Development of Bio-Based Materials: Synthesis, Characterization, and Applications II

Development of Bio-Based Materials: Synthesis, Characterization, and Applications II

Guest Editors

Antonio M. Borrero-López
Concepción Valencia-Barragán
Esperanza Cortés Triviño
Adrián Tenorio-Alfonso
Clara Delgado-Sánchez

MDPI

Basel • Beijing • Wuhan • Barcelona • Belgrade • Novi Sad • Cluj • Manchester

Guest Editors

Antonio M. Borrero-López
Chemical Engineering,
Physical Chemistry and
Science of Materials
University of Huelva
Huelva
Spain

Concepción Valencia-Barragán
Chemical Engineering,
Physical Chemistry and
Science of Materials
University of Huelva
Huelva
Spain

Esperanza Cortés Triviño
Chemical Engineering,
Physical Chemistry and
Science of Materials
University of Huelva
Huelva
Spain

Adrián Tenorio-Alfonso
Chemical Engineering,
Physical Chemistry and
Science of Materials
University of Huelva
Huelva
Spain

Clara Delgado-Sánchez
Chemical Engineering,
Physical Chemistry and
Science of Materials
University of Huelva
Huelva
Spain

Editorial Office
MDPI AG
Grosspeteranlage 5
4052 Basel, Switzerland

This is a reprint of the Special Issue, published open access by the journal *Polymers* (ISSN 2073-4360), freely accessible at: www.mdpi.com/journal/polymers/special_issues/T8F1Z2OS62.

For citation purposes, cite each article independently as indicated on the article page online and using the guide below:

Lastname, A.A.; Lastname, B.B. Article Title. *Journal Name* **Year**, *Volume Number*, Page Range.

ISBN 978-3-7258-3680-2 (Hbk)
ISBN 978-3-7258-3679-6 (PDF)
https://doi.org/10.3390/books978-3-7258-3679-6

Contents

About the Editors . . . . **ix**

Preface . . . . **xi**

Tongsai Jamnongkan, Nitchanan Intraramongkol, Wesarach Samoechip, Pranut Potiyaraj, Rattanaphol Mongkholrattanasit and Porntip Jamnongkan et al.
Towards a Circular Economy: Study of the Mechanical, Thermal, and Electrical Properties of Recycled Polypropylene and Their Composite Materials
Reprinted from: *Polymers* **2022**, *14*, 5482, https://doi.org/10.3390/polym14245482 . . . . **1**

Vojtěch Jašek, Jan Fučík, Lucia Ivanová, Dominik Veselý, Silvestr Figalla and Ludmila Mravcova et al.
High-Pressure Depolymerization of Poly(lactic acid) (PLA) and Poly(3-hydroxybutyrate) (PHB) Using Bio-Based Solvents: A Way to Produce Alkyl Esters Which Can Be Modified to Polymerizable Monomers
Reprinted from: *Polymers* **2022**, *14*, 5236, https://doi.org/10.3390/polym14235236 . . . . **17**

Luisa García-Fuentevilla, Gabriela Domínguez, Raquel Martín-Sampedro, Manuel Hernández, María E. Arias and José I. Santos et al.
Enzyme-Catalyzed Polymerization of Kraft Lignin from *Eucalyptus globulus*: Comparison of Bacterial and Fungal Laccases Efficacy
Reprinted from: *Polymers* **2023**, *15*, 513, https://doi.org/10.3390/polym15030513 . . . . **35**

Piotr Koczoń, Bartłomiej Bartyzel, Anna Iuliano, Dorota Klensporf-Pawlik, Dorota Kowalska and Ewa Majewska et al.
Chemical Structures, Properties, and Applications of Selected Crude Oil-Based and Bio-Based Polymers
Reprinted from: *Polymers* **2022**, *14*, 5551, https://doi.org/10.3390/polym14245551 . . . . **55**

Rongrong An, Chengguo Liu, Jun Wang and Puyou Jia
Wood–Derived Polymers from Olefin–Functionalized Lignin and Ethyl Cellulose via Thiol–Ene Click Chemistry
Reprinted from: *Polymers* **2023**, *15*, 1923, https://doi.org/10.3390/polym15081923 . . . . **76**

Guillem Ferreres, Sílvia Pérez-Rafael, Angela Gala Morena, Tzanko Tzanov and Liudmyla Gryshchuk
Influence of Enzymatically Hydrophobized Hemp Protein on Morphology and Mechanical Properties of Bio-Based Polyurethane and Epoxy Foams
Reprinted from: *Polymers* **2023**, *15*, 3608, https://doi.org/10.3390/polym15173608 . . . . **94**

Sadia Munir, Wei Yue, Jinling Li, Xiaoyue Yu, Tianhao Ying and Ru Liu et al.
Effects of Phenolics on the Physicochemical and Structural Properties of Collagen Hydrogel
Reprinted from: *Polymers* **2023**, *15*, 4647, https://doi.org/10.3390/polym15244647 . . . . **106**

Agustín Maceda, Marcos Soto-Hernández and Teresa Terrazas
Cellulose in Secondary Xylem of Cactaceae: Crystalline Composition and Anatomical Distribution
Reprinted from: *Polymers* **2022**, *14*, 4840, https://doi.org/10.3390/polym14224840 . . . . **125**

Xiaoli Liang, Shan Wei, Yanpeng Xu, Liang Yin, Ruiming Wang and Piwu Li et al.
Construction and Characterization of Fitting Equations for a New Wheat Straw Pulping Method
Reprinted from: *Polymers* **2023**, *15*, 4637, https://doi.org/10.3390/polym15244637 . . . . **142**

Carolina Villegas, Sara Martínez, Alejandra Torres, Adrián Rojas, Rocío Araya and Abel Guarda et al.
Processing, Characterization and Disintegration Properties of Biopolymers Based on Mater-Bi® and Ellagic Acid/Chitosan Coating
Reprinted from: *Polymers* **2023**, *15*, 1548, https://doi.org/10.3390/polym15061548 . . . . **154**

Marina G. Holyavka, Svetlana S. Goncharova, Andrey V. Sorokin, Maria S. Lavlinskaya, Yulia A. Redko and Dzhigangir A. Faizullin et al.
Novel Biocatalysts Based on Bromelain Immobilized on Functionalized Chitosans and Research on Their Structural Features
Reprinted from: *Polymers* **2022**, *14*, 5110, https://doi.org/10.3390/polym14235110 . . . . **170**

Nathaly Vasquez-Martínez, Daniel Guillen, Silvia Andrea Moreno-Mendieta, Sergio Sanchez and Romina Rodríguez-Sanoja
The Role of Mucoadhesion and Mucopenetration in the Immune Response Induced by Polymer-Based Mucosal Adjuvants
Reprinted from: *Polymers* **2023**, *15*, 1615, https://doi.org/10.3390/polym15071615 . . . . **189**

Sergio Alcalá-Alcalá, José Eduardo Casarrubias-Anacleto, Maximiliano Mondragón-Guillén, Carlos Alberto Tavira-Montalvan, Marcos Bonilla-Hernández and Diana Lizbeth Gómez-Galicia et al.
Melanin Nanoparticles Obtained from Preformed Recombinant Melanin by *Bottom-Up* and *Top-Down* Approaches
Reprinted from: *Polymers* **2023**, *15*, 2381, https://doi.org/10.3390/polym15102381 . . . . **215**

Milena Ignatova, Ina Anastasova, Nevena Manolova, Iliya Rashkov, Nadya Markova and Rositsa Kukeva et al.
Bio-Based Electrospun Fibers from Chitosan Schiff Base and Polylactide and Their Cu^{2+} and Fe^{3+} Complexes: Preparation and Antibacterial and Anticancer Activities
Reprinted from: *Polymers* **2022**, *14*, 5002, https://doi.org/10.3390/polym14225002 . . . . **234**

Jingjing Mai, Steven Pratt, Bronwyn Laycock and Clement Matthew Chan
Synthesis and Characterisation of Poly(3-hydroxybutyrate-*co*-3-hydroxyvalerate)-*b*-poly(3-hydroxybutyrate-*co*-3-hydroxyvalerate) Multi-Block Copolymers Produced Using Diisocyanate Chemistry
Reprinted from: *Polymers* **2023**, *15*, 3257, https://doi.org/10.3390/polym15153257 . . . . **258**

Emmanuelle Richely, Johnny Beaugrand, Michel Coret, Christophe Binetruy, Pierre Ouagne and Alain Bourmaud et al.
In Situ Tensile Testing under High-Speed Optical Recording to Determine Hierarchical Damage Kinetics in Polymer Layers of Flax Fibre Elements
Reprinted from: *Polymers* **2023**, *15*, 2794, https://doi.org/10.3390/polym15132794 . . . . **273**

Elsa Veronica Flores-Vela, Alain Salvador Conejo-Dávila, Claudia Alejandra Hernández-Escobar, Rocio Berenice Dominguez, David Chávez-Flores and Lillian V. Tapia-Lopez et al.
Silk Fibroin-*g*-Polyaniline Platform for the Design of Biocompatible-Electroactive Substrate
Reprinted from: *Polymers* **2022**, *14*, 4653, https://doi.org/10.3390/polym14214653 . . . . **292**

Goretti Arias-Ferreiro, Aurora Lasagabáster-Latorre, Ana Ares-Pernas, Pablo Ligero, Sandra María García-Garabal and María Sonia Dopico-García et al.
Lignin as a High-Value Bioaditive in 3D-DLP Printable Acrylic Resins and Polyaniline Conductive Composite
Reprinted from: *Polymers* **2022**, *14*, 4164, https://doi.org/10.3390/polym14194164 . . . . **305**

About the Editors

Antonio M. Borrero-López

Antonio María Borrero-López completed a degree in Chemical Engineering at the University of Huelva (Spain) in 2013 and later obtained a Master's degree in Product Formulation and Design from the International University of Andalusia (Spain). Afterwards, he received his PhD degree from the University of Huelva, with a thesis entitled "Development of new lignocellulosic-based thickening agents for biodegradable oleogel formulations with several industrial applications". This thesis has been awarded the Best 2021–2022 Doctoral Thesis in the branch of Engineering and Architecture of the University of Huelva and received an honourable mention at the Annual European Rheology Conference 2022 (AERC 2022). Since completing his PhD, he has been working as a postdoc at the Institut Jean Lamour, University of Lorraine (France), where new greener approaches for the performance of bio-based materials are being targeted. He has participated in a total of eight research projects, including the European Project UCGWATERplus, which aims to remediate waters polluted with organic and inorganic contaminants by the formulation of different products via the valorisation of the residues from underground coal gasification and other processes. As a result of his research, he has published a total of 28 articles and contributed to more than 30 international and national congresses.

Concepción Valencia-Barragán

Concepción Valencia Barragán received her PhD degree from the University of Extremadura. She is a Full Professor at the University of Huelva. She has carried out extensive research activity for more than 23 years, mainly focused on Chemical Product Engineering, specifically concerning the rheology of complex materials, the processing of non-Newtonian fluids, lubricants, adhesives and coatings, polymers and biopolymers, food colloids, and emulsion technologies. In total, she has participated in more than 30 research projects with public funding, obtained in competitive calls, with some of them as a lead researcher and more than 40 contracts with private sector companies. Her main scientific and technical achievements have been focused on the modification of the rheological properties of lubricating greases by means of reactive and recycled polymeric additives as well as on the development of biodegradable oleogels capable of replacing traditional lubricating greases formulated from non-renewable resources. As a result of this research activity, she is the author of more than 130 papers in peer-reviewed journals and the inventor of six patents. She has also presented more than 130 communications at national and international scientific conferences.

Esperanza Cortés Triviño

Esperanza Cortés-Triviño pursued her PhD in the development of new thickening agents of vegetable oil from different chemically modified lignocellulosic fractions, with support from a contract grant under project TEP-1499, awarded by the Junta de Andalucía. Her doctoral research, conducted within the Doctoral Programme in Industrial and Environmental Science and Technology at the Chemical Process and Product Technology Research Centre (Pro2TecS), was recognized as the best 2019 Doctoral Thesis in Engineering at the University of Huelva and received an honourable mention at the Iberian Meeting on Rheology (IBEREO 2019). Following her PhD and as a member of the Complex Fluid Engineering Research Group (TEP-185) at Pro2TecS, she has contributed to multiple research projects, collaborating with multinational companies like Procter & Gamble over the past 5 years.

Her research focused on the development of bio-lubricants, bio-adhesives, bioplastics, and emulsions for energy storage, along with the modification of polymeric materials for functional applications. Dr. Cortés-Triviño has published 15 articles in high-impact international journals and has presented numerous contributions at national and international conferences.

Adrián Tenorio-Alfonso

Adrián Tenorio-Alfonso conducted his PhD degree on the development of polyurethane formulations based on cellulose acetate and castor oil with the aid of a national grant for a PhD contract "Ayuda a la Formación de Profesorado Universitario (FPU13/01114)" by el Ministerio de Educación, Cultura y Deporte, affiliated with the Doctoral Programme in Industrial and Environmental Science and Technology at the Chemical Process and Product Technology Research Centre (Pro2TecS) from the University of Huelva. Afterwards, he has taken part in several research projects, not only aimed at developing and processing advanced polymeric materials but mainly focusing on efficient energy storage and management. In this respect, he was awarded a prestigious postdoctoral fellowship in the development of novel phase change materials. As a result of his research experience, he has published 19 research articles in international journals with high impact factors, accounting for more than 50 contributions to national and international congresses.

Clara Delgado-Sánchez

Clara Delgado-Sánchez holds two PhD degrees, first in Solid-State Chemistry from the University of Lorraine (France) and later in Chemical Engineering from the University of Huelva (Spain). Her initial doctoral research, conducted within the Biosourced Materials group at the Institut Jean Lamour (IJL - UMR CNRS 7198), focused on the development and characterization of tannin-based foams as sustainable thermal insulation materials for buildings. Following her PhD in France, she continued her research at the Pro2Tec research center (University of Huelva), where she obtained her second PhD and contributed to multiple competitive national and regional projects on advanced materials, thermal energy storage, and circular economy strategies. Currently, she leads a research line as a principal investigator under the prestigious EMERGIA postdoctoral program: Bio-Based Insulating Materials for Advanced Thermoregulation of Environments with High Energy Demand. This project integrates biopolymers, circular economy principles, and energy efficiency to develop innovative materials for sustainable thermal management. Her work has resulted in over 29 publications in international journals and numerous contributions to national and international conferences

Preface

The shift towards a sustainable and circular economy is one of today's most pressing challenges. Growing global demand for materials, along with concerns about environmental impact and resource depletion, has fuelled interest in innovative polymeric materials that support sustainability, recyclability, and advanced functionality.

Given the increasing interest in this field, this reprint brings together a selection of 19 innovative research articles originally published in the Special Issue "Development of Bio-Based Materials: Synthesis, Characterization, and Applications II". These studies highlight recent advances in polymer science, with a strong focus on sustainability, recycling, bio-based materials, biomedical applications, and advanced functional materials.

The first set of articles in this volume presents groundbreaking advancements in polymer recycling and sustainability, showcasing innovative strategies aimed at enhancing the circular economy. These contributions explore the chemical and structural modifications of natural products, industrial waste, and natural proteins to develop enhanced biopolymers for diverse industrial applications. Furthermore, these articles propose cutting-edge approaches for the reutilization of both synthetic polymers and biopolymers, offering sustainable and environmentally friendly alternatives to conventional methods.

Subsequent studies examine the role of polymeric materials in drug delivery, immune response modulation, and biocompatible coatings, as well as the induction of biopolymer gelation, highlighting innovative approaches to biodegradable and bioactive medical materials. The reprint also covers advancements in sustainable electronics, including biocompatible electroactive materials and 3D-printable bioadditives, demonstrating the potential of renewable materials to replace conventional synthetic polymers.

Designed as a valuable resource for researchers, engineers, and professionals in polymer science, materials engineering, and sustainability, this collection aims to inspire further innovation in bio-based materials and their applications.

Antonio M. Borrero-López, Concepción Valencia-Barragán, Esperanza Cortés Triviño,
Adrián Tenorio-Alfonso, and Clara Delgado-Sánchez
Guest Editors

polymers MDPI

Article

Towards a Circular Economy: Study of the Mechanical, Thermal, and Electrical Properties of Recycled Polypropylene and Their Composite Materials

Tongsai Jamnongkan [1,*], Nitchanan Intraramongkol [1], Wesarach Samoechip [2], Pranut Potiyaraj [2], Rattanaphol Mongkholrattanasit [3], Porntip Jamnongkan [4], Piyada Wongwachirakorn [5], Masataka Sugimoto [6], Hiroshi Ito [6] and Chih-Feng Huang [7,*]

1 Department of Fundamental Science and Physical Education, Faculty of Science at Sriracha, Kasetsart University, Chonburi 20230, Thailand
2 Department of Materials Science, Faculty of Science, Chulalongkorn University, Bangkok 10330, Thailand
3 Faculty of Industrial Textiles and Fashion Design, Rajamangala University of Technology Phra Nakhon, Bangkok 10110, Thailand
4 Department of Public Health and Environment, Saensuk Municipality, Chonburi 20130, Thailand
5 Department of Environmental Science, Faculty of Science and Technology, Pibulsongkarm Rajabhat University, Phitsanulok 65000, Thailand
6 Graduated School of Organic Materials Science, Faculty of Engineering, Yamagata University, Yonezawa 32000, Yamagata, Japan
7 Department of Chemical Engineering, i-Center for Advanced Science and Technology (iCAST), National Chung Hsing University, Taichung 40227, Taiwan
* Correspondence: jamnongkan.t@ku.ac.th (T.J.); huangcf@dragon.nchu.edu.tw (C.-F.H.)

Abstract: This research focuses on the mechanical properties of polypropylene (PP) blended with recycled PP (rPP) at various concentrations. The rPP can be added at up to 40 wt% into the PP matrix without significantly affecting the mechanical properties. MFI of blended PP increased with increasing rPP content. Modulus and tensile strength of PP slightly decreased with increased rPP content, while the elongation at break increased to up to 30.68% with a 40 wt% increase in rPP content. This is probably caused by the interfacial adhesion of PP and rPP during the blending process. The electrical conductivity of materials was improved by adding carbon black into the rPP matrices. It has a significant effect on the mechanical and electrical properties of the composites. Stress-strain curves of composites changed from ductile to brittle behaviors. This could be caused by the poor interfacial interaction between rPP and carbon black. FTIR spectra indicate that carbon black did not have any chemical reactions with the PP chains. The obtained composites exhibited good performance in the electrical properties tested. Finally, DSC results showed that rPP and carbon black could act as nucleating agents and thus increase the degree of crystallinity of PP.

Keywords: polypropylene; plastic waste; mechanical properties; carbon black; sustainability

check for updates

Citation: Jamnongkan, T.; Intraramongkol, N.; Samoechip, W.; Potiyaraj, P.; Mongkholrattanasit, R.; Jamnongkan, P.; Wongwachirakorn, P.; Sugimoto, M.; Ito, H.; Huang, C.-F. Towards a Circular Economy: Study of the Mechanical, Thermal, and Electrical Properties of Recycled Polypropylene and Their Composite Materials. *Polymers* **2022**, *14*, 5482. https://doi.org/10.3390/polym14245482

Academic Editors: Antonio M. Borrero-López, Concepción Valencia-Barragán, Esperanza Cortés Triviño, Adrián Tenorio-Alfonso and Clara Delgado-Sánchez

Received: 29 November 2022
Accepted: 12 December 2022
Published: 14 December 2022

Publisher's Note: MDPI stays neutral with regard to jurisdictional claims in published maps and institutional affiliations.

1. Introduction

In recent times, the production of petroleum-based plastic materials has rapidly increased because they exhibit several advantages for various uses over other materials such as a light weight, high flexibility, resilience, resistance to corrosion, excellent chemical resistance, transparency, and ease of processing [1–4]. It is forecasted that the global annual production of plastic will grow further due to the continued expansion of world population, and therefore the demand for consumables will also increase. Thus, waste management of plastic post-consumption is very important [5,6]. It is well-known that plastic waste has numerous damaging environmental effects, including increased energy consumption and greenhouse gas exhaustion [7–9]. Particularly, fossil-based plastics such as polyethylene (PE), polypropylene (PP), polystyrene (PS), polyethylene terephthalate

(PET), and many others are used for daily purposes which increases the amounts being disposed of in landfills day by day [10–16]. In consequence, it has been a challenge to establish a cost-effective method to solve these problems. One effective approach could be the use of natural resources, called biomaterials, to replace their corresponding functional products [17]. Upcycling or recycling plastic waste based on the circular economy (CE) concept is an effective alternative to manage and reduce the effects of plastic waste on the environment [18,19]. This concept could be managed waste from the plastic production and consumption system by imported and consumed as raw materials for recycling, reuse, reduction and repairing or refurbishing in the process of product production. Recently, the concept of CE has received great attention from several scientists across the world because it has directly responded to one of the seventeen targets of the Sustainable Development Goals (SDGs). CE is a manufacturing method for waste management that allows the remanufacturing, reprocessing, and recycling of materials in the production process [20,21]. Therefore, the recycling method not only reduces poisonous materials and waste pollution on the environment compared to landfilling and incineration [22–24], but can also create new products with good mechanical and physical properties. However, it is not easy to recycle these materials because they have been combined with different types of plastics and additives during processing [25]. In general, the mechanical properties of recycled plastics do not remain the same because of several parameters such as degradation from heat, mechanical stress, and oxidation during reprocessing. Recently, several research groups have studied and focused on the influence of multiple recycling cycles on the quality of the reprocessed plastics. For example, Canevarolo [26] reported the thermo-mechanical properties of PP when subjected to multi-extrusion with different processing conditions and screw profiles. They found that the material was reprocessed five times and found that the PP chain scission process occurred during multiple processes, particularly at higher molecular weights of PP. Jin et al. [27] have investigated and reported the effects of the recycling process of low-density polyethylene (LDPE) on its mechanical properties. They found that LDPE could be extruded for up to 40 times without significantly changing its processability and long-time mechanical properties. Tominaga and his colleagues worked on the mechanical properties of recycled PP and found that they are closely related to the inner structure of materials [28]. They also mentioned that the molecular weight of recycled PP did not change significantly during the recycling process, and they were able to improve its mechanical properties and durability by adopting an appropriate molding condition of the mechanical recycling process. In Thailand, it was also found that the demand and production of PP have increased tremendously due to its applications in various areas, which has caused an increase in PP waste in landfills. PP is a semicrystalline thermoplastic that can be remelted and used again, and also has a low cost, processability, and excellent chemical stability [29]. Over last decade, it has been attempted to solve the problem of recycled plastics through several methods, both physical (i.e., blending, reprecipitation, composite, etc. [30–35]) and chemical (such as reaction processes and catalytic cracking methods [36–39]), leading to products ready to be used in specific applications. The easiest way is to recycle the material directly by a physical process such as blending it with the virgin material. As we have mentioned earlier, it is still difficult to keep consistency of mechanical properties and performance of raw materials during the recycling of post-consumer materials. In addition, the idea of recycling as any technology that converts post-use plastics into specialty polymers or new materials has been increasingly recognized as a potential solution to the recycling of plastic waste. Therefore, the performance of value-added specialty polymers should be paramount concerns. The electrical conductivity of a polymer is one of the important properties that have recently attracted attention from researchers because of its relevance to various practical fields, such as electronic devices and sensors [40–42]. Carbon black has attracted attention due to its numerous desirable properties, such as good electrical conductivity, low cost, and good thermal and chemical stability [43,44]. Accordingly, we have chosen to use carbon black as a conductive filler within rPP matrices. Our aim is to study the reprocessing of problematic industrial PP scrap or rPP by using the melt-extrusion process. The effects of the concentrations of rPP on their mechanical properties were investigated. In addition, the chemical, mechanical, thermal, and surface resistivity properties of the prepared rPP/carbon black composites were also examined. This research will advance the knowledge and help to promote the use of rPP in some material engineering applications.

2. Materials and Methods

2.1. Materials

Commercial-grade virgin polypropylene (vPP), 1100NK grade, was purchased from IRPC Co., Ltd., Rayong, Thailand. Recycled PP (rPP) was obtained from Panich Parts and Mold Co., Ltd., Chonburi, Thailand. Distilled glycerol monostearate (DMG) and carbon black were received from Suntor Chemical Co., Ltd., Samutprakarn, Thailand. Other chemical agents were of analytical-grade purity and used as received.

2.2. Preparation of vPP/rPP Blends

At the first stage, rPP was sorted and crushed into 0.5 cm of granular form by using a pulverizer milling machine (Siamlab, Nonthaburi, Thailand) for enhancing the surface area of the material. After that, it was dried at a temperature of 80 °C for 24 h in a hot air oven to remove the moisture content. Different ratios of vPP/rPP blends were prepared by the extrusion method as follows. rPP and vPP pellets were first dried at 80 °C for 24 h and then the different ratios of rPP/vPP were fed into a lab-scale, twin-screw extruder (Thermo PRISM, Bangkok, Thailand), with a screw diameter of 28 mm and a length/diameter ratio of 25:1, to mix the compounds. During the extrusion process, the fed temperature in the first zone of the extruder and the die temperature were set to 160 and 180 °C, respectively. The extruder temperature profile was set at 170 °C with a screw speed of 40–50 rpm. The extrudate was cooled down in a water bath at a temperature of 30–35 °C. Then, the extrudate was cut into pellets nominally 3 mm long by using a pelletizer. After extrusion, the vPP/rPP compounds were dried and indirectly heated for dehumidifying in the oven at a temperature of 80 °C for 4 h, and then stored in moisture-barrier bags before injection molding. An injection molding machine (Model SG50M, Sumitomo, Tokyo, Japan) was used to fabricate the specimens for the mechanical test. This machine had a single screw of 40 mm in diameter with a length/diameter ratio of 18:1. The temperature of the feeding, compression, and metering sections were maintained at 160, 170, and 175 °C, respectively. During the injection process, the holding step was set to be three steps with pressure and time as follows: (i) 90.0 bar for 3 s, (ii) 95.0 bar for 5 s, and (iii) 90.0 bar for 3 s. Mold temperature was maintained at 40 °C, which was controlled by water cooling. In this paper, the recycling PP samples are referred to as vPP, rPP20, rPP30, and rPP40, corresponding to the virgin PP and the PP blended with rPP at the concentrations of 20, 30, and 40 wt%, respectively. The melt flow index (MFI) of all blended samples was determined with the melt flow index instrument (Kayeness, Model 7053, Morgantown, PA, USA) in accordance with the ASTM D1238 standard. The same sample was tested at least five times at 230 °C with a load of 2.10 kg. The MFI was calculated by using the following equation:

$$\text{MFI (g/10 min)} = 600\ \text{m/t}$$

where m and t refer to the average weight of extrudates and the time of extrudate in seconds, respectively.

In addition, rPP30 was chosen as a representative of recycled PP composited with carbon black at the concentration of 40 wt%. This composite was prepared through the same procedure for vPP/rPP blends preparation, as follows: rPP30 was milled and mixed with carbon black and DMG at the concentration of 40 wt% and 0.5 wt%, respectively, by using a high-speed mixer machine. Then, the mixture powder was transferred to the twin screw extruder. During the extrusion process, the temperature in the mixing chamber was set to a range of 160–180 °C and the screw rotation rate was maintained at 210 rpm. In this process, polyethylene (PE) wax was used as a plasticizer to improve the processibility of composites for comparison. After obtaining the composite pellets, the specimens for mechanical testing were fabricated using an injection molding machine following ASTM D638 [45]. In this study, the samples are referred to with codes rPP30CBW and rPP30CBNW, corresponding to the rPP30 composited with 40 wt% of carbon black with and without PE wax, respectively.

2.3. Tensile Test

Tensile tests of all the samples were performed on the universal testing machine (Model 5560, Instron, MA, USA). A load cell of 25 kN was employed for testing all the samples. All the specimens were fabricated to a rectangular shape ($30 \times 10 \times 0.4$ mm^3), according to the ASTM D638 [45], by using an injection molding machine. Crosshead speed and gauge length were set at 5 mm/min and 25 mm, respectively. At least five samples were tested in each sample and the average of results were reported.

2.4. FTIR-ATR Analysis

FTIR was performed to examine the presence of chemical reactions within the structure of compounding molecules. An FTIR spectrometer (Invenio, Bruker, MA, USA) in ATR mode equipped with a diamond crystal was used for the tests. All the spectra were recorded in transmittance mode with 4 cm^{-1} resolutions in the wavenumber range of 4000–400 cm^{-1} under ambient conditions.

2.5. Surface Resistivity Test

Electrostatic discharge is one of the key issues that cause losses in the properties of products containing sensitive electronics during production, storage, and transportation. Therefore, the surface resistivity of rPP30 composited with carbon black was measured at room temperature using a probe digital multimeter (Model KS-385B, Kingsom, Shenzhen, China), according to ASTM D257 [46]. The specimen of vPP was also investigated for comparison. The experimental procedure was as follows: two copper electrodes were pressed on the surface of the composites at a distance of 1.5 cm from each other.

2.6. Differential Scanning Calorimetry Test

The crystallization behavior and melting characteristics of vPP, rPP30, and its composites were analyzed by using a differential scanning calorimetry (DSC) technique (Model DSC 200 F3, Netzsch, Selb, Wunsiedel, Germany). All the samples were in a sealed aluminum crucible and each sample used was approximately 5 mg. DSC analysis was run at a temperature range from 25 to 200 °C, with heating and cooling rates of 10 °C/min. All runs were carried out under inert atmosphere with a nitrogen flow of 50 mL/min to prevent the thermal degradation of samples. After that, the degree of crystallinity (χ_c, %) was determined from the melting enthalpy values using the following equation:

$$\chi_c/\% = \frac{\Delta H_m}{w \times \Delta H_m^{\circ}} \times 100$$

where ΔH_m, w, and ΔH_m° are the melting enthalpy of the specimens, the weight fraction of filler in PP composites, and the enthalpy value for a theoretically 100% crystalline PP (207 J/g) [47], respectively.

3. Results and Discussion

3.1. Feasibility of the Obtained vPP/rPP Blends

Figure 1 shows optical images of the compound pellets of vPP and vPP blended with rPP at different concentrations. The images show that the extrusion process can produce homogeneous pellets composed of vPP and its blends, with roughly the same external surface in the four samples. Furthermore, the images suggest that the color of the vPP/rPP compounds became gray in the case of the vPP containing rPP when compared to those without rPP. Additionally, the shade of gray color of vPP/rPP compounds is directly proportional to the rPP content within the compounds, as can be seen in Figure 1. This result could indicate that the rPP content will affect mechanical properties. Therefore, it is worthwhile to further examine the effects of rPP content on mechanical properties. We begin our examination of the effect of rPP concentration on the melt flow behavior (results shown in Figure 2). The MFI of PP blends is proportional to the amount of rPP concentration. The MFI of blended vPP increased with increasing rPP content. It was found that vPP and its blend with 20, 30, and 40 wt% of rPP content could be extruded evidently and their MFI values were 11.72, 11.98, 12.98, and 16.74 g/10 min, respectively. Obviously, rPP has extremely affected the melt flow property of PP. This suggested that the high MFI of PP-blended compounds led to poor interfacial adhesion between the filler particles and matrices, which is in agreement with the research results of Xu et al. [48].

Figure 1. Feasibility images of different compounds. (**A**) vPP, (**B**) rPP20, (**C**) rPP30, and (**D**) rPP40.

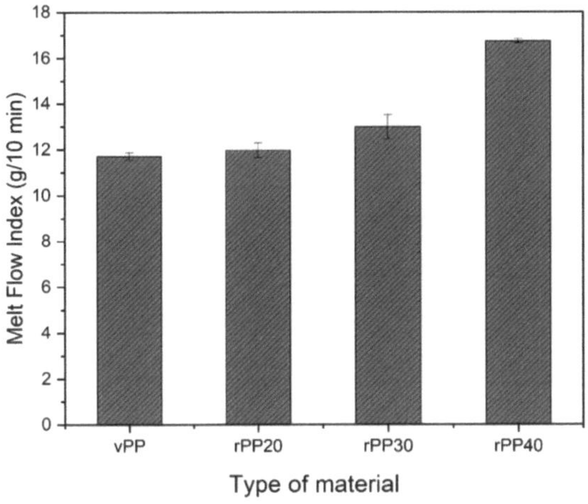

Figure 2. Melt flow index of all prepared samples.

3.2. Tensile Test

The fracture profile of tensile stress-strain curves and optical feasibility fracture for all specimens were obtained and are illustrated in Figures 3 and 4, respectively. We found that all samples exhibited the fracture ductility behavior.

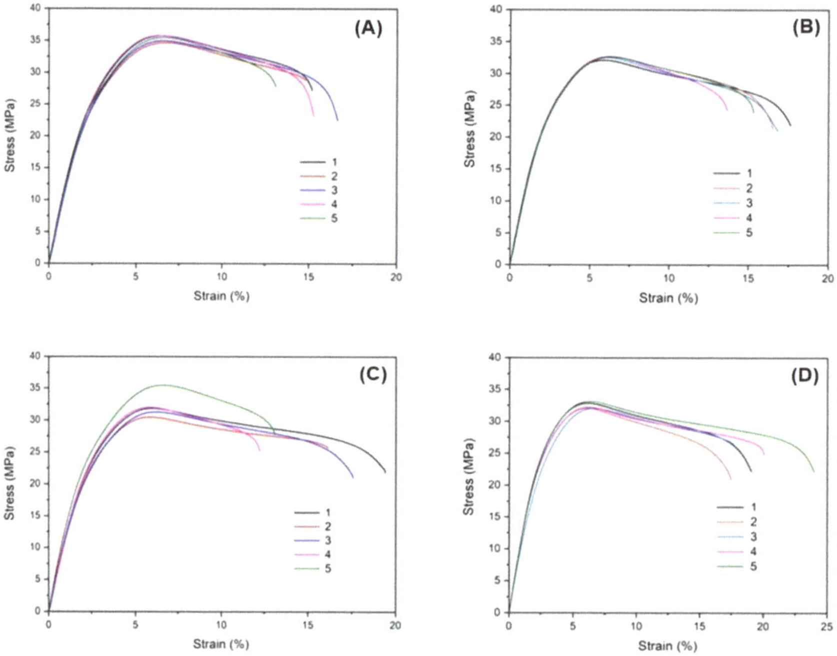

Figure 3. Tensile stress-strain curves of vPP/rPP blended with different rPP concentrations. (**A**) vPP, (**B**) rPP20, (**C**) rPP30, and (**D**) rPP40. The number of each curve indicates the testing numbers.

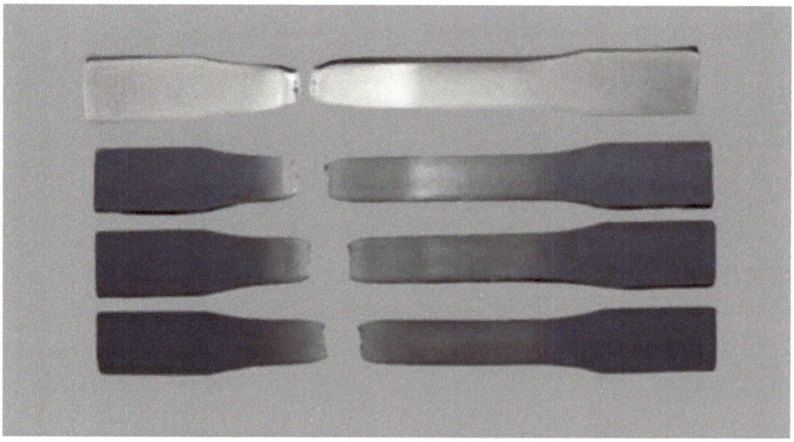

Figure 4. Optical images of all specimens after tensile test at room temperature.

The tensile strength, modulus, and elongation at the break of all samples were obtained and presented in Figure 5 and Table 1. In general, the interfacial adhesion between the fillers and the matrices has directly affected the mechanical properties of blends and composite materials [49,50]. We found that the modulus and tensile strength of the vPP/rPP blends slightly decreased with increasing rPP concentration. In addition, the elongation at the breaks of all samples increased with increasing rPP concentration, as enumerated in Table 1. This phenomenon is probably caused by the interfacial adhesion of vPP and rPP during the melt extrusion process.

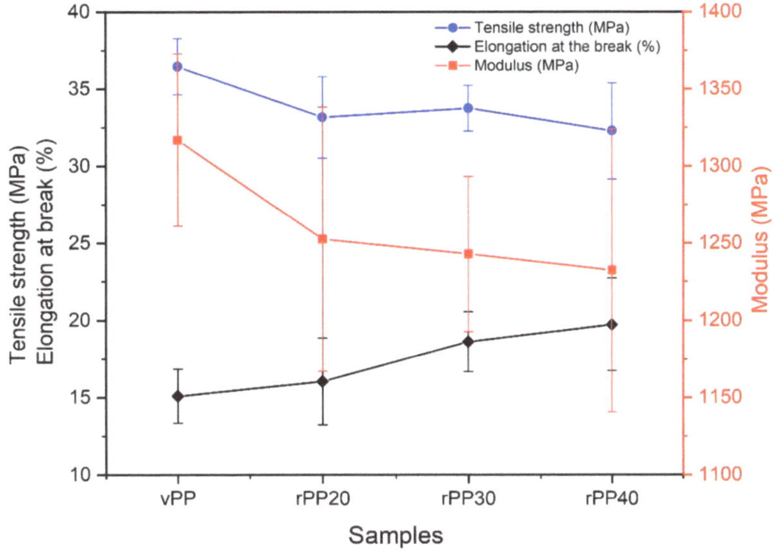

Figure 5. Effect of varying rPP contents on the mechanical properties of PP blends.

Table 1. Mechanical properties of the vPP specimens blended with rPP at various concentrations.

Sample Code	Modulus (MPa)	Tensile Strength (MPa)	Elongation at the Break (%)
vPP	1317 ± 55.85	36.47 ± 1.83	15.09 ± 1.75
rPP20	1252 ± 85.63	33.17 ± 2.65	16.04 ± 2.82
rPP30	1243 ± 50.15	33.75 ± 1.48	18.62 ± 1.93
rPP40	1232 ± 91.62	32.28 ± 3.12	19.72 ± 2.99

In Figure 5, we found that vPP exhibited the highest modulus (1317 MPa) and tensile strength (36.47 MPa), while it has the lowest elongation at the break (15.09%) among the four specimens. This result has the same tendency of earlier reports [51–53]. In this study, however, we found that the mechanical properties of the recycled PP samples are not insignificantly different when compared with the virgin PP. The experimental results seem to indicate that rPP can be added to a blend with vPP up to a 40 wt% concentration without significantly affecting the mechanical properties.

To develop innovative materials and increase the value addition of PP waste, we attempted to improve the electrical conductivity properties of the rPPs by compositing with carbon black. rPP30 was chosen as a representative of vPP/rPP blends for mixing with carbon black at the concentration of 40 wt%. We successfully compounded these composite materials by using the melt-extrusion process. It was found that the color of the obtained composite compound changed from gray to black, as depicted in Figure 6, coinciding with the natural color of carbon black. Then, we continued to investigate their mechanical properties and found that the stress-strain curves of all composite samples exhibited brittle profile curves, as illustrated in Figure 7.

Figure 6. Optical images of pellets of compound (**A**) rPP30 and (**B**) rPP30 composited with carbon black.

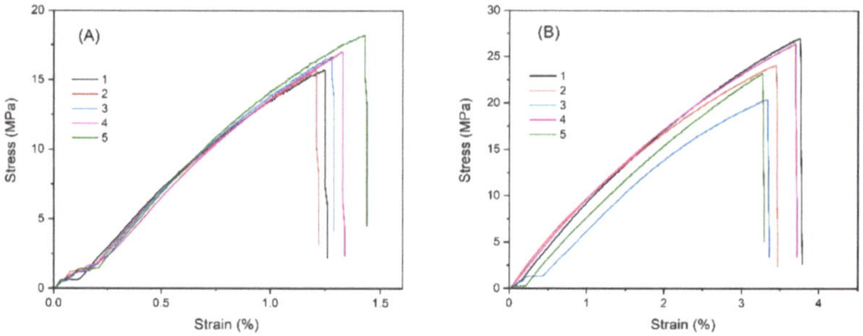

Figure 7. The stress-strain curves of (**A**) rPP30CBNW and (**B**) rPP30CBW composites. The number of each curve indicates the testing numbers.

Mechanical properties such as modulus, tensile strength, and elongation at the break of these composite materials were analyzed and calculated from the results in Figure 7. We found that carbon black significantly affected the modulus, tensile strength, and elongation at the break of composite materials, as shown in detail in Table 2.

Table 2. Mechanical properties of rPP30, rPP30CBNW, and rPP30CBW composites.

Sample Codes	Modulus (MPa)	Tensile Strength (MPa)	Elongation at the Break (%)
rPP30	1243 ± 50.15	33.75 ± 1.48	18.62 ± 1.93
rPP30CBNW	1676 ± 87.06	17.16 ± 1.18	1.27 ± 0.08
rPP30CBW	1038 ± 22.80	26.08 ± 2.11	3.73 ± 0.47

Figure 8 shows the mechanical properties of all rPP30/carbon black composite materials compared to rPP30. The rPP30CBNW displayed the highest modulus (1676 MPa) as well as the lowest tensile strength (17.16 MPa) and elongation at the break (1.27%) among the three composite materials. The elongation at the break of rPP30CBW was greater than that of rPP30CBNW. This might be attributed to the addition of a plasticizer that makes the polymer chain move easily, thereby causing the material to have less stiffness. Thus, the materials will have lower modulus and higher percentage of elongation at the break than those composites without the plasticizer [54,55]. Additionally, the interfacial adhesion between filler particles and matrices is also a key parameter that affects their mechanical properties [56], as described earlier.

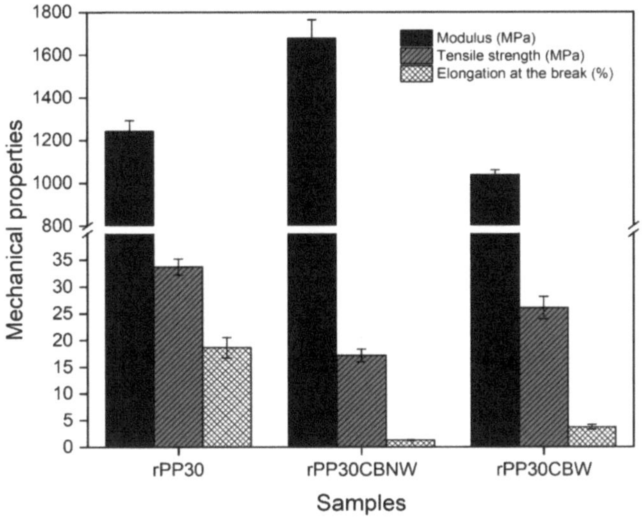

Figure 8. Mechanical properties of the rPP30 and the obtained rPP30 composited with and without the plasticizer content.

3.3. FTIR–ATR Analysis

FTIR was employed to analyze the chemical interaction of the vPP/rPP blends and rPP30/carbon black composites. We began by investigating the IR spectra of rPP in comparison to vPP and the result was depicted in the Supplementary Material, as shown in Figure S1. Figure 9 depicts the FTIR spectrum of rPP30 and rPP30 composited with carbon black samples. IR spectra of the obtained composites and vPP were then compared. We found that the IR spectra of vPP and rPP30 show almost the same characteristic peaks of polypropylene, as shown in spectra A and B, respectively. A strong band around 2950 cm^{-1} revealed the -CH_3 asymmetrical stretching vibrations of the surface of PP [57]. Moreover, two sharp signals at 1375 and 1458 cm^{-1} indicate the presence of C-H bonds in the chemical structure [58]. In addition, we also found the identified characteristic peaks of the partially crystalline PP such as the transmittance peaks occurring at around wavelengths 1165, 972, and 843 cm^{-1}, corresponding to the asymmetrical -C-H stretching vibration, -CH_3C-C rocking, and -CH_2- rocking vibration of polypropylene, respectively [57,59].

The IR spectra of rPP30CBW and rPP30CBNW similarly indicated the presence of the functional group of carbon black within the matrices of polypropylene, as illustrated in spectra C and D, respectively. The spectra contain information about both the organic and inorganic parts of the materials. A peak at around 2120 cm^{-1} reveals the presence of a triple bond between C-C atoms (alkynes group) with stretching vibration, as reported in a previous publication [60]. The peak at around 2348–2350 cm^{-1} is ascribed to asymmetrical stretching vibrations of carbon dioxide molecules. Its presence could be associated with the porous nature of carbon black samples [61]. In addition, we found that the weak peaks around 840 cm^{-1} is attributed to out-of-plane deformation vibrations of C-H groups in aromatic structures [62]. The spectrum displays all the characteristic peaks of both carbon black and polypropylene. Thus, we deduced that no chemical bonding occurred between the carbon black and rPP30 molecules. Even though the spectra of the composites contain all the characteristic absorption bands of the polypropylene molecule (i.e., -CH_3, -CH_2, -CH, -C-C- stretching), but they are slightly shifted to lower wavelengths and exhibit predominantly low intensities. This phenomenon is probably caused by the effect of the intermolecular interaction between polypropylene chains and carbon black particles.

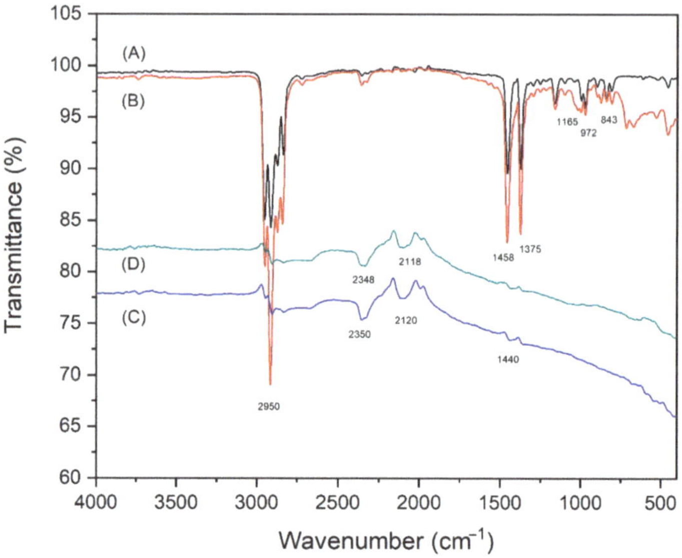

Figure 9. FTIR spectra of (**A**) vPP, (**B**) rPP30, (**C**) rPP30CBW, and (**D**) rPP30CBNW composites.

3.4. Surface Resistivity Test

We expected these obtained composite materials to have electrical conductivity properties to use as an antistatic material. Thus, the effect of carbon black on the surface resistance of PP was studied. The surface resistivity of rPP30CB composites as a function of plasticizer content is shown in Figure 10 and also enumerated in detail in Table 3. We found that vPP displayed high surface resistivity of approximately 10^{12} Ohm/sq, while this value was reduced to 10^{4} Ohm/sq when carbon black was added into the rPP matrix. Obviously, this indicates that carbon black can improve the conductivity property of polypropylene. It is well known that PP is a natural electrical insulator with high surface resistivity. This property means that the polymer surface can resist the generation of charges [63]. We added conductive fillers such as carbon black into those materials to effectively improve this property by reducing the surface resistivity. In general, when the composite material becomes conductive enough, electrons are able to transfer from one surface to another, resulting in the electrostatic shielding property of materials [64,65]. Additionally, we found that the PE wax, used as a plasticizer for processing, is not affected by the surface resistivity of polypropylene. These results indicate that the surface resistivity of the obtained composites could be suitable for use in engineering materials applications.

Table 3. The surface resistivity of vPP, rPP30CBNW and rPP30CBW composites.

Sample Codes	Surface Resistivity (Ohm/sq)
vPP	10^{12}
rPP30CBNW	10^{4}
rPP30CBW	10^{4}

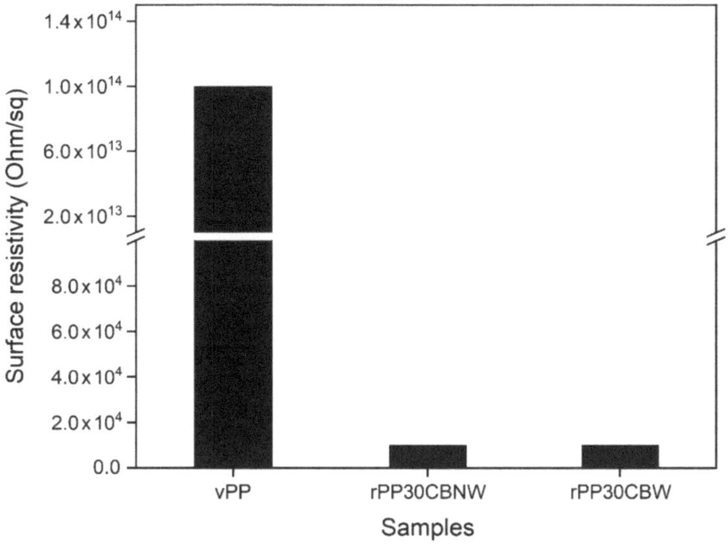

Figure 10. Comparison of the surface resistivity of the neat PP and the obtained rPP composites with and without the plasticizer content.

3.5. Differential Scanning Calorimetry Test

It is well known that the crystalline structure and the degree of crystallinity of polymers can affect the physical and mechanical properties of the material. Thus, DSC was performed in order to detect the crystallization and melting temperatures of all the specimens. The thermal behaviors of vPP, rPP30 and its composites are presented in Figure 11. The numerical values of the DSC thermograms were extracted and the results are summarized in Table 4. No significant difference was observed between the values measured before and after recycling for the vPP and rPP30 samples. The melting temperature of the rPP30 higher than the of vPP caused by the shorted of rPP30 can be reacted with the surface of neat PP. However, a small difference was noticed between the values of the vPP and the waste sample based on this polymer for polypropylene. Besides, it was found that rPP30 showed the small exothermic peak at 125.8 °C. This is probably attributed to admixtures of additives in the commercial waste products, which was also observed in the melting thermograms as smaller and broader curves. In addition, we found that the melting temperature peaks of rPP30 composites, both in rPP30CBNW and rPP30CBW, were higher than those of rPP30 and vPP, respectively. This result shows that the addition of carbon black particles acts as a nucleating agent, which increases the melting temperature of PP [66]. According to our investigation, in the rPP30 composites with and without PE wax (rPP30CBW and rPP30CBNW) within the rPP matrix, the melting peak temperature raised from 166.6 °C (for rPPCBW) to 167.4 °C, respectively. This indicated that the PE wax can act as a plasticizer agent, resulting in the reduction of the melting temperature of the rPP30 composite. However, the PE wax content did not show a significant influence on the thermal properties of the composites. Additionally, the small broader exothermic peaks of rPP30CBNW and rPP30CBW were found to be at 93.1 and 91.2 °C, respectively. This is probably attributed to admixtures of additives (such as carbon black and carbon black with PE wax, respectively) into the recycled composites. Interestingly, we found that the crystalline temperature enormously increased when adding rPP into the vPP matrix, as can been see in Figure 11a. rPP30 showed the highest values of crystalline temperature (approximately 121 °C) among the four samples. As we mentioned earlier, this is probably because rPP30 can be reacted with the surface of polypropylene and thus increases the crystalline temperature and the degree of crystallinity (57.10%). It was found that the degree of crystallinity increased up to approximately 26.27% when compared to the degree of crystallinity of vPP. However, in these cases, carbon black and PE wax slightly affected the crystalline temperature and the low-intensity and broader endothermic peaks were at 88.7 and 88.0 °C, respectively. The degree of crystallinity of rPP30CBNW and rPP30CBW did not significantly change and these values were calculated to be 40.36% and 39.03%, respectively. In addition, we found that the melting enthalpy of vPP was higher than that of rPP30, rPP30CBNW and rPP30CBW. As presented in Table 4, the melting enthalpy of PP decreased from 93.60 to 56.59 J/g while the fillers, such as rPP, carbon black, and PE wax, were added into PP matrices. This could indicate that the thermal stability of PP increases when adding fillers because these fillers absorb more heat energy in the melting of the composites. This finding is in agreement with the results of other literatures [67–69].

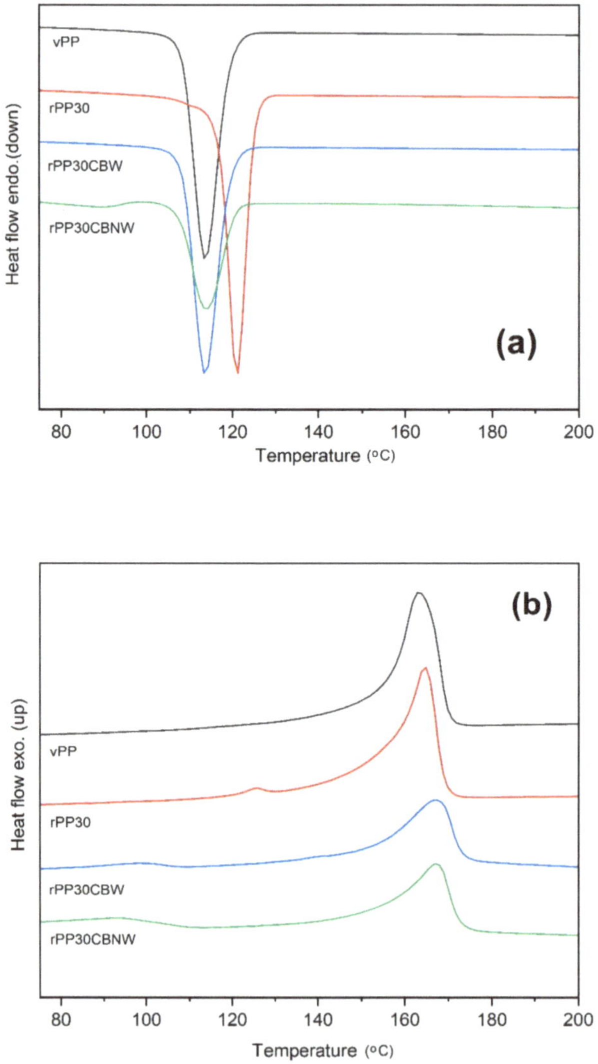

Figure 11. Crystallization temperature (**a**) and melting temperature (**b**) of vPP, rPP30, and its composites.

Table 4. Summarized results of DSC analysis of vPP, rPP30 and its composites.

Sample Codes	T_c (°C)	T_m (°C)	ΔH_m(J/g)	χ_c (%)
vPP	113.6	163.2	93.60	45.22
rPP30	121.0	164.6	82.74	57.10
rPP30CBNW	113.8	167.4	58.48	40.36
rPP30CBW	113.1	166.6	56.59	39.03

Abbreviations: PP, polypropylene; CB, carbon black; NW, non-PE wax; W, PE wax.

4. Conclusions

This study aimed at developing alternative upcycling polypropylene waste and anti-static composite materials with high performance and acceptable mechanical properties for various applications. The vPP/rPP blends and rPP/carbon black composites were successfully fabricated using a melt-extrusion process. All samples showed good performance of mechanical properties. However, the modulus and tensile strength of PP slightly decreased with increased recycled PP concentrations. The modulus and tensile strength of the vPP were 1317 and 36.47 MPa, respectively. These values were slightly higher than those of the vPP blended with rPP. On the other hand, the elongation at break of vPP was 15.09%. Our studies indicated that rPP can be blended with vPP at a concentration of up to 40 wt% without changing the mechanical properties significantly. In addition, it was found that MFI of the blended vPP increased with increasing rPP content.

With value-added materials in mind, we improved the electrical conductivity of recycled PP by using carbon black as a conductive filler. We found that carbon black affected the mechanical and electrical properties. The stress-strain curves of the PP composites changed from ductile to brittle behaviors. This might be caused by the poor interfacial intermolecular interaction between recycled PP and carbon black. FTIR spectra indicates that carbon black did not have any chemical reactions or bonding with the PP chains. Finally, as expected, the obtained PP-based composites exhibited good performance on the electrical properties tested. We found that vPP displayed high surface resistivity of approximately 10^{12} Ohm/sq, while this value was reduced to 10^4 Ohm/sq when adding carbon black into rPP matrices. DSC results showed that the rPP and carbon black powder could act as nucleating agents and thus increase the degree of crystallinity of PP. The obtained rPP30/carbon black composites have potential to serve as a conductive material for various applications in the future.

Supplementary Materials: The following supporting information can be downloaded at: https://www.mdpi.com/article/10.3390/polym14245482/s1, Figure S1: FTIR spectra of vPP and rPP.

Author Contributions: Conceptualization, T.J., P.P. and C.-F.H.; methodology and investigation, P.W., W.S., P.J. and R.M.; physical tests and analysis, N.I., W.S. and H.I.; Writing—Original Draft, T.J. and N.I.; Writing—Review & Editing, M.S., H.I. and C.-F.H. All authors have read and agreed to the published version of the manuscript.

Funding: The present work was financially co-supported by National Research Council of Thailand (NRCT), Thailand, and the two-institution co-research scholarship provided by Kasetsart University and National Chung Hsing University, grant number 00072022.

Institutional Review Board Statement: Not applicable.

Informed Consent Statement: Not applicable.

Data Availability Statement: Not applicable.

Acknowledgments: We gratefully acknowledge the Faculty of Science at Sriracha, Kasetsart University and the Faculty of Science, Chulalongkorn University for providing funding and convenient laboratories, respectively. The authors also would like to thank the anonymous referees for their helpful comments on this article.

Conflicts of Interest: The authors declare no conflict of interest.

References

1. Amrullah, A.; Farobie, O.; Septarini, S.; Satrio, J.A. Synergetic biofuel production from co-pyrolysis of food and plastic waste: Reaction kinetics and product behavior. *Heliyon* **2022**, *8*, e10278. [CrossRef]
2. Kameel, N.I.A.; Daud, W.M.A.W.; Patah, M.F.A.; Zulkifli, N.W.M. Influence of reaction parameters on thermal liquefaction of plastic wastes into oil: A review. *Energy Convers. Manag. X* **2022**, *14*, 100196. [CrossRef]
3. Sangroniz, A.; Zhu, J.-B.; Tang, X.; Etxeberria, A.; Chen, E.Y.-X.; Sardon, H. Packaging materials with desired mechanical and barrier properties and full chemical recyclability. *Nat. Commun.* **2019**, *10*, 3559. [CrossRef]
4. Râpă, M.; Spurcaciu, B.N.; Ion, R.-M.; Grigorescu, R.M.; Darie-Niță, R.N.; Iancu, L.; Nicolae, C.-A.; Gabor, A.R.; Matei, E.; Predescu, C. Valorization of Polypropylene Waste in the Production of New Materials with Adequate Mechanical and Thermal Properties for Environmental Protection. *Materials* **2022**, *15*, 5978. [CrossRef]
5. Zhao, X.; Korey, M.; Li, K.; Copenhaver, K.; Tekinalp, H.; Celik, S.; Kalaitzidou, K.; Ruan, R.; Ragauskas, A.J.; Ozcan, S. Plastic waste upcycling toward a circular economy. *Chem. Eng. J.* **2021**, *428*, 131928. [CrossRef]
6. Takarada, W.; Barique, M.A.; Kunimitsu, T.; Kameda, T.; Kikutani, T. Verification of the Influence of Processing History through Comparing High-Speed Melt Spinning Behavior of Virgin and Recycled Polypropylene. *Polymers* **2022**, *14*, 3238. [CrossRef]
7. Cabernard, L.; Pfister, S.; Oberschelp, C.; Hellweg, S. Growing environmental footprint of plastics driven by coal combustion. *Nat. Sustain.* **2021**, *5*, 139–148. [CrossRef]
8. Huang, Q.; Chen, G.; Wang, Y.; Chen, S.; Xu, L.; Wang, R. Modelling the global impact of China's ban on plastic waste imports. *Resour. Conserv. Recycl.* **2019**, *154*, 104607. [CrossRef]
9. Morales, M.A.; Maranon, A.; Hernandez, C.; Porras, A. Development and Characterization of a 3D Printed Cocoa Bean Shell Filled Recycled Polypropylene for Sustainable Composites. *Polymers* **2021**, *13*, 3162. [CrossRef]
10. Alhazmi, H.; Almansour, F.; Aldhafeeri, Z. Plastic Waste Management: A Review of Existing Life Cycle Assessment Studies. *Sustainability* **2021**, *13*, 5340. [CrossRef]
11. Aboulkas, A.; El Harfi, K.; El Bouadili, A. Thermal degradation behaviors of polyethylene and polypropylene. Part I: Pyrolysis kinetics and mechanisms. *Energy Convers. Manag.* **2010**, *51*, 1363–1369. [CrossRef]
12. Fletes, R.C.V.; López, E.O.C.; Sánchez, F.J.M.; Mendizábal, E.; Núñez, R.G.; Rodrigue, D.; Gudiño, P.O. Morphological and Mechanical Properties of Bilayers Wood-Plastic Composites and Foams Obtained by Rotational Molding. *Polymers* **2020**, *12*, 503. [CrossRef]
13. Oh, S.; Stache, E.E. Chemical Upcycling of Commercial Polystyrene via Catalyst-Controlled Photooxidation. *J. Am. Chem. Soc.* **2022**, *144*, 5745–5749. [CrossRef]
14. Lewandowski, K.; Skórczewska, K. A Brief Review of Poly(Vinyl Chloride) (PVC) Recycling. *Polymers* **2022**, *14*, 3035. [CrossRef]
15. Assaad, J.J.; Khalil, M.; Khatib, J. Alternatives to Enhance the Structural Performance of PET-Modified Reinforced Concrete Beams. *Environments* **2022**, *9*, 37. [CrossRef]
16. Assaad, J.J.; Khatib, J.M.; Ghanem, R. Bond to Bar Reinforcement of PET-Modified Concrete Containing Natural or Recycled Coarse Aggregates. *Environments* **2022**, *9*, 8. [CrossRef]
17. Dziuba, R.; Kucharska, M.; Madej-Kiełbik, L.; Sulak, K.; Wiśniewska-Wrona, M. Biopolymers and Biomaterials for Special Applications within the Context of the Circular Economy. *Materials* **2021**, *14*, 7704. [CrossRef]
18. Soares, C.T.D.M.; Ek, M.; Östmark, E.; Gällstedt, M.; Karlsson, S. Recycling of multi-material multilayer plastic packaging: Current trends and future scenarios. *Resour. Conserv. Recycl.* **2021**, *176*, 105905. [CrossRef]
19. Benyathiar, P.; Kumar, P.; Carpenter, G.; Brace, J.; Mishra, D.K. Polyethylene Terephthalate (PET) Bottle-to-Bottle Recycling for the Beverage Industry: A Review. *Polymers* **2022**, *14*, 2366. [CrossRef]
20. Hahladakis, J.N.; Velis, C.A.; Weber, R.; Iacovidou, E.; Purnell, P. An overview of chemical additives present in plastics: Migration, release, fate and environmental impact during their use, disposal and recycling. *J. Hazard. Mater.* **2018**, *344*, 179–199. [CrossRef]
21. Gharde, S.; Kandasubramanian, B. Mechanothermal and chemical recycling methodologies for the Fibre Reinforced Plastic (FRP). *Environ. Technol. Innov.* **2019**, *14*, 100311. [CrossRef]
22. He, P.; Chen, L.; Shao, L.; Zhang, H.; Lü, F. Municipal solid waste (MSW) landfill: A source of microplastics? -Evidence of microplastics in landfill leachate. *Water Res.* **2019**, *159*, 38–45. [CrossRef] [PubMed]
23. Kim, Y.; Lim, J.; Cho, H.; Kim, J. Novel mechanical vapor recompression-assisted evaporation process for improving energy efficiency in pulp and paper industry. *Int. J. Energy Res.* **2021**, *46*, 3409–3427. [CrossRef]
24. Lee, J.; Ahn, Y.; Cho, H.; Kim, J. Economic performance assessment of elemental sulfur recovery with carbonate melt desulfurization process. *Process Saf. Environ. Prot.* **2022**, *158*, 123–133. [CrossRef]
25. Fico, D.; Rizzo, D.; De Carolis, V.; Montagna, F.; Corcione, C.E. Sustainable Polymer Composites Manufacturing through 3D Printing Technologies by Using Recycled Polymer and Filler. *Polymers* **2022**, *14*, 3756. [CrossRef]
26. Canevarolo, S.V. Chain scission distribution function for polypropylene degradation during multiple extrusions. *Polym. Degrad. Stab.* **2000**, *70*, 71–76. [CrossRef]
27. Jin, H.; Gonzalez-Gutierrez, J.; Oblak, P.; Zupancic, B.; Emri, I. The effect of extensive mechanical recycling on the properties of ow density polyethylene. *Polym. Degrad. Stab.* **2012**, *97*, 2262–2272. [CrossRef]
28. Tominaga, A.; Sekiguchi, H.; Nakano, R.; Yao, S.; Takatori, E. Thermal process-dependence of the mechanical properties and inner structures of pre-consumer recycled polypropylene. *AIP Conf. Proc.* **2015**, *1664*, 150011. [CrossRef]

29. Vidakis, N.; Petousis, M.; Velidakis, E.; Mountakis, N.; Fischer-Griffiths, P.E.; Grammatikos, S.A.; Tzounis, L. Fused Filament Fabrication 3D printed polypropylene/alumina nanocomposites: Effect of filler loading on the mechanical reinforcement. *Polym. Test.* **2022**, *109*, 107545. [CrossRef]
30. Kolek, Z. Recycled polymers from food packaging in relation to environmental protection. *Pol. J. Environ. Stud.* **2001**, *10*, 73–76.
31. Jiun, Y.L.; Tze, C.T.; Moosa, U.; Tawawneh, M.A. Effects of Recycling Cycle on Used Thermoplastic Polymer and Thermoplastic Elastomer Polymer. *Polym. Polym. Compos.* **2016**, *24*, 735–740. [CrossRef]
32. Reis, J.; Pacheco, L.; Mattos, H.D.C. Tensile behavior of post-consumer recycled high-density polyethylene at different strain rates. *Polym. Test.* **2013**, *32*, 338–342. [CrossRef]
33. Żenkiewicz, M.; Dzwonkowski, J. Effects of electron radiation and compatibilizers on impact strength of composites of recycled polymers. *Polym. Test.* **2007**, *26*, 903–907. [CrossRef]
34. Yam, K.L.; Gogoi, B.K.; Lai, C.C.; Selke, S.E. Composites from compounding wood fibers with recycled high density polyethylene. *Polym. Eng. Sci.* **1990**, *30*, 693–699. [CrossRef]
35. Poulakis, J.; Papaspyrides, C. Recycling of polypropylene by the dissolution/reprecipitation technique: I. A model study. *Resour. Conserv. Recycl.* **1997**, *20*, 31–41. [CrossRef]
36. Achilias, D.S.; Roupakias, C.; Megalokonomos, P.; Lappas, A.A.; Antonakou, E.V. Chemical recycling of plastic wastes made from polyethylene (LDPE and HDPE) and polypropylene (PP). *J. Hazard. Mater.* **2007**, *149*, 536–542. [CrossRef]
37. Deka, B.; Maji, T. Study on the properties of nanocomposite based on high density polyethylene, polypropylene, polyvinyl chloride and wood. *Compos. Part A Appl. Sci. Manuf.* **2011**, *42*, 686–693. [CrossRef]
38. Okan, M.; Aydin, H.M.; Barsbay, M. Current approaches to waste polymer utilization and minimization: A review. *J. Chem. Technol. Biotechnol.* **2018**, *94*, 8–21. [CrossRef]
39. Cardona, S.C.; Corma, A. Tertiary recycling of polypropylene by catalytic cracking in a semibatch stirred reactor: Use of spent equilibrium FCC commercial catalyst. *Appl. Catal. B Environ.* **2000**, *25*, 151–162. [CrossRef]
40. Ragab, H.; Algethami, N.; Elamin, N.Y.; Asnag, G.; Rajeh, A.; Alzahrani, H.S. An insight into the influence of Ag/Se nanoparticles on the structural, optical, and electrical properties of Cs/PAM nanocomposites films as application in electrochemical devices. *J. Mol. Struct.* **2022**, *1267*, 133619. [CrossRef]
41. Xu, Z.; Wang, N.; Li, N.; Zheng, G.; Dai, K.; Liu, C.; Shen, C. Liquid sensing behaviors of conductive polypropylene composites containing hybrid fillers of carbon fiber and carbon black. *Compos. Part B Eng.* **2016**, *94*, 45–51. [CrossRef]
42. Orozco, F.; Salvatore, A.; Sakulmankongsuk, A.; Gomes, D.R.; Pei, Y.; Araya-Hermosilla, E.; Pucci, A.; Moreno-Villoslada, I.; Picchioni, F.; Bose, R.K. Electroactive performance and cost evaluation of carbon nanotubes and carbon black as conductive fillers in self-healing shape memory polymers and other composites. *Polymer* **2022**, *260*, 125365. [CrossRef]
43. Yang, H.; Guan, Y.; Ye, L.; Wang, S.; Li, S.; Wen, X.; Chen, X.; Mijowska, E.; Tang, T. Synergistic effect of nanoscale carbon black and ammonium polyphosphate on improving thermal stability and flame retardancy of polypropylene: A reactive network for strengthening carbon layer. *Compos. Part B Eng.* **2019**, *174*, 107038. [CrossRef]
44. Ali, A.; Azeem, M.; Noman, M.T.; Amor, N.; Militky, J.; Petru, M.; Wang, Y.; Masin, I. Development of silver plated electrically conductive elastomers embedded with carbon black particles obtained from Kevlar waste source. *Polym. Test.* **2022**, *116*, 107793. [CrossRef]
45. *ASTM D638*; Standard Test Methods for Tensile Properties of Plastics. ASTM International: West Conshohocken, PA, USA, 2010.
46. *ASTM D257*; Standard Test Method for DC Resistance of Conductance of Insulating Materials. ASTM International: West Conshohocken, PA, USA, 2010.
47. Abadchi, M.R.; Jalali-Arani, A. Crystallization and melting behavior of polypropylene (PP) in (vulcanized nanoscale polybutadiene rubber powder/PP) polymer-nanocomposites. *Thermochim. Acta* **2015**, *617*, 120–128. [CrossRef]
48. Xu, J.; Hao, X.; Tang, W.; Zhou, H.; Chen, L.; Guo, C.; Wang, Q.; Ou, R. Mechanical properties, morphology, and creep resistance of ultra-highly filled bamboo fiber/polypropylene composites: Effects of filler content and melt flow index of polypropylene. *Constr. Build. Mater.* **2021**, *310*, 125289. [CrossRef]
49. Wu, C.L.; Zhang, M.Q.; Rong, M.Z.; Friedrich, K. Silica nanoparticles filled polypropylene: Effects of particle surface treatment, matrix ductility and particle species on mechanical performance of the composites. *Compos. Sci. Technol.* **2005**, *65*, 635–645. [CrossRef]
50. Jamnongkan, T.; Jaroensuk, O.; Khankhuean, A.; Laobuthee, A.; Srisawat, N.; Pangon, A.; Mongkholrattanasit, R.; Phuengphai, P.; Wattanakornsiri, A.; Huang, C.-F. A Comprehensive Evaluation of Mechanical, Thermal, and Antibacterial Properties of PLA/ZnO Nanoflower Biocomposite Filaments for 3D Printing Application. *Polymers* **2022**, *14*, 600. [CrossRef]
51. Anosike-Francis, E.N.; Obianyo, I.I.; Salami, O.W.; Ihekweme, G.O.; Ofem, M.I.; Olorunnisola, A.O.; Onwualu, A.P. Physical-Mechanical properties of wood based composite reinforced with recycled polypropylene and cowpea (Vigna unguiculata Walp.) husk. *Clean. Mater.* **2022**, *5*, 100101. [CrossRef]
52. Faraj, R.H.; Sherwani, A.F.H.; Daraei, A. Mechanical, fracture and durability properties of self-compacting high strength concrete containing recycled polypropylene plastic particles. *J. Build. Eng.* **2019**, *25*, 100808. [CrossRef]
53. Lin, T.A.; Lin, J.-H.; Bao, L. Polypropylene/thermoplastic polyurethane blends: Mechanical characterizations, recyclability and sustainable development of thermoplastic materials. *J. Mater. Res. Technol.* **2020**, *9*, 5304–5312. [CrossRef]
54. Xiao, H.; Lu, W.; Yeh, J.-T. Effect of plasticizer on the crystallization behavior of poly(lactic acid). *J. Appl. Polym. Sci.* **2009**, *113*, 112–121. [CrossRef]

55. Yang, K.; Yang, Q.; Li, G.; Zhang, Y.; Zhang, P. Mechanical properties and morphologies of polypropylene/single-filler or hybrid-filler calcium carbonate composites. *Polym. Eng. Sci.* **2007**, *47*, 95–102. [CrossRef]
56. Abd El-Fattah, A.; Youssef, H.; Gepreel, M.A.H.; Abbas, R.; Kandil, S. Surface Morphology and Mechanical Properties of Polyether Ether Ketone (PEEK) Nanocomposites Reinforced by Nano-Sized Silica (SiO_2) for Prosthodontics and Restorative Dentistry. *Polymers* **2021**, *13*, 3006. [CrossRef]
57. Caban, R. FTIR-ATR spectroscopic, thermal and microstructural studies on polypropylene-glass fiber composites. *J. Mol. Struct.* **2022**, *1264*, 133181. [CrossRef]
58. Shao, Y.; Guizani, C.; Grosseau, P.; Chaussy, D.; Beneventi, D. Biocarbons from microfibrillated cellulose/lignosulfonate precursors: A study of electrical conductivity development during slow pyrolysis. *Carbon* **2018**, *129*, 357–366. [CrossRef]
59. Lanyi, F.J.; Wenzke, N.; Kaschta, J.; Schubert, D.W. A method to reveal bulk and surface crystallinity of Polypropylene by FTIR spectroscopy—Suitable for fibers and nonwovens. *Polym. Test.* **2018**, *71*, 49–55. [CrossRef]
60. Jamnongkan, T.; Intaramongkol, N.; Kanjanaphong, N.; Ponjaroen, K.; Sriwiset, W.; Mongkholrattanasit, R.; Wongwachirakorn, P.; Lin, K.-Y.A.; Huang, C.-F. Study of the Enhancements of Porous Structures of Activated Carbons Produced from Durian Husk Wastes. *Sustainability* **2022**, *14*, 5896. [CrossRef]
61. Gómez-Hernández, R.; Panecatl-Bernal, Y.; Méndez-Rojas, M. High yield and simple one-step production of carbon black nanoparticles from waste tires. *Heliyon* **2019**, *5*, e02139. [CrossRef]
62. Moreno-Castilla, C.; Lopez-Ramon, M.V.; Carrasco-Marín, F. Changes in surface chemistry of activated carbons by wet oxidation. *Carbon* **2000**, *38*, 1995–2001. [CrossRef]
63. Liu, W.; Cheng, L.; Li, S. Review of electrical properties for polypropylene based nanocomposite. *Compos. Commun.* **2018**, *10*, 221–225. [CrossRef]
64. Zhang, R.-Y.; Zhao, Q.; Ge, M.-L. The effect of electrostatic shielding using invisibility cloak. *AIP Adv.* **2011**, *1*, 042126. [CrossRef]
65. Srihata, W.; Jamnongkan, T.; Rattanasak, U.; Boonsang, S.; Kaewpirom, S. Enhanced electrostatic dissipative properties of chitosan/gelatin composite films filled with reduced graphene oxide. *J. Mater. Sci. Mater. Electron.* **2016**, *28*, 999–1010. [CrossRef]
66. Su, Z.; Dong, M.; Guo, Z.; Yu, J. Study of Polystyrene and Acrylonitrile–Styrene Copolymer as Special β-Nucleating Agents To Induce the Crystallization of Isotactic Polypropylene. *Macromolecules* **2007**, *40*, 4217–4224. [CrossRef]
67. Beate, K.; Regine, B.; Liane, H.; Petra, P. Ultralow percolation threshold in polyamide 6.6/MWCNT composites. *Compos. Sci. Technol.* **2015**, *114*, 119–125.
68. Yetgin, S.H. Effect of multi walled carbon nanotube on mechanical, thermal and rheological properties of polypropylene. *J. Mater. Res. Technol.* **2019**, *8*, 4725–4735. [CrossRef]
69. Alghyamah, A.A.; Elnour, A.Y.; Shaikh, H.; Haider, S.; Poulose, A.M.; Al-Zahrani, S.; Almasry, W.A.; Park, S.Y. Biochar/polypropylene composites: A study on the effect of pyrolysis temperature on crystallization kinetics, crystalline structure, and thermal stability. *J. King Saud Univ. Sci.* **2021**, *33*, 101409. [CrossRef]

polymers

MDPI

Article

High-Pressure Depolymerization of Poly(lactic acid) (PLA) and Poly(3-hydroxybutyrate) (PHB) Using Bio-Based Solvents: A Way to Produce Alkyl Esters Which Can Be Modified to Polymerizable Monomers

Vojtěch Jašek [1,*], Jan Fučík [2], Lucia Ivanová [2], Dominik Veselý [2], Silvestr Figalla [1], Ludmila Mravcova [2], Petr Sedlacek [3], Jozef Krajčovič [2] and Radek Přikryl [1]

[1] Institute of Materials Chemistry, Faculty of Chemistry, Brno University of Technology, 61200 Brno, Czech Republic

[2] Institute of Environmental Chemistry, Faculty of Chemistry, Brno University of Technology, 61200 Brno, Czech Republic

[3] Institute of Physical and Applied Chemistry, Faculty of Chemistry, Brno University of Technology, 61200 Brno, Czech Republic

* Correspondence: xcjasekv@vutbr.cz

Citation: Jašek, V.; Fučík, J.; Ivanová, L.; Veselý, D.; Figalla, S.; Mravcova, L.; Sedlacek, P.; Krajčovič, J.; Přikryl, R. High-Pressure Depolymerization of Poly(lactic acid) (PLA) and Poly(3-hydroxybutyrate) (PHB) Using Bio-Based Solvents: A Way to Produce Alkyl Esters Which Can Be Modified to Polymerizable Monomers. *Polymers* **2022**, *14*, 5236. https://doi.org/10.3390/polym14235236

Academic Editors: Antonio M. Borrero-López, Concepción Valencia-Barragán, Esperanza Cortés Triviño, Adrián Tenorio-Alfonso and Clara Delgado-Sánchez

Received: 8 November 2022
Accepted: 29 November 2022
Published: 1 December 2022

Abstract: The polyesters poly(lactic acid) (PLA) and poly(3-hydroxybutyrate) (PHB) used in various applications such as food packaging or 3D printing were depolymerized by biobased aliphatic alcohols—methanol and ethanol with the presence of *para*-toluenesulphonic acid (*p*-TSA) as a catalyst at a temperature of 151 °C. It was found that the fastest depolymerization is reached using methanol as anucleophile for the reaction with PLA, resulting in the value of reaction rate constant (k) of 0.0425 min^{-1} and the yield of methyl lactate of 93.8% after 120 min. On the other hand, the value of constant k for the depolymerization of PHB in the presence of ethanol reached 0.0064 min^{-1} and the yield of ethyl 3-hydroxybutyrate was of 76.0% after 240 min. A kinetics study of depolymerization was performed via LC–MS analysis of alkyl esters of lactic acid and 3-hydroxybutanoic acid. The structure confirmation of the products was performed via FT-IR, MS, ^{1}H NMR, and ^{13}C NMR. Synthesized alkyl lactates and 3-hydroxybutyrates were modified into polymerizable molecules using methacrylic anhydride as a reactant and potassium 2-ethylhexanoate as a catalyst at a temperature of 80 °C. All alkyl esters were methacrylated for 24 h, guaranteeing the quantitative yield (which in all cases reached values equal to or of more than 98%). The methacrylation rate constants (k') were calculated to compare the reaction kinetics of each alkyl ester. It was found that lactates reach afaster rate of reaction than 3-hydroxybutyrates. The value of k' for themethacrylated methyl lactate reached 0.0885 $dm^3/(mol{\cdot}min)$. Opposite to this result, methacrylated ethyl 3-hydroxybutyrate's constant k' was 0.0075 $dm^3/(mol{\cdot}min)$. The reaction rate study was conducted by the GC-FID method and the structures were confirmed via FT-IR, MS, ^{1}H NMR, and ^{13}C NMR.

Keywords: poly(lactic acid); poly(3-hydroxybutyrate); depolymerization; alcoholysis; methacrylation; polymerizable monomers; kinetics

1. Introduction

In recent years, processes producing plastic materials mostly use fossil-based polymers due to the fact that these molecules are cheap to obtain and their properties can be determined efficiently. However, this type of manufacturing is dependent on non-renewable resources which might lead to potential supply risks [1–4]. Therefore, various biopolymers based on bio-source inputs such as PLA or PHB attract a significant amount of attention [5,6]. Poly(lactic acid) (PLA) is a particular biopolymer used for numerous applications such as food packaging or 3D filament printing [2,7,8]. The increasing usage of this polymer leads to considerable problems—the rate of degradation of PLA in moderate environmental conditions is very slow; therefore, the accumulation of waste can occur [1,9,10]. There are ways to handle PLA waste, such as composting, incinerating, or mechanical recycling [7,11,12]. The first two processes do not generate any usable material and the mechanical recycling of PLA results in the production of a polymer with considerably worse properties than the original one. The reason for this outcome is that the thermal and photochemical degradation takes place during the mechanical processing of the PLA polymer, which leads to the decrease in molecular weight of the material [2,7,13,14]. Poly(3-hydroxybutyrate) (PHB) is a biopolymer that is produced by various microorganisms and can be degraded by living systems as well. The biodegradability of PHB is aconsiderable benefit of this polyester [2,15,16]. Nevertheless, PHB suffers rapid thermal degradation during the potential mechanical recycling, similarly to PLA, particularly at temperatures starting at 170–180 °C [2,17]. These two biopolymers are miscible, which is beneficial for a potential 3D printing usage, for example [18,19]. However, the fact that mixtures of numerous polymers and plasticizers are used complicates their potential recycling [20].

The chemical recycling of polyesters such as PLA or PHB is another way of handling leftover polymers or those with bad quality or properties such as low values of MW. This type of process results in forming monomers out of the polyester polymer chain. The produced monomers can either be carboxylic acids or esters depending on the chosen nucleophile. When water is used as a nucleophile, hydrolysis takes the place of the depolymerization process. Alcoholysis, on the other hand, requires alcohol to undergo the reaction, resulting in the formation of an ester via the transesterification mechanism [21–27]. These processes involve either specific reaction conditions (high pressure, high temperature) or the presence of a particular catalyst and appropriate organic solvents. Extreme conditions such as high temperature (above 120 °C) and high pressure (depending on the chosen nucleophile) can be complicated in terms of up-scaling or energy consumption [28–31]. On the other hand, specific catalysts can be expensive, and particularly organic solvents may be inappropriate for regeneration in the production process and lower temperatures could result in a decrease in the reaction rate [30,32]. The hydrolysis of PLA was experimentally verified using a temperature of 250 °C and high pressure. The molar ratio of 1:20 (PLA:H_2O) and a reaction time of 10–20 min resulted in a 90% conversion to L-lactic acid [33–35]. Adding microwave irradiation to the process of hydrolysis with aratio of 1:3 (PLA:H_2O) provided a 45% conversion to l-lactic acid after 120 min [34,35]. PHB alcoholysis was observed using ionic liquids as catalysts and methanol as a nucleophile. The highest yield of methyl 3-hydroxybutyrate (83.75%) was reached with the mixture of the molar ratio 5:1 (MeOH:PHB) and of the mass ratio 1:0.03 (PHB:cat.) with the conditions of a temperature of 140 °C and a reaction time of 3 h [36–38].

The main aim of this work is to describe a depolymerization process of the polyesters PLA and PHB in high-pressure and high-temperature conditions. The nucleophiles chosen for the reactions are methanol and ethanol and the reaction takes place in ahigh-pressure reactor. *Para*-toluensulphonic acid serves as an acidic catalyst for all alcoholyses. The monoester product structures are verified via numerous analyses (MS, FT-IR, ^{1}H NMR, ^{13}C NMR) and their reaction kinetics are studied. All synthesized monoesters (lactates and 3-hydroxybutyrates) undergo a methacrylation process to produce methacrylated esters which can be polymerized. The methacrylation reaction uses methacrylic anhydride as areagent and the chosen catalyst for this reaction is potassium 2-ethylhexanoate. The polymerizable products' structures are verified by numerous methods (MS, FT-IR, ^{1}H NMR, ^{13}C NMR) as well.

2. Materials and Methods

2.1. Materials

PLA granulate was supplied from Fillamentum Manufacturing Czech s.r.o., Hulín, Czech Republic. Measured polymer parameters were as follows: number-average molecular weight (M_n), 123,500 g/mol; weight-average molecular weight (MW), 235,300 g/mol; dispersity, 1.90. PHB powder was acquired from NAFIGATE Corporation a.s., Prague,

Czech Republic. Measured polymer parameters were as follows: number-average molecular weight (M_n), 85,040 g/mol; weight-average molecular weight (MW), 211,400 g/mol; dispersity, 2.49. All measurements of polymer properties were measured via GPC (Agilent 1100, Santa Clara, CA, USA) in chloroform ($CHCl_3$) and the analysis parameters were as follows: mobile phase flow 1 mL/min; column temperature 30 °C, used column: PLgel 5 µm MIXED-C (300 × 7.5 mm). Aliphatic alcohols for depolymerization (methanol 99%, ethanol 99%) were supplied by Honeywell Research Chemicals, Charlotte, NC, USA (used alcohols were not claimed either synthetic or bio-source by the supplier). The catalyst for alcoholyses (*p*-toluensulphonic acid monohydrate), methacrylic anhydride (94%), potassium hydroxide (p.a.), d-chloroform ($CDCl_3$; 99.8%), and 2-ethylhexanoic acid (for synthesis) were all acquired from Sigma-Aldrich, Prague, Czech Republic.

2.2. *Methods for the Characterization of Products*

2.2.1. Fourier-Transform Infrared Spectrometry (FT-IR)

Infrared spectrometry was used as one of the structure verification methods, but itwas mainly supposed to serve as a confirmation of –OH hydroxyl functional groups in alkyl esters of either lactic acid or 3-hydroxybutanoic acid. Analyses were performed on the infrared spectrometer Bruker Tensor 27 (Billerica, MA, USA) by the attenuated total reflectance (ATR) method using diamond as a dispersion component. The irradiation source in this type of spectrometer is a diode laser. Due to the fact that instrumentation uses Fourier transformation, the Michelson interferometer was used for the quantification of the signal. Spectra were composed out of 32 total scans with a measurement resolution of 2 cm^{-1}.

2.2.2. Mass Spectrometry (MS)

MS conditions were as follows: ESI in positive mode; spray voltage: 3500 V; cone temperature 350 °C; cone gas flow: 35 a.u.; heated probe temperature: 650 °C; probe gas flow: 40 a.u., nebulizer gas flow: 55 a.u., and exhaust gas: ON. For quantification, MRM mode was used with the following MRM transitions for MeLa (RT 1.24 min; 105.1 > 84.6 with CE 0.25 eV;105.1 > 93.9 with CE 0.25 eV and 105.1 > 45.0 with CE 2 eV), for EtLa (RT 2.35 min; 119.1 > 47.3 with CE 2.0 eV and 119.1 > 91.1 with CE 2.0 eV) for M3HB (RT 1.91 min; 119.1 > 59.0 with CE 10 eV; 119.1 > 87.2 with CE 2.5 eV; and 119.1 > 101.2 with CE 1.0 eV), and for E3HB (RT 3.58 min; 133.1 > 73.3 CE with 5 eV and 133.1 > 87.2 with CE 5 eV). The collision gas Argon was used ata pressure of 1.5 mTorr.

Additionally, by the same MS method, newly synthesized monomers methacrylated methyl lactate (MeLaMMA), methacrylated ethyl lactate (EtLaMMA), methacrylated methyl 3-hydroxybutyrate (M3HBMMA), and methacrylated ethyl 3-hydroxybutyrate (E3HBMMA) were qualitatively characterized by product scan; therefore, themass spectra of these compounds were obtained. The precursor of MeLAaMMA (*m/z* 173.1) was fragmented (CE 10 eV) and the following product ions were obtained: 57.3 and 74.2. The precursor of EtLaMMA (*m/z* 187.0) was fragmented (CE 2.5 eV) and the following product ions were obtained: 69.2; 113.1 and 141.1. The precursor of M3HBMMA (*m/z* 187.1) was fragmented (CE 2.5 eV) and the following product ions were obtained: 59.3; 69.2; 101.1 and 155.1. The precursor of E3HBMMA (*m/z* 201.0) was fragmented (CE 2.5 eV) and the following product ions were obtained: 69.0; 73.3; 115.1; and 155.0.

2.2.3. Nuclear Magnetic Resonance (NMR)

Nuclear magnetic resonance was used to obtain ^{1}H and ^{13}C spectra to confirm the structure of synthesized molecules. The measurements were conducted by instrument Bruker Avance III 500 MHz (Bruker, Billerica, MA, USA) with the measuring frequency of 500 MHz for ^{1}H NMR and 126 MHz for ^{13}C NMR at the temperature of 30 °C using d-chloroform ($CDCl_3$) as a solvent with tetramethylsilane (TMS) as an internal standard. The chemical shifts (δ) are expressed in part per million (ppm) units which are referenced by a solvent. Coupling constant *J* has (Hz) unit with coupling expressed as s—singlet, d—doublet, t—triplet, q—quartet, p—quintet, m—multiplet.

2.3. Alcoholyses of Polyesters

All depolymerization reactions took place in the high-pressure reactor of a volume reservoir of 1.8 L. Mixtures consisting of the polyester (PLA or PHB) and the alcohol (methanol or ethanol) in amolar ratio 1:4 (PLA/PHB:MeOH/EtOH) with the presence of dissolved catalyst p-toluensulphonic acid in a molar ratio 1:0.01 (PLA/PHB:*p*-TSA) were transferred into the reactor (see Scheme 1 for PLA and Scheme 2 for PHB). The reactor was heated up to 151 °C and the pressure increased regarding the particular type of suspense according to the vapor-pressure characteristics of the alcohol and forming ester product (ranging from 7 to 15 bar). The samples for kinetics analysis were taken directly from the reactor and cooled down immediately. The kinetics of forming monoesters was monitored via LC–MS analysis. The conversion of the particular polyester was calculated from the leftover polymer acquired by mixing the reaction solution sample with water. The unreacted polymer was precipitated and weighed. After the alcoholyses were stopped (depending on the particular combination of reagents) the leftover alcohol and the formed monoester were distilled from the solution. Products of the depolymerization were analyzed via MS, FT-IR, ^{1}H NMR, and^{13}C NMR for structure verification. Yields of monoester (Yield) and polymer conversions (*X*) were calculated as follows:

$$X = \frac{\text{Starting polymer weight} - \text{leftover polymer weight}}{\text{Starting polymer weight}} \times 100\% \quad (1)$$

$$\text{Yield} = \frac{\text{Measured quantity of ester}}{\text{Theoretical quantity of ester}} \times 100\% \quad (2)$$

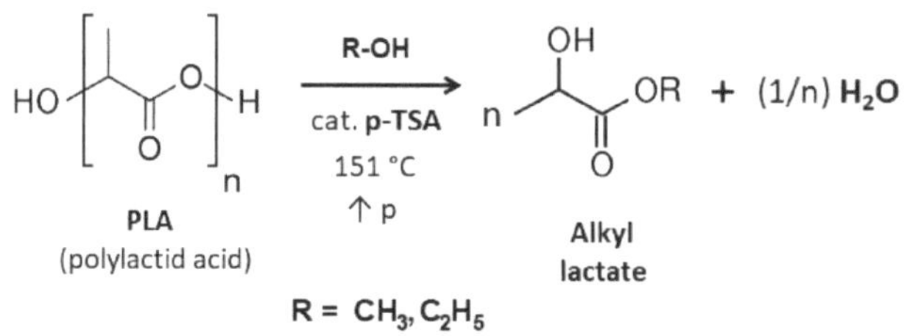

Scheme 1. Alcoholysis of polylactic acid (PLA) catalyzed by *p*-TSA (↑ p means pressure's increase).

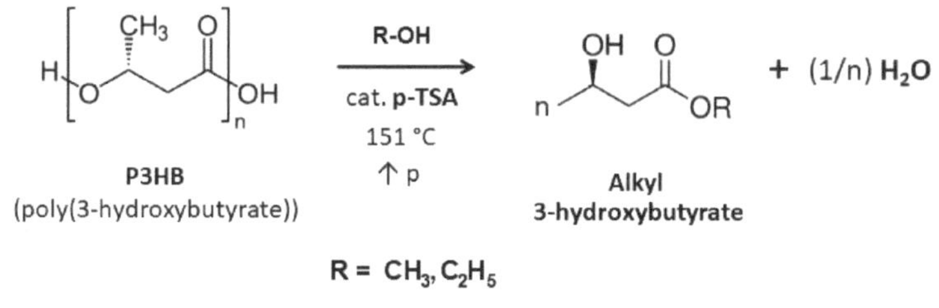

Scheme 2. Alcoholysis of poly(3-hydroxybutyrate) (PHB) catalyzed by p-TSA (↑ p means pressure's increase).

Methyl lactate (MeLa): ^{1}H NMR (Figure S1) ($CDCl_3$, 500 MHz): δ(ppm) 4.30–4.26 (q; *J* = 6.9 Hz; 1H), 3.78 (s; 3H), 2.80 (s; 1H), 1.42–1.41 (d; *J* = 6.9 Hz; 3H). ^{13}C NMR (Figure S5) ($CDCl_3$, 126 MHz): δ(ppm) 176.26; 66.90; 52.65; 20.50.

Ethyl lactate (EtLa): ^{1}H NMR (Figure S2) ($CDCl_3$, 500 MHz): δ (ppm) 4.27–4.21 (m; 3H), 2.84 (s; 1H), 1.41–1.40 (d; *J* = 6.9 Hz; 3H), 1.31–1.28 (t; *J* = 7.16; 7.16 Hz; 3H). ^{13}C NMR (Figure S6) ($CDCl_3$, 126 MHz): δ (ppm) 175.72; 66.76; 61.64; 20.38; 14.15.

Methyl 3-hydroxybutyrate (M3HB): ^{1}H NMR (Figure S3) ($CDCl_3$, 500 MHz): δ (ppm) 4.22–4.16 (qd; *J* = 8.54; 6.30; 6.27; 6.27 Hz; 1H), 3.70 (s; 3H), 2.93 (s; 1H), 2.51–2.40 (m; 2H), 1.23–1.22 (d; *J* = 6.30 Hz 3H). ^{13}C NMR (Figure S7) ($CDCl_3$, 126 MHz): δ (ppm) 173.29; 64.28; 51.71; 42.63; 22.49.

Ethyl 3-hydroxybutyrate (E3HB): ^{1}H NMR (Figure S4) ($CDCl_3$, 500 MHz): δ 4.22–4.15 (q; *J* = 7.16; 7.12; 7.12 Hz; 3H), 2.98 (s; 1H), 2.50–2.46 (dd; *J* = 16.38; 3.48 Hz; 1H), 2.43–2.38 (dd; *J* = 16.40; 8.68 Hz; 1H), 1.28–1.25 (t; *J* = 6.52; 6.52 Hz; 3H), 1.23–1.22 (d; *J* = 6.93 Hz; 3H). ^{13}CNMR (Figure S8) ($CDCl_3$, 126 MHz): δ (ppm) 172.93; 64.30; 60.67; 42.83; 22.45, 14.19.

2.4. Methacrylation of Alkyl Esters

Synthesized alkyl esters of either lactic acid or 3-hydroxybutanoic acid did undergo a reaction with methacrylic anhydride (MAA) in order to form a polymerizable monomer. The reaction mixtures were prepared in a molar ratio 1:1 (ester:MAA) (see Scheme 3 for lactates and Scheme 4 for 3-hydroxybutyrates). The reaction mixture was poured into a three-necked round bottom flask and placed in an oil bath tempered at 80 °C and stirred via a magnetic stirrer. The reactions were catalyzed by a 50% solution of potassium 2-ethylhexanoate in 2-ethylhexynoic acid (2-EHA) that was prepared via a neutralization reaction of 2-ethylhexanoic acid with potassium hydroxide in a mass ratio 1:2 (KOH:acid) while the reaction water was evaporated. The catalyst was added to the mixture in a molar ratio 1:0.02 (ester:catalyst). The reaction started the moment the catalyst was added and all reactions took 24 h of reaction time. The conversions of reactants and the yields of forming products were monitored via GC-FID analysis. The formed methacrylic acid was neutralized by potassium hydroxide aqueous solution and separated from the metracrylated product. Methacrylated alkyl ester structures were verified via MS, FT-IR, ^{1}H NMR, and^{13}C NMR methods.

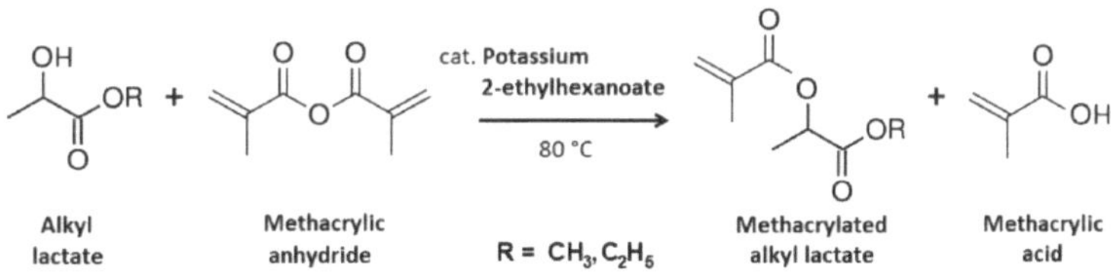

Scheme 3. Methacrylation of alkyl lactate esters by methacrylic anhydride (MAA).

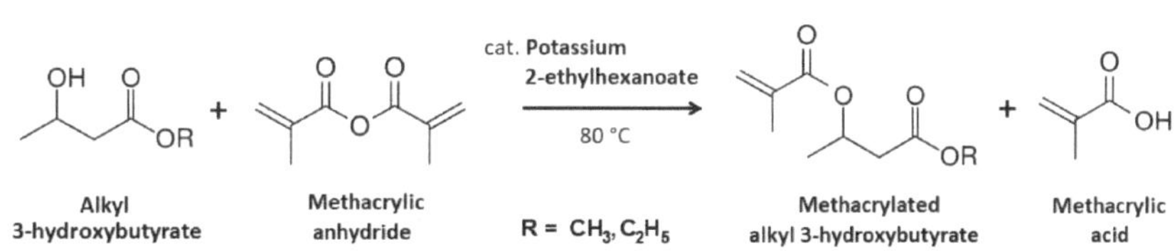

Scheme 4. Methacrylation of alkyl 3-hydroxybutyrate esters by methacrylic anhydride (MAA).

Methacrylated methyl lactate (MeLaMMA): ^{1}H NMR (Figure S9) ($CDCl_3$, 500 MHz): δ(ppm) 6.20–6.19 (dd; *J* = 1.54; 0.94 Hz; 1H), 5.63–5.62 (p; *J* = 1.53; 1.53; 1.52; 1.52 Hz; 1H), 5.17–5.13 (q; *J* = 7.08; 7.07; 7.07 Hz; 1H), 3.75 (s; 3H), 1.97–1.96 (dd; *J* = 1.6; 1.00 Hz; 3H), 1.53(d; *J* = 7.00 Hz; 3H). ^{13}C NMR (Figure S13) ($CDCl_3$, 126 MHz): δ (ppm)171.30; 166.68; 135.63; 126.44; 68.77; 52.27; 18.13; 16.96.

Methacrylated ethyl lactate (EtLaMMA): ^{1}H NMR (Figure S10) ($CDCl_3$, 500 MHz): δ(ppm) 6.20–6.19 (p; *J* = 1.07; 1.07; 1.07; 1.07 Hz; 1H), 5.63–5.61 (p; *J* = 1.57; 1.57; 1.57; 1.57Hz; 1H), 5.14–5.10 (q; *J* = 7.07; 7.07; 7.03 Hz; 1H), 4.23–4.18 (q; *J* = 7.16; 7.16; 7.15 Hz; 2H), 1.97–1.96 (dd; *J* = 1.58; 1.01 Hz; 3H), 1.53–1.52 (d; *J* = 7.05 Hz; 3H), 1.28–1.26 (t; *J* = 7.15; 7.15 Hz; 3H); 1.53–1.52 (d; *J* = 7.05 Hz; 3H), 1.28–1.26 (t; *J* = 7.15; 7.15 Hz; 3H). ^{13}C NMR (Figure S14) ($CDCl_3$, 126 MHz): δ (ppm) 170.86; 166.76; 135.74; 126.35; 68.93; 61.31; 18.18; 16.97; 14.11.

Methacrylated methyl 3-hydroxybutyrate (M3HBMMA): ^{1}H NMR (Figure S11) ($CDCl_3$, 500 MHz): δ (ppm) 6.07–6.06 (dq; *J* = 1.96; 1.02; 0.98; 0.98 Hz; 1H), 5.56–5.53 (p; *J* = 1.60; 1.60; 1.58; 1.58 Hz; 1H), 5.35–5.29 (dp; *J* = 7.32; 6.26; 6.26; 6.25; 6.25 Hz; 1H), 3.68 (s; 3H), 2.72–2.67 (dd; *J* = 15.34; 7.29 Hz; 1H), 2.57–2.53 (dd; *J* = 15.35; 5.79 Hz; 1H), 1.92 (dd; *J* = 1.63; 1.01 Hz; 3H), 1.35–1.34 (d; *J* = 6.36 Hz; 3H). ^{13}C NMR (Figure S15) ($CDCl_3$, 126MHz): δ (ppm) 170.70; 166.59; 136.47; 125.41; 67.68; 51.73; 40.74; 19.89; 18.23.

Methacrylated ethyl 3-hydroxybutyrate (E3HBMMA): ^{1}HNMR (Figure S12) ($CDCl_3$, 500 MHz): δ (ppm) 6.07–6.06 (dd; *J* = 1.75; 0.97 Hz; 1H), 5.54–5.53 (q; *J* = 1.63; 1.63; 1.63 Hz; 1H), 5.35–5.29 (dp; *J* = 7.50; 6.24; 6.24; 6.24; 6.24 Hz; 1H), 4.16–4.10 (qd; *J* = 7.11; 7.06; 7.06; 0.96Hz; 2H), 2.69–2.65 (dd; *J* = 15.28; 7.42 Hz; 1H), 2.55–2.51 (dd; *J* = 15.29; 5.75 Hz; 1H), 1.94–1.91 (m; 3H), 1.34–1.33 (d; *J* = 6.28 Hz; 3H), 1.25–1.22 (t; *J* = 7.13; 7.13 Hz; 3H). ^{13}C NMR (Figure S16) ($CDCl_3$, 126 MHz): δ (ppm) 170.21; 166.54; 136.46; 125.33; 67.71; 60.57; 41.01; 19.85; 18.19; 14.14.

2.5. *Methods for the Reaction Kinetics Study*

2.5.1. LC–MS Method for Depolymerization Kinetics

Samples were obtained from the reactor during organic synthesis, followed by quantification of products methyl lactate (MeLa), ethyl lactate(EtLa), methyl 3-hydroxybutyrate(M3HB), and ethyl 3-hydroxybutyrate(E3HB) by ultra-performance liquid chromatography (UHPLC Agilent 1290 Infinity LC) in tandem with triple quadruple (Bruker EVOQ LC-TQ) (Billerica, MA, USA) with atmospheric pressure electrospray ionization (ESI). An external generator of gases was used as the source of nitrogen and air (Peak Scientific—Genius 3045) (Inchinnan, UK). As a stationary phase column, Luna® Omega Polar C18 Phenomenex (100 × 2.1 mm, 1.6 μm) was used. The optimum column temperature was adjusted to 40 °C and the flow rate was set to 0.5 mL/min. The mobile phases were as follows: (A) 0.1% HCOOH in H_2O and (B) ACN were used with the following gradient program of An eluent (%): t (0 min) = 90, t (0.5 min) = 85, t (3.5 min) = 5, t (4.5 min) = 95. Stop time was set to 6.0 min and re-equilibration time was set to 2.0 min. The injection volume applied in all analyses was 7 μL.

MS conditions were as follows: ESI in positive mode; spray voltage: 3500 V; cone temperature 350 °C; cone gas flow: 35 a.u.; heated probe temperature: 650 °C; probe gas flow: 40 a.u., nebulizer gas flow: 55a.u. and exhaust gas: ON. For quantification, MRM mode was used with the following MRM transitions for MeLa (RT 1.24 min; 105.1 > 84.6 with CE 0.25 eV; 105.1 > 93.9 with CE 0.25 eV and 105.1 > 45.0 with CE 2 eV), for EtLa (RT 2.35 min; 119.1 > 47.3 with CE 2.0 eV and 119.1 > 91.1 with CE 2.0 eV), for M3HB (RT 1.91 min; 119.1 > 59.0 with CE 10eV; 119.1 > 87.2 with CE 2.5 eV and 119.1 > 101.2 with CE 1.0 eV), and for E3HB (RT 3.58 min; 133.1 > 73.3 CE with 5eV and 133.1 > 87.2 with CE 5 eV). The collision gas Argon was used ata pressure of 1.5 mTorr.

2.5.2. GC-FID Method for Methacrylation Kinetics

Samples were obtained from the reactor during organic synthesis, followed by quantification of reactants (alkyl esters, methacrylic anhydride) by gas chromatography (Hewlett Packard 5890 Series II) (Palo Alto, CA, USA) with a flame ionization detector (FID). Gas bottles of nitrogen (as auxiliary gas for FID), air (as an oxidizer for FID), and hydrogen (as carrier gas and fuel for FID) were used. Capillary GC column ZB-624 (60 m × 0.32 mm, 1.8 μm) served as a stationary phase. The temperature of the inlet was set to 200 °C and the temperature of the detector to 260 °C. Substances were separated with atemperature gradient, with an initial temperature of 60 °C (held for 1 min) followed by atemperature rate of 20 °C/min with a final temperature of 250 °C (held for 15 min). The column flow rate set for the analyses was 3 mL/min and the split ratio was 1:40. The injection volume applied in all analyses was 1 μL. The retention time of peaks: MeLa (RT6.15 min); EtLa (RT 7.05 min); M3HB (RT 7.76 min); E3HB (RT 8.53 min); MAA (RT 9.88 min); MeLaMMA (RT 9.80); EtLaMMA (RT10.45 min); M3HBMMA (RT 10.85); E3HBMMA (RT 14.48 min).

3. Results

3.1. Depolymerization of PLA and PHB via Alcoholysis

The reaction mixtures for the depolymerization were prepared according to the mass proportion shown in Table 1. The mass of a particular polymer in every mixture was constant and the amount of reacting nucleophile (alcohol) changed depending on the molar ratio of the treactants. The amounts of catalyst (*p*-TSA) for each reaction solution are written in the table as well. Table 1 also contains information on the boiling points of each alkyl ester that was synthesized.

Table 1. Depolymerization mixture components and boiling point values of synthesized monoesters.

		Methyl Lactate	Ethyl Lactate	Methyl 3-Hydroxybutyrate	Ethyl 3-Hydroxybutyrate
Reaction mixture	Polymer	200 g PLA	200 g PLA	200 g PHB	200 g PHB
	Alcohol	356 g MeOH	511 g EtOH	298 g MeOH	428 g EtOH
	Catalyst	5.28 g *p*-TSA	5.28 g *p*-TSA	4.42 g *p*-TSA	4.42 g *p*-TSA
Boiling point		145 °C [39]	154 °C [40]	159 °C [41]	185 °C [42]

The results of the LC–MS analysis in Figure 1a show that methyl esters of each lactic or 3-hydroxybutanoic acid reach their reaction equilibria after about 90 min of depolymerization (MeLawas slightly faster than M3HB), resulting in yields of 93.8% for MeLa and 91.6% for M3HB. On the other hand, the ethyl esters of both carboxylic acids did not reach their total yield value after 4 h of reaction. The reaction rate of ethyl 3-hydroxybutyrate seems to be the slowest, reaching a yield of 76.0% after 240 min. The ethyl lactate's yield after the same reaction time reached 85.1%. The differences between the product yields could be caused by the steric effects of the particular molecules involved in the reaction.

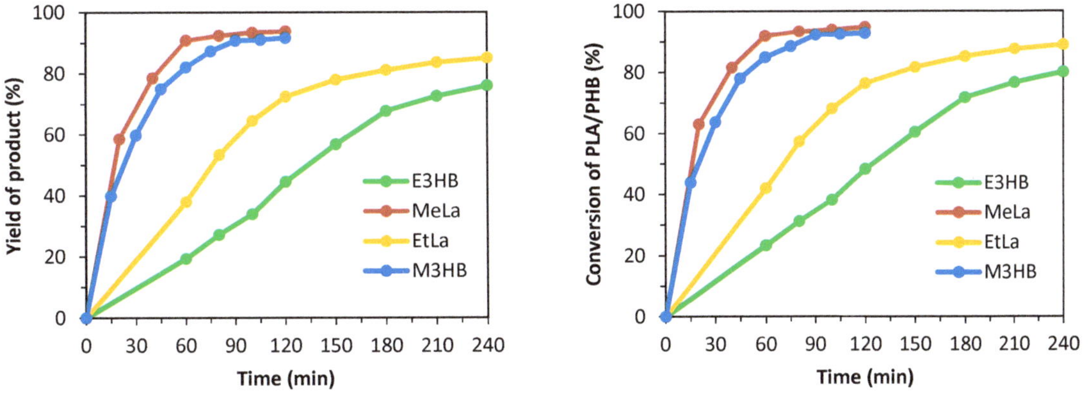

(**a**) (**b**)

Figure 1. (**a**) The increase in particular yields of monoesters during the depolymerization reaction of PLA or PHB; (**b**) the increase in particular conversions of polymers during the depolymerization reaction of PLA or PHB.

The conversion signs of progress of either PLA or PHB during the depolymerization reactions are displayed in Figure 1b. They have been acquired by weighing the residual polymer from the taken sample. The precipitated content of the unreacted polymer was measured and the percentage of conversion was calculated. The results have very similar data curves as the values of the products' yields due to the fact that these values are connected. The conversion of PLA is 94.8% for methanolysis (after 120 min) and the conversion of PLA is 89.1% for ethanolysis (after 240 min). PHB reached conversions values of 92.8% (MeOH, 120 min) and 80.0% (EtOH, 240 min).

The pressure of the reacting mixture was monitored during every alcoholysis. The results shown in Figure 2 confirm the progressing depolymerization for each mixture. Due to the incorporation of alcohol into the structure of the alkyl ester, the pressure in the system should decrease as a result of decreasing the presence of evaporating alcohol. These expectations are fulfilled except for the methyl lactate. The pressure of the system for the methanolysis of PLA increases after 45 min of reaction. This elevation is caused by the forming MeLa since the reaction temperature was 151 °C and the boiling point value of MeLa is below this temperature (shown in Table 1). Therefore, the occurring monoester participates in the pressure increase.

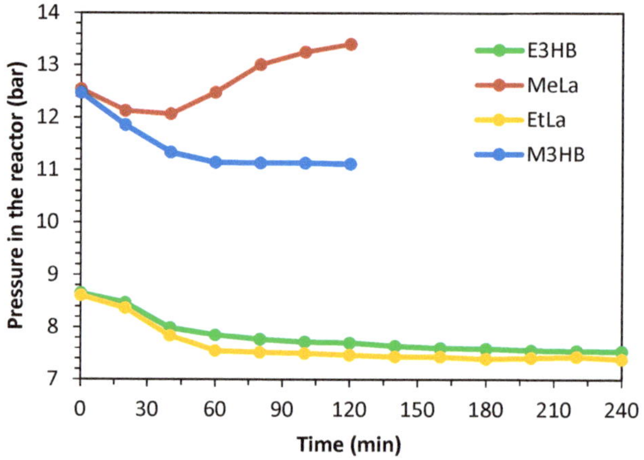

Figure 2. Pressure values dependence on time during depolymerization.

3.2. Kinetics of the Depolymerization of PLA and PHB via Alcoholysis

The first-order reaction order is mostly used to describe the alcoholysis depolymerization due to the fact that if the excess of alcohol is added to the mixture, the concentration of polymer molecules affects the reaction kinetics directly due to its lesser molar amount of mixture [43,44]. If the first-order reaction is used for the calculation, the equations are the following:

$$r = \frac{dc_{polymer}}{dt} = -kc_{polymer}, \tag{3}$$

where r is the reaction rate (mol/(dm^3·min)), k represents the reaction constant (min^{-1}), and $c_{polymer}$ stands for the molar concentration of the polymer (mol/dm^3) at time t (min). The molar concentration of the polymer as a reagent can be substituted by the following conversion values:

$$c_{polymer} = (1 - X), \tag{4}$$

where the molar concentration ($c_{polymer}$) is expressed by conversion (X). This equation is applied for the values of conversion from the number interval of <0,1>. If the molar concentration is replaced with the conversion, the first-order equation has to be rewritten:

$$\frac{dX}{dt} = k(1 - X) \tag{5}$$

If Equation (5) is calculated generally, the steps are as follows:

$$\frac{dX}{(1 - X)} = kdt \tag{6}$$

$$\int_0^X \frac{dX}{(1-X)} = k\int_0^t dt \tag{7}$$

$$-[\ln(1-X) - \ln(1)] = kt \tag{8}$$

$$\ln\frac{1-X}{1} = -kt \tag{9}$$

$$\ln\frac{1}{1-X} = kt \tag{10}$$

Equation (10) was applied to the kinetics values measured by LC–MS analysis. The conversion was transferred into a modified form using the logarithm of a fraction and the dependence on time, as shown in Figure 3. There is evidence that the highest reaction rate constant comprises the depolymerization of PLA in methanol (producing MeLa) with the value of approximately 0.0425 min^{-1}. On the other hand, the depolymerization of PHB in ethanol provided the lowest value of reaction rate constant, reaching approx. 0.0064 min^{-1}. The rest of the reaction rate constants (k) are shown in Table 2.

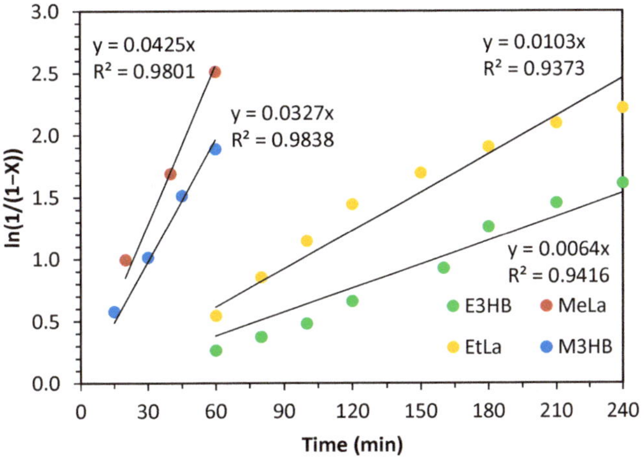

Figure 3. Kinetics of depolymerization of PLA and PHB polymers providing reaction rate constants (k).

Table 2. The values of reaction rate constants (k) for depolymerization of PLA and PHB via alcoholysis.

	Methyl Lactate	**Ethyl Lactate**	**Methyl 3-Hydroxybutyrate**	**Ethyl 3-Hydroxybutyrate**
Reaction rate constant k (min^{-1})	0.0425	0.0327	0.0103	0.0064

3.3. Structural Characterization of Synthesized Alkyl Esters

3.3.1. FT-IR Analyses of Alkyl Esters

Infrared spectrometry using Fourier transformation was used as the first method to confirm the distilled synthesized product. The main peaks shown in Figure 4 lay in the intervals of wave numbers of either 3700–3200 cm^{-1} (–OH stretching) or the values of 1210–1163 cm^{-1} (C=O stretching) and 1750–1735 cm^{-1} (C–O stretching). In particular, the ester bond stretching signals are important since the initial suspension mixture would contain aliphatic alcohol, which provides the signal for the hydroxyl functional groups as well.

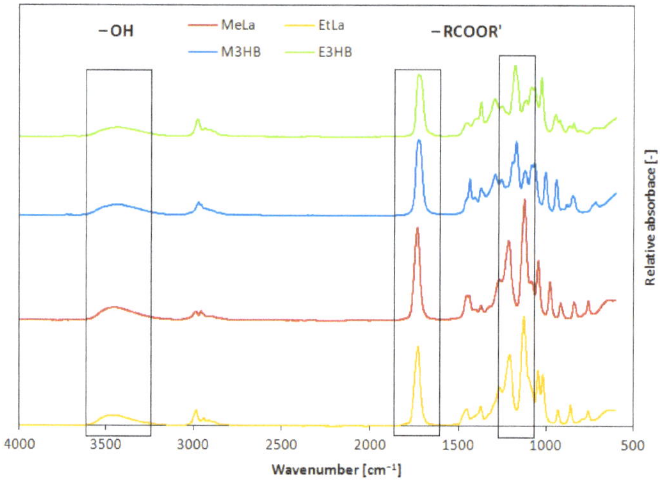

Figure 4. FT-IR spectra of the synthesized alkyl esters as products of the depolymerization reaction of PLA and PHB.

3.3.2. MS Analyses of Alkyl Esters

The analytic procedure to obtain all MS spectra of synthesized alkyl esters of lactic and 3-hydroxybutyric acid is described above (see Section 2.2.2). All molecular precursor ions have been measured as products of the ESI ionization process. Other signals are particular product ions occurring due to the fragmentation of the molecules apart during the MS/MS analysis. All spectra are shown in Figure 5.

3.4. Methacrylation of the Alkyl Esters of Lactic and 3-Hydroxybutanoic Acid

The reaction mixtures for the syntheses of methacrylated alkyl esters of carboxylic acids were prepared according to the mass proportion shown in Table 3. The mass of a particular alkyl ester in every mixture was constant and the amount of reacting methacrylic anhydride changed depending on the molar ratio of the reactants. The amount of catalyst (50% solution of potassium 2-ethylhexanoate in 2-ethylhexanoic acid) was calculated according to the particular reacting alkyl ester.

Table 3. Methacrylation mixture components for the synthesis of methacrylated alkyl monoesters.

		Methacrylated Methyl Lactate	**Methacrylated Ethyl Lactate**	**Methacrylated Methyl 3-Hydroxybutyrate**	**Methacrylated Ethyl 3-Hydroxybutyrate**
Reaction mixture	Ester	30 g MeLa	30 g EtLa	30 g M3HB	30 g E3HB
	Anhydride	44.4 g MAA	39.2 g MAA	39.2 g MAA	35.0 g MAA
	Catalyst	2.1 g solution	1.85 g solution	1.85 g solution	1.66 g solution
Mixture volume		70.42 cm^3	66.92 cm^3	66.26 cm^3	63.31 cm^3

The results of the methacrylation reaction that forms methacryled alkyl esters of lactic or 3-hydroxybutanoic acid are shown in Figure 6. Both reactants which were monitored (alkyl ester and methacrylic anhydride) via GC-FID analysis have similar time progressions of their conversion values due to the fact that their molar ratio in the reacting mixtures was 1:1 mol in all cases. The conversion values of methacrylic anhydride are slightly higher, likely due to the fact that the anhydride participated in secondary reactions in the reaction mixture (water hydrolysis, etc.). It is evident that the esters of lactic acid (MeLa and EtLa), formed into the products MeLaMMA and EtLaMMA, respectively, progressed faster in time than the alkyl esters of 3-hydroxybutanoic acid. These results may have been determined by the steric effects of each ester and due to their varying polarity which could have affected the effectiveness of the catalyst.

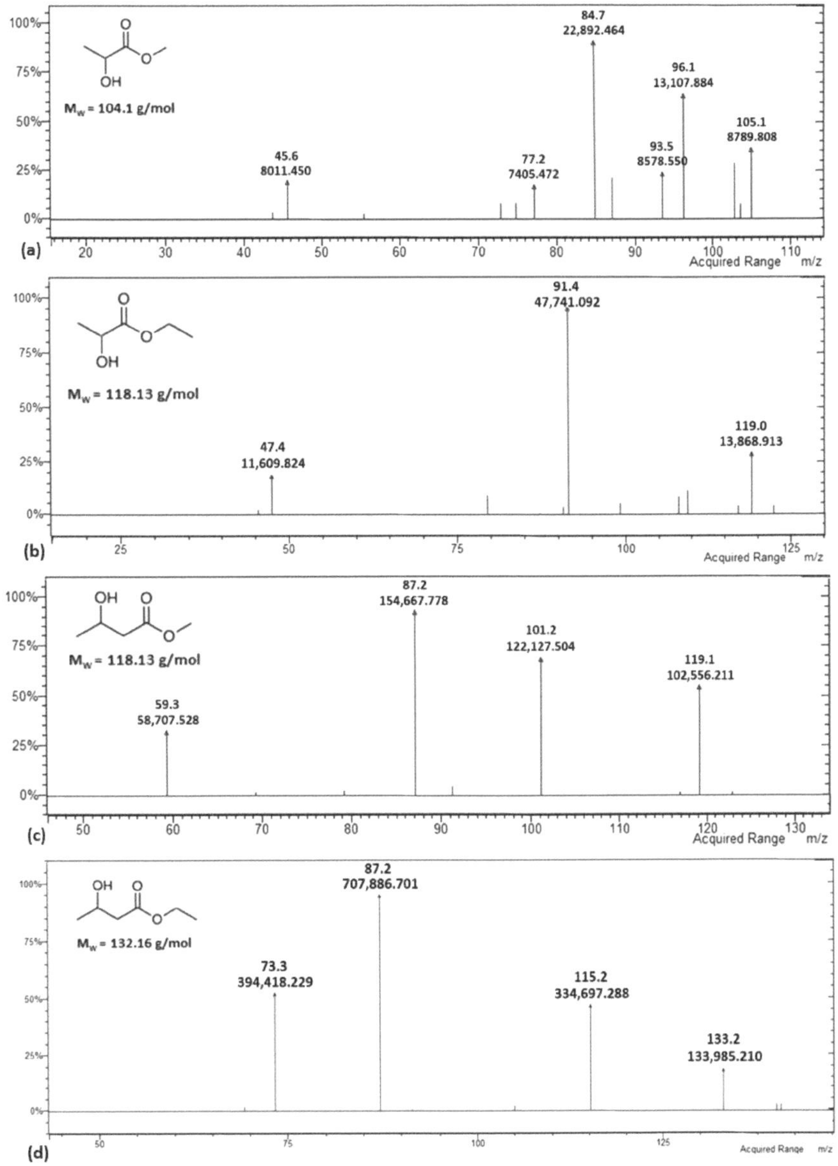

Figure 5. MS spectra of the synthesized alkyl monoesters based on PLA and PHB. (**a**) Methyl lactate (MeLa); (**b**) Ethyl lactate (EtLa); (**c**) Methyl 3-hydroxybutyrate (M3HB); (**d**) Ethyl 3-hydroxybutyrate (E3HB).

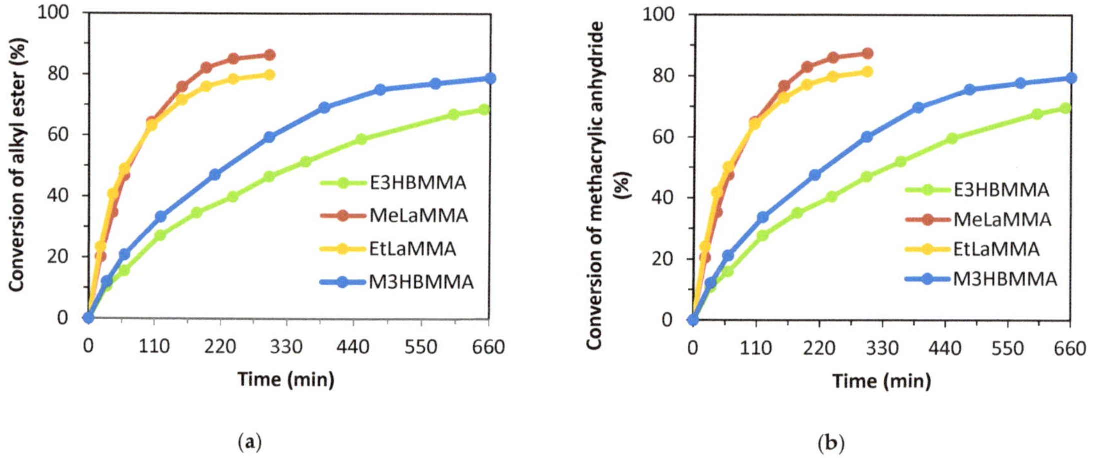

Figure 6. (**a**) The increase in particular conversions of alkyl esters of lactic and 3-hydroxybutanoic acid in time during the methacrylation reactions; (**b**) The increase in conversion of methacrylic anhydride in time during the methacrylation reactions.

The methacrylation reactions were performed for 24 h to obtain the highest yields of the methacrylated product. The graphs below are presented in order to compare the rates of each particular reaction which served for the calculation of reaction rate constants. The yields of methacrylated alkyl esters products increasing in time are shown in Table 4. The yield quantified after 5 h of reaction is shown for comparison.

Table 4. Methacrylated alkyl ester yields compared after 5 h and 24 h of methacrylation reaction.

		Methacrylated Methyl Lactate	Methacrylated Ethyl Lactate	Methacrylated Methyl 3-Hydroxybutyrate	Methacrylated Ethyl 3-Hydroxybutyrate
Yield of product	5 h	86.2%	80.2%	59.5%	45.6%
	24 h	99.5%	99.4%	98.3%	98.0%

3.5. Kinetics of the Methacrylation Reactions

It is assumed that when equimolar amounts of both reactants (alkyl ester and methacrylic anhydride) are used, the rate of methacrylation is dependent on the concentration of both reactants. The acylation of hydroxyl functional groups using homogeneous catalysis was considered as potential reaction kinetics (methacrylation reaction is a type of the acylation) [45]. The Equations defining the dependence of the concentration of both reactants (alkyl ester and methacrylic anhydride) on time leading to acquiring the reaction rate constant are as follows:

$$r' = \frac{dc_{ester}}{dt} = \frac{dc_{MAA}}{dt} = -k' c_{ester} c_{MAA}, \tag{11}$$

Equation (11) defines the conventional second-order rate where r' is the reaction rate (mol/(dm^3·min)), k' represents the reaction constant (dm^3/(mol·min)), c_{ester} stands for the molar concentration of a particular alkyl ester (mol/dm^3), and c_{MAA} stands for the molar concentration of methacrylic anhydride (mol/dm^3) at time t (min). To solve the second-order rate of reaction, several mathematical adjustments need to be made:

$$c^0{}_{ester} = a \tag{12}$$

$$c^0{}_{MAA} = b \tag{13}$$

$$c_{ester} = a - x \tag{14}$$

$$c_{MAA} = b - x, \tag{15}$$

where x stands for the concentration of each reactant in particular time t (min), $c^0{}_{ester}$ (mol/dm^3) is the initial concentration of alkyl ester, and $c^0{}_{MAA}$ (mol/dm^3) is the initial concentration of methacrylic anhydride. Considering these additional defined quantities, Equation (11) can be rearranged and solved:

$$-\frac{dx}{dt} = -k'\left(c^0{}_{ester} - x\right)\left(c^0{}_{MAA} - x\right) \tag{16}$$

$$\int_0^x \frac{dx}{(c^0{}_{ester} - x)(c^0{}_{MAA} - x)} = k' \int_0^t dt \tag{17}$$

$$\frac{1}{b-a}(\ln\frac{1}{a-x} - \ln\frac{1}{b-x}) = k't \tag{18}$$

$$\frac{1}{c^0{}_{MAA} - c^0{}_{ester}}(\ln\frac{c^0{}_{ester}}{c^0{}_{ester} - x} - \ln\frac{c^0{}_{MAA}}{c^0{}_{MAA} - x}) = k't, \tag{19}$$

The left side of Equation (19) can be simplified by applying the rule of logarithm, and when the simplified equation has been rearranged, the dependence of the actual concentration during the reaction on time can be formed:

$$\frac{1}{c^0{}_{MAA} - c^0{}_{ester}} \ln\frac{c^0{}_{ester} c_{MAA}}{c^0{}_{MAA} c_{ester}} = k't \tag{20}$$

$$\ln\frac{c^0{}_{ester} c_{MAA}}{c^0{}_{MAA} c_{ester}} = k'\left(c^0{}_{MAA} - c^0{}_{ester}\right)t \tag{21}$$

Equation (21) can be used to obtain the reaction rate constant k'. If the graphic solution is applied, the slope of the linear curve acquired from the graph contains the constant k'. All data gathered during the reaction progress in time recalculated for mathematical purposes are shown in Figure 7. All reaction rate constants are written in Table 5. It is evident from the results that, in general, the methacrylation reactions of alkyl esters of lactic acid have higher reaction rate constants than the alkyl esters of 3-hydroxybutanoic acid. It is assumed that the availability of the hydroxyl functional group of lactates is better than for 3-hydroxybutyrates.

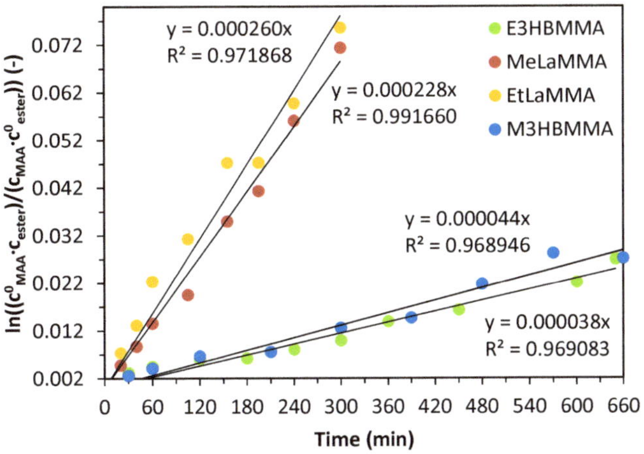

Figure 7. The graphic solution of second-order reaction rate of methacrylation reaction of alkyl esters.

Table 5. The calculated values of reaction rate constants (k') for methacrylation of alkyl esters of lactic and 3-hydroxybutyric acid from the slope of the graphic solutions.

	Methacrylated Methyl Lactate	**Methacrylated Ethyl Lactate**	**Methacrylated Methyl 3-Hydroxybutyrate**	**Methacrylated Ethyl 3-Hydroxybutyrate**
Reaction rate constant k' ($dm^3/(mol \cdot min)$)	0.0885	0.0554	0.0092	0.0079

3.6. *Structural Characterization of Synthesized Methycrylated Alkyl Esters*

3.6.1. FT-IR Analyses of Methacrylated Alkyl Esters

Infrared spectrometry using Fourier transformation's results for the confirmation of the structures of the synthesized methacrylated alkyl esters are displayed in Figure 8. Peaks showing the presence of signals, which belong to ester bonds, lay in the intervals of wave numbers of either 1210–1163 cm^{-1} (C=O stretching) or the values of 1750–1735 cm^{-1} (C–O stretching). The signals referring to C–O stretching are split in every spectrum. The reason for the splitting of the peak is the presence of two different types of ester bonding in molecules. One bond belongs to the ester of lactic or 3-hydroxybutanoic acid and aliphatic alcohol. The other signal refers to the ester bond between the formed alkyl ester of carboxylic acid and the methacrylic acid. The second type of signal that can be found in FT-IR spectra reaches the values of wave numbers of either 1670–1600 cm^{-1} (C=C stretching) or 1000–650 cm^{-1} (C=C bending). These peaks uncover the presence of unsaturated double bonds within the structures of synthesized products that belong to methacrylates. Another confirmation of the successful methacrylation is the absence of signal in the area of 3700–3200 cm^{-1} (–OH stretching). These functional groups were supposed to react with methacrylic anhydride. Therefore, their peaks are missing in comparison with FT-IR spectra in Figure 4.

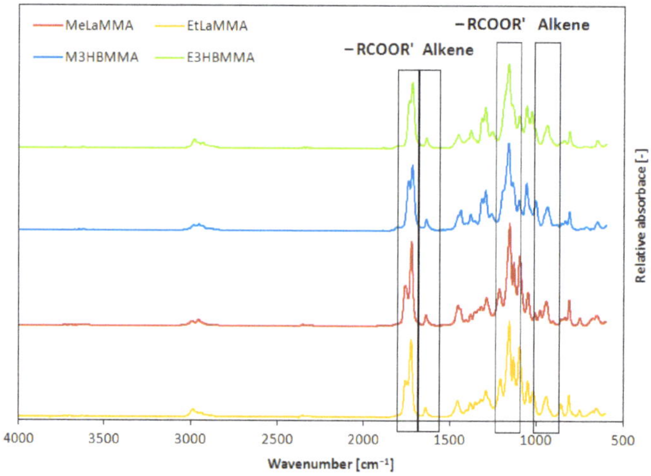

Figure 8. FT-IR spectra of the synthesized methacryled alkyl esters as products of the methacrylation reaction of alkyl esters of lactic acid or 3-hydroxybutanoic acid and methacrylic anhydride.

3.6.2. MS Analyses of Methacrylated Alkyl Esters

The analytic procedure to obtain all MS spectra of synthesized methacrylated alkyl esters is described above (see Section 2.2.2). All molecular precursor ions have been measured as products of the ESI ionization process. Other signals are particular product ions occurring due to the fragmentation of the molecules during the MS/MS analysis. All spectra are shown in Figure 9.

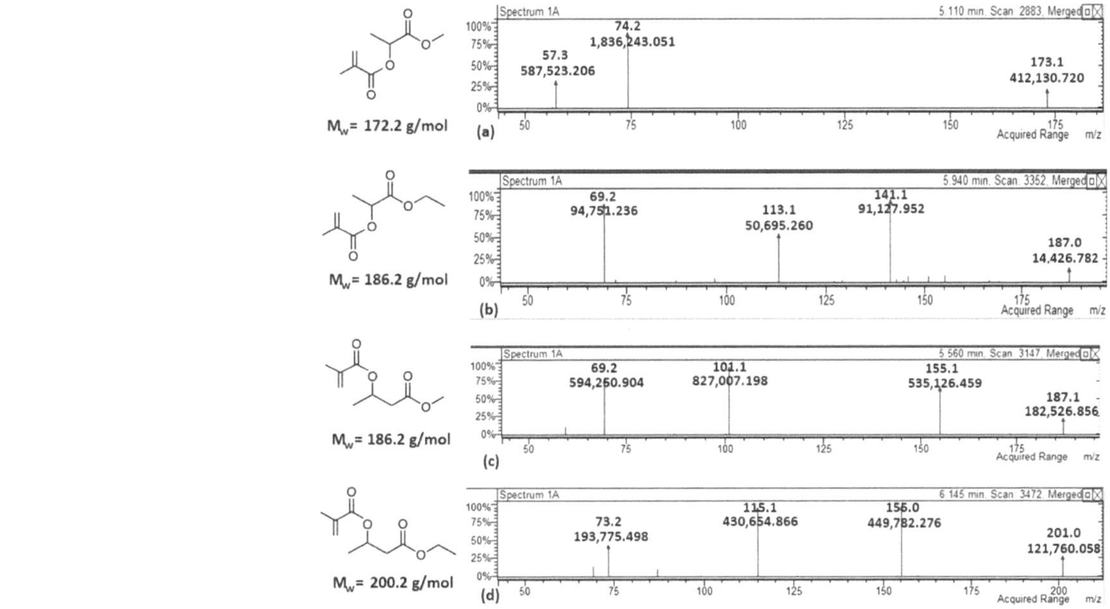

Figure 9. MS spectra of the synthesized methacrylated alkyl esters of lactic acid or 3-hydroxybutanoic acid. (**a**) Methacrylated methyl lactate (MeLaMMA); (**b**) methacrylated ethyl lactate (EtLaMMA); (**c**) methacrylated methyl 3-hydroxybutyrate (M3HB); (**d**) methacrylated ethyl 3-hydroxybutyrate (E3HB).

4. Discussion

This work was focused on the experimental confirmation of the depolymerization of poly(lactic acid) (PLA) and poly(3-hydroxybutyrate) (PHB) via alcoholysis. All reactions were performed in identical conditions, which were a temperature of 151 °C, the presence of the constant amount of molar alcohol access (4:1) and the catalyst *para*-toluenesulphonic acid, particularly 1% mol. of the particular polyester. The pressure of the reaction solution differed regarding the used alcohol and the particular reaction combination. In all cases, the pressure in the reactor decreased over time except for the mixture containing methanol and poly(lactic acid). This mixture's pressure elevated from 12.54 bar to 13.40 due to the boiling point value of methyl lactate being 145 °C, which means this ester evaporated in the reactor as well. Generally, lower pressure values were measured for theethanolyses of both PLA and PHB, which decreased from an approximate value of 8.63 bar to 7.54 (E3HB) and 7.39 (M3HB). The decrease is the consequence of lowering the volatility of the reacting solution due to the forming of alkyl esters. It was also found that the rates of methanolyses are faster than the rates of ethanolyses. Depolymerization reaction rate constants (k) were calculated for all experimental reactions, resulting in the highest one of 0.0425 (min^{-1}) for the methanolysis of PLA and the lowest constant related to ethanolysis of PHB, reaching a value of 0.0064 (min^{-1}). These differences are probably caused by the steric effects of particular reactants. The yields of each product were: 93.79% MeLa (120 min); 91.64% M3HB (120 min); 85.08% EtLa (240 min); and 76.03% E3HB (240 min).

The synthesized alkyl esters from the depolymerization of PLA and PHB were modified by the methacrylation reaction with methacrylic anhydride (MAA) to methacrylated alkyl esters, which are polymerizable. The reaction mixtures were composed of equal mass amounts of alkyl esters and different amounts of methacrylic anhydride in equimolar ratio to the esters. The catalyst used for the methacrylation was potassium 2-ethylheaxoate (50% mass solution) in 2-ethylhexanoic acid. The amount of catalyst was 2% mol. The rates of the methacrylation at 80 °C of each alkyl ester define the calculated reaction rate constants (k').

It has been observed that in general, the methacrylation of alkyl lactates was faster than in the case of alkyl 3-hydroxybutyrates. The highest value of methacrylation rate constant comprises the mixture containing methyl lactate ($k' = 0.0885$ dm^3/(mol·min)).On the other hand, the lowest one belongs to ethyl 3-hydroxybutyrate ($k' = 0.0079$ dm^3/(mol·min)). These constants have been calculated according to the reactions' progress in time to evaluate a comparison for all reactants, but the methacrylation reactions of all alkyl esters were performed for 24 h to ensure the biggest possible yield. All yields reached values equal to or higher than 98%.

5. Conclusions

Several conclusions from the performed experiments can be summarized. The type of aliphatic alcohol plays a major role in the rate of depolymerization of the polyesters PLA and PHB. Methanol ensures a faster depolymerization process than ethanol. However, using methanol can be problematic regarding the used equipment due to the fact that these mixtures produce a higher overpressure. The methacrylation of all alkyl esters of lactic or 3-hydroxybutanoic acid reaches high yields after 24 h, close to 100%. Nevertheless, the lactic esters undergo methacrylation reaction at a faster rate than the 3-hydroxybutyrates. The distillation of the forming methacrylic acid could be performed instead of neutralizing and washing the acid; however, the appropriate stabilization against spontaneous polymerization has to be ensured.

Supplementary Materials: The following supporting information can be downloaded at: https://www.mdpi.com/article/10.3390/polym14235236/s1, Figure S1: ^{1}H NMR spectrum of methyl lactate (MeLa). ^{1}H NMR ($CDCl_3$, 500 MHz): δ (ppm) 4.30–4.26 (q; *J* = 6.9 Hz; 1H), 3.78 (s; 3H), 2.80 (s; 1H), 1.42–1.41 (d; *J* = 6.9 Hz; 3H). Figure S2: ^{1}H NMR spectrum of ethyl lactate (EtLa). ^{1}H NMR ($CDCl_3$, 500 MHz): δ (ppm) 4.27–4.21 (m; 3H), 2.84 (s; 1H), 1.41–1.40 (d; *J* = 6.9 Hz; 3H), 1.31–1.28 (t; *J* = 7.16; 7.16 Hz; 3H). Figure S3: ^{1}H NMR spectrum of methyl 3-hydroxybutyrate (M3HB). ^{1}H NMR ($CDCl_3$, 500 MHz): δ (ppm) 4.22–4.16 (qd; *J* = 8.54; 6.30; 6.27; 6.27 Hz; 1H), 3.70 (s; 3H), 2.93 (s; 1H), 2.51–2.40 (m; 2H), 1.23–1.22 (d; *J* = 6.30 Hz 3H). Figure S4: ^{1}H NMR spectrum of ethyl 3-hydroxybutyrate (E3HB). ^{1}H NMR ($CDCl_3$, 500 MHz): δ 4.22–4.15 (q; *J* = 7.16; 7.12; 7.12 Hz; 3H), 2.98 (s; 1H), 2.50–2.46 (dd; *J* = 16.38; 3.48 Hz; 1H), 2.43–2.38 (dd; *J* = 16.40; 8.68 Hz; 1H), 1.28–1.25 (t; *J* = 6.52; 6.52 Hz; 3H), 1.23–1.22 (d; *J* = 6.93 Hz; 3H). Figure S5: ^{13}C NMR spectrum of methyl lactate (MeLa). ^{13}C NMR ($CDCl_3$, 126 MHz): δ (ppm) 176.26; 66.90; 52.65; 20.50. Figure S6: ^{13}C NMR spectrum of ethyl lactate (EtLa). ^{13}C NMR ($CDCl_3$, 126 MHz): δ (ppm) 175.72; 66.76; 61.64; 20.38; 14.15. Figure S7: ^{13}C NMR spectrum of methyl 3-hydroxybutyrate (M3HB). ^{13}C NMR ($CDCl_3$, 126 MHz): δ (ppm) 173.29; 64.28; 51.71; 42.63; 22.49. Figure S8: ^{13}C NMR spectrum of ethyl 3-hydroxybutyrate (E3HB). ^{13}C NMR ($CDCl_3$, 126 MHz): δ (ppm) 172.93; 64.30; 60.67; 42.83; 22.45, 14.19. Figure S9: ^{1}H NMR spectrum of methacrylated methyl lactate (MeLaMMA). ^{1}H NMR ($CDCl_3$, 500 MHz): δ (ppm) 6.20–6.19 (dd; *J* = 1.54; 0.94 Hz; 1H), 5.63–5.62 (p; *J* = 1.53; 1.53; 1.52; 1.52 Hz; 1H), 5.17-5.13 (q; *J* = 7.08; 7.07; 7.07 Hz; 1H), 3.75 (s; 3H), 1.97–1.96 (dd; *J* = 1.6; 1.00 Hz; 3H), 1.53 (d; *J* = 7.00 Hz; 3H). Figure S10: ^{1}H NMR spectrum of methacrylated ethyl lactate (EtLaMMA). ^{1}H NMR ($CDCl_3$, 500 MHz): δ (ppm) 6.20–6.19 (p; *J* = 1.07; 1.07; 1.07; 1.07 Hz; 1H), 5.63–5.61 (p; *J* = 1.57; 1.57; 1.57; 1.57 Hz; 1H), 5.14–5.10 (q; *J* = 7.07; 7.07; 7.03 Hz; 1H), 4.23–4.18 (q; *J* = 7.16; 7.16; 7.15 Hz; 2H), 1.97–1.96 (dd; *J* = 1.58; 1.01 Hz; 3H), 1.53–1.52 (d; *J* = 7.05 Hz; 3H), 1.28–1.26 (t; *J* = 7.15; 7.15 Hz; 3H). Figure S11: ^{1}H NMR spectrum of methacrylated methyl 3-hydroxybutyrate (M3HBMMA). ^{1}H NMR ($CDCl_3$, 500 MHz): δ (ppm) 6.07–6.06 (dq; *J* = 1.96; 1.02; 0.98; 0.98 Hz; 1H), 5.56–5.53 (p; *J* = 1.60; 1.60; 1.58; 1.58 Hz; 1H), 5.35–5.29 (dp; *J* = 7.32; 6.26; 6.26; 6.25; 6.25 Hz; 1H), 3.68 (s; 3H), 2.72–2.67 (dd; *J* = 15.34; 7.29 Hz; 1H), 2.57–2.53 (dd; *J* = 15.35; 5.79 Hz; 1H), 1.92 (dd; *J* = 1.63; 1.01 Hz; 3H), 1.35-1.34 (d; *J* = 6.36 Hz; 3H). Figure S12: ^{1}H NMR spectrum of methacrylated ethyl 3-hydroxybutyrate (E3HBMMA). ^{1}H NMR ($CDCl_3$, 500 MHz): δ (ppm) 6.07–6.06 (dd; *J* = 1.75; 0.97 Hz; 1H), 5.54–5.53 (q; *J* = 1.63; 1.63; 1.63 Hz; 1H), 5.35–5.29 (dp; *J* = 7.50; 6.24; 6.24; 6.24; 6.24 Hz; 1H), 4.16–4.10 (qd; *J* = 7.11; 7.06; 7.06; 0.96 Hz; 2H), 2.69–2.65 (dd; *J* = 15.28; 7.42 Hz; 1H), 2.55–2.51 (dd; *J* = 15.29; 5.75 Hz; 1H), 1.94–1.91 (m; 3H), 1.34–1.33 (d; *J* = 6.28 Hz; 3H), 1.25–1.22 (t; *J* = 7.13; 7.13 Hz; 3H). Figure S13: ^{13}C NMR spectrum of methacrylated methyl lactate (MeLaMMA). ^{13}C NMR ($CDCl_3$, 126 MHz): δ (ppm) 171.30; 166.68; 135.63; 126.44; 68.77; 52.27; 18.13; 16.96. Figure S14: ^{13}C NMR spectrum of methacrylated ethyl lactate (EtLaMMA). ^{13}C NMR ($CDCl_3$, 126 MHz): δ (ppm) 170.86; 166.76; 135.74; 126.35; 68.93; 61.31; 18.18; 16.97; 14.11.

Figure S15: ^{13}C NMR spectrum of methacrylated methyl 3-hydroxybutyrate (M3HBMMA). ^{13}C NMR ($CDCl_3$, 126 MHz): δ (ppm) 170.70; 166.59; 136.47; 125.41; 67.68; 51.73; 40.74; 19.89; 18.23. Figure S16: ^{13}C NMR spectrum of methacrylatedethyl 3-hydroxybutyrate (E3HBMMA). ^{13}C NMR ($CDCl_3$, 126 MHz): δ (ppm) 170.21; 166.54; 136.46; 125.33; 67.71; 60.57; 41.01; 19.85; 18.19; 14.14.

Author Contributions: V.J.: conceptualization, experimental work, mathematical calculation, gas chromatography analysis, and infrared analysis; J.F.: liquid chromatography analysis, mass spectrometry analysis, and conceptualization; L.I.: nuclear magnetic resonance analysis; D.V.: nuclear magnetic resonance analysis; L.M.: gas chromatography supervision; S.F., R.P., P.S. and J.K.: supervision; J.K. and P.S.: investigation. All authors have read and agreed to the published version of the manuscript.

Funding: This work was supported by the Materials Research Centre. L.I.: Appreciation to Brno City Municipality (Brno Ph.D. Talent Scholarship) for the financial support. V.J., J.F., L.I. and D.V.: Appreciation to the Ministry of Education, Youth and Sport of the Czech Republic (project No. FCH-S-22-8012) and (project No. FCH-S-22-8001).

Institutional Review Board Statement: Not applicable.

Informed Consent Statement: Not applicable.

Data Availability Statement: Not applicable.

Acknowledgments: We would like to thank Nafigate Corporation a.s. for the support in the form of supplies of PHB material.

Conflicts of Interest: The authors declare no conflict of interest.

References

1. Beltrán, F.R.; Barrio, I.; Lorenzo, V.; Del Río, B.; Martínez Urreaga, J.; de La Orden, M.U. Valorization of poly(lactic acid) wastes via mechanical recycling: Improvement of the properties of the recycled polymer. *Waste Manag. Res. J. A Sustain. Circ. Econ.* **2019**, *37*, 135–141. [CrossRef]
2. Arrieta, M.P.; Samper, M.D.; Aldas, M.; López, J. On the Use of PLA-PHB Blends for Sustainable Food Packaging Applications. *Materials* **2017**, *10*, 1008. [CrossRef] [PubMed]
3. Ray, S.S.; Bousmina, M. Biodegradable polymers and their layered silicate nanocomposites: In greening the 21st century materials world. *Prog. Mater. Sci.* **2005**, *50*, 962–1079.
4. Hottel, T.; Bilec, M.; Landis, A. Sustainability assessments of bio-based polymers. *Polym. Degrad. Stab.* **2013**, *98*, 1898–1907. [CrossRef]
5. Reddy, M.M.; Vivekanandhan, S.; Misra, M.; Bhatia, S.K.; Mohanty, A.K. Biobased plastics and bionanocomposites: Current status and future opportunities. *Prog. Polym. Sci.* **2013**, *38*, 1653–1689. [CrossRef]
6. Gough, C.R.; Callaway, K.; Spencer, E.; Leisy, K.; Jiang, G.; Yang, S.; Hu, X. Biopolymer-Based Filtration Materials. *ACS Omega* **2021**, *6*, 11804–11812. [CrossRef]
7. Beltrán, F.; Arrieta, M.; Moreno, E.; Gaspar, G.; Muneta, L.; Carrasco-Gallego, R.; Yáñez, S.; Hidalgo-Carvajal, D.; de la Orden, M.; Urreaga, J.M. Evaluation of the Technical Viability of Distributed Mechanical Recycling of PLA 3D Printing Wastes. *Polymers* **2021**, *13*, 1247. [CrossRef] [PubMed]
8. Moreno, E.; Beltrán, F.R.; Arrieta, M.P.; Gaspar, G.; Muneta, L.M.; Carrasco-Gallego, R.; Yáñez, S.; Hidalgo-Carvajal, D.; de la Orden, M.U.; Martínez Urreaga, J. Technical Evaluation of Mechanical Recycling of PLA 3D Printing Wastes. In Proceedings of the First International Conference on "Green" Polymer Materials, Basel, Switzerland, 4 November 2020; MDPI: Basel, Switzerland, 2021; p. 19.
9. Leejarkpai, T.; Mungcharoen, T.; Suwanmanee, U. Comparative assessment of global warming impact and eco-efficiency of PS (polystyrene), PET (polyethylene terephthalate) and PLA (polylactic acid) boxes. *J. Clean. Prod.* **2016**, *125*, 95–107. [CrossRef]
10. Piemonte, V.; Gironi, F. Kinetics of Hydrolytic Degradation of PLA. *J. Polym. Environ.* **2013**, *21*, 313–318. [CrossRef]
11. Castro-Aguirre, E.; Iniguez-Franco, F.; Samsudin, H.; Fang, X.; Auras, R. Poly(lactic acid)—Mass production, processing, industrial applications, and end of life. *Adv. Drug Deliv. Rev.* **2016**, *107*, 333–366. [CrossRef] [PubMed]
12. Papong, S.; Malakul, P.; Trungkavashirakun, R.; Wenunun, P.; Chom-in, T.; Nithitanakul, M.; Sarobol, E. Comparative assessment of the environmental profile of PLA and PET drinking water bottles from a life cycle perspective. *J. Clean. Prod.* **2014**, *65*, 539–550. [CrossRef]
13. Chen, H.; Chen, F.; Chen, H.; Liu, H.; Chen, L.; Yu, L. Thermal degradation and combustion properties of most popular synthetic biodegradable polymers. *Waste Manag. Res. J. A Sustain. Circ. Econ.* **2022**, 0734242X221129054. [CrossRef] [PubMed]
14. Zhou, Q.; Xanthos, M. Nanosize and microsize clay effects on the kinetics of the thermal degradation of polylactides. *Polym. Degrad. Stab.* **2009**, *94*, 327–338. [CrossRef]
15. Uğur, A.; Şahin, N.; Beyatli, Y. Accumulation of Poly-\beta-Hydroxybutyrate in Streptomyces Species During Growth with Different Nitrogen Sources. *Turk. J. Biol.* **2002**, *26*, 171–174.

16. Lenz, R.W.; Marchessault, R.H. Bacterial Polyesters: Biosynthesis, Biodegradable Plastics and Biotechnology. *Biomacromolecules* **2005**, *6*, 1–8. [CrossRef]
17. Arrieta, M.P.; Fortunati, E.; Dominici, F.; Rayón, E.; López, J.; Kenny, J.M. Multifunctional PLA–PHB/cellulose nanocrystal films: Processing, structural and thermal properties. *Carbohydr. Polym.* **2014**, *107*, 16–24. [CrossRef]
18. Kervran, M.; Vagner, C.; Cochez, M.; Ponçot, M.; Saeb, M.R.; Vahabi, H. Thermal degradation of polylactic acid (PLA)/polyhydroxybutyrate (PHB) blends: A systematic review. *Polym. Degrad. Stab.* **2022**, *201*, 109995. [CrossRef]
19. Kontárová, S.; Přikryl, R.; Melčová, V.; Menčík, P.; Horálek, M.; Figalla, S.; Plavec, R.; Feranc, J.; Sadílek, J.; Pospíšilová, A. Printability, Mechanical and Thermal Properties of Poly(3-Hydroxybutyrate)-Poly(lactic acid)-Plasticizer Blends for Three-Dimensional (3D) Printing. *Materials* **2020**, *13*, 4736. [CrossRef]
20. Plavec, R.; Hlaváčiková, S.; Omaníková, L.; Feranc, J.; Vanovčanová, Z.; Tomanová, K.; Alexy, P. Recycling possibilities of bioplastics based on PLA/PHB blends. *Polym. Test.* **2020**, *92*, 106880. [CrossRef]
21. Elsawy, M.A.; Kim, K.H.; Park, J.W.; Deep, A. Hydrolytic degradation of polylactic acid (PLA) and its composites. *Renew. Sustain. Energy Rev.* **2017**, *79*, 1346–1352. [CrossRef]
22. Piemonte, V.; Sabatini, S.; Gironi, F. Chemical Recycling of PLA: A Great Opportunity Towards the Sustainable Development? *J. Polym. Environ.* **2013**, *21*, 640–647. [CrossRef]
23. Coszach, P.; Bogaert, J.C.; Willocq, J. Chemical Recycling of PLA by Hydrolysis. U.S. Patent No 8,431,683, 2013.
24. Cosate de Andrade, M.F.; Souza, P.; Cavalett, O.; Morales, A.R. Life Cycle Assessment of Poly(lactic acid) (PLA): Comparison Between Chemical Recycling, Mechanical Recycling and Composting. *J. Polym. Environ.* **2016**, *24*, 372–384. [CrossRef]
25. McKeown, P.; Jones, M.D. The Chemical Recycling of PLA: A Review. *Sustain. Chem.* **2020**, *1*, 1–22. [CrossRef]
26. Cristina, A.M.; Rosaria, A.; Sara, F.; Fausto, G. PLA Recycling by Hydrolysis at High Temperature. In *AIP Conference Proceedings*; AIP Publishing LLC: Melville, NY, USA, 2016; p. 020011.
27. Tsuji, H.; Daimon, H.; Fujie, K. A New Strategy for Recycling and Preparation of Poly(l-lactic acid): Hydrolysis in the Melt. *Biomacromolecules* **2003**, *4*, 835–840. [CrossRef] [PubMed]
28. Li, S.M.; Rashkov, I.; Espartero, J.L.; Manolova, N.; Vert, M. Synthesis, characterization, and hydrolytic degradation of PLA/PEO/PLA triblock copolymers with long poly (l-lactic acid) blocks. *Macromolecules* **1996**, *29*, 57–62. [CrossRef]
29. Piemonte, V.; Gironi, F. Lactic Acid Production by Hydrolysis of Poly(Lactic Acid) in Aqueous Solutions: An Experimental and Kinetic Study. *J. Polym. Environ.* **2013**, *21*, 275–279. [CrossRef]
30. Majgaonkar, P.; Hanich, R.; Malz, F.; Brüll, R. Chemical Recycling of Post-Consumer PLA Waste for Sustainable Production of Ethyl Lactate. *Chem. Eng. J.* **2021**, *423*, 129952. [CrossRef]
31. Grewell, D.; Srinivasan, G.; Cochran, E. Depolymerization of Post-Consumer Polylactic Acid Products. *J. Renew. Mater.* **2014**, *2*, 157–165. [CrossRef]
32. Lee, S.H.; Song, W.S. Enzymatic Hydrolysis of Polylactic Acid Fiber. *Appl. Biochem. Biotechnol.* **2011**, *164*, 89–102. [CrossRef]
33. Payne, J.; McKeown, P.; Jones, M.D. A circular economy approach to plastic waste. *Polym. Degrad. Stab.* **2019**, *165*, 170–181. [CrossRef]
34. Hirao, K.; Ohara, H. Synthesis and Recycle of Poly(L-lactic acid) using Microwave Irradiation. *Polym. Rev.* **2011**, *51*, 1–22. [CrossRef]
35. Hirao, K.; Nakatsuchi, Y.; Ohara, H. Alcoholysis of Poly(l-lactic acid) under microwave irradiation. *Polym. Degrad. Stab.* **2010**, *95*, 925–928. [CrossRef]
36. Song, X.; Liu, F.; Wang, H.; Wang, C.; Yu, S.; Liu, S. Methanolysis of microbial polyester poly(3-hydroxybutyrate) catalyzed by Brønsted-Lewis acidic ionic liquids as a new method towards sustainable development. *Polym. Degrad. Stab.* **2018**, *147*, 215–221. [CrossRef]
37. Song, X.; Wang, H.; Liu, F.; Yu, S. Kinetics and mechanism of monomeric product from methanolysis of poly (3-hydroxybutyrate) catalyzed by acidic functionalized ionic liquids. *Polym. Degrad. Stab.* **2016**, *130*, 22–29. [CrossRef]
38. Siddiqui, M.N.; Redhwi, H.H.; Al-Arfaj, A.A.; Achilias, D.S. Chemical Recycling of PET in the Presence of the Bio-Based Polymers, PLA, PHB and PEF: A Review. *Sustainability* **2021**, *13*, 10528. [CrossRef]
39. National Center for Biotechnology Information. Methyl Lactate. PubChem Compound Database. Available online: https://pubchem.ncbi.nlm.nih.gov/compound/Methyl-lactate (accessed on 30 October 2022).
40. National Center for Biotechnology Information. Ethyl Lactate. PubChem Compound Database. Available online: https://pubchem.ncbi.nlm.nih.gov/compound/Ethyl-lactate (accessed on 30 October 2022).
41. National Center for Biotechnology Information. Methyl 3-Hydroxybutyrate. PubChem Compound Database. Available online: https://pubchem.ncbi.nlm.nih.gov/compound/Methyl-3-hydroxybutyrate (accessed on 30 October 2022).
42. National Center for Biotechnology Information. Ethyl 3-Hydroxybutyrate. PubChem Compound Database. Available online: https://pubchem.ncbi.nlm.nih.gov/compound/Ethyl-3-hydroxybutyrate (accessed on 30 October 2022).
43. Amarasekara, A.S.; Owereh, O.S. Synthesis of a sulfonic acid functionalized acidic ionic liquid modified silica catalyst and applications in the hydrolysis of cellulose. *Catal. Commun.* **2010**, *11*, 1072–1075. [CrossRef]
44. Codari, F.; Lazzari, S.; Soos, M.; Storti, G.; Morbidelli, M.; Moscatelli, D. Kinetics of the hydrolytic degradation of poly(lactic acid). *Polym. Degrad. Stab.* **2012**, *97*, 2460–2466. [CrossRef]
45. Hill, C.A.S.; Jones, D.; Strickland, G.; Cetin, N.S. Kinetic and Mechanistic Aspects of the Acetylation of Wood with Acetic Anhydride. *Holzforschung* **1998**, *52*, 623–629. [CrossRef]

polymers

MDPI

Article

Enzyme-Catalyzed Polymerization of Kraft Lignin from *Eucalyptus globulus*: Comparison of Bacterial and Fungal Laccases Efficacy

Luisa García-Fuentevilla [1], Gabriela Domínguez [2], Raquel Martín-Sampedro [1], Manuel Hernández [2], María E. Arias [2], José I. Santos [3], David Ibarra [1] and María E. Eugenio [1,*]

1 Forest Sciences Institute (ICIFOR-INIA), CSIC, Ctra. de la Coruña Km 7.5, 28040 Madrid, Spain
2 Department of Biomedicine and Biotechnology, University of Alcalá, 28805 Alcalá de Henares, Spain
3 General Services of Research SGIKER, University of the Basque Country (UPV/EHU), Edificio Joxe Mari Korta, Avda. Tolosa 72, 20018 San Sebastian, Spain
* Correspondence: mariaeugenia@inia.csic.es; Tel.: +34-913-473-948

Abstract: Kraft lignin, a side-stream from the pulp and paper industry, can be modified by laccases for the synthesis of high added-value products. This work aims to study different laccase sources, including a bacterial laccase from *Streptomyces ipomoeae* (SiLA) and a fungal laccase from *Myceliophthora thermophila* (MtL), for kraft lignin polymerization. To study the influence of some variables in these processes, a central composite design (CCD) with two continuous variables (enzyme concentration and reaction time) and three levels for each variable was used. The prediction of the behavior of the output variables (phenolic content and molecular weight of lignins) were modelled by means of response surface methodology (RSM). Moreover, characterization of lignins was performed by Fourier-transform infrared (FTIR) spectroscopy and different nuclear magnetic resonance (NMR) spectroscopy techniques. In addition, antioxidant activity was also analyzed. Results showed that lignin polymerization (referring to polymerization as lower phenolic content and higher molecular weight) occurred by the action of both laccases. The enzyme concentration was the most influential variable in the lignin polymerization reaction within the range studied for SiLA laccase, while the most influential variable for MtL laccase was the reaction time. FTIR and NMR characterization analysis corroborated lignin polymerization results obtained from the RSM.

Keywords: bacterial laccase; central composite design; characterization; eucalypt; fungal laccase; kraft lignin; polymerization; response surface methodology

Citation: García-Fuentevilla, L.; Domínguez, G.; Martín-Sampedro, R.; Hernández, M.; Arias, M.E.; Santos, J.I.; Ibarra, D.; Eugenio, M.E. Enzyme-Catalyzed Polymerization of Kraft Lignin from *Eucalyptus globulus*: Comparison of Bacterial and Fungal Laccases Efficacy. *Polymers* **2023**, *15*, 513. https://doi.org/10.3390/polym15030513

Academic Editors: Esperanza Cortés Triviño, Adrián Tenorio-Alfonso and Clara Delgado-Sánchez

Received: 20 December 2022
Revised: 11 January 2023
Accepted: 13 January 2023
Published: 18 January 2023

1. Introduction

After cellulose, lignin is the second most abundant biopolymer on the planet [1]. It is synthetized from *p*-coumaryl alcohol (H), coniferyl alcohol (G), and sinapyl alcohol (S) monomers by enzymatic polymerization, in which oxidoreductase enzymes such as laccases and peroxidases are involved [2]. As a result, a heterogeneous complex tridimensional macromolecule is formed, containing different types of both ether (e.g., β-O-4′, α-O-4′, 5-O-4′, etc.) and carbon–carbon bonds (e.g., β-β′, β-5′, β-1′, 5-5′, etc.), and a wide variety of reactive groups depending on the biomass source. Accordingly, dicots (e.g., eucalypt, birch, poplar) contain around 20–25% of lignin, which is mostly composed of G and S units and traces of H units. On the other hand, gymnosperms (e.g., pine, spruce) with approximately 20–35% of lignin, are mostly composed of G units and very low proportions of H units. Finally, monocot grasses (e.g., flax, hemp, sisal) have lower lignin content (9–20%) composed of G and S units, together with high levels of H units [3]. Moreover, along with their source, the lignin isolation technology used strongly affects their features and properties and, therefore, the valorization ways [4].

Actually, the main source of lignin is the pulp and paper industry. Approximately 100 million tons per year of lignin were produced in 2015 with a value of roughly USD 732.7 million. Moreover, it is expected to increase to USD 913.1 million in 2025 [5]. Among the different pulp and paper processes, the kraft process is the most extended pulping technology, with an average lignin production estimated at 55–90 million tons per year [6]. Most of this kraft lignin is normally combusted, due to their high calorific value, to produce energy that is partially used in the same pulp and paper mills. However, this process generates an excess of energy, making the valorization of this waste lignin more interesting as high-added-value chemicals and materials that guarantee the sustainability and competitiveness of these mills [7]. In addition, kraft lignin valorization is also expected to benefit the future circular bioeconomy, which aims to maximize the usage and value of all raw materials, products, and wastes.

Although the inherent heterogeneity of kraft lignin (i.e., chemical composition, molecular structure, and molecular weight distribution) makes this waste material extraordinarily interesting, these features may be an obstacle for certain applications. To overcome this fact, the modification of the lignin structure is often a necessary step to produce the right lignin for each possible application [8]. The oxidative enzymes, such as laccases and peroxidases, involved in lignin biosynthesis in nature, can accomplish this modification [9]. Laccases (EC 1.10.3.2) are multicopper-containing oxidases with phenoloxidase activity, being widely expressed in nature, mainly in plants, insects, fungi, and bacteria [10]. The biological role of these enzymes is determined by their source and the phase of life of the organism producing them. For instance, fungal laccases participate in stress defense, morphogenesis, fungal plant–pathogen/host interactions, and lignin degradation, while bacterial laccases are involved in pigmentation, morphogenesis, toxin oxidation, and protection against oxidizing agents, and ultraviolet light [11]. These enzymes catalyze the oxidation of an extensive variety of phenolic and non-phenolic molecules, using oxygen as the final electron acceptor and releasing water as a by-product [11]. The catalytic site of laccases contains four copper ions. On the one hand, type-T1 copper, responsible for the characteristic blue color of the enzyme, is involved in the oxidation of the reducing substrate, acting as the primary electron acceptor. On the other hand, type-T2 copper together with two type-T3 coppers form a tri-nuclear copper cluster where oxygen is reduced to water [12]. The electrochemical potential of type-T1 copper is one of the most important properties of laccases, fluctuating between 0.4 and 0.8 V. Bacterial and plant laccases have low redox potential, whereas medium and high values are usually reported for fungal laccases [12].

The oxidative versatility, low catalytic requirements, and capacity of laccases to catalyze degradation or polymerization reactions make these enzymes suitable for a wide range of applications in different sectors, including lignocellulosic biorefinery, pulp and paper industry, food and textile sectors, bioremediation, and biosensor applications, among others [13]. More specifically, the enzymatic polymerization of lignin by laccases has been applied in the synthesis of new lignin-based polymeric materials [14] in, for example, the manufacture of green binders for fiberboard manufacturing [15], nanocomposite films formed by coating lignin nanoparticles along the microfibrilled cellulose fiber network [16], controlled-delivery fertilizer systems [17], and a pesticide release system [18]. In most of the studies on the enzymatic polymerization of lignin, fungal laccases are commonly used [15,17–23]. Only in recent years, bacterial laccases have also gained attention for this purpose [16,24,25]. Hence, there is a necessity to explore the potential of novel laccases, including bacterial enzymes, for kraft lignin polymerization.

This work aims to study the oxidative polymerization of *Eucalyptus globulus* kraft lignin by using different laccase sources, such as a bacterial laccase isolated from *Streptomyces ipomoeae* (SiLA) and a commercial fungal laccase from the ascomycete *Myceliophthora thermophila* (MtL). Laccase dosage and reaction time were the input variables evaluated at three levels to study laccase polymerization reactions, using a central composite design (CCD). The prediction of the behavior of the output variables (phenolic content and molecular weight of the resulting laccase-treated lignins) was modelled by means of response surface methodology (RSM). Moreover, structural characterization of the resulting lignins by Fourier-transform infrared spectroscopy (FTIR) and nuclear magnetic resonance (NMR) spectroscopy, as well as their antioxidant activity, were also evaluated.

2. Materials and Methods

2.1. Raw Material, Enzymes and Chemicals

Eucalypt (*Eucalyptus globulus*) residual lignin was isolated from kraft black liquor provided by La Montañanesa pulp mill (Lecta, Zaragoza, Spain). Then, the lignin was precipitated (pH of liquor lowered to 2.5 with concentrated sulfuric acid), centrifuged and washed with acid water (pH 2.5), dried and finally homogenized.

A recombinant bacterial laccase (SiLA) from *Streptomyces ipomoeae* CECT 3341.16 used for this study was overproduced and purified according to Guijarro et al. [26]. A commercial fungal laccase (MtL) from *Myceliophtora thermophila* (Novozym® 51003) was also used, being kindly supplied by Novozymes (Bagsvaerd, Denmark). Enzyme activities were determined by oxidation of 5 mM 2,2′-azino-bis(3-ethylbenzothiazoline-6-sulphonic acid) (ABTS) to its cation radical ($\varepsilon_{436\ nm}$ = 29,300 M^{-1} cm^{-1}) in 0.1 mM sodium acetate (pH 5) at 24 °C.

All the other reagents used were of analytical grade purchased either from Sigma–Aldrich (Madrid, Spain) or Merck (Barcelona, Spain).

2.2. Kraft Lignin Enzymatic Polymerization

Lignin was solubilized in phosphate buffer at pH 7.0 and 8.0 (100 mM), according to optimal pH for MtL and SiLA laccase activities, respectively [26,27], obtaining a solution of 1.5 g/L. The reactions took place at 60 °C and 45 °C, according also to optimal temperature for MtL and SiLA laccases, respectively [26,27]. At the end of reactions, the pH was lowered to 2.5 causing lignin precipitation. The reaction product was filtered and washed twice with acidified water (pH 2.5) and oven-dried at 40 °C under vacuum.

To study and optimize the influence of some variables on lignin polymerization, a central composite design (CCD) was used, considering two input variables (laccase dosage and reaction time) at three levels (−1, 0, +1). The design consisted of thirteen runs (experiments) from which five were center points and four were axial points (alpha = ±1.414). The values of laccase dosages (minimum value, 40 IU/g of kraft lignin; maximum value, 160 IU/g of kraft lignin) and reaction times (minimum value, 90 min; maximum value, 390 min) for the CCD were established based on previous experiments (Table 1) [28,29].

Table 1. Experimental conditions of the central composite design for investigation of kraft lignin polymerization by SiLA and MtL laccases considering two input variables (laccase dosage and reaction time) at three levels (−1, 0, +1).

	Coded Levels		Experimental Values	
Run	**Reaction Time**	**Laccase Dosage**	**Reaction Time (Min)**	**Laccase Dosage (IU/g)**
1	−1	−1	90.0	40.0
2	1	−1	390.0	40.0
3	−1	1	90.0	160.0
4	1	1	390.0	160.0
5	−1.414	0	27.87	100.0
6	1.414	0	452.13	100.0
7	0	−1.414	240.0	15.15
8	0	1.414	240.0	184.85
9	0	0	240.0	100.0
10	0	0	240.0	100.0
11	0	0	240.0	100.0
12	0	0	240.0	100.0
13	0	0	240.0	100.0

The output variables considered in this study were: the total phenolic content and the molecular weight of the resulting enzyme-treated lignins. Then, for each of the output variables, the obtained data from the CCD were used to fit a quadratic regression equation representing the influence of the two input variables. Each equation was plotted by means of response surface methodology (RSM). The design of the experiments of the CCD, the quadratic regression equations, and the plots obtained from modelling the equations by RSM were obtained using Minitab 19.1 software (Minitab Ltd., Coventry, United Kingdom).

2.3. *Enzyme-Treated Lignins Characterization*

2.3.1. Total Phenolic Content

The total phenolic content of kraft lignins was determined following the Folin–Ciocalteu method with some modifications, according to Jiménez-López et al. [30]. Firstly, kraft lignin samples were dissolved in dimethylsulfoxide (DMSO). Then, 500 μL of Folin–Ciocalteau reagent was added to 100 μL of the sample dilution, followed by the addition of 400 μL of Na_2CO_3. The reaction mixture was incubated at 50 °C for 10 min, and absorbance at 760 nm were measured after cooling, using a UV–Vis spectrophotometer (Lambda 365, PerkinElmer, Boston, MA, USA). The total phenolic content of samples was quantified using a calibration curve prepared from a standard solution of gallic acid (1–200 mg/L) and expressed as mg gallic acid equivalent (GAE)/g of lignin (on a dry basis).

2.3.2. Size Exclusion Chromatography (SEC)

SEC analysis (weight-average (Mw), number-average (Mn) molecular weights, and polydispersity (Mw/Mn)) of kraft lignins was carried out in an Agilent Technologies 1260 HPLC. The samples were analyzed at 254 nm (G1315D DAD detector, Agilent, Waldbronn, Germany) using two columns (Phenomenex) coupled in series (GPC P4000 and P5000, both 300 × 7.8 mm) and a safeguard column (35 × 7.8 mm). NaOH (0.05 M), pumped at a rate of 1 mL min^{-1}, was employed as mobile phase at 25 °C for 30 min. Samples were dissolved at a final concentration of 0.5 g/L in NaOH (0.05 M). Polystyrene sulfonated standard (peak average molecular weights of 4210, 9740, 65,400, 470,000, PSS-Polymer Standards Service) were used for the calibration curve [31].

2.3.3. Fourier-Transform Infrared (FTIR) and Ultraviolet–Visible (UV–Vis) Spectroscopy

FTIR spectra of kraft lignins were acquired by a JASCO FT/IR 460 Plus spectrometer (Jasco, Japan), equipped with an accessory single-reflection diamond, working with a resolution of 1 cm^{-1}, 400 scans, and a spectral range of 600–2000 cm^{-1} [32].

UV–Vis analysis of kraft lignins, dissolved in 0.1 M NaOH to a final concentration of 50 μg mL^{-1}, was carried out using a UV–Vis spectrophotometer (Lambda 365, PerkinElmer, Boston, MA, USA), and the absorbances were measured between λ 200 and 800 nm.

2.3.4. Nuclear Magnetic Resonance Spectroscopy (NMR)

Solid-state ^{13}C nuclear magnetic resonance (13C NMR) spectroscopy of kraft lignins was performed at 25 °C in a Bruker Avance III 400 MHz (Bruker, Billerica, MA, USA) at 100.64 MHz with the cross polarization/magic angle spinning (CP/MAS) technique. Lignin samples were prepared in 4 mm rotors and the spinning rate was 10 KHz. The contact time was 2 ms and the delay between scans was 5 s. The number of scans was 10,240 [32].

^{13}C–^{1}H two-dimensional nuclear magnetic resonance (2D NMR) spectra of kraft lignins were recorded at 25 °C in a Bruker AVANCE 500 MHz (Bruker, Billerica, MA, USA) equipped with a z-gradient double resonance probe. Around 40 mg of each lignin was dissolved in 0.75 mL of deuterated dimethylsulfoxide (DMSO-d6) and an HSQC (heteronuclear single quantum correlation) experiment was recorded. The spectral widths for the HSQC were 5000 Hz and 13,200 Hz for the ^{1}H and ^{13}C dimensions. The number of collected complex points was 2048 for ^{1}H-dimension with a recycle delay of 5 s. The number of transients for the HSQC spectra was 64, and 256 time increments were always recorded in ^{13}C-dimension. The J-coupling evolution delay was set to 3.2 ms. Squared cosine-bell apodization function was applied in both dimensions. Prior to Fourier transform the data matrices were zero filled up to 1024 points in the ^{13}C-dimension. Residual DMSO (from DMSO-d6) was used as an internal reference (δ_C/δ_H 39.6/2.5 ppm) [32].

2.3.5. Antioxidant Activity of Kraft Enzyme-Treated Lignins

The antioxidant activity of kraft lignin samples was estimated following the $ABTS^{+\bullet}$ methods according to Re et al. [33] and Ratanasumarn et al. [34]. $ABTS^{+\bullet}$ was produced by the reaction between 2,2′-azino-bis(3-ethylbenzthiazoline-6-sulphonic acid) diammonium salt (ABTS) stock solution (7 mM) and potassium persulfate (2.45 mM). Prior to use, the $ABTS^{+\bullet}$ stock solution was diluted with ethanol to obtain an absorbance of 0.7 ± 0.02 at 734 nm. Then, 1 mL of the $ABTS^{+\bullet}$ stock solution was mixed with 10 µL sample (32 mg mL^{-1}) or control and allowed to react for 6 min. The change in absorbance of the reaction mixture was measured at 734 nm and the antioxidant activity was expressed as trolox equivalent antioxidant capacity (TEAC). To calculate TEAC capacity, the gradient of the plot of the percentage inhibition of absorbance vs. concentration for lignin samples was divided by the gradient of the plot for trolox.

3. Results and Discussion

There is an increasing interest to evaluate the potential of novel laccases, including those from bacterial origin, for residual lignin polymerization. Presently, the fungal *M. thermophila* (MtL) laccase, with a high pH range and noticeable thermal stability, is widely employed for lignin polymerization [15,20,21,35]. *S. ipomoeae* (SiLA) laccase, with similar properties to MtL laccase, has been also assessed for this purpose [36], although to a lesser extent.

3.1. *Effect of Laccase Dosage and Reaction Time on Total Phenolic Content*

As it can be observed in Figure 1, both MtL and SiLA laccases showed the ability to reduce the phenolic content of the treated kraft lignins obtained from all experiments of the CCD. Nevertheless, the bacterial laccase achieved a higher reduction of total phenolic content compared to the fungal laccase, in spite of the similar low redox potential showed by both laccases (0.450 mV) [27,37].

The quadratic regression equation obtained from the phenolic content values (Figure 1a,b) when using MtL laccase (Equation (1)) and SiLA laccase (Equation (2)) over kraft lignin, and their regression coefficients are the following:

$$\text{mg GAE/g lignin} = 434.8 - 0.306 \cdot \text{Time (min)} - 0.786 \cdot \text{Dosage (IU/g)} - 0.000339 \cdot \text{Time (min)} \cdot \text{Time (min)} + 0.00089 \cdot \text{Dosage (IU/g)} \cdot \text{Dosage (IU/g)} + 0.002381 \cdot \text{Time (min)} \cdot \text{Dosage (IU/g)}$$
$$\text{R-squared} = 91.24\% \quad (1)$$

$$\text{mg GAE/g lignin} = 222.5 + 0.1458 \cdot \text{Time (min)} - 0.588 \cdot \text{Dosage (IU/g)} - 0.000343 \cdot \text{Time (min)} \cdot \text{Time (min)} + 0.00157 \cdot \text{Dosage (UA/g)} \cdot \text{Dosage (IU/g)} - 0.000776 \cdot \text{Time (min)} \cdot \text{Dosage (IU/g)}$$
$$\text{R-squared} = 92.93\% \quad (2)$$

Figure 2a,b show the response surface for lignin phenolic content as a function of enzyme concentration and reaction time using MtL or SiLA laccases, respectively. As it can be seen, the total phenolic content of lignin decreased with increasing reaction time in the studied region when MtL laccase was used. The maximum phenolic content reduction (53%) was obtained at 452 min using 100 IU/g of MtL laccase. In the case of SiLA laccase, the most influential variable on the total phenolic content reduction was the laccase dosage, observing a maximum decrease (77%) using 160 IU/g of SiLA laccase at 390 min.

As it is widely known, laccases can oxidize free phenolic lignin units, yielding resonance-stabilized phenoxyl radicals via a single electron transfer process [38]. Thus, the establishment of new linkages between the formed phenoxyl radicals leads to a lower content of free phenols in lignin. Different studies have already described the ability of MtL laccase to reduce the phenolic content of both kraft lignin and lignosulfonates. Then, the phenolic content was decreased by this fungal laccase by around 66% in the case of eucalypt kraft lignin [15], whereas a reduction of 52% was described using lignosulfonates [35]. On the other hand, the phenolic content reduction in lignin by bacterial laccases has been also reported. In this regard, Mayr et al. [25] showed the ability of CotA laccase to decrease the phenolic content between 30% and 65% in different kraft lignins of softwood and hardwood origin, respectively. Similar results were described by Wang et al. [16] when a commercial bacterial laccase (Metzyme®) was used to polymerize alkali lignins from birch and spruce materials.

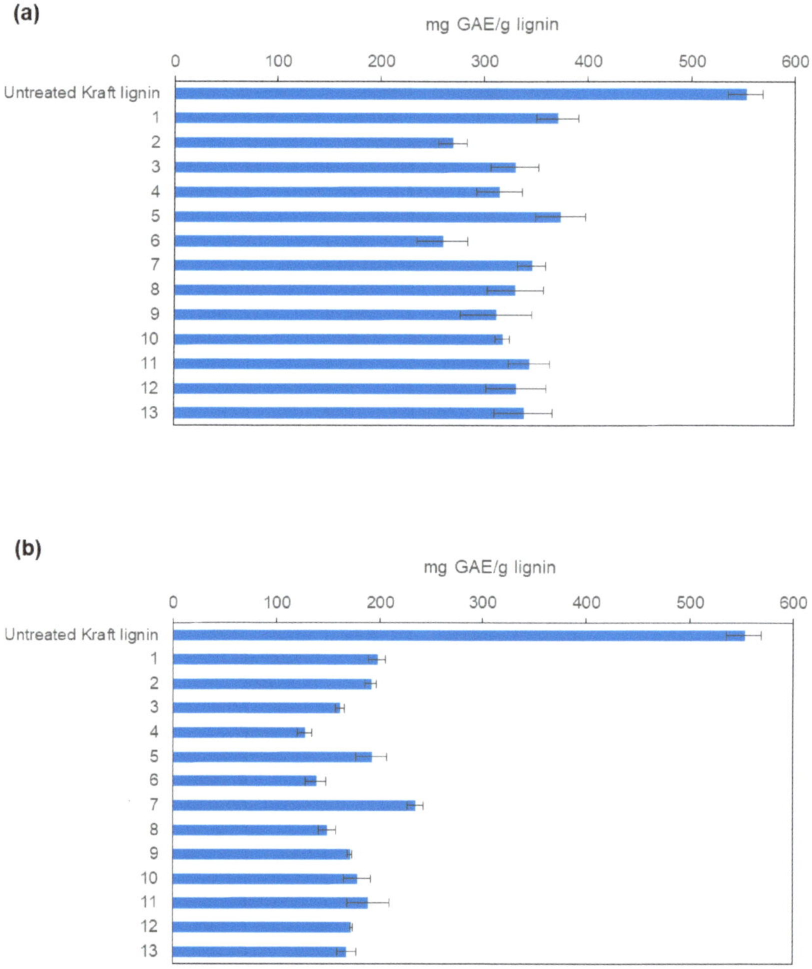

Figure 1. Total phenolic content of the resulting treated lignins with MtL (**a**) and SiLA (**b**) laccases expressed as mg GAE/g lignin (experiments resulting from CCD).

Contrary to this study, the effects of laccase dosage and reaction time on the phenolic content of lignin are usually evaluated separately. Gillgren et al. [19] showed that longer reaction time resulted in higher reductions of phenolic content of both organosolv lignin and lignosulfonates when they were treated with a fungal laccase from *Trametes* (syn. *Coriolus polyporus*). A similar trend was reported by Huber et al. [35], using a laccase from

MtL to polymerize both eucalypt kraft lignin and lignosulfonates. Moreover, these authors also observed a higher decrease in the phenolic content using higher MtL laccase dosages, indicating that the amount of enzyme used, together with reaction time, are important factors for the lignin polymerization process. Finally, Mayr et al. [25] also reported the influence of reaction time on the phenolic content decrease when a bacterial CotA laccase was used to polymerize different kraft lignins, observing a decrease in phenolic content by extending the reaction time.

(a)

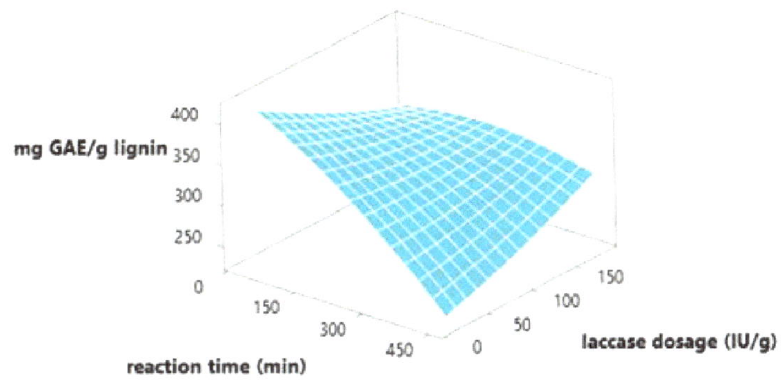

(b)

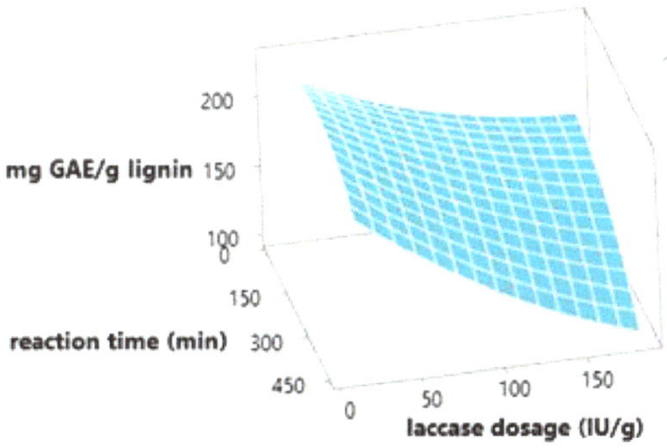

Figure 2. Response surface plot of total phenolic content expressed as mg GAE/g lignin to evaluate MtL (**a**) and SiLA (**b**) laccases dosage and reaction time.

3.2. Effect of Laccase Dosage and Reaction Time on Molecular Weight

The molecular weight distributions of laccase-treated lignins are displayed in Figure S1. From them, weight-average (Mw) and number-average (Mn) molecular weights, as well as polydispersity (Mw/Mn) values were calculated (Table S1). In general, both MtL and SiLA laccases produced an increment in the Mw values of the treated kraft lignins obtained from all experiments (Figure 3a,b). Polydispersity values also showed higher values compared to the untreated lignin (Table S1).

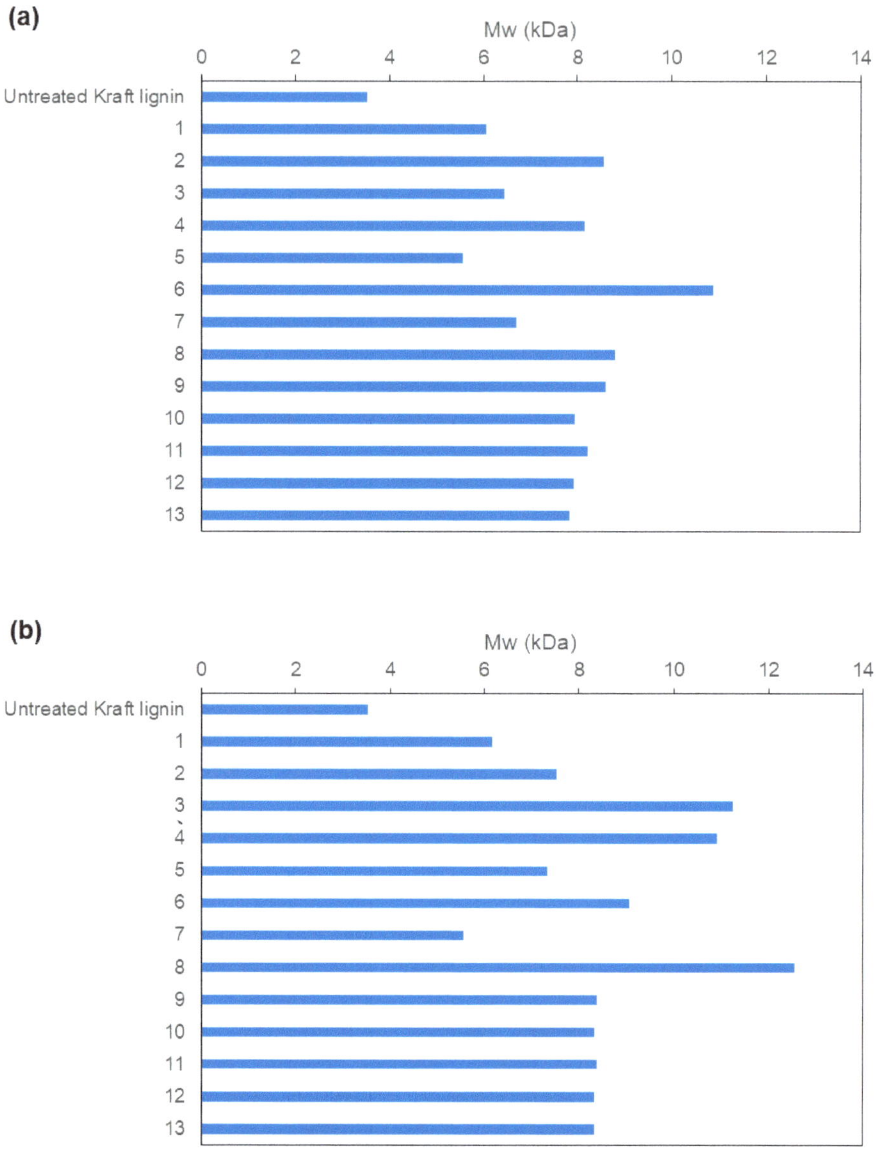

Figure 3. Weight-average (Mw) molecular weight values (KDa) of the resulting treated lignins with MtL (**a**) and SiLA (**b**) laccases (experiments resulting from CCD).

The quadratic regression equation obtained from the Mw values (Figure 3a,b) when using MtL laccase (Equation (3)) and SiLA laccase (Equation (4)) over kraft lignin, and their regression coefficients are the following:

$$\text{Mw (Da)} = 3371 + 14.40 \cdot \text{Time (min)} + 30.7 \cdot \text{Dosage (IU/g)} - 0.0050 \cdot \text{Time (min)} \cdot \text{Time (min)} - 0.0959 \cdot \text{Dosage (IU/g)} \cdot \text{Dosage (UA/g)} - 0.0223 \cdot \text{Time (min)} \cdot \text{Dosage (IU/g)}$$

$$\text{R-squared} = 83.75\% \quad (3)$$

$$\text{Mw (Da)} = 3936 + 7.42 \cdot \text{Time (min)} + 25.46 \cdot \text{Dosage (IU/g)} + 0.00040 \cdot \text{Time (min)} \cdot \text{Time (min)} + 0.1201 \cdot \text{Dosage (IA/g)} \cdot \text{Dosage (IU/g)} - 0.0471 \cdot \text{Time (min)} \cdot \text{Dosage (IU/g)}$$

$$\text{R-squared} = 98.38\% \quad (4)$$

Figure 4a,b show the response surface for lignin Mw as a function of enzyme concentration and reaction time of MtL and SiLA laccases, respectively. As can be observed, the molecular weight of the lignin increased by increasing the reaction time in the studied interval when MtL laccase was used, in agreement with the observed reduction in phenolic content (Section 3.1). The maximum molecular weight increment (3.0-fold, which correspond to the value of Mw 10,865 Da) was obtained at 452 min using 100 IU/g of MtL laccase. In the case of SiLA laccase, the most influential variable on the molecular weight was the laccase dosage used, as also observed for the phenolic content of lignin (Section 3.1). The maximum increment (3.5-fold, which corresponds to the value of Mw 12,545 Da) was obtained using 184.85 IU/g of SiLA laccase at 240 min.

As previously commented, the stabilized phenoxyl radicals, generated from lignin by laccase oxidation, undergo radical–radical coupling through phenyl ether–carbon and carbon–carbon linkages, yielding the observed increase in Mw values of kraft lignin by both laccases. Moreover, the polydispersity increase is also expected due to the non-selective radical–radical coupling reactions, which link lignin end groups to each other spontaneously with low or no control and, consequently leading to higher polydispersity values [20]. MtL laccase has already shown its capability to increase the molecular weight of both kraft lignin and lignosulfonates. Thus, Gouveia et al. [15] reported a strong increase (17.0-fold) in the average molecular weight of laccase-treated eucalypt kraft lignin (80,000 Da) compared to the untreated lignin sample (4700 Da). Huber et al. [35] also described a 12.0-fold increase in Mw (22,400 Da) for enzymatic polymerization of lignosulfonates (1900 Da for untreated lignin), and only a 1.4-fold increase (2300 Da) when kraft lignin (1600 Da for untreated lignin) was used. On the other hand, the molecular weight increase by bacterial laccases has been also described. Thus, Wang et al. [16] reported a 2.9-fold increase (from 17,750 Da for untreated lignin to 52,000 Da for laccase treated lignin) when a Metzyme® laccase was used to polymerize alkali spruce lignin. Mayr et al. [25] achieved 6.0-fold increases in molecular weight for softwood (from 21,600 Da for untreated lignin to 130,000 Da for laccase-treated lignin) and 19.2-fold for hardwood kraft lignins (from 3150 Da for untreated lignin to 60,000 Da for laccase-treated lignin) when they were treated with a CotA laccase.

Similarly to phenolic content, the effects of laccase dosage and reaction time on the molecular weight of lignin are generally studied separately, reporting different results in function of both laccase and lignin sources. Gouveia et al. [21] observed that the major changes in molecular weight of kraft lignin treated with the fungal MtL laccase occurred during the first 2 h, although longer reaction time resulted in higher Mw values of the resulting treated lignins. These authors also showed a molecular weight increase as the enzyme dosage was augmented. In this regard, Areskogh et al. [39] determined that no significant increments in the molecular weight of lignosulfonates were observed at low MtL enzyme dosage, while the molecular weight increased by augmenting the enzyme concentration. Huber et al. [35] also demonstrated that the amount of biocatalyst used strongly influences the polymerization process. When 50 mg of MtL laccase was used, 4.0-fold and 1.7-fold molecular weight increments were determined for lignosulfonates and kraft lignin, respectively. However, when the MtL laccase was augmented to 100 mg, a 12.0-fold increase in the molecular weight was measured for enzymatic polymerization of lignosulfonates, and only a 1.4-fold increase was seen for kraft lignin. Finally, Mayr et al. [25] also achieved higher increases in the molecular weight of softwood and hardwood kraft lignins at longer reaction times using a bacterial CotA laccase.

(a)

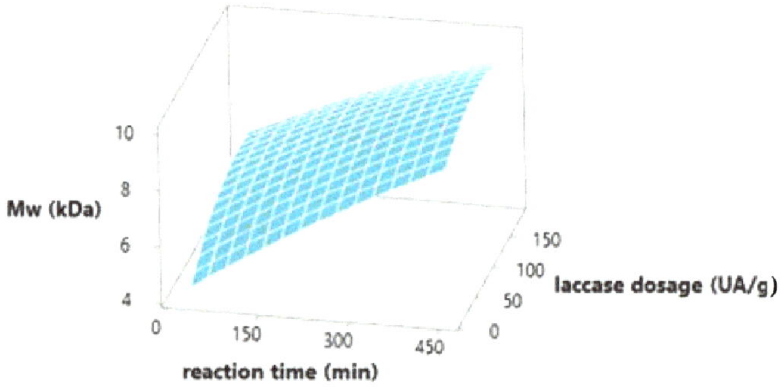

(b)

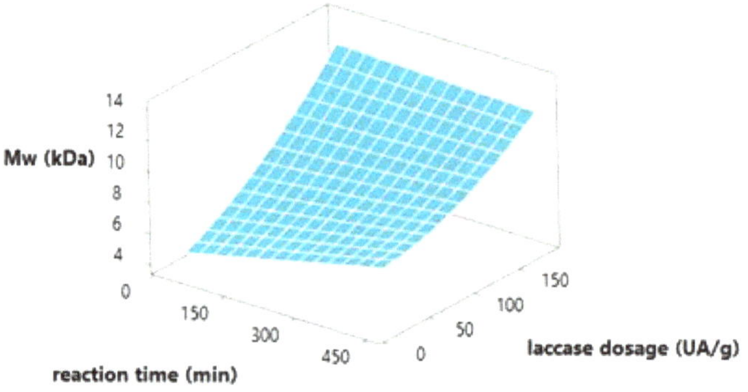

Figure 4. Response surface plot of weight-average (Mw) molecular weight values (KDa) to evaluate MtL (**a**) and SiLA (**b**) laccases dosage and reaction time.

While the significant phenolic content of kraft lignin can translate into good reactivity for producing phenol-formaldehyde resins, epoxy resins, polyester systems, and polyurethanes, among others, the molecular weight increase of kraft lignin by laccase enzymes enables new applications as lignin-based dispersants providing better adsorption properties, stabilizer for emulsions, and in thermoplastic blends or copolymers enhancing thermal and mechanical performance [32].

3.3. Antioxidant Activity

The antioxidant ability of lignins (i.e., their capacity to act as radical scavengers) promotes their use as natural additives in food, cosmetics, pharmaceuticals, and polymeric formulations as an alternative to synthetic compounds such as butylated hydroxyanisole (BHA) and butylated hydroxytoluene (BHT), among others [40]. It is widely known that having a high phenolic content, low molecular weight, and narrow distribution seem to be favorable for the antioxidant capacity of lignin [40]. Nevertheless, in spite of the phenolic content decrease and molecular weight increase observed herein for the treated kraft lignins by both laccases, they still showed some antioxidant capacity, expressed as TEAC, (0.02–0.18) compared to the untreated sample (0.2) (Figure 5a,b for MtL and SilA laccases, respectively).

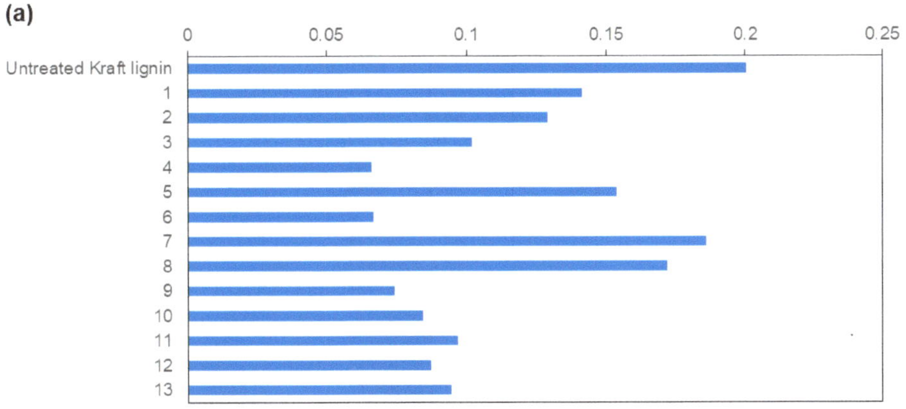

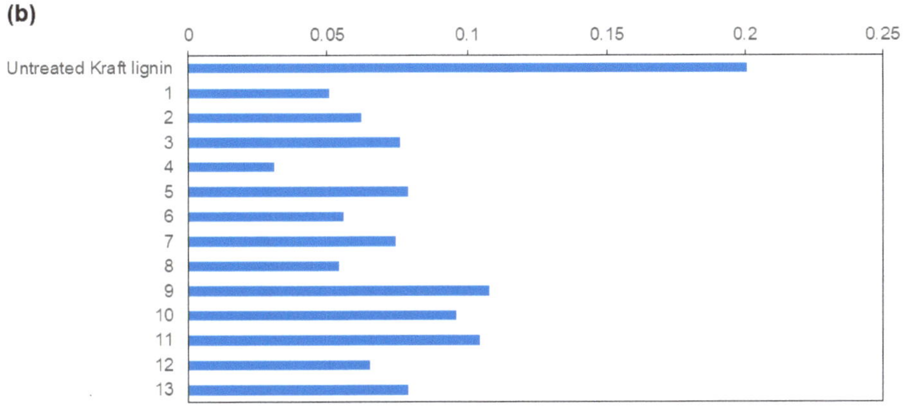

Figure 5. Antioxidant activity of the resulting treated lignins with MtL (**a**) and SiLA (**b**) laccases expressed as TEAC (experiments resulting from CCD).

3.4. FTIR Characterization

FTIR spectra of untreated and laccase-treated kraft lignins (MtL-KL and SiLA-KL resulting from experiments with maximum decrease in phenolic content and increase in molecular weight achieved) are displayed in Figure 6. The observed bands were assigned in comparison with others previously reported in the literature [41,42] and are displayed in Table S2. FTIR spectrum of kraft lignin showed the characteristic bands of lignin, which include those observed at 1610, 1515 and 1415 cm^{-1} associated with aromatic ring vibrations, and at 1455 cm^{-1} attributed to C−H asymmetric vibrations and deformations (Figure 6a). Bands attributed to syringyl (S) and guaiacyl (G) units were also identified, including those at 1315 cm^{-1} (S and G units), a shoulder at 1270 cm^{-1} (G units), 1220 cm^{-1} (G units), 1115 cm^{-1} (S units), 1025 cm^{-1} (G units) and 820 cm^{-1} (S units).

The major change in the FTIR spectra of MtL-KL and SiLA-KL samples compared to untreated kraft lignin spectra was observed at the bands corresponding to the C=O stretching for conjugated (1650 cm^{-1}) and unconjugated (1715 cm^{-1}) linkages (Figure 6b,c), as a consequence of the lignin oxidation caused by both laccases, being more noticeable in the case of the bacterial laccase. This effect was supported by UV–Vis, observing a decrease in the two absorption maxima at λ 230–240 and 280 nm, attributed to non-conjugated phenolic groups, in both MtL-KL and SiLA-KL samples due to lignin oxidation (Figure S2). Comparable results have been previously described by Gouveia et al. [15,21], when a laccase from *M. thermophila* was used for eucalypt kraft lignin polymerization, and Gillgren et al. [19], when a laccase from the white-fungus *C. polyporus* was employed to polymerize organosolv lignin and lignosulfonates. Moreover, both MtL-KL and SiLA-KL samples kept their characteristic triplet at 1610, 1515 and 1415 cm^{-1}, which is indicative of no modification of lignin aromatic backbone, as previously observed by Areskogh et al. [43] during polymerization of lignosulfonates by *M. thermophila* laccase.

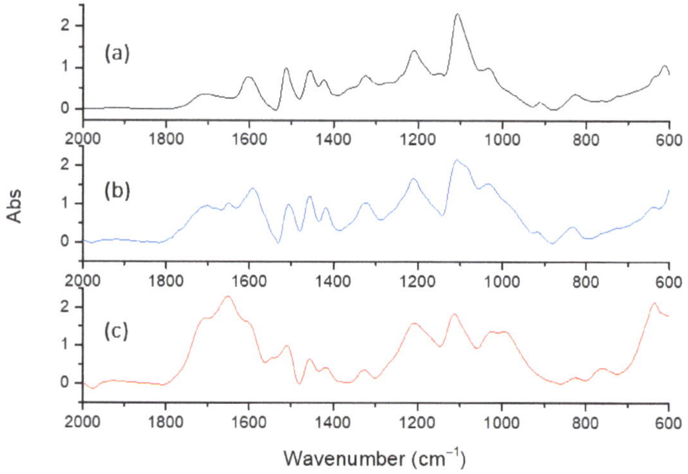

Figure 6. FTIR spectra, 2000–600 cm^{-1} region of the untreated lignin ((**a**), black line) and of the resulting treated lignins with MtL ((**b**), blue line) and SiLA ((**c**), red line) laccases. The bands in each spectrum are normalized with regard to the band at 1515 cm^{-1}.

3.5. NMR Characterization

The HSQC spectra of untreated and laccase-treated kraft lignins (MtL-KL and SiLA-KL resulting from experiments with maximum decrease in phenolic content and increase in molecular weight achieved) are shown. They included the whole spectra (δ_C/δ_H 0.0–150.0/0.0–9.0) in Supplementary Figure S3, and the spectra corresponding to the oxygenated aliphatic (δ_C/δ_H 45.0–95.0/2.5–6.0 ppm) and the aromatic (δ_C/δ_H 90.0–150.0/5.0–9.0 ppm) regions in Figures 7 and 8, respectively. The main ^{13}C–^{1}H lignin correlation signals identified in HSQC spectra are displayed in Table S3, endorsed according to those described by the literature [42,44–47]. The lignin substructures identified are depicted in Figures S4 and S5.

The oxygenated aliphatic region of the kraft lignin spectrum exhibited information about the different interunit linkages present (Figure 7a), including those from native and kraft-derived linkages. Despite the well-known lignin degradation under alkaline conditions during kraft pulping [48], several remaining signals from native β-O-4′ and β-β′ resinol substructures were observed, as well as correlation signals for spirodienones and cinnamyl alcohol end-groups. Signals from kraft-derived lignin linkages could also be recognized. Among them, signals from epiresinols and diaresinol, both diastereomers from the transformation of the native resinol substructure during kraft pulping [47,49]. An aryl-glycerol substructure could also be hesitantly identified, produced from the non-phenolic β-aryl ether linkage under alkaline conditions during kraft pulping [50]. Finally, a correlation signal of lignin terminal structures with a carboxyl group in C_α (Ar–CHOH–COOH; F_α), overlapping with aryl-glycerol, could also be found.

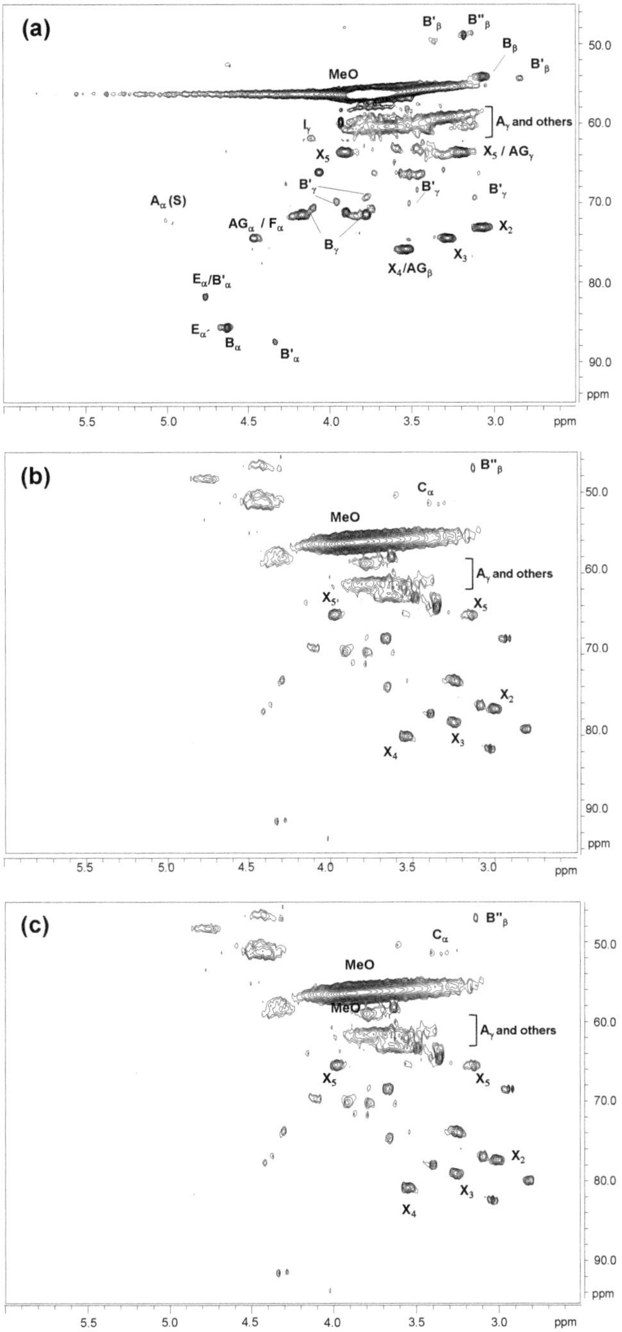

Figure 7. HSQC 2D-NMR spectra, δ_C/δ_H 45.0–95.0/2.5–6.0 ppm aliphatic oxygenated region, of the untreated lignin (**a**) and of the resulting treated lignins with MtL (**b**) and SiLA (**c**) laccases. Correspondences between lignin structures and assigned letters in the figure: A, β-O-4′ alkyl-aryl ether; AG, aryl-glycerol; B, resinols; B′, epiresinols; B″, diaresinol; C, α-5′; E, spirodienones; F, Ar–CHOH–COOH; I, cinnamyl alcohol end-groups; X, xylopyranose (R, OH).

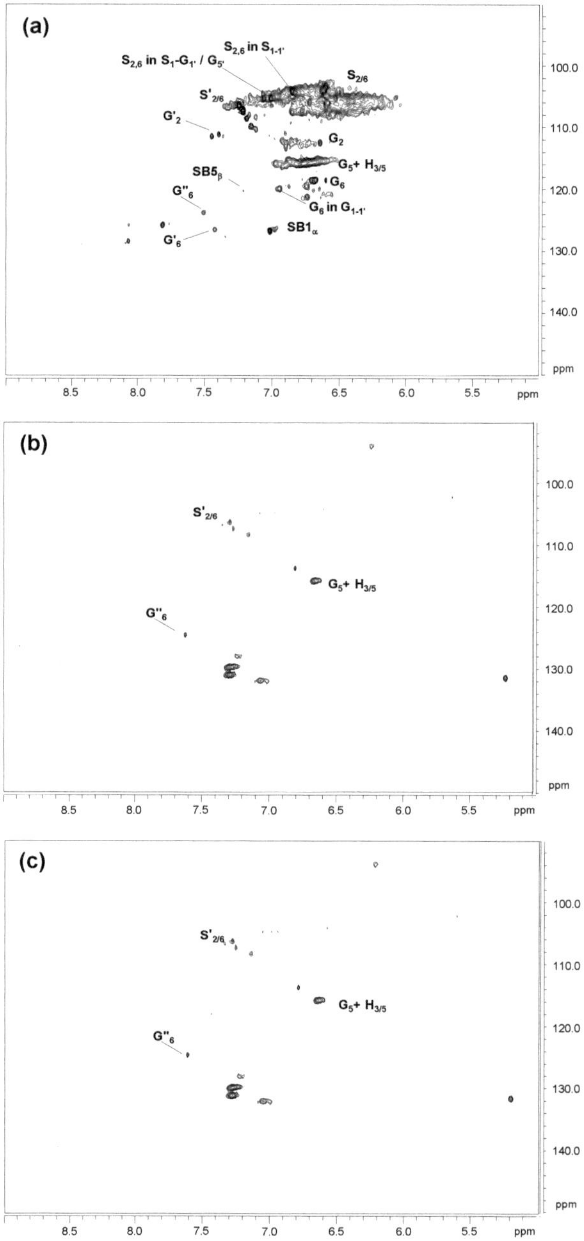

Figure 8. HSQC 2D-NMR spectra, δ_C/δ_H 90.0–150.0/5.0–9.0 ppm aromatic region, of the untreated lignin (**a**) and of the resulting treated lignins with MtL (**b**) and SiLA (**c**) laccases. Correspondences between lignin structures and assigned letters in the figure: G, guaiacyl unit; G′, vanillin; G″, acetovanillona; H, *p*-hydroxyphenyl unit; S, syringyl unit; S′, syringaldehyde (R=H) or acetosyringone (R=CH_3); $S_{1-1'}$, 3,5-tetramethoxy-para-diphenol; $G_{1-1'}$, 3-dimethoxy-para-diphenol; S_1-$G_{1'}$/$G_{5'}$; SB1, stilbene-β-1′; SB5, stilbene-β-5′.

The aromatic region of kraft lignin spectra displayed the typical correlation signals of S, G, and H lignin units (Figure 8a), the usual pattern of hardwood lignins [51]. Moreover, a group of signals from lignin oxidation, such as oxidized S units, corresponding to syringaldehyde or acetosyringone, and oxidized G units, attributed to vanillin and acetovanillone, could also be detected. Signals from kraft-derived lignin linkages were also found in the aromatic region. Among them, correlation signals endorsed to β1 and β5 stilbene, derived from degradation of spirodienone and β-5′ phenylcoumaran during kraft pulping, respectively, were identified [46,47]. Finally, correlation signals from $S_{1\text{-}1'}$ (3,5-tetramethoxy-para-diphenol), $G_{1\text{-}1'}$ (3-dimethoxy-para-diphenol) and S_1-$G_{1'}$/$G_{5'}$ were tentatively identified as a result of C_α-C_1 breakdown in a retro-aldol reaction, followed by a radical coupling reaction, during kraft pulping [32,45,46].

Significant changes in the oxygenated aliphatic region of MtL-KL and SiLA-KL spectra (Figure 7b,c for MtL-KL and SiLA-KL samples, respectively), compared to the kraft lignin spectrum (Figure 7a), were found. In general, a complete disappearance of signals assigned to native and kraft-derived linkages was observed for both MtL-KL and SiLA-KL lignin samples, probably due to the cleavage of interunit linkages by treatment with both laccases. Nevertheless, some signals from β-O-4′ and diaresinol were still found. In this sense, Prasetyo et al. [23] reported a decrease in the intensity signals of β-O-4′ linkages when lignosulfonates were treated with *Trametes villosa* and *Trametes hirsuta* laccases. Wang et al. [16] also described a cleaved of β-aryl ether and β-β′ resinol substructures during the treatment of alkali lignins with a commercial (MetZyme®) bacterial laccase. In addition to the cleavage of interunit linkages, a new signal could also be observed in the aliphatic oxygenated region of both MtL-KL and SiLA-KL spectra, which was tentatively attributed to α-5′ condensed structures. The appearance of this structure is probably due to lignin condensation/polymerization reactions by laccases action. In this sense, Wang et al. [16] already described the formation of this condensed structure during the treatment of alkali lignins with the MetZyme bacterial laccase.

MtL-KL and SiLA-KL spectra also showed a near complete disappearance of the aromatic ^{13}C–^1H correlation signals (Figure 8b,c) compared to the kraft lignin spectrum (Figure 8a). Nevertheless, some intensity of signals corresponding to G and H units as well as to lignin oxidation, such as oxidized S and G units, could still be found. This important loss of aromatic correlation signals observed in the HSQC spectra after the enzymatic treatment with both laccases could indicate a significant modification of the lignin aromatic backbone. However, when laccase-treated lignins were analyzed by 1D-NMR, it could be inferred that the loss of aromatic correlation signals observed by 2D-NMR was due to deprotonation of the lignin benzene rings, as revealed by the ^1H NMR MtL-KL and SiLA-KL spectra (Figure S6). Meanwhile strong signals of aromatic carbons could be seen in the ^{13}C-NMR MtL-KL and SiLA-KL spectra (Figure S7), proving that benzene rings were not degraded by laccase treatment, as previously seen by FTIR analysis (Section 3.4). The decrease or complete disappearance of aromatic proton signals was not entirely unexpected, as it has been previously described in lignosulfonates by the action of *T. villosa* and *T. hirsuta* laccases [23], in eucalypt kraft lignin treated with *M. thermophila* laccase [15], in alkali lignins by the action of MetZyme bacterial laccase [16], and in organosolv lignin and lignosulfonates treated with *C. polyporus* laccase [19]. All these authors have related this effect with the formation of condensed structures such as 5-5′ or 4-O-5′.

^{13}C-NMR of laccase treated lignins (Figure S7b,c) also showed a remarkable increase in the signal at δ_C 176 ppm (carbonyl groups), especially in the SiLA lignin sample as previously described by FTIR analysis (Section 3.4), resulting from lignin oxidation caused by laccases action. A significant decrease in the signals at δc 147 ppm, corresponding to C_3 and C_5 of phenolic S units and C_3 and C_5 in phenolic G units, and at δc 134 ppm, endorsed to C_1 and C_4 in phenolic S units and C_1 in phenolic G units, was also observed, resulting from the phenolic lignin units' oxidation by laccase treatment, which supports the phenolic content decrease observed in Section 3.1. At a time, an increase in the shoulders at δc 152 ppm and at δc 130 ppm, from non-phenolic lignin units were also visible. Santos et al. [52] assigned the signal at δc 152 ppm to C_3 in new 5-5′ or C_3 (and C_4/C_5) in new 4-O-5′ structures formed during laccase (*T. villosa*) treatment of lignosulfonates, whereas Magina et al. [53] endorsed the signal at δc 130 ppm to C_5 in 5-5′ structure during MtL laccase treatment of lignosulfonates.

Then, these observations arising from NMR analysis suggest lignin condensation/ polymerization reactions by laccases action, supporting the molecular weight increment observed by SEC (Section 3.2). The phenoxy radicals formed by the action of the laccase enzymes on the phenolic units present in the initial kraft lignin together with those derived from the cleavage of interunit linkages underwent radical–radical coupling through phenyl ether–carbon and carbon–carbon links resulting in new condensed structures such as α-5′, 5-5′, and 4-O-5′ (Figure 9).

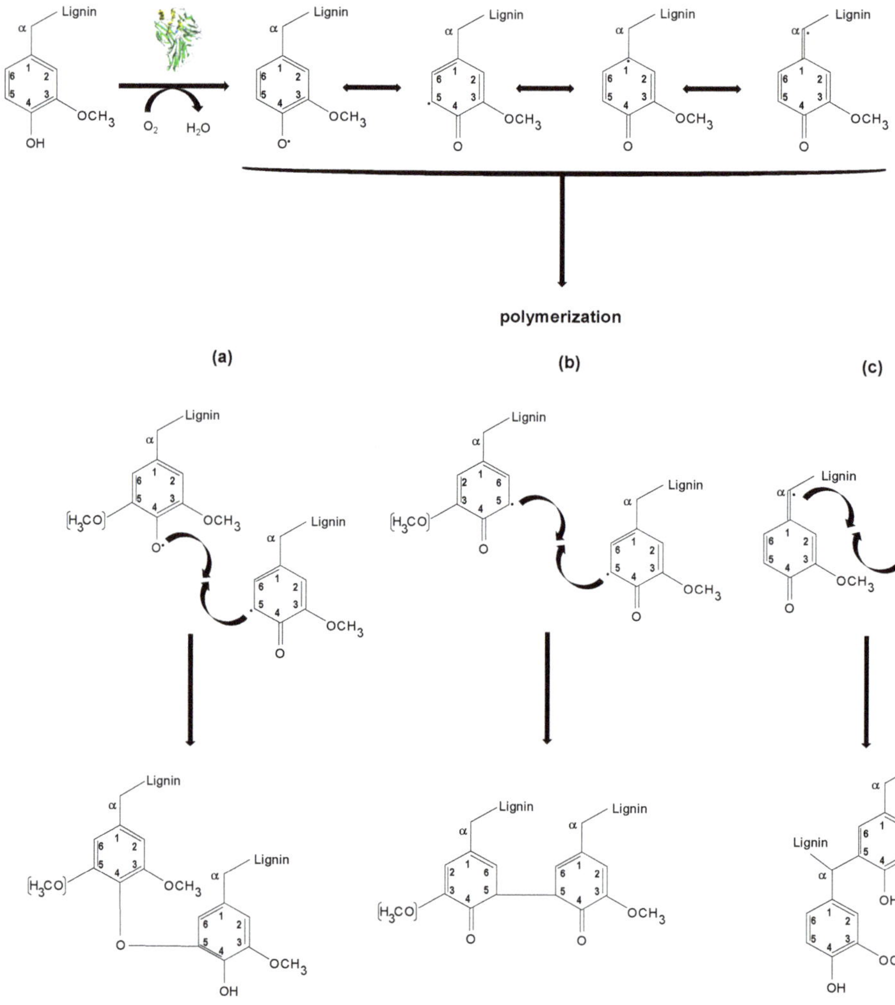

Figure 9. Schematic representation of oxidative lignin polymerization catalyzed by laccase. Radical coupling leads to formation of new condensed structures such as 4-O-5′ (**a**), 5-5′ (**b**), and α-5′ (**c**).

4. Conclusions

In order to produce the most appropriate lignin for each possible application, structural modification of this molecule is needed. A possible way to achieve this goal is by using laccase enzymes, which is considered a sustainable and environmentally friendly approach. Concerning this, this study confirmed the ability of a bacterial laccase from *S. ipomoeae* (SiLA) and a commercial fungal laccase from *M. thermophila* (MtL) to polymerize kraft lignin (referring to polymerization as lower phenolic content and higher molecular weight). Specifically, the enzyme dosage was the most influential variable in the kraft lignin polymerization reaction, within the range studied, when the bacterial SiLA laccase was used, while for MtL fungal laccase the most influential variable was the reaction time. FTIR and NMR characterization spectra verified lignin polymerization, observing new condensed structures such as α-5′, 5-5′, and 4-O-5′.

Supplementary Materials: The following supporting information can be downloaded at: https://www.mdpi.com/article/10.3390/polym15030513/s1, Figure S1. Molecular weight distributions of the resulting treated lignins with MtL (a) and SiLA (b) laccases. Figure S2. UV–Vis spectra, λ 200–800 nm, of the untreated lignin (a) and of the resulting treated lignins with MtL (b) and SiLA (c) laccases. Figure S3. HSQC 2D-NMR whole spectra, δ_C/δ_H 45.0–95.0/2.5–6.0 ppm, of the untreated lignin (a) and of the resulting treated lignins with MtL (b) and SiLA (c) laccases. Figure S4. Main lignin and carbohydrate substructures identified in aliphatic oxygenated region of the untreated kraft lignin and of the resulting treated lignins with MtL and SiLA laccases. Figure S5. Main lignin substructures identified in aromatic region of the untreated kraft lignin and of the resulting treated lignins with MtL and SiLA laccases. Figure S6. ^{1}H NMR spectra, δ_H 0.0–9.0 ppm, of the untreated kraft lignin (a) and of the resulting treated lignins with MtL (b) and SiLA (c) laccases. Figure S7. ^{13}C NMR spectra, δ_C 0.0–200.0 ppm, of the untreated kraft lignin (a) and of the resulting treated lignins with MtL (b) and SiLA (c) laccases. Table S1. Weight-average (Mw) and number-average (Mn) molecular weights and polidispersity (Mw/Mn) of the untreated kraft lignin and of the resulting treated lignins with MtL and SilA laccases. Mw and Mn are given in Da. Table S2. Main assignments of untreated and laccase-treated kraft lignins FTIR bands. Table S3. Assignment of main lignin and carbohydrates ^{13}C–^{1}H correlation signals in the HSQC spectra of untreated kraft lignin and the resulting treated lignins with MtL and SiLA laccases.

Author Contributions: Conceptualization, D.I. and M.E.E.; methodology, L.G.-F., G.D. and R.M.-S.; software, L.G.-F., G.D., R.M.-S. and J.I.S.; validation, L.G.-F., G.D., R.M.-S. and J.I.S.; formal analysis, L.G.-F., G.D., R.M.-S. and J.I.S.; investigation, L.G.-F., G.D., R.M.-S., M.H., M.E.A., J.I.S., D.I. and M.E.E.; resources, M.H., M.E.A., D.I. and M.E.E.; data curation, M.H., M.E.A., D.I. and M.E.E.; writing—original draft preparation, D.I. and M.E.E.; writing—review and editing, L.G.-F., G.D., R.M.-S., M.H., M.E.A., J.I.S., D.I. and M.E.E.; visualization, L.G.-F., G.D., R.M.-S., M.H., M.E.A., J.I.S., D.I. and M.E.E.; supervision, D.I. and M.E.E.; project administration, M.H., M.E.A., D.I. and M.E.E.; funding acquisition, M.H., M.E.A., D.I. and M.E.E. All authors have read and agreed to the published version of the manuscript.

Funding: This research was funded by Comunidad de Madrid via Project SUSTEC-CM S2018/EMT-4348; MICINN/AEI/10.13039/501100011033 and by "ERDF A way of making Europe" via Project RTI2018-096080-B-C22; and MICINN via Project TED2021-132122B-C21.

Institutional Review Board Statement: Not applicable.

Informed Consent Statement: Not applicable.

Data Availability Statement: Not applicable.

Acknowledgments: L.G.-F., R.M.-S., D.I. and M.E.E. are grateful for the support of the Interdisciplinary Platform for Sustainable Plastics towards a Circular Economy (SusPlast-CSIC), Madrid, Spain; Interdisciplinary Platform for Sustainability and Circular Economy (SosEcoCir-CSIC), Madrid, Spain; and Interdisciplinary Platform Horizonte Verde (CSIC). The contribution of COST Action LignoCOST(CA17128), supported by COST (European Cooperation in Science and Technology), in promoting interaction, exchange of knowledge and collaborations in the field of lignin valorization is also gratefully acknowledged.

Conflicts of Interest: The authors declare no conflict of interest.

References

1. Ralph, J.; Lapierre, C.; Boerjan, W. Lignin structure and its engineering. *Curr. Opin. Biotechnol.* **2019**, *56*, 240–249. [CrossRef] [PubMed]
2. Tobimatsu, Y.; Schuetz, M. Lignin polymerization: How do plants manage the chemistry so well? *Curr. Opin. Biotechnol.* **2019**, *56*, 75–81. [CrossRef] [PubMed]
3. Lourenço, A.; Pereira, H. Compositional Variability of Lignin in Biomass. In *Lignin—Trends and Applications*; InTech: Rang-Du-Fliers, France, 2018.
4. Sun, R. Lignin Source and Structural Characterization. *ChemSusChem* **2020**, *13*, 4385–4393. [CrossRef] [PubMed]
5. Rodrigues, A.E.; de Oliveira Rodrigues Pinto, P.C.; Barreiro, M.F.; Esteves da Costa, C.A.; Ferreira da Mota, M.I.; Fernandes, I. *An Integrated Approach for Added-Value Products from Lignocellulosic Biorefineries*; Springer International Publishing: Cham, Switzerland, 2018; ISBN 978-3-319-99312-6.
6. Liao, J.J.; Latif, N.H.A.; Trache, D.; Brosse, N.; Hussin, M.H. Current advancement on the isolation, characterization and application of lignin. *Int. J. Biol. Macromol.* **2020**, *162*, 985–1024. [CrossRef] [PubMed]
7. Sethupathy, S.; Murillo Morales, G.; Gao, L.; Wang, H.; Yang, B.; Jiang, J.; Sun, J.; Zhu, D. Lignin valorization: Status, challenges and opportunities. *Bioresour. Technol.* **2022**, *347*, 126696. [CrossRef]
8. Glasser, W.G. About Making Lignin Great Again—Some Lessons from the Past. *Front. Chem.* **2019**, *7*, 565. [CrossRef]
9. Munk, L.; Sitarz, A.K.; Kalyani, D.C.; Mikkelsen, J.D.; Meyer, A.S. Can laccases catalyze bond cleavage in lignin? *Biotechnol. Adv.* **2015**, *33*, 13–24. [CrossRef]
10. Janusz, G.; Pawlik, A.; Świderska-Burek, U.; Polak, J.; Sulej, J.; Jarosz-Wilkołazka, A.; Paszczyński, A. Laccase Properties, Physiological Functions, and Evolution. *Int. J. Mol. Sci.* **2020**, *21*, 966. [CrossRef] [PubMed]
11. Mate, D.M.; Alcalde, M. Laccase: A multi-purpose biocatalyst at the forefront of biotechnology. *Microb. Biotechnol.* **2017**, *10*, 1457–1467. [CrossRef]
12. Khatami, S.H.; Vakili, O.; Movahedpour, A.; Ghesmati, Z.; Ghasemi, H.; Taheri-Anganeh, M. Laccase: Various types and applications. *Biotechnol. Appl. Biochem.* **2022**, *69*, 2658–2672. [CrossRef]
13. Moreno, A.D.; Ibarra, D.; Eugenio, M.E.; Tomás-Pejó, E. Laccases as versatile enzymes: From industrial uses to novel applications. *J. Chem. Technol. Biotechnol.* **2020**, *95*, 481–494. [CrossRef]
14. Agustin, M.B.; Carvalho, D.M.; Lahtinen, M.H.; Hilden, K.; Lundell, T.; Mikkonen, K.S. Laccase as a Tool in Building Advanced Lignin-Based Materials. *ChemSusChem* **2021**, *14*, 4615–4635. [CrossRef] [PubMed]
15. Gouveia, S.; Otero, L.; Fernández-Costas, C.; Filgueira, D.; Sanromán, Á.; Moldes, D. Green Binder Based on Enzymatically Polymerized Eucalypt Kraft Lignin for Fiberboard Manufacturing: A Preliminary Study. *Polymers* **2018**, *10*, 642. [CrossRef]
16. Wang, L.; Tan, L.; Hu, L.; Wang, X.; Koppolu, R.; Tirri, T.; van Bochove, B.; Ihalainen, P.; Seleenmary Sobhanadhas, L.S.; Seppälä, J.V.; et al. On Laccase-Catalyzed Polymerization of Biorefinery Lignin Fractions and Alignment of Lignin Nanoparticles on the Nanocellulose Surface via One-Pot Water-Phase Synthesis. *ACS Sustain. Chem. Eng.* **2021**, *9*, 8770–8782. [CrossRef]
17. Legras-Lecarpentier, D.; Stadler, K.; Weiss, R.; Guebitz, G.M.; Nyanhongo, G.S. Enzymatic Synthesis of 100% Lignin Biobased Granules as Fertilizer Storage and Controlled Slow Release Systems. *ACS Sustain. Chem. Eng.* **2019**, *7*, 12621–12628. [CrossRef]
18. Weiss, R.; Ghitti, E.; Sumetzberger-Hasinger, M.; Guebitz, G.M.; Nyanhongo, G.S. Lignin-Based Pesticide Delivery System. *ACS Omega* **2020**, *5*, 4322–4329. [CrossRef]
19. Gillgren, T.; Hedenström, M.; Jönsson, L.J. Comparison of laccase-catalyzed cross-linking of organosolv lignin and lignosulfonates. *Int. J. Biol. Macromol.* **2017**, *105*, 438–446. [CrossRef]
20. Gouveia, S.; Fernández-Costas, C.; Sanromán, M.A.; Moldes, D. Enzymatic polymerisation and effect of fractionation of dissolved lignin from Eucalyptus globulus Kraft liquor. *Bioresour. Technol.* **2012**, *121*, 131–138. [CrossRef]
21. Gouveia, S.; Fernández-Costas, C.; Sanromán, M.A.; Moldes, D. Polymerisation of Kraft lignin from black liquors by laccase from *Myceliophthora thermophila*: Effect of operational conditions and black liquor origin. *Bioresour. Technol.* **2013**, *131*, 288–294. [CrossRef]
22. Mattinen, M.-L.; Tapani, S.; Gosselink, R.; Argyropoulos, D.; Evtuguin, D.; Suurnäkki, A.; de Jong, E.; Tamminen, T. Polymerization of different lignins by laccase. *Bioresources* **2008**, *32*, 549–555.
23. Nugroho Prasetyo, E.; Kudanga, T.; Østergaard, L.; Rencoret, J.; Gutiérrez, A.; del Río, J.C.; Ignacio Santos, J.; Nieto, L.; Jiménez-Barbero, J.; Martínez, A.T. Polymerization of lignosulfonates by the laccase-HBT (1-hydroxybenzotriazole) system improves dispersibility. *Bioresour. Technol.* **2010**, *101*, 5054–5062. [CrossRef] [PubMed]
24. Braunschmid, V.; Binder, K.; Fuerst, S.; Subagia, R.; Danner, C.; Weber, H.; Schwaiger, N.; Nyanhongo, G.S.; Ribitsch, D.; Guebitz, G.M. Comparison of a fungal and a bacterial laccase for lignosulfonate polymerization. *Process Biochem.* **2021**, *109*, 207–213. [CrossRef]
25. Mayr, S.A.; Subagia, R.; Weiss, R.; Schwaiger, N.; Weber, H.K.; Leitner, J.; Ribitsch, D.; Nyanhongo, G.S.; Guebitz, G.M. Oxidation of Various Kraft Lignins with a Bacterial Laccase Enzyme. *Int. J. Mol. Sci.* **2021**, *22*, 13161. [CrossRef]
26. Molina-Guijarro, J.M.; Pérez, J.; Muñoz-Dorado, J.; Guillén, F.; Moya, R.; Hernández, M.; Arias, M.E. Detoxification of azo dyes by a novel pH-versatile, salt-resistant laccase from *Streptomyces ipomoea*. *Int. Microbiol.* **2009**, *12*, 13–21.

27. Ibarra, D.; Romero, J.; Martínez, M.J.; Martínez, A.T.; Camarero, S. Exploring the enzymatic parameters for optimal delignification of eucalypt pulp by laccase-mediator. *Enzyme Microb. Technol.* **2006**, *39*, 1319–1327. [CrossRef]
28. Antúnez-Argüelles, E.; Herrera-Bulnes, M.; Torres-Ariño, A.; Mirón-Enríquez, C.; Soriano-García, M.; Robles-Gómez, E. Enzymatic-assisted polymerization of the lignin obtained from a macroalgae consortium, using an extracellular laccase-like enzyme (Tg-laccase) from *Tetraselmis gracilis*. *J. Environ. Sci. Health Part A* **2020**, *55*, 739–747. [CrossRef]
29. Pardo, I.; Rodríguez-Escribano, D.; Aza, P.; de Salas, F.; Martínez, A.T.; Camarero, S. A highly stable laccase obtained by swapping the second cupredoxin domain. *Sci. Rep.* **2018**, *8*, 15669. [CrossRef] [PubMed]
30. Jiménez-López, L.; Martín-Sampedro, R.; Eugenio, M.E.; Santos, J.I.; Sixto, H.; Cañellas, I.; Ibarra, D. Co-production of soluble sugars and lignin from short rotation white poplar and black locust crops. *Wood Sci. Technol.* **2020**, *54*, 1617–1643. [CrossRef]
31. Martín-Sampedro, R.; Santos, J.I.; Eugenio, M.E.; Wicklein, B.; Jiménez-López, L.; Ibarra, D. Chemical and thermal analysis of lignin streams from *Robinia pseudoacacia* L. generated during organosolv and acid hydrolysis pre-treatments and subsequent enzymatic hydrolysis. *Int. J. Biol. Macromol.* **2019**, *140*, 311–322. [CrossRef]
32. Eugenio, M.E.; Martín-Sampedro, R.; Santos, J.I.; Wicklein, B.; Martín, J.A.; Ibarra, D. Properties versus application requirements of solubilized lignins from an elm clone during different pre-treatments. *Int. J. Biol. Macromol.* **2021**, *181*, 99–111. [CrossRef]
33. Re, R.; Pellegrini, N.; Proteggente, A.; Pannala, A.; Yang, M.; Rice-Evans, C. Antioxidant activity applying an improved ABTS radical cation decolorization assay. *Free Radic. Biol. Med.* **1999**, *26*, 1231–1237. [CrossRef] [PubMed]
34. Ratanasumarn, N.; Chitprasert, P. Cosmetic potential of lignin extracts from alkaline-treated sugarcane bagasse: Optimization of extraction conditions using response surface methodology. *Int. J. Biol. Macromol.* **2020**, *153*, 138–145. [CrossRef] [PubMed]
35. Huber, D.; Pellis, A.; Daxbacher, A.; Nyanhongo, G.; Guebitz, G. Polymerization of Various Lignins via Immobilized *Myceliophthora thermophila* Laccase (MtL). *Polymers* **2016**, *8*, 280. [CrossRef] [PubMed]
36. Domínguez, G.; Blánquez, A.; Borrero-López, A.M.; Valencia, C.; Eugenio, M.E.; Arias, M.E.; Rodríguez, J.; Hernández, M. Eco-Friendly Oleogels from Functionalized Kraft Lignin with Laccase SilA from *Streptomyces ipomoeae*: An Opportunity to Replace Commercial Lubricants. *ACS Sustain. Chem. Eng.* **2021**, *9*, 4611–4616. [CrossRef]
37. De La Torre, M.; Martín-Sampedro, R.; Fillat, Ú.; Eugenio, M.E.; Blánquez, A.; Hernández, M.; Arias, M.E.; Ibarra, D. Comparison of the efficiency of bacterial and fungal laccases in delignification and detoxification of steam-pretreated lignocellulosic biomass for bioethanol production. *J. Ind. Microb. Biotechnol.* **2017**, *44*, 1561–1573. [CrossRef] [PubMed]
38. Fillat, Ú.; Ibarra, D.; Eugenio, M.E.; Moreno, A.D.; Tomás-Pejó, E.; Martín-Sampedro, R. Laccases as a Potential Tool for the Efficient Conversion of Lignocellulosic Biomass: A Review. *Fermentation* **2017**, *3*, 17. [CrossRef]
39. Areskogh, D.; Li, J.; Gellerstedt, G.; Henriksson, G. Investigation of the Molecular Weight Increase of Commercial Lignosulfonates by Laccase Catalysis. *Biomacromolecules* **2010**, *11*, 904–910. [CrossRef]
40. Espinoza-Acosta, J.L.; Torres-Chávez, P.I.; Ramírez-Wong, B.; López-Saiz, C.M.; Montaño-Leyva, B. Antioxidant, antimicrobial, and antimutagenic properties of technical lignins and their applications. *BioResources* **2016**, *11*, 5452–5481. [CrossRef]
41. Fodil Cherif, M.; Trache, D.; Brosse, N.; Benaliouche, F.; Tarchoun, A.F. Comparison of the Physicochemical Properties and Thermal Stability of Organosolv and Kraft Lignins from Hardwood and Softwood Biomass for Their Potential Valorization. *Waste Biomass Valoriz.* **2020**, *11*, 6541–6553. [CrossRef]
42. García-Fuentevilla, L.; Rubio-Valle, J.F.; Martín-Sampedro, R.; Valencia, C.; Eugenio, M.E.; Ibarra, D. Different Kraft lignin sources for electrospun nanostructures production: Influence of chemical structure and composition. *Int. J. Biol. Macromol.* **2022**, *214*, 554–567. [CrossRef]
43. Areskogh, D.; Li, J.; Gellerstedt, G.; Henriksson, G. Structural modification of commercial lignosulphonates through laccase catalysis and ozonolysis. *Ind. Crops Prod.* **2010**, *32*, 458–466. [CrossRef]
44. Wang, J.; Zhao, C.; Zhang, T.; Yang, L.; Chen, H.; Yue, F. In-Depth Identification of Phenolics Fractionated from Eucalyptus Kraft Lignin. *Adv. Sustain. Syst.* **2022**, *6*, 2100406. [CrossRef]
45. Giummarella, N.; Lindén, P.A.; Areskogh, D.; Lawoko, M. Fractional Profiling of Kraft Lignin Structure: Unravelling Insights on Lignin Reaction Mechanisms. *ACS Sustain. Chem. Eng.* **2020**, *8*, 1112–1120. [CrossRef]
46. Giummarella, N.; Pylypchuk, I.V.; Sevastyanova, O.; Lawoko, M. New Structures in Eucalyptus Kraft Lignin with Complex Mechanistic Implications. *ACS Sustain. Chem. Eng.* **2020**, *8*, 10983–10994. [CrossRef]
47. Lancefield, C.S.; Wienk, H.L.J.; Boelens, R.; Weckhuysen, B.M.; Bruijnincx, P.C.A. Identification of a diagnostic structural motif reveals a new reaction intermediate and condensation pathway in kraft lignin formation. *Chem. Sci.* **2018**, *9*, 6348–6360. [CrossRef]
48. Jardim, J.M.; Hart, P.W.; Lucia, L.; Jameel, H. Insights into the Potential of Hardwood Kraft Lignin to Be a Green Platform Material for Emergence of the Biorefinery. *Polymers* **2020**, *12*, 1795. [CrossRef]
49. Zhao, C.; Hu, Z.; Shi, L.; Wang, C.; Yue, F.; Li, S.; Zhang, H.; Lu, F. Profiling of the formation of lignin-derived monomers and dimers from Eucalyptus alkali lignin. *Green Chem.* **2020**, *22*, 7366–7375. [CrossRef]
50. Zhao, C.; Huang, J.; Yang, L.; Yue, F.; Lu, F. Revealing Structural Differences between Alkaline and Kraft Lignins by HSQC NMR. *Ind. Eng. Chem. Res.* **2019**, *58*, 5707–5714. [CrossRef]
51. Suota, M.J.; da Silva, T.A.; Zawadzki, S.F.; Sassaki, G.L.; Hansel, F.A.; Paleologou, M.; Pereira, L.R. Chemical and structural characterization of hardwood and softwood LignoForce™ lignins. *Ind. Crops Prod.* **2021**, *173*, 114138. [CrossRef]

52. Santos, J.I.; Nieto, L.; Jiménez-Barbero, J.; Rencoret, J.; Suárez, A.G.; del Río Andrade, J.C.; Martínez, Á.T. NMR study on enzymatic polymerization of spruce lignosulfonate. In Proceedings of the Oxidative Enzymes as Sustainable Industrial Biocatalyst, Santiago de Compostela, Spain, 14–15 September 2010.
53. Magina, S.; Barros-Timmons, A.; Evtuguin, D.V. Laccase-catalyzed oxidative modification of lignosulfonates from acidic sulfite pulping of eucalyptus wood. *Holzforschung* **2020**, *74*, 589–596. [CrossRef]

polymers

MDPI

Review

Chemical Structures, Properties, and Applications of Selected Crude Oil-Based and Bio-Based Polymers

Piotr Koczoń [1], Bartłomiej Bartyzel [2], Anna Iuliano [3], Dorota Klensporf-Pawlik [4], Dorota Kowalska [1], Ewa Majewska [1], Katarzyna Tarnowska [1], Bartłomiej Zieniuk [1] and Eliza Gruczyńska-Sękowska [1,*]

1 Department of Chemistry, Institute of Food Sciences, Warsaw University of Life Sciences, 02-776 Warsaw, Poland
2 Department of Morphological Sciences, Institute of Veterinary Medicine, Warsaw University of Life Sciences, 02-776 Warsaw, Poland
3 Faculty of Chemistry, Warsaw University of Technology, 00-664 Warsaw, Poland
4 Department of Food Quality and Safety, Poznan University of Economics and Business, 61-875 Poznan, Poland
* Correspondence: eliza_gruczynska@sggw.edu.pl

Abstract: The growing perspective of running out of crude oil followed by increasing prices for all crude oil-based materials, e.g., crude oil-based polymers, which have a huge number of practical applications but are usually neither biodegradable nor environmentally friendly, has resulted in searching for their substitutes—namely, bio-based polymers. Currently, both these types of polymers are used in practice worldwide. Owing to the advantages and disadvantages occurring among plastics with different origin, in this current review data on selected popular crude oil-based and bio-based polymers has been collected in order to compare their practical applications resulting from their composition, chemical structure, and related physical and chemical properties. The main goal is to compare polymers in pairs, which have the same or similar practical applications, regardless of different origin and composition. It has been proven that many crude oil-based polymers can be effectively replaced by bio-based polymers without significant loss of properties that ensure practical applications. Additionally, biopolymers have higher potential than crude oil-based polymers in many modern applications. It is concluded that the future of polymers will belong to bio-based rather than crude oil-based polymers.

Keywords: crude oil-based; bio-based polymers; chemical structure; properties; application

check for updates

Citation: Koczoń, P.; Bartyzel, B.; Iuliano, A.; Klensporf-Pawlik, D.; Kowalska, D.; Majewska, E.; Tarnowska, K.; Zieniuk, B.; Gruczyńska-Sękowska, E. Chemical Structures, Properties, and Applications of Selected Crude Oil-Based and Bio-Based Polymers. *Polymers* **2022**, *14*, 5551. https://doi.org/10.3390/polym14245551

Academic Editors: Antonio M. Borrero-López, Concepción Valencia-Barragán, Esperanza Cortés Triviño, Adrián Tenorio-Alfonso and Clara Delgado-Sánchez

Received: 31 October 2022
Accepted: 12 December 2022
Published: 19 December 2022

1. Introduction

The increasing demand for crude oil-based materials has contributed to the depletion of natural reserves of petroleum. It is estimated that a serious shortage in crude oil and significant increase in its costs will be noticed as early as 2040 [1], and petroleum supplies will have been consumed until the end of this century. As much as 11–12% of crude oil is used in the production of polymers [2,3]. Fossil fuel-based polymeric materials show a variety of desirable physical properties such as durability, light weight, and resistance to corrosion and chemical reagents. For this reason the spectrum of the applications of crude oil-based polymers is extremely wide, from packaging [4] to constructional materials [5] and medical equipment [6]. Unfortunately, the properties that make fossil-based polymers so attractive trigger enormous environmental problems, since the vast majority of petroleum-based polymers do not decompose and continue to remain almost untouched for centuries. Environmental issues along with the upcoming deficiency in crude oil supplies have gained a great concern among the polymer industry and researchers who try to find new alternatives to crude oil-based polymers. Polymers from renewable resources have recently attracted the attention of global scientists as a future perspective in replacing crude oil-based polymers or reducing their usage.

Bio-based polymers are polymers derived (at least in part) from renewable raw materials such as plant and animal biomass, and organic waste. This group of materials can be biodegradable (such as polylactic acid—PLA) or nondegradable (such as biopolyethylene—Bio-PE). However, bio-based polymers are considered environmentally responsible, since they do not depend on finite fossil fuel reserves and can be obtained with a lower carbon footprint than their crude oil counterparts. Renewable raw materials such as starch, vegetable oils, and proteins are nearly inexhaustible natural resources, the supply of which can be restored in a short period of time. Renewables are inexpensive and readily available, ensuring sustainable carbon transfer from biomass and similar materials to bio-based polymers or intermediates used in their production.

Natural bio-based polymers of industrial application can be directly extracted from biomass. Such polymers are produced in large quantities by plants (e.g., cellulose, hemicellulose, starch, inulin, and pectin) or by animals (e.g., chitin and chitosan), and they often need to undergo special modification to meet requirements for their use [7]. Molecular biology and genetic engineering can be successfully incorporated into the process of creating agricultural crops of desirable properties or facilitating the procedure for subsequent recovery of biopolymers. Bio-based polymers can also be synthesized de novo by various bacterial strains (e.g., *Pseudomonas putida, Aeromonas hydrophila, Bacillus subtilis*) during fermentation processes using low molecular weight metabolism products. Another group of bio-based polymers includes synthetic polymers from bio-derived monomers (e.g., PLA and other polyesters). Several prominent building blocks such as succinic, itaconic, muconic, and lactic acids can be effectively produced from biomass [8].

The physicochemical properties of bio-based polymers are often similar to crude oil-based polymers, which makes them a potential substitute for their crude oil-based counterparts. Moreover, bio-based polymers can exhibit a wide range of new features, which can enable novel applications in many technological fields. Not only should bio-based production be cost-effective but also more sustainable in comparison to crude oil-based production.

The paper discusses the possibility of replacing selected popular crude oil-based polymers such as polyethylene, low density polyethylene, polystyrene, polyethylene terephthalate, and polyvinyl chloride with bio-based polymers made from renewable resources such as polylactic acid (PLA), derived from an animal origin, such as chitosan, and received on biotechnological pathways such as polyhydroxybutyrate (PHB) and pullulan. The choice of the petroleum-based polymers was dictated by their versatile application in a multitude of industries, whereas PLA and PHB are the major industry players that are bringing bio-based polymeric materials to the market, and polysaccharides such as pullulan and chitosan are fully appreciated by researchers and industrialists worldwide for their high biocompatibility and biodegradability. The properties and practical applications of both groups of polymers have been collected in order to compare them and indicate the possible future perspectives for bio-based polymers.

2. Polyethylene vs. Polyhydroxybutyrate

Shopping bags, including those from supermarkets, plastic packaging, various lids, industrial foils, milk containers, and others, including thin-wall containers, wire insulators, pipes, injection, and blow molding, are all made from different types of chemically the same material—polyethylene (PE). In addition to the products listed that are in common use, clearly visible in modern world, PE has many other important, yet not so commonly known applications, e.g., parts of various machines used in the food and paper industry, veterinary and medicine tools and materials, and in the construction industry [9]. Although there are currently many voices pointing out PE disadvantages, above all its inability to be naturally degradable and having a share in the dramatic increase in the volume of litter, one must admit that PE is a companion in everyday life [10].

The history of PE can be considered serendipitous both in terms of laboratory discovery and industrial synthesis. It goes back to 1898 when Hans von Pechmann accidentally obtained a white waxy substance by heating diazomethane. Later on in 1900, a similarly obtained waxy substance was proved to contain a long carbon chain and was given the name "polymethylene" by Eugen Bamberger and Friedrich Tschirner. In both cases, the waxy substance was polyethylene. Laboratory discovery was followed by industrial synthesis of PE at the works of Imperial Chemical Industries (ICI), Northwich, the UK, in 1933. The very first patent referring to production of PE from ethylene was granted on 6 September 1937. Between first information in 1898 and the patent in 1937, several reactions were tested to produce the waxy substance first obtained by Pechmann. Numerous chemical pathways to obtain PE were worked out, which included several reactions, e.g., reaction of decamethylene dibromide with sodium in a Wurtz-type reaction done by Carothers and Van Natta in 1930 [11], hydrogenation of polybutadiene, modified Fischer–Tropsch reduction of carbon monoxide with hydrogen, or reduction of poly(vinyl chloride) with lithium aluminum hydride. Every pathway has advantages and disadvantages, e.g., inability to obtain a polymer with molecular mass greater than 1300 with use of Wurtz-type reaction or formation of branched-only or unbranched-only products. Currently, ethylene and other materials generated from crude oil are primary source materials to produce PE commercially [12–14].

The monomer of PE, ethylene, is chemically unsaturated and contains a double bond, while PE is saturated containing only single bonds between carbon atoms (Figure 1). Carbon chains can be straight or branched. In general, if the chains are straight, the basic form of PE is considered, while branched chains are associated with low-density PE (LDPE). Both forms are produced in polymerization reactions taking place under high pressure and temperature. Ethylene is heated and pressed to break double bonds and form long chains. Pressure is applied to make less space for ethylene molecules that are in gaseous form. Although no equilibrium is considered in the polymerization reaction, the Le Chatelier principle can be used to explain favoring of the forward direction: solid state products (right-hand side of reaction) occupy much less space than gaseous substrates (left-hand side of reaction) [12,15].

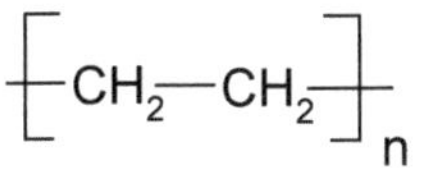

Figure 1. The fragment of polyethylene structure.

The production process starts with preparation of ethylene with appropriate properties. The monomer is sourced by dehydration of ethyl alcohol fermented from molasses. Another source for ethylene is a selected fraction of crude oil fractional distillation. Before starting the polymerization process, all undesired contaminants such as carbon monoxide, oxygen, water, and acetylene must be removed. The use of inadequately prepared reactant yields product of undesired properties, especially in terms of insulation or heat resistance [10,12].

The polymerization route requires applying high pressure of 100–300 MPa and temperature of 350–600 K. This is the most common process, known as high-pressure polymerization. There are also other well-known technological methods of PE production, e.g., Standard Oil Company Indiana process, high-density PE (HDPE) Phillips or Ziegler process, and metallocene catalyzed process. Application of different conditions, namely pressure, temperature, or catalyst, leads to production of PE, LDPE, or HDPE [10,12,16].

Currently PE is divided into classes that consider chemical structure and resultant properties [10,12]. Structural classes are:

- Linear PE—with a macromolecule made of many monomer units arranged in a straight line;
- Cross-linked PE (PEX)—where macromolecule has covalent bonds between the polymer molecules.

PE can also be classified according to material density. The following classes are described:

- Low density PE (LDPE);
- Linear low density PE (LLDPE);
- Middle density PE (MDPE);
- High density PE (HDPE);
- Ultra-high molecular weight PE (UHMWPE).

In summary, PE with a general molecular formula of $-[C_2H_4]_n-$, a degree of polymerization of 1500–9000, and a melting point at 395–400 K [15] has very useful and desirable mechanical and chemical properties, including being soft, transparent, tasteless and odorless, but not resistant to high temperature. The main advantage of all PE types is low price, as costs of production are low. PE from each class has different properties; hence, it is used in different areas of human life, for example, LLDPE has the highest impact strength, tensile strength, and extensibility of all PE classes [15,17–19].

Without specific treatment, decomposition of PE takes hundreds of years, hence many scientific and industrial interdisciplinary teams work on methods to increase its decomposition rate or reuse [20–23]. Promising investigations cover the use of bacteria, yeasts, and enzymes present in microorganisms to recover material that can be used for the next synthesis of plastic exhibiting the quality equal to the one obtained in petrochemical processes. Investigated microorganisms can also be used for remediation of plastic waste present in soil and landfills [24–26]. However, there is no similar research on PE/LDPE biomedical wastes.

Polyhydroxybutyrate—PHB—was discovered in 1925 by Maurice Lemoigne [27,28]. Chemically, it is polyester. There are several slightly different forms of this polymer, namely poly-4-hydroxybutyrate (P4HP), polyhydroxyvalerate (PHV), polyhydroxyhexanoate (PHH), and the most common poly-3-hydroxybutyrate (P3HB).

With the molecular formula $[OCH(CH_3)CH_2CO]_n$ (Figure 2), PHB has similar mechanical properties to PE, while its greatest advantage over PE is biodegradability [29]. However, the biggest disadvantage is its high cost of production, and currently there is intensive research focused on decreasing these costs [24,30–32].

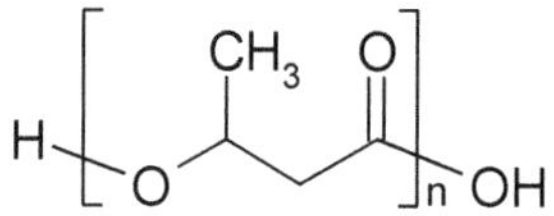

Figure 2. The fragment of polyhydroxybutyrate structure.

PHB that belongs to a polyhydroxyalkanoate (PHA) family is a semicrystalline thermoplastic polyester. Its glass transition temperature is 278–282 K; the melting point is 440–450 K. PHB has a very low Young's modulus (3–3.5 GPa) compared to other biodegradable biopolymers. Its decomposition lasts up to several years [15,33]. PHB is used as a component in medical implants, surgical sutures, and elements of artificial tissues [34,35].

PHB can be obtained from its monomer—butyric acid—by microorganisms on biological pathways [29]. Production of PHB is more expensive than any other PE. Therefore, sources of cheap organic matter are required. One way is food-originated residuals that undergo fermentation to form volatile fatty acids, which are in turn converted by specific bacteria, e.g., *Cupriavidus necator* to form PHB. Another method of PHB production is the use of the strain *R. piridinivorans* BSRT1-1. This strain is isolated from soil, and it has been stated that fructose and KNO_3 are the best sources of carbon and nitrogen for bacteria to produce PHB. Under optimal conditions, 3.60 g of PHB can be formed from 1 dm^3 of biomass. Details are provided in [36]. *Bacillus cereus* SH-02 (OM992297) is also considered to be a good producer of PHB [37].

Both PE/LDPE and PHB are widely used in medicine and the veterinary and agricultural industries. Although both are polymers, their physical, chemical, and biological properties, together with their formation and methods of utilization, differ significantly. Both plastics are commonly used in production of medicinal and veterinary materials [12,29,38,39]. Materials and products obtained from both polymers can be in direct touch with human organs. They can be used for production of bone tunnels [40], surgical threads [41], containers, foils, slices, packaging [42], and antibacterial foils [43]. Owing to relatively low costs, PE single-use products are manufactured and applied [21]. Orthopedics is a rapidly developing area for the use of polymers, including PE and PHB applications. In the most common surgery worldwide, i.e., total knee arthroplasty (TKA), endoprostheses used are made of PE. A specific type of PE used for this type of endoprostheses provides physiological kinematics of the knee joint [44,45]. Additional advantage is low specific weight, mechanical resistance, and transparency to X-rays [46].

A characteristic and practical important feature of this biopolymer is its significantly high biodegradability and biocompatibility. As biopolymers generally, PHB can be relatively easily decomposed by enzymes produced by living species. In the biological environment, biodegradation can occur by oxidation, including photo-oxidation or hydrolysis [47]. PHB can be easily decomposed inside human tissues by enzymatic or hydrolytic decomposition, with pH kept at a constant level [48,49]. Products formed are easy metabolized by human or animal organisms [50]. This ability to be degradable, especially in the human body, allows use of PHB as a composite for implants. Additionally, PHB can promote cell growth in a near-natural biological environment [28,49,51,52].

3. Polystyrene vs. Polylactide

Polystyrene (PS) is a thermoplastic polymer (Figure 3) made of aromatic hydrocarbon monomer styrene that is derived from fossil-fuels [53].

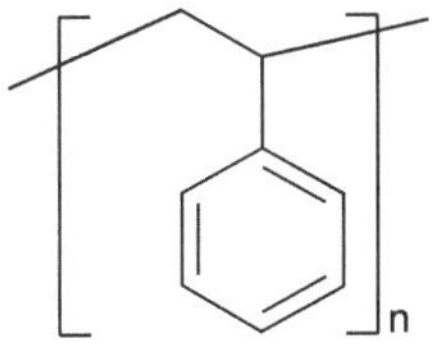

Figure 3. The fragment of polystyrene structure.

The synthesis of PS is based on the free radical polymerization of styrene using free-radical initiators. It is mostly used in solid (high impact and general purpose PS), foam and expanded PS forms. The main advantages of PS are low-cost, easy processing ability, and resistance to ethylene oxide, as well as radiation sterilization. It is, however, not resistant to organic solvents such as cyclic ethers, ketones, acids, and bases. The most popular general purpose PS (GPPS or unmodified PS) is transparent, brittle, and rigid, which makes this kind of material suitable for laboratory purposes, such as diagnostic and analytical, and medical packaging (e.g., Petri dishes, tissue culture trays, pipettes, test tubes). For high-strength products, high-impact PS (HIPS) is competitive with polypropylene and PVC [54]. It is typically used in thermoformed products, such as catheters, heart pumps, and epidural trays, and toys, packaging, and electronic appliances. Owing to its high dimensional stability and easy processing, it is often chosen for the preproduction prototypes in 3D-printing technique [55]. As a result of strong C-C and C-H bonds present in the structure, PS is resistant to biodegradation without special treatment such as copolymerization and fictionalization. However, it was proved that some bacterial species are able to form biofilm on the PS surface, which leads to its partial degradation [56]. PS can be recycled using several methods. Mechanical recycling is the one with lowest cost, but it has many limitations. The main obstacle is efficient separation of PS from the plastic waste stream. Currently, PS is sorted using near-infrared technologies and complementary sorting methods, including density, electrostatics, selective dissolution, and flotation [57]. The latter—froth flotation—is the most common due to its low cost and the possibility to separate polymers with similar density. To increase the flotation effectiveness, surface modification can be performed, e.g., in the presence of $KMnO_4$ [58] or K_2FeO_4 [59]. Recycled

PS exhibits worse mechanical properties than neat polymer, and reduction in molecular mass is also observed. Nevertheless, many products made of recycled PS can be found on the market, e.g., pencils, doors, window frames, cups, plates, and bottles; some of them even approved for food contact [60]. Chemical recycling of PS is less common due to the high cost. It leads to the production of styrene, and other useful chemicals such as benzene, toluene, indan, ethylbenzene, and benzoic acid, via pyrolysis and oxidation. Recently, a novel simple and low-cost method has been reported that enables the oxidative cleavage of PS to benzoic acid, formic acid, and acetophenone by singlet oxygen at ambient temperature and pressure [61]. PS waste can also be converted to biodegradable PHAs [62]. To summarize, PS is one of the most important polymers present in our daily life. However, once it has fulfilled its designed purpose, it is not easily degradable. Chemical recycling is not economically convenient since the feedstocks are cheaper than the process itself; additionally, mechanical recycling is limited due to the low separation efficiency of PS from the plastic waste stream. That is why there is an urgent need to find sustainable alternatives that can at least partially replace petroleum-based PS in use. The most popular green substitutes for PS are cellulose and thermoplastic starch used as thermal insulation materials (foams) [63,64], and poly(vinyl alcohol) for bead-foaming process [65] and polylactide [66].

Polylactide (PLA)—biodegradable and compostable aliphatic polyester (Figure 4)—is one of the key biopolymers with the largest market significance. The global volume of PLA production was around 457,000 metric tons in 2021, which accounted for 29% of the total biodegradable bioplastics production worldwide [67]. The PLA production on industrial scale is either based on the ring-opening polymerization (ROP) of lactide, (method applied by NatureWorks LLC, Plymouth, MN, United States, and Corbion N.V., Amsterdam, the Netherlands) or direct polycondensation of lactic acid in an azeotropic solution (applied by Mitsui Toatsu Chemicals, Inc., Tokyo, Japan) [68]. In both cases, high molecular mass PLA is obtained; however, solvent-free ROP is preferable for production in large scale. In this case, optically pure *L*-lactic or *D*-lactic acid is produced as a monomer of PLA by microbial fermentation from renewable resources such as molasses, whey, sugar cane, and plants with high starch content [69]. Next, LA is condensed to form low molecular mass prepolymer PLA, which undergoes a controlled depolymerization to a cyclic dimer of lactate–lactide. The polymerization of lactide is generally catalyzed by tin octanoate and requires short reaction time at a temperature of about 440–460 K [70].

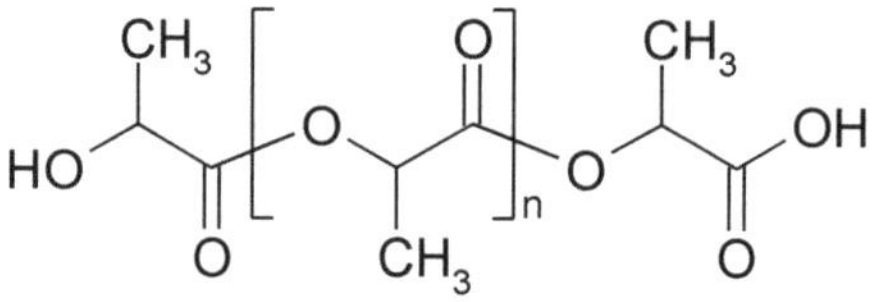

Figure 4. The fragment of polylactide structure.

The mechanical properties of PLA are similar to those of PS and polyethylene terephthalate (PET) [71], also described in detail in Section 4 of this paper. It can also be a sustainable alternative to polypropylene (PP) and PVC. PLA is as rigid and brittle as PS, and its resistance to fats and oils resembles PET [71]. Although CO_2, O_2, N_2, and H_2O permeabilities for PLA are higher than for PET, but lower than for PS [72], therefore many attempts to improve the PLA barrier properties have been reported, e.g., by introducing nanofillers with a lamellar structure [73]. In addition, it is characterized by a high tensile modulus and resistance to UV radiation. Good mechanical and optical properties allow PLA to compete with the existing crude oil-based thermoplastics. PLA containing approx. 5% of *D*-repeating units is a transparent, colorless, and relatively rigid material resembling PS [74]. An extra advantage of PLA is its easy processing ability through conventional melt processes such as extrusion, injection molding, compression molding, or blow molding, which are also used for other commercial polymers, namely PS and PET [75].

The properties of PLA depend on the polymer molecular mass and the degree of crystallinity [76]. Stereochemistry also plays an important role. The stereochemical composition and distribution of monomer units along the polyester chain affect the properties of PLA [74]. *L*-PLA (PLLA) and *D*-PLA (PDLA) are composed of lactic acid units of the same chirality [77]. They are isotactic, stereoregular, and partially crystalline polymers (degree of crystallinity up to 60%), the glass-transition temperature (T_g) is approx. 320–330 K, and the melting point is 440–470 K [29,74,78]. On the other hand, *D,L*-PLA is an amorphous polymer with a T_g of about 330 K. It shows worse mechanical properties and degrades faster than PLLA and PDLA. The highest melting point, about 500 K, shows a racemic mixture of PLLA and PDLA, in which chains of different chirality form a densely packed network. Compared to the parent polymers, the resulting racemic PLA (PDLLA) has enhanced functional properties, such as mechanical strength, durability, and thermal and hydrolytic stability [79].

Desirable properties allow PLA to compete with PS in several application fields, described below.

3.1. Packaging Application

Low toxicity, strong flavor and aroma barrier, and high transparency make PLA an ideal material for fresh food packaging, especially fruit and vegetables [80]. Auras et al. [81] tested and compared oriented PLA (OPLA) with PET and oriented PS (OPS) films intended for production of fresh fruit and vegetables storage containers. According to these results, mechanical, physical, and barrier properties of OPLA were comparable and, in some cases, better than standard OPS and PET containers. Similar studies were performed for the shelf life of blackberries [82] and blueberries [83] under retail conditions closed in the OPS and OPLA containers. In both cases the shelf life was extended, proving that PLA can be a good replacement for PS. PLA can be used also as trays for storage of mangoes, melons, and other tropical fruit. The shelf life of the fruit packed in such a way was the same as of the fruit packed in PET trays [84]. However, the PLA packaging is more susceptible to cracking and breaking during transport when compared with OPS or PET. Neither the sheet nor the finished product can be stored at temperatures above 313 K or relative humidity greater than 50% [85].

3.2. Three-Dimensional Printing

The filaments used in 3D printing are primarily thermoplastics. The most popular are PLA, acrylonitrile butadiene styrene (ABS) and HIPS [86]. In all three cases, filament can also be produced from recycled plastic, which can significantly reduce its price. It is worth mentioning that commercial filaments for 3D printing are 20 to 200 times more expensive than those of raw plastics [87]. The source for PLA waste is food containers and bottles, ABS filaments originating from car dashboards, and HIPS derived from refrigerators or automotive parts [88]. The advantages of PLA as filament for 3D printing are ease of printing, glossiness, and multicolor appearance. The dimensional accuracy of the parts printed from PLA is high since it poses less warp behavior than the other filaments. Compared to HIPS, PLA filament does not require a heated bed, it is odorless, and what is more important, it releases many fewer volatile organic compounds and exhibits lower particle emission during printing [89]. PLA prints have wider application than HIPS due to biocompatibility and susceptibility to biodegradation, which are important in biomedical application and tissue engineering [90]. Moreover, the price of 1 kg of PLA filament is comparable to that for HIPS. This is why PLA can be a good alternative to HIPS in rapid manufacturing of packaging prototypes using 3D printing technology [91].

3.3. Medical Application/Drug Delivery

Medical plastic has to be biocompatible, stable under different sterilization conditions, and robust to surface modification. While PLA fulfils all these requirements, PS is not applicable because of the cancerogenic properties of styrene and its very moderate biocompatibility [54]. However, there are several studies on improving the biocompatibility of PS, e.g., by nonequilibrium gaseous plasma treatment [92]. Both polymers can be sterilized by ethylene oxide, gamma radiation, and electron-beam radiation, however, due to the presence of a benzene ring in its structure, PS is more resistant to high radiation doses than PLA. PLA exhibits strong resistance to sterilization processes with use of an autoclave or dry heat [93]; standard PS is not autoclavable, but syndiotactic PS is excellent [94]. The main application of PS in the laboratory field is the production of different containers for a variety of liquids, cells, and bacteria, together with microspheres used as drug carriers and magnetic particles. The biocompatibility of PLA makes this material an excellent application as scaffolds for bone regeneration, implants, stents, along with bioresorbable surgical and orthopedic threads and dental implants. Owing to the good mechanical properties of PLA, it can be used in catheters, heart pumps, and epidural trays to replace PS [54]. PLA and PS are also used as a surface for adhesion and proliferation of fibroblast and osteoblast cell lines [95].

PLA is a promising bioplastic with mechanical properties comparable to those of PS. In addition to its established position as a material for biomedical applications, it can replace mass production plastics from petroleum. However, there are still challenges that need to be addressed, e.g., improvement of barrier properties, which play a very important role in maintaining food quality and safety [96]. Moreover, the cost of PLA manufacturing is still too high to compete with PS. That is why there is a need to find low-cost substrates and high-performance microorganisms to increase the efficiency of LA production and obtain low-cost, high-quality PLA. Another concern is the recycling of PLA. PLA can be easily degraded in the natural environment or in compost; however, the idea of introducing a large amount of waste for biodegradation is unreasonable and its transformation into chemical products more valuable than simply carbon dioxide and water should be considered. Currently, several attempts of PLA recycling have been made but an industrially feasible chemical recycling concept, in adherence to the fundamental principles of closed-loop recycling within a Circular Economy, has not yet been developed [97]. Other than PLA, products made from PS can be recycled, but the high cost of the recycling process and the segregation problem make the technology inefficient. Moreover, the production of biopolymers is considered more sustainable than petroleum-based materials due to the reduced net carbon footprint [98].

4. Polyethylene Terephthalate vs. Chitosan

The abbreviation "PET" is well-known to all consumers worldwide and stands for a petroleum-based synthetic polymer of terephthalic acid and ethylene glycol, i.e., polyethylene terephthalate (Figure 5). The history of this polymer production dates back to the 1940s when Whinfield and Dickson [99] patented terephthalic esters in the form of linear polymers. The second most important patent relating to the described plastic is the invention of Wyeth and Roseveare [100], i.e., a plastic soda bottle. In recent years, the global production of this plastic exceeded 30 million metric tons and a 4% market growth rate is expected in the coming years [101].

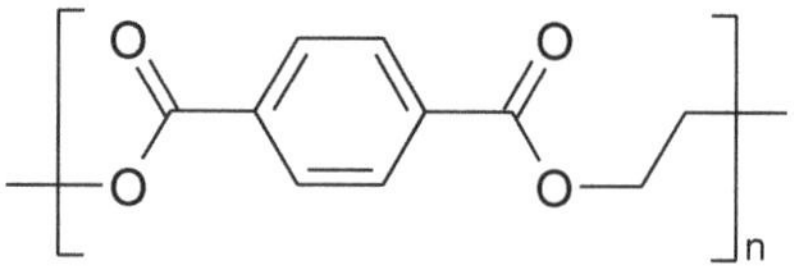

Figure 5. The fragment of polyethylene terephthalate structure.

The technology of PET production has been developed over the years, and now the substrates for the synthesis of this polymer, i.e., terephthalic acid or dimethyl terephthalate and ethylene glycol, are obtained from fossil-based resources. Consequently, the first two compounds are produced from *p*-xylene and the diol is made through the oxidation of ethene. One of the major industrial-scale synthesis methods of polyalkylene terephthalates is the so-called melt polycondensation, including a two-step process in the presence of catalysts for (trans)esterification of substrates in an inert atmosphere at 460–500 K, and subsequently, polycondensation of the intermediates at reduced pressure and increased temperature (520–550 K) until obtaining enough viscous mixture resulting in the end product [102].

The technology affects the appearance and the final form of PET, wherein the amorphous state the polymer is transparent, while in the semicrystalline state PET appears opaque [103]. According to the ASTM (American Society for Testing and Materials) International Resin Identification Coding System the number 1 was assigned to PET, meaning low production costs together with high recycling possibilities [104,105]. PET materials are characterized by high strength, rigidity, and hardness. High thermal stability is observed, which is connected to the presence of para-substituted aromatic rings in the polymer structure. Such polyesters have a melting point above 520 K, and even a theoretical value of 565 K was calculated for PET if an accurate process of annealing would be used [103]. Moreover, both the T_g and density are strictly dependent on the final form of the polymer; hence, T_g for the amorphous state is 340 K and for the semicrystalline PET is about 350 K [104]. Amorphous PET has a density of 1.30–1.34 g/cm^3 and semicrystalline PET is about 1.50 g/cm^3. Crystalline PET is much denser than the amorphous form because polymer chains of the former are closely packed and parallel, while they are disordered in the latter [106]. Interestingly, exceeding the T_g point, thus the temperature above 340–350 K, may lead to the decreased resistance of PET to hydrolysis [107].

Polyethylene terephthalate has found its application in single-use food packaging, mainly for beverage bottles, but another major market is the textile industry and the production of clothes, shoes, and carpet fibers [101,108]. Features such as transparency, lightness, strength, and durability made PET bottles favored over those made of glass, and the results of the Coca-Cola Company life-cycle assessment study in 1969 revealed that bottles obtained from PET affected the environment less than their glass analogues [104]. The possibility of using PET in packaging material for food or pharmaceuticals results from low permeability for gases and solvents, together with low moisture absorption [109]. Chemical stability of polymers is also extremely important. PET's stability is observed in weak acids and is also inert to several organic solvents from the alcohol, halogen, and ketone groups [104,105]. Strong acids, bases, hydrocarbons, and aromatic compounds are examples of chemicals that influence the stability of PET, and moreover, according to Lepoittevin and Roger [103], PET may be soluble in a mixture of phenol and trichloroethane, and in trifluoroacetic acid, *o*-chlorophenol, or hexafluoroisopropanol.

It is estimated that the degradation of plastic bottles lasts around 450 years, and 8% of solid waste weight is contributed to PET [110]. Owing to the fact that almost 60% of PET is discarded into landfills, solutions are being sought to reuse this plastic. Approaches such as bottle reuse systems, bans on plastic bags, and the introduction of taxes and deposit systems have been successfully applied in many geographical regions, e.g., Europe, America, and South Asia [111]. Currently PET waste is subjected to recycling processes using both mechanical and chemical methods. Reactive extrusion is often used because of its simplicity and the multiple extrusions carried out at 540–550 K lead to a decrease in molecular mass of the polymer [110]. Among chemical recycling methods, the following reactions are distinguished: hydrolysis at higher temperature and pressure, methanolysis, glycolysis, aminolysis, and ammonolysis, which lead to the depolymerization of PET. Products obtained from chemical depolymerization found their application in various industries, e.g., as cement replacement, corrosion inhibitors, paints, etc. [112]. Unfortunately, chemical recycling of polymers seems to be unprofitable because the polymerization of fossil-based substrates is much cheaper than reprocessing of polyesters [108].

PET is considered resistant to hydrolysis and enzymatic treatment, but the current prospects for its biodegradation are very promising. Numerous studies have shown the usefulness of bacteria of the genera *Ideonella*, *Bacillus*, and *Streptomyces*, along with the enzymes produced from the esterase and cutinase group, the so-called PETases [101,104,108]. The discovery of such microorganisms with their ability to produce PET hydrolyzing enzymes, and their possible application in polymer biodegradation may be an innovative approach in waste management in line with the circular economy system, which meets the goals of sustainable development [108].

Currently, new packaging materials, mainly those which are bio-based and biodegradable, are gaining interest among the consumers and food manufacturers. An interesting example of such a polymer is chitosan (Figure 6).

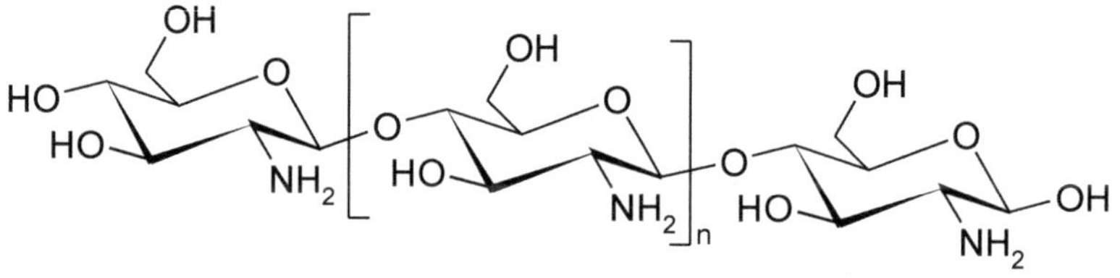

Figure 6. The fragment of chitosan structure.

Chitosan is a biopolymer obtained after chitin deacetylation, the discovery of which is attributed to the French physiologist Charles Rouget, who in 1859 heated chitin (a glycan consisting of *N*-acetyl-D-glucosamine molecules) in an alkaline solution. From the chemical point of view, this polysaccharide is a linear polymer consisting of D-glucosamine units linked with β 1–4 glycosidic bonds. The process of removing acetyl groups is not often entirely performed; therefore there are several chitosan preparations available on the market with different degrees of deacetylation [113,114]. Interestingly, the industrial production of chitosan started in Japan in 1971 [115].

It is believed that chitin, along with cellulose, is one of the most common biopolymers found worldwide. The main sources of chitin and chitosan are crustaceans (shrimps and prawns, krills, or crabs), and insects, together with microorganisms, mainly fungi, algae, and some yeasts [113,116]. Shells and other inedible parts, i.e., crustacean waste, are a good source for these polysaccharides, especially since their contents in this arthropod taxon reach up to 20% dry weight and seafood consumption will grow increasingly [113,117]. In the last few years, several biotechnology companies, such as the Mycodev Group (Fredericton, NB, Canada), Chibio Biotech Co., Ltd. (Qingdao City, Shandong Province, China), and KitoZyme (Herstal, Belgium), have launched and have been producing the so-called "vegetal chitosan" or "mycochitosan", a fully nonanimal source for this polymer obtained through the fermentation process with the use of filamentous fungi; hence, these preparations may be an alternative for people with shellfish allergies and for vegetarians [113].

The properties of chitosan and PET differ, especially in terms of strength and hardness. Therefore, these polymers cannot be replaced one-to-one, but other attributes of the former allow the use of plastics to be limited, as mentioned below. One of the main advantages of this polysaccharide is that it can be consumed as opposed to petroleum-based synthetic polymers. Chitosan has been considered as material for food packaging, but also as a dietary supplement, biofertilizer, and biopesticide. Moreover, hydrogels and wound-healing bandages based on chitosan are currently obtainable on the market [117]. This polymer could be used for bodyweight reduction, but there are some concerns that it may interact with fat-soluble vitamins. The lowest observed adverse effect level (LOAEL) for chitosan is relatively high and it accounts for 450 mg/kg for men and 6000 mg/kg for women [117]. In the case of oral median lethal dose (LD_{50}) in mice, it was revealed that the value was higher than that of sucrose and it exceeded 16 g/day/kg body mass [114].

Furthermore, its potential widespread application comes from both biological and physical properties, such as antimicrobial activity, biodegradability, and film-forming ability. Chitosan is not soluble in water and organic solvents but dissolves in dilute acetic or hydrochloric acids. Chitosan-based edible coatings and films are able to extend the shelf life of food products, prolong their quality, improve nutritional and antioxidant properties, and prevent the growth of food-spoilage microorganisms. In order to improve some features of such films, various additives are added, such as plasticizers in the form of polyols (e.g., glycerol) and emulsifiers. Chitosan can also be blended with other biopolymers, such as alginate, pectin, starch, or caseinate, or enriched with different valuable compounds (e.g., polyphenol extracts) to enhance its applicability [118].

Among various polymers, chitosan is distinguished by its antimicrobial activity. Interestingly, two different mechanisms may be responsible for this activity, but in each case the chemical structure is the clue. The first mechanism is associated with chitosan binding to DNA molecules causing bacterial cell death. In the second one, chitosan and more specifically the amino groups in the chain of this polysaccharide in the acidic medium generate a cationic charge that binds to negatively charged bacterial cell walls and membranes, resulting in permeability disturbances and cell leakage. In addition, by lowering the pH value, the improvement in the antibacterial activity is observed due to the increase in the protonation of the amino groups [119].

Chitosan shows remarkable potential for biomedical applications. Apart from already-mentioned bioactive dressings for healing wounds and burns, this polysaccharide can be used in preparing drug delivery systems of different forms (tablets, gels, granules, films, or microcapsules) or can serve as an artificial kidney membrane impermeable for serum proteins, and potential material for contact lenses, specifically due to its good tolerance by living tissues [116]. Over the past few decades, scientists were also interested in the use of chitosan in tissue engineering and regenerative medicine. Both PET and expanded polytetrafluoroethylene (ePTFE) are usually standard materials for prosthetic vascular grafts, but the need for small-diameter (<4 mm) application becomes problematic with the use of these two synthetic polymers [120]. Chupa et al. [121] indicated the meaningful potential of chitosan and its complexes with glycosaminoglycans or dextran sulfate in the preparation of newly bioactive materials that exhibited activity both in vitro and in vivo and modulated the activity of smooth muscle and vascular endothelial cells.

Pure chitosan films are not thermoplastic and thus cannot be softened by heating, thus their extrusion, molding, or heat-sealing is limited [122]. Some properties of chitosan-based films, e.g., water vapor and oxygen permeabilities, are thus important parameters influencing the applicability of the coating or film for food packaging (since the moisture and oxygen levels can lead to lowering the quality of food), can change depending on the addition of some materials incorporated into chitosan films, as well as the characteristics of chitosan, i.e., the degree of deacetylation and the methodology of coatings and films manufacturing. Wang et al. [118] reported that the use of organic compounds such as polyphenols in chitosan films led to a decrease in water vapor permeability due to the interactions between compounds and limiting interactions with water molecules. Moreover, the use of nanoparticles led to similar observations because good dispersion in the film and filling the spaces resulted in hindered migration of water. Furthermore, it is believed that the tightly packed structure of chitosan with plenty of hydrogen bonds may have limited oxygen permeability, and the incorporation of chitosan-based films with graphene oxide nanosheets or silver nanoparticles caused a significant reduction in oxygen permeability [119].

In the near future, chitosan-based packaging may find its niche in producing active and intelligent packaging. The first type, in addition to the features typically assigned to packaging, is characterized by carrying additional properties maintaining the quality and increasing the safety of the products, which in the case of chitosan relates to its antimicrobial and antioxidant activities [119]. The term "intelligent" in the case of the latter refers to the possibility of using packaging for real-time food quality monitoring. This approach arouses the interest of scientists, thereupon extensive research is undertaken, but also industry attention is attracted. Recently colorimetric and pH-indicating films are under special examination, where both their manufacturing and specific properties on the selected food products are assessed. Methylene blue is an example of a dye applied in chitosan-based films, whereby its color depends on oxygen concentration in the atmosphere surrounding the product and decreasing oxygen content may indicate microbiological contamination [118]. As pH-indicating compounds, both synthetic and natural substances are used in chitosan films. The studies on the use of alizarin, along with anthocyanins from purple potato or grapes were conducted and confirmed the applicability of these compounds to monitor changes in food, where pH alternations may suggest that the product is stale or contaminated [118,119].

Concluding, it cannot be clearly stated which polymer is better, PET or chitosan. Similarly it cannot be confirmed with certainty that the latter, i.e., a polymer of natural origin, will replace PET because they differ significantly in their properties. PET is primarily a packaging material used in the beverage industry with high hardness and rigidity. Inversely, chitosan is used to create coatings and films, and the possibility of producing edible packaging is an additional advantage. Finally, the prospect of producing active and intelligent packaging based on chitosan, which will be also edible, biodegradable, and compostable, points in favor of its use. Therefore, the production of chitosan-based packaging will be a limitation in the use of PET, replacing it in selected applications, rather than a complete replacement.

5. Polyvinyl Chloride vs. Pullulan

Polyvinyl chloride—PVC—is a long-chain thermoplastic polymer produced by a free-radical polymerization of vinyl chloride monomer. Industrial synthesis of PVC is dated to the early 1930s, and recent estimates account production volume as third among plastics after polypropylene and polyethylene, and up to 25% of total plastic production [21]. The basic raw materials for the PVC synthesis come from crude oil and sodium chloride and hence only 43% of this polymer mass is petroleum-based [123]. The chemical formula of PVC is $(C_2H_3Cl)_n$, and the chemical structure is a long chain with the repeating unit of vinyl chloride, as shown in Figure 7.

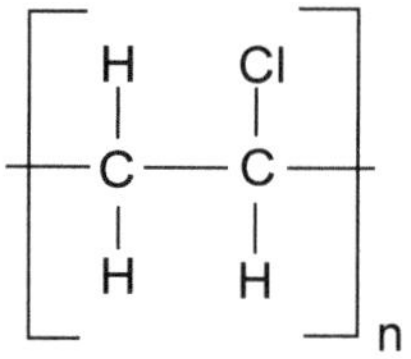

Figure 7. The fragment of polyvinyl chloride structure.

As a thermoplastic material with excellent chemical and mechanical properties, PVC has widespread uses. These properties together with low production cost affect the economic significance of PVC worldwide. The most important properties are polymer durability, high chemical resistance, resistance to water and weather conditions, and adhesiveness [124,125]. Moreover, because of its high polarity, it has the ability to accommodate a wide range of additives such as stabilizers, plasticizers, lubricants, or pigments [126]. The addition of different chemicals to PVC resins modifies its properties and may change the possible way of use. The unplasticized PVC is hard and rigid and can be used in plumbing, construction, fencing, or drainage systems, whereas plasticizers incorporated into the polymer make it softer and more flexible, thus useful in electrical cable insulation, medical devices, or inflatable toys [127]. These hard and soft variants of PVC compounds differ considerably in terms of the T_g and flexibility at specific temperatures [128]. Additionally, high transparency of this polymer is useful in film production or light-transmitting panels. It is a lightweight material, with high strength-to-weight ratio, which imparts no taint or taste. Notwithstanding, almost 70% of PVC compounds is used in the construction industry [129], next in medicine (flexible blood containers or inhalation masks) [130,131], and the packaging industry (wrap films with good oxygen barrier properties) [132].

Despite the outstanding position of PVC mainly among medical polymers, it is considered to be harmful to both human and environment because of various chemicals and dangerous degradation products released during its life cycle [123]. Its complex composition together with low thermal stability makes this polymer difficult to recycle, but there are available techniques for the management of PVC waste, both mechanical [133] and feedstock [134].

Pullulan is an extra-cellular, unbranched, water soluble, neutral, nonionic, nontoxic, nonmutagenic, noncarcinogenic exopolisaccharide. This biopolymer is obtained from fermentation medium of the fungus-like yeast *Aureobasidium pullulans*, generally referred to as "black yeast". The final yield of pullulan is highly affected by media composition and culture conditions [135,136]. The large scale production started in 1976 by Hayashibara Co., Ltd. (Okayama, Japan), which is still the leading commercial producer worldwide together with Shandong Jinmei Biotechnology Co., Ltd. (Zhucheng, China). Pullulan is generally marketed as white or off-white dry powder or capsules [137]. The chemical formula of pullulan has been suggested to be $(C_6H_{10}O_5)_n$, and its chemical structure consists of repeating units of maltotriose linked with each other by the α 1–6 glycosidic bonds [138–140]. Pullulan structure is often seen as an intermediate between amylose and dextran structures. The chemical structure is shown in Figure 8.

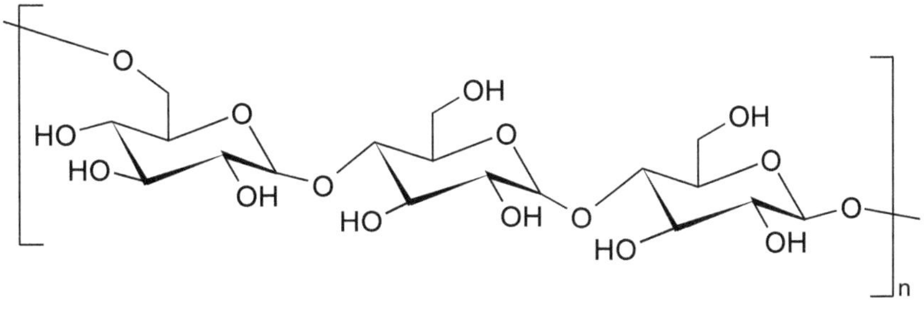

Figure 8. The fragment of pullulan structure.

The peculiar structure of pullulan (mainly the α 1–6 linkages) affects strongly its high solubility in both cold and hot water, and its lack of ability to form gels. It is responsible for the high structural flexibility, but is also reflected in the lack of crystalline regions within the polymer, which has a completely amorphous organization [141]. Pullulan compounds have a high heat resistance and are biodegradable in biologically active environments, therefore it can be utilized in many different ways. It has a significant mechanical strength, adhesiveness, thick film, and fiber formability, stability of aqueous solution over a broad range of pH, low viscosity, and good oxygen- and moisture-barrier properties [136,142]; it also inhibits fungal growth in food [143]. Pullulan can be formed into compression moldings that can resemble PS and PVC in transparency, gloss, hardness, and strength, but which are far more elastic [135]. Additionally, the capacity to form thin layers, nanoparticles, flexible coatings, and standalone films means it can successfully replace other synthetic polymers derived from petroleum, such as polyvinyl alcohol [138].

This polysaccharide is also colorless, tasteless, odorless, and, what is more important, edible, however, not attacked by the digestive enzymes in human gut [136,142]. It has a "generally recognized as safe" status in the United States [144] and in the European Union is a food additive of microbial origin E1204 [145].

As a nonpolluting "plastic", the biodegradable and biocompatible biopolymer pullulan could be used in different sectors, especially pharmaceutical (hard and soft capsules, drug delivery systems, anticancer nanoparticles) [146,147], biomedical (wound healing) [137], environmental, food, and cosmetics (body and skin application) [148]. However, despite all the pullulan advantages, its high cost is the limiting factor for wide scale applications.

The most emerging area of interest in pullulan application is still the food packaging sector. Because of the nonbiodegradable character of synthetic polymers and their great impact on the environment, research remain focused on developing ecofriendly and biodegradable food packaging systems obtained from natural sources. Such applications are edible coatings and active films, which are known to protect food, extend the shelf life of the product, and improve its quality. A thin layer of pullulan is formed directly on the surface of a product and can be safely eaten with a protected food. Pullulan was used as an edible coating on, e.g., highbush blueberries [149], Fuji apples [150], and white asparagus [151]. Moreover, pullulan is also an excellent medium for different compounds, which play an important role in prolonging shelf life: chitosan [152], thymol [153], pectin [154], or propolis [142].

6. Summary

The most important properties of the polymers discussed earlier in this paper have been collected and are presented below (Table 1).

Table 1. Selected properties of all the polymers discussed in this paper.

Property	Pairs of Polymers (Fossil–Bio)							
	PE–PHB		PS–PLA		PET–Chitosan		PVC–Pullulan	
Density (kg/m^3)	950 [c]	1262 [d]	1111–1127 [d]	1248–1290 [d]	1300–1500 [c]	1000 [e]	1330–1380 [c]	1500 ± 100 [e]
Young's Modulus (GPa)	0.7 [c]	3–3.5 [a]	3.4 [c]	3.0 [f]	1.7 [d]	0.63 [g]	3.4 [c]	N/A *
Tensile Strength (MPa)	30–40 [a]	20–40 [a]	34.5–68.9 [b]	50–70 [f]	62 [a]	9.7 [g]	48 [c]	40.2 [e]
Elongation at Break (%)	200–700 [a]	5–10 [a]	1–2.3 [b]	4.0 [f]	30–80 [b]	2.6 [g]	85–104 [d]	N/A *
Flexural Strength (MPa)	40 [d]	33–40 [d]	68.9–103 [b]	100 [f]	96.5–124.1 [b]	N/A *	72 [d]	N/A *
Izod Notched Impact Strength (J/m)	60–80 [d]	35–50 [d]	21 [b]	26 [b]	59 [b]	N/A *	21–53 [d]	N/A *
Degree of Crystallinity (%)	25–80 [a]	50–60 [a]	N/A *	3.5–14 [a]	7.97 [a]	57 [h]	11–15 [d]	N/A *
Melting Temp. (K)	390–410 [a]	440–450 [a]	513 [d]	440–470 [a]	530 [a]	360 [e]	485–583 [d]	193 [e]
Glass Transition Temp. (K)	140–370 [a]	278–282 [a]	373 [d]	320–330 [a]	340–354 [a]	N/A *	353–370 [d]	N/A *

[a] [29], [b] [53], [c] [155], [d] [156], [e] [157], [f] [158], [g] [159], [h] [160], * N/A—not available.

The total replacement of crude oil-based polymers with bio-based polymers is not yet possible, mainly because of functional properties of synthetic polymers, namely durability, mechanical, and water vapor barrier. Therefore, to partially eliminate synthetic polymers from packaging systems and to reduce environmental impact, biopolymers combined with crude oil-based polymers are of interest. Chemical modification helps in combining specific characteristics of different polymers, resulting in a unique and suitable material with interesting properties [161].

7. Conclusions

Bio-based polymers can replace crude oil-based polymers in a variety of applications from everyday products to medical materials or advanced technology. Yet a noticeable difference in the physicochemical parameters between both groups of polymers prevents the petroleum-based synthetic polymers from being substituted directly by their bio-based counterparts. Nevertheless, the replacement of crude oil-based polymers by bio-based polymers in only selected applications would considerably diminish the net carbon footprint and create sustainable solutions in polymeric materials management. Many bio-based polymers such as PHB and PLA, which are commonly adapted in the production of biocompatible devices extensively used in surgery, drug delivery, and cardiovascular systems, or dentistry, can successfully compete with their hydrocarbon polymeric analogues, namely PE and PS. PLA meets all the requirements to be a perfect biomaterial, whereas PS, which may be contaminated with residual styrene, might be considered cancerogenic. Similarly, PVC, used primarily in the construction, medical, and packaging industry, is regarded as toxic to humans and the environment due to harmful compounds released during its life cycle.

The food packaging sector, which is dominated by PET, a polymer resistant to biodegradation, is currently searching for a bio-based and biodegradable replacement. Chitosan, one of the most widespread biopolymers found in nature, can at least in part rival PET and contribute to a considerable reduction in plastics. This polysaccharide, which has a film-forming ability, is edible and can be consumed together with the food product; additionally, its antimicrobial and antioxidant properties can increase the safety and stability of the protected foodstuff. Edible coatings formed by pullulan are reported to prolong the shelf-life of numerous fruit and vegetables; these properties can be enhanced by the addition of other bioactive molecules. All of these ecofriendly and innovative advantages may constitute a springboard for creating active and intelligent packaging in the future. However, PET is still the best option for the beverage industry, which needs hard and rigid polymeric materials for packaging.

The implementation of bio-based polymers in many industrial sectors is currently limited mainly due to the high costs of their production. The manufacturing of crude oil-based polymers remains still more cost-effective, although such polymers are resistant to biodegradation, and their recycling is as yet uneconomical.

A significant challenge for the future is to synthesize novel, sustainable bio-based polymers with such functionalities that could enable the substitution of their conventional analogues. Biotechnology is a strategic factor, which can notably contribute to the transition from fossil-derived plastics to bioplastics acquired from renewable resources. Technological advancements and new biotechnological findings may allow for better development of bio-based materials and reduce production expenditures.

Author Contributions: Conceptualization, P.K. and E.G.-S.; resources, P.K., E.G.-S., B.B., A.I., D.K.-P., D.K., E.M., K.T. and B.Z.; writing—original draft preparation, P.K., E.G.-S., B.B., A.I., D.K.-P., D.K., E.M., K.T. and B.Z.; writing—review and editing, P.K., E.G.-S., A.I. and E.M.; visualization, D.K. and K.T.; supervision, P.K. and E.G.-S. All authors have read and agreed to the published version of the manuscript.

Funding: This research received no external funding.

Institutional Review Board Statement: Not applicable.

Informed Consent Statement: Not applicable.

Data Availability Statement: Not applicable.

Conflicts of Interest: The authors declare no conflict of interest.

References

1. Bhunia, H.P.; Basak, A.; Chaki, T.K.; Nando, G.B. Synthesis and characterization of polymers from cashew nut shell liquid: A renewable resource V. Synthesis of copolyester. *Eur. Polym. J.* **2000**, *36*, 1157–1165. [CrossRef]
2. Clark, J.H. Green chemistry: Today (and tomorrow). *Green Chem.* **2006**, *8*, 17–21. [CrossRef]
3. Scott, G. "Green" polymers. *Polym. Degrad. Stab.* **2000**, *68*, 1–7. [CrossRef]
4. Chaudhary, V.; Punia Bangar, S.; Thakur, N.; Trif, M. Recent Advancements in Smart Biogenic Packaging: Reshaping the Future of the Food Packaging Industry. *Polymers* **2022**, *14*, 829. [CrossRef] [PubMed]
5. Hou, W.J.; Xiao, Y.; Han, G.Y.; Lin, J.Y. The Applications of Polymers in Solar Cells: A Review. *Polymers* **2019**, *11*, 143. [CrossRef] [PubMed]
6. Zaharescu, T.; Varca, G.H.C. Radiation modified polymers for medical applications. *Radiat. Phys. Chem.* **2022**, *194*, 110043. [CrossRef]
7. Babu, R.P.; O'Connor, K.; Seeram, R. Current progress on bio-based polymers and their future trends. *Prog. Biomater.* **2013**, *2*, 8. [CrossRef] [PubMed]

8. Rorrer, N.A.; Vardon, D.R.; Dorgan, J.R.; Gjersing, E.J.; Beckham, G.T. Biomass-derived monomers for performance-differentiated fiber reinforced polymer composites. *Green Chem.* **2017**, *19*, 2812–2825. [CrossRef]
9. Ferreira, T.; Mendes, G.A.; de Oliveira, A.M.; Dias, C.G.B.T. Manufacture and Characterization of Polypropylene (PP) and High-Density Polyethylene (HDPE) Blocks for Potential Use as Masonry Component in Civil Construction. *Polymers* **2022**, *14*, 2463. [CrossRef]
10. Jeremic, D. Polyethylene. In *Ullmann's Encyclopaedia of Industrial Chemistry*; Wiley-VCH: Weinheim, Germany, 2014; pp. 1–34.
11. Carothers, W.H.; Van Natta, F.J. Studies on polymerization and ring formation. III. Glycol esters of carbonic acid. *J. Am. Chem. Soc.* **1930**, *52*, 314–326. [CrossRef]
12. Brydson, J.A. Polyethylene. In *Plastics Materials*, 7th ed.; Butterworth-Heinemann: Oxford, UK, 1999; pp. 205–246.
13. Chen, L.; Pelton, R.E.O.; Smith, T.M. Comparative life cycle assessment of fossil and bio-based polyethylene terephthalate (PET) bottles. *J. Clean. Prod.* **2016**, *137*, 667–676. [CrossRef]
14. Siracusa, V.; Blanco, I. Bio-Polyethylene (Bio-PE), Bio-Polypropylene (Bio-PP) and Bio-Poly(ethylene terephthalate) (Bio-PET): Recent Developments in Bio-Based Polymers Analogous to Petroleum-Derived Ones for Packaging and Engineering Applications. *Polymers* **2020**, *12*, 1641. [CrossRef]
15. Rabek, J.F. *Polimery. Otrzymywanie, Metody Badawcze, Zastosowanie*; Wydawnictwo Naukowe PWN: Warszawa, Polska, 2013; p. 15.
16. Zhong, X.; Zhao, X.; Qian, Y.; Zou, Y. Polyethylene plastic production process. *Insight Mater. Sci.* **2018**, *1*, 1–8. [CrossRef]
17. Fico, D.; Rizzo, D.; Casciaro, R.; Esposito Corcione, C. A Review of Polymer-Based Materials for Fused Filament Fabrication (FFF): Focus on Sustainability and Recycled Materials. *Polymers* **2022**, *14*, 465. [CrossRef]
18. Salakhov, I.I.; Shaidullin, N.M.; Chalykh, A.E.; Matsko, M.A.; Shapagin, A.V.; Batyrshin, A.Z.; Shandryuk, G.A.; Nifant'ev, I.E. Low-Temperature Mechanical Properties of High-Density and Low-Density Polyethylene and Their Blends. *Polymers* **2021**, *13*, 1821. [CrossRef]
19. Rana, S.K. Blend of high-density polyethylene and a linear low-density polyethylene with compositional-invariant mechanical properties. *J. Appl. Polym. Sci.* **2002**, *83*, 2604–2608. [CrossRef]
20. Coates, G.W.; Getzler, Y.D.Y.L. Chemical recycling to monomer for an ideal, circular polymer economy. *Nat. Rev. Mater.* **2020**, *5*, 501–516. [CrossRef]
21. Geyer, R.; Jambeck, J.R.; Law, K.L. Production, use, and fate of all plastics ever made. *Sci. Adv.* **2017**, *3*, e1700782. [CrossRef]
22. Vollmer, I.; Jenks, M.J.F.; Roelands, M.C.P.; White, R.J.; van Harmelen, T.; de Wild, P.; van der Laan, G.P.; Meirer, F.; Keurentjes, J.T.F.; Weckhuysen, B.M. Beyond Mechanical Recycling: Giving New Life to Plastic Waste. *Angew. Chem. Int. Ed.* **2020**, *59*, 15402–15423. [CrossRef]
23. Wu, W.M.; Criddle, C.S. Characterization of biodegradation of plastics in insect larvae. In *Enzymatic Plastic Degradation*; Weber, G., Bornscheuer, U.T., Wei, R., Eds.; Elsevier Inc.: Amsterdam, The Netherlands, 2021; Volume 648, pp. 95–120. [CrossRef]
24. Filiciotto, L.; Rothenberg, G. Biodegradable Plastics: Standards, Policies, and Impacts. *ChemSusChem* **2021**, *14*, 56–72. [CrossRef]
25. Conk, R.J.; Hanna, S.; Shi, J.X.; Yang, J.; Ciccia, N.R.; Qi, L.; Bloomer, B.J.; Heuvel, S.; Wills, T.; Su, J.; et al. Catalytic deconstruction of waste polyethylene with ethylene to form propylene. *Science* **2022**, *377*, 1561–1566. [CrossRef] [PubMed]
26. Uddin, M.A.; Koizumi, K.; Murata, K.; Sakata, Y. Thermal and catalytic degradation of structurally different types of polyethylene into fuel oil. *Polym. Degrad. Stab.* **1997**, *56*, 37–44. [CrossRef]
27. Lemoigne, M. Etudes sur L'autolyse microbienne acidification par formation D'acide β-oxybutyrique. *Ann. Inst. Pasteur.* **1925**, *39*, 144–173.
28. Degli Esposti, M.; Chiellini, F.; Bondioli, F.; Morselli, D.; Fabbri, P. Highly porous PHB-based bioactive scaffolds for bone tissue engineering by in situ synthesis of hydroxyapatite. *Mater. Sci. Eng. C* **2019**, *100*, 286–296. [CrossRef] [PubMed]
29. McAdam, B.; Brennan Fournet, M.; McDonald, P.; Mojicevic, M. Production of Polyhydroxybutyrate (PHB) and Factors Impacting Its Chemical and Mechanical Characteristics. *Polymers* **2020**, *12*, 2908. [CrossRef]
30. Janczak, K.; Dąbrowska, G.B.; Raszkowska-Kaczor, A.; Kaczor, D.; Hrynkiewicz, K.; Richert, A. Biodegradation of the plastics PLA and PET in cultivated soil with the participation of microorganisms and plants. *Int. Biodeterior. Biodegrad.* **2020**, *155*, 105087. [CrossRef]
31. Tournier, V.; Topham, C.M.; Gilles, A.; David, B.; Folgoas, C.; Moya-Leclair, E.; Kamionka, E.; Desrousseaux, M.L.; Texier, H.; Gavalda, S.; et al. An engineered PET depolymerase to break down and recycle plastic bottles. *Nature* **2020**, *580*, 216–219. [CrossRef]
32. Belgacem, M.N.; Gandini, A. *Monomers, Polymers and Composites from Renewable Resources*, 1st ed.; Elsevier Science: Amsterdam, The Netherlands, 2008.
33. Kosior, E.; Braganca, R.M.; Fowler, P. Lightweight compostable packaging, literature review. In *Milestone Report: A*; The Waste & Resources Action Programme: Banbury, UK, 2006; Volume 26, pp. 1–48.
34. Celarek, A.; Kraus, T.; Tschegg, E.K.; Fischerauer, S.F.; Stanzl-Tschegg, S.; Uggowitzer, P.J.; Weinberg, A.M. PHB, crystalline and amorphous magnesium alloys: Promising candidates for bioresorbable osteosynthesis implants? *Mater. Sci. Eng. C* **2012**, *32*, 1503–1510. [CrossRef]
35. Chen, G.Q.; Wu, Q. The application of polyhydroxyalkanoates as tissue engineering materials. *Biomaterials* **2005**, *26*, 6565–6578. [CrossRef]

36. Trakunjae, C.; Boondaeng, A.; Apiwatanapiwat, W.; Kosugi, A.; Arai, T.; Sudesh, K.; Vaithanomsat, P. Enhanced polyhydroxybutyrate (PHB) production by newly isolated rare actinomycetes *Rhodococcus* sp. strain BSRT1-1 using response surface methodology. *Sci. Rep.* **2021**, *11*, 1896. [CrossRef]
37. Hamdy, S.M.; Danial, A.W.; Gad El-Rab, S.M.F.; Shoreit, A.A.M.; Hesham, A.E.-L. Production and optimization of bioplastic (Polyhydroxybutyrate) from *Bacillus cereus* strain SH-02 using response surface methodology. *BMC Microbiol.* **2022**, *22*, 183. [CrossRef]
38. Kök, F.N.; Hasirci, V. Polyhydroxybutyrate and its copolymers: Applications in the medical field. In *Tissue Engineering and Novel Delivery Systems*; Yaszemski, M.J., Trantolo, D.J., Lewandrowski, K.-U., Hasirci, V., Altobelli, D.E., Wise, D.L., Eds.; Marcel Dekker Inc.: New York, NY, USA, 2004; pp. 543–561.
39. Todros, S.; Todesco, M.; Bagno, A. Biomaterials and Their Biomedical Applications: From Replacement to Regeneration. *Processes* **2021**, *9*, 1949. [CrossRef]
40. Luescher, M.; Schmierer, P.A.; Park, B.H.; Pozzi, A.; Gutbrod, A.; Evans, R.; Knell, S.C. Biomechanical comparison of knotted and knotless stabilization techniques of the tarsal medial collateral ligament in cats: A cadaveric study. *Vet. Surg.* **2020**, *49*, 390–400. [CrossRef]
41. Manfredini, M.; Ferrario, S.; Beretta, P.; Farronato, D.; Poli, P.P. Evaluation of Breaking Force of Different Suture Materials Used in Dentistry: An In Vitro Mechanical Comparison. *Materials* **2022**, *15*, 1082. [CrossRef]
42. McKeen, L.W. Plastics Used in Medical Devices. In *Handbook of Polymer Applications in Medicine and Medical Devices*; Ebnesajjad, S., Modjarrad, K., Eds.; Elsevier Inc.: Oxford, UK, 2014; pp. 21–53. [CrossRef]
43. Ordon, M.; Zdanowicz, M.; Nawrotek, P.; Stachurska, X.; Mizielińska, M. Polyethylene Films Containing Plant Extracts in the Polymer Matrix as Antibacterial and Antiviral Materials. *Int. J. Mol. Sci.* **2021**, *22*, 13438. [CrossRef]
44. Heckmann, N.D.; Steck, T.; Sporer, S.M.; Meneghini, R.M. Conforming Polyethylene Inserts in Total Knee Arthroplasty: Beyond the Posterior-Stabilized and Cruciate-Retaining Debate. *J. Am. Acad. Orthop. Surg.* **2021**, *29*, e1097–e1104. [CrossRef] [PubMed]
45. Indelli, P.F.; Risitano, S.; Hall, K.E.; Leonardi, E.; Migliore, E. Effect of polyethylene conformity on total knee arthroplasty early clinical outcomes. *Knee Surg. Sports Traumatol. Arthrosc.* **2019**, *27*, 1028–1034. [CrossRef]
46. Mano, J.F.; Sousa, R.A.; Boesel, L.F.; Neves, N.M.; Reis, R.L. Bioinert, biodegradable and injectable polymeric matrix composites for hard tissue replacement: State of the art and recent developments. *Compos. Sci. Technol.* **2004**, *64*, 789–817. [CrossRef]
47. Kunduru, K.R.; Basu, A.; Domb, A.J. Biodegradable Polymers: Medical Applications. In *Encyclopedia of Polymer Science and Technology*, 4th ed.; John Wiley & Sons: Hoboken, NJ, USA, 2016; pp. 1–22. [CrossRef]
48. Reddy, C.S.K.; Ghai, R.; Rashmi; Kalia, V.C. Polyhydroxyalkanoates: An overview. *Bioresour. Technol.* **2003**, *87*, 137–146. [CrossRef] [PubMed]
49. Wei, S.; Ma, J.X.; Xu, L.; Gu, X.S.; Ma, X.L. Biodegradable materials for bone defect repair. *Mil. Med. Res.* **2020**, *7*, 54. [CrossRef]
50. Goonoo, N.; Bhaw-Luximon, A.; Passanha, P.; Esteves, S.R.; Jhurry, D. Third generation poly(hydroxyacid) composite scaffolds for tissue engineering. *J. Biomed. Mater. Res. Part B* **2017**, *105*, 1667–1684. [CrossRef] [PubMed]
51. Ao, C.; Niu, Y.; Zhang, X.; He, X.; Zhang, W.; Lu, C. Fabrication and characterization of electrospun cellulose/nano-hydroxyapatite nanofibers for bone tissue engineering. *Int. J. Biol. Macromol.* **2017**, *97*, 568–573. [CrossRef] [PubMed]
52. Hosseini, E.S.; Dervin, S.; Ganguly, P.; Dahiya, R. Biodegradable Materials for Sustainable Health Monitoring Devices. *ACS Appl. Bio Mater.* **2021**, *4*, 163–194. [CrossRef] [PubMed]
53. Freeland, B.; McCarthy, E.; Balakrishnan, R.; Fahy, S.; Boland, A.; Rochfort, K.D.; Dabros, M.; Marti, R.; Kelleher, S.M.; Gaughran, J. A Review of Polylactic Acid as a Replacement Material for Single-Use Laboratory Components. *Materials* **2022**, *15*, 2989. [CrossRef] [PubMed]
54. Jiang, D.-H.; Satoh, T.; Tung, S.H.; Kuo, C.-C. Sustainable Alternatives to Nondegradable Medical Plastics. *ACS Sustain. Chem. Eng.* **2022**, *10*, 4792–4806. [CrossRef]
55. Singh, R.; Kumar, R.; Farina, I.; Colangelo, F.; Feo, L.; Fraternali, F. Multi-Material Additive Manufacturing of Sustainable Innovative Materials and Structures. *Polymers* **2019**, *11*, 62. [CrossRef]
56. Ho, B.T.; Roberts, T.K.; Lucas, S. An Overview on Biodegradation of Polystyrene and Modified Polystyrene: The Microbial Approach. *Crit. Rev. Biotechnol.* **2018**, *38*, 308–320. [CrossRef]
57. Schyns, Z.O.G.; Shaver, M.P. Mechanical Recycling of Packaging Plastics: A Review. *Macromol. Rapid Commun.* **2021**, *42*, 2000415. [CrossRef]
58. Cui, Y.; Li, Y.; Wang, W.; Wang, X.; Lin, J.; Mai, X.; Song, G.; Naik, N.; Guo, Z. Flotation Separation of Acrylonitrile-Butadienestyrene (ABS) and High Impact Polystyrene (HIPS) from Waste Electrical and Electronic Equipment (WEEE) by Potassium Permanganate Surface Modification. *Sep. Purif. Technol.* **2021**, *269*, 118767. [CrossRef]
59. Wang, J.; Liu, W.; Wang, H.; Wang, C.; Huang, W. Separation of Acrylonitrile-Butadiene-Styrene and Polystyrene Waste Plastics after Surface Modification Using Potassium Ferrate by Froth Flotation. *Waste Manag.* **2018**, *78*, 829–840. [CrossRef]
60. Styrolution-eco. Available online: https://styrolution-eco.com.html (accessed on 23 September 2022).
61. Huang, Z.; Shanmugam, M.; Liu, Z.; Brookfield, A.; Bennett, E.L.; Guan, R.; Vega Herrera, D.E.; Lopez-Sanchez, J.A.; Slater, A.G.; McInnes, E.J.L.; et al. Chemical Recycling of Polystyrene to Valuable Chemicals via Selective Acid-Catalyzed Aerobic Oxidation under Visible Light. *J. Am. Chem. Soc.* **2022**, *144*, 6532–6542. [CrossRef] [PubMed]

62. Johnston, B.; Radecka, I.; Hill, D.; Chiellini, E.; Ilieva, V.I.; Sikorska, W.; Musioł, M.; Zięba, M.; Marek, A.A.; Keddie, D.; et al. The Microbial Production of Polyhydroxyalkanoates from Waste Polystyrene Fragments Attained Using Oxidative Degradation. *Polymers* **2018**, *10*, 957. [CrossRef] [PubMed]
63. Debiagi, F.; Mali, S.; Grossmann, M.V.E.; Yamashita, F. Biodegradable Foams Based on Starch, Polyvinyl Alcohol, Chitosan and Sugarcane Fibers Obtained by Extrusion. *Braz. Arch. Biol. Technol.* **2011**, *54*, 1043–1052. [CrossRef]
64. Wang, P.; Aliheidari, N.; Zhang, X.; Ameli, A. Strong Ultralight Foams Based on Nanocrystalline Cellulose for High-Performance Insulation. *Carbohydr. Polym.* **2019**, *218*, 103–111. [CrossRef] [PubMed]
65. Wang, Q.; Yang, J.; Liu, P.; Li, L. Facile One-Step Approach to Manufacture Environmentally Friendly Poly(Vinyl Alcohol) Bead Foam Products. *Ind. Eng. Chem. Res.* **2021**, *60*, 2962–2970. [CrossRef]
66. Halloran, M.W.; Danielczak, L.; Nicell, J.A.; Leask, R.L.; Marić, M. Highly Flexible Polylactide Food Packaging Plasticized with Nontoxic, Biosourced Glycerol Plasticizers. *ACS Appl. Polym. Mater.* **2022**, *4*, 3608–3617. [CrossRef]
67. European Bioplastics. Available online: https://www.european-bioplastics.org/market/ (accessed on 11 October 2022).
68. Castro-Aguirre, E.; Iñiguez-Franco, F.; Samsudin, H.; Fang, X.; Auras, R. Poly(Lactic Acid)—Mass Production, Processing, Industrial Applications, and End of Life. *Adv. Drug Deliv. Rev.* **2016**, *107*, 333–366. [CrossRef]
69. Czajka, A.; Bulski, R.; Iuliano, A.; Plichta, A.; Mizera, K.; Ryszkowska, J. Grafted Lactic Acid Oligomers on Lignocellulosic Filler towards Biocomposites. *Materials* **2022**, *15*, 314. [CrossRef]
70. Florjańczyk, Z.; Jóźwiak, A.; Kundys, A.; Plichta, A.; Dębowski, M.; Rokicki, G.; Parzuchowski, P.; Lisowska, P.; Zychewicz, A. Segmental Copolymers of Condensation Polyesters and Polylactide. *Polym. Degrad. Stab.* **2012**, *97*, 1852–1860. [CrossRef]
71. Gołębiewski, J.; Gibas, E.; Malinowski, R. Wybrane polimery biodegradowalne–otrzymywanie, właściwości, zastosowanie. *Polimery* **2008**, *53*, 799–807.
72. Lim, L.-T.; Auras, R.; Rubino, M. Processing Technologies for Poly(Lactic Acid). *Prog. Polym. Sci.* **2008**, *33*, 820–852. [CrossRef]
73. Svagan, A.J.; Åkesson, A.; Cárdenas, M.; Bulut, S.; Knudsen, J.C.; Risbo, J.; Plackett, D. Transparent Films Based on PLA and Montmorillonite with Tunable Oxygen Barrier Properties. *Biomacromolecules* **2012**, *13*, 397–405. [CrossRef] [PubMed]
74. Duda, A.; Penczek, S. Polilaktyd [poli(kwas mlekowy)]: Synteza, właściwości i zastosowania. *Polimery* **2003**, *48*, 16–27. [CrossRef]
75. Kosmalska, D.; Janczak, K.; Raszkowska-Kaczor, A.; Stasiek, A.; Ligor, T. Polylactide as a Substitute for Conventional Polymers—Biopolymer Processing under Varying Extrusion Conditions. *Environments* **2022**, *9*, 57. [CrossRef]
76. Södergård, A.; Stolt, M. Properties of Lactic Acid Based Polymers and Their Correlation with Composition. *Prog. Polym. Sci.* **2002**, *27*, 1123–1163. [CrossRef]
77. Piórkowska, E.; Kuliński, Z.; Gadzinowska, K. Plastyfikacja polilaktydu. *Polimery* **2009**, *54*, 83–90.
78. Pang, X.; Zhuang, X.; Tang, Z.; Chen, X. Polylactic Acid (PLA): Research, Development and Industrialization. *Biotechnol. J.* **2010**, *5*, 1125–1136. [CrossRef]
79. Dimonie, D.; Mathe, S.; Iftime, M.M.; Ionita, D.; Trusca, R.; Iftimie, S. Modulation of the PLLA Morphology through Racemic Nucleation to Reach Functional Properties Required by 3D Printed Durable Applications. *Materials* **2021**, *14*, 6650. [CrossRef]
80. Marano, S.; Laudadio, E.; Minnelli, C.; Stipa, P. Tailoring the Barrier Properties of PLA: A State-of-the-Art Review for Food Packaging Applications. *Polymers* **2022**, *14*, 1626. [CrossRef]
81. Auras, R.A.; Singh, S.P.; Singh, J.J. Evaluation of Oriented Poly(Lactide) Polymers vs. Existing PET and Oriented PS for Fresh Food Service Containers. *Packag. Technol. Sci.* **2005**, *18*, 207–216. [CrossRef]
82. Joo, M.; Lewandowski, N.; Auras, R.; Harte, J.; Almenar, E. Comparative Shelf Life Study of Blackberry Fruit in Bio-Based and Petroleum-Based Containers under Retail Storage Conditions. *Food Chem.* **2011**, *126*, 1734–1740. [CrossRef] [PubMed]
83. Almenar, E.; Samsudin, H.; Auras, R.; Harte, B.; Rubino, M. Postharvest Shelf Life Extension of Blueberries Using a Biodegradable Package. *Food Chem.* **2008**, *110*, 120–127. [CrossRef] [PubMed]
84. Chonhenchob, V.; Chantarasomboon, Y.; Singh, S.P. Quality Changes of Treated Fresh-Cut Tropical Fruits in Rigid Modified Atmosphere Packaging Containers. *Packag. Technol. Sci.* **2007**, *20*, 27–37. [CrossRef]
85. Jamshidian, M.; Tehrany, E.A.; Imran, M.; Jacquot, M.; Desobry, S. Poly-Lactic Acid: Production, Applications, Nanocomposites, and Release Studies. *Compr. Rev. Food Sci. Food Saf.* **2010**, *9*, 552–571. [CrossRef] [PubMed]
86. Anderson, I. Mechanical Properties of Specimens 3D Printed with Virgin and Recycled Polylactic Acid. *3D Print. Addit. Manuf.* **2017**, *4*, 110–115. [CrossRef]
87. Cruz Sanchez, F.A.; Boudaoud, H.; Hoppe, S.; Camargo, M. Polymer Recycling in an Open-Source Additive Manufacturing Context: Mechanical Issues. *Addit. Manuf.* **2017**, *17*, 87–105. [CrossRef]
88. Mikula, K.; Skrzypczak, D.; Izydorczyk, G.; Warchoł, J.; Moustakas, K.; Chojnacka, K.; Witek-Krowiak, A. 3D Printing Filament as a Second Life of Waste Plastics—A Review. *Environ. Sci. Pollut. Res.* **2021**, *28*, 12321–12333. [CrossRef]
89. Davis, A.Y.; Zhang, Q.; Wong, J.P.S.; Weber, R.J.; Black, M.S. Characterization of Volatile Organic Compound Emissions from Consumer Level Material Extrusion 3D Printers. *Build. Environ.* **2019**, *160*, 106209. [CrossRef]
90. Tümer, E.H.; Erbil, H.Y. Extrusion-Based 3D Printing Applications of PLA Composites: A Review. *Coatings* **2021**, *11*, 390. [CrossRef]
91. Wojciechowska, P.; Wolek, P. Design and Properties of Packaging Prototypes Made from PLA and HIPS Using 3D Printing Technology. *Pol. J. Commod. Sci.* **2020**, *62*, 41–49.
92. Vesel, A.; Mozetic, M.; Jaganjac, M.; Milkovic, L.; Cipak, A.; Zarkovic, N. Biocompatibility of Oxygen-Plasma-Treated Polystyrene Substrates. *Eur. Phys. J. Appl. Phys.* **2011**, *56*, 24024. [CrossRef]

93. Iuliano, A.; Nowacka, M.; Rybak, K.; Rzepna, M. The Effects of Electron Beam Radiation on Material Properties and Degradation of Commercial PBAT/PLA Blend. *J. Appl. Polym. Sci.* **2020**, *137*, 48462. [CrossRef]
94. Rogers, W.J. Sterilisation Techniques for Polymers. In *Sterilisation of Biomaterials and Medical Devices*, 1st ed.; Lerouge, S., Simmons, A., Eds.; Woodhead Publishing: Cambridge, UK, 2012; pp. 151–211. [CrossRef]
95. Tilkin, R.G.; Régibeau, N.; Lambert, S.D.; Grandfils, C. Correlation between Surface Properties of Polystyrene and Polylactide Materials and Fibroblast and Osteoblast Cell Line Behavior: A Critical Overview of the Literature. *Biomacromolecules* **2020**, *21*, 1995–2013. [CrossRef] [PubMed]
96. Iuliano, A.; Fabiszewska, A.; Kozik, K.; Rzepna, M.; Ostrowska, J.; Dębowski, M.; Plichta, A. Effect of Electron-Beam Radiation and Other Sterilization Techniques on Structural, Mechanical and Microbiological Properties of Thermoplastic Starch Blend. *J. Polym. Environ.* **2021**, *29*, 1489–1504. [CrossRef] [PubMed]
97. Majgaonkar, P.; Hanich, R.; Malz, F.; Brüll, R. Chemical Recycling of Post-Consumer PLA Waste for Sustainable Production of Ethyl Lactate. *Chem. Eng. J.* **2021**, *423*, 129952. [CrossRef]
98. Durkin, A.; Taptygin, I.; Kong, Q.; Gunam Resul, M.F.M.; Rehman, A.; Fernández, A.M.L.; Harvey, A.P.; Shah, N.; Guo, M. Scale-up and Sustainability Evaluation of Biopolymer Production from Citrus Waste Offering Carbon Capture and Utilisation Pathway. *ChemistryOpen* **2019**, *8*, 668–688. [CrossRef]
99. Whinfield, J.R.; Dickson, J.T. Polymeric Linear Terephthalic Esters. U.S. Patent 2465319-A, 22 March 1949. Available online: https://patents.google.com/patent/US2465319 (accessed on 26 September 2022).
100. Wyeth, N.; Roseveare, R. Biaxially Oriented Poly(ethylene terephthalate) Bottle. U.S. Patent 3733309-A, 15 May 1973. Available online: https://patents.google.com/patent/US3733309A (accessed on 26 September 2022).
101. Kim, N.K.; Lee, S.H.; Park, H.D. Current biotechnologies on depolymerization of polyethylene terephthalate (PET) and repolymerization of reclaimed monomers from PET for bio-upcycling: A critical review. *Bioresour. Technol.* **2022**, *363*, 127931. [CrossRef]
102. De Vos, L.; van de Voorde, B.; van Daele, L.; Dubruel, P.; van Vlierberghe, S. Poly(alkylene terephthalate)s: From current developments in synthetic strategies towards applications. *Eur. Polym. J.* **2021**, *161*, 110840. [CrossRef]
103. Lepoittevin, B.; Roger, P. Chapter 4—Poly(ethylene terephthalate). In *Handbook of Engineering and Speciality Thermoplastics: Polyethers and Polyesters*; Sabu, T., Visakh, P.M., Eds.; Scrivener Publishing LLC: Beverly, MA, USA, 2011; Volume 3, pp. 97–126.
104. Dhaka, V.; Singh, S.; Anil, A.G.; Naik, T.S.S.K.; Garg, S.; Samuel, J.; Kumar, M.; Ramamurthy, P.C.; Singh, J. Occurrence, toxicity and remediation of polyethylene terephthalate plastics. A review. *Environ. Chem. Lett.* **2022**, *20*, 1777–1800. [CrossRef]
105. Nanda, S.; Berruti, F. Thermochemical conversion of plastic waste to fuels: A review. *Environ. Chem. Lett.* **2021**, *19*, 123–148. [CrossRef]
106. Demirel, B.; Yaraş, A.; Elçiçek, H. Crystallization Behavior of PET Materials. *BAÜ Fen Bil. Enst. Derg. Cilt* **2011**, *13*, 26–35.
107. Thomsen, T.B.; Hunt, C.J.; Meyer, A.S. Influence of substrate crystallinity and glass transition temperature on enzymatic degradation of polyethylene terephthalate (PET). *N. Biotechnol.* **2022**, *69*, 28–35. [CrossRef] [PubMed]
108. Hiraga, K.; Taniguchi, I.; Yoshida, S.; Kimura, Y.; Oda, K. Biodegradation of waste PET: A sustainable solution for dealing with plastic pollution. *EMBO Rep.* **2019**, *20*, e49365. [CrossRef] [PubMed]
109. Soong, Y.-H.V.; Sobkowicz, M.J.; Xie, D. Recent Advances in Biological Recycling of Polyethylene Terephthalate (PET) Plastic Wastes. *Bioengineering* **2022**, *9*, 98. [CrossRef] [PubMed]
110. Suhaimi, N.A.S.; Muhamad, F.; Razak, N.A.A.; Zeimaran, E. Recycling of polyethylene terephthalate wastes: A review of technologies, routes, and applications. *Polym. Eng. Sci.* **2022**, *62*, 2355–2375. [CrossRef]
111. Benyathiar, P.; Kumar, P.; Carpenter, G.; Brace, J.; Mishra, D.K. Polyethylene Terephthalate (PET) Bottle-to-Bottle Recycling for the Beverage Industry: A Review. *Polymers* **2022**, *14*, 2366. [CrossRef]
112. Jeya, G.; Dhanalakshmi, R.; Anbarasu, M.; Vinitha, V.; Sivamurugan, V. A short review on latest developments in catalytic depolymerization of Poly (ethylene terephathalate) wastes. *J. Indian Chem. Soc.* **2022**, *99*, 100291. [CrossRef]
113. Crognale, S.; Russo, C.; Petruccioli, M.; D'Annibale, A. Chitosan Production by Fungi: Current State of Knowledge, Future Opportunities and Constraints. *Fermentation* **2022**, *8*, 76. [CrossRef]
114. Raafat, D.; Sahl, H.G. Chitosan and its antimicrobial potential–a critical literature survey. *Microb. Biotechnol.* **2009**, *2*, 186–201. [CrossRef]
115. Crini, G. Historical review on chitin and chitosan biopolymers. *Environ. Chem. Lett.* **2019**, *17*, 1623–1643. [CrossRef]
116. Jiménez-Gómez, C.P.; Cecilia, J.A. Chitosan: A Natural Biopolymer with a Wide and Varied Range of Applications. *Molecules* **2020**, *25*, 3981. [CrossRef] [PubMed]
117. Amiri, H.; Aghbashlo, M.; Sharma, M.; Gaffey, J.; Manning, L.; Masoud, S.; Basri, M.; Kennedy, J.F.; Gupta, V.K.; Tabatabaei, M. Chitin and chitosan derived from crustacean waste valorization streams can support food systems and the UN Sustainable Development Goals. *Nat. Food* **2022**, *3*, 822–828. [CrossRef]
118. Wang, H.; Qian, J.; Ding, F. Emerging Chitosan-Based Films for Food Packaging Applications. *J. Agric. Food Chem.* **2018**, *66*, 395–413. [CrossRef] [PubMed]
119. Oladzadabbasabadi, N.; Nafchi, A.M.; Ariffin, F.; Wijekoon, M.M.J.O.; Al-Hassan, A.A.; Dheyab, M.A.; Ghasemlou, M. Recent advances in extraction, modification, and application of chitosan in packaging industry. *Carbohydr. Polym.* **2022**, *277*, 118876. [CrossRef] [PubMed]
120. Xue, L.; Greisler, H.P. Biomaterials in the development and future of vascular grafts. *J. Vasc. Surg.* **2003**, *37*, 472–480. [CrossRef]

121. Chupa, J.M.; Foster, A.M.; Sumner, S.R.; Madihally, S.V.; Matthew, H.W. Vascular cell responses to polysaccharide materials: In vitro and in vivo evaluations. *Biomaterials* **2000**, *21*, 2315–2322. [CrossRef]
122. Van den Broek, L.A.M.; Knoop, R.J.I.; Kappen, F.H.J.; Boeriu, C.G. Chitosan films and blends for packaging material. *Carbohydr. Polym.* **2015**, *116*, 237–242. [CrossRef]
123. Jones, M.P.; Archodoulaki, V.-M.; Kock, B.-M. The power of good decisions: Promoting eco-informed design attitudes in plastic selection and use. *Resour. Conserv. Recycl.* **2022**, *182*, 106324. [CrossRef]
124. IPEX. *Chemical Resistance Guide, Polyvinyl Chloride (PVC)*, 1st ed.; IPEX: Oakville, ON, Canada, 2020.
125. Wypych, G. *PVC Degradation and Stabilization*; ChemTec Publishing: Toronto, ON, Canada, 2020.
126. Leadbitter, J. *Polyvinyl Chloride (PVC) for Food Packaging Applications*; International Life Science Institute report; ILSI Europe: Brussels, Belgium, 2003.
127. Turner, A.; Fillella, M. Polyvinyl chloride in consumer and environmental plastics, with a particular focus on metal-based additives. *Environ. Sci. Process. Impacts* **2021**, *23*, 1376–1384. [CrossRef]
128. Wypych, G. *PVC Formulary*; ChemTec Publishing: Toronto, ON, Canada, 2020.
129. Cousins, K. *Polymers in Building and Construction*; Rapra Technology LTD: Shawbury, UK, 2002.
130. Abdel-Monem, R.A.; Rabie, S.T.; El-Liethy, M.A.; Hemdan, B.A.; El-Nazer, H.A.; Gaballah, S.T. Chitosan-PVC conjugates/metal nanoparticles for biomedical applications. *Polym. Adv. Technol.* **2022**, *33*, 514–523. [CrossRef]
131. Zhong, R.; Wang, H.; Wu, X.; Cao, Y.; He, Z.; He, Y.; Liu, J. In vitro investigation of the effect of plasticizers on the blood compatibility of medical grade plasticized poly(vinyl chloride). *J. Mater. Sci. Mater. Med.* **2013**, *24*, 1985–1992. [CrossRef] [PubMed]
132. Marsh, K.; Bugusu, B. Food packaging roles, materials, and environmental issues. *J. Food Sci.* **2007**, *72*, 39–55. [CrossRef] [PubMed]
133. Janajreh, I.; Alshrah, M.; Zamzam, S. Mechanical recycling of PVC plastic waste streams from cable industry: A case study. *Sustain. Cities Soc.* **2015**, *18*, 13–20. [CrossRef]
134. Buhl, R. Progress in PVC feedstock recycling. *Polimery* **2003**, *48*, 263–267. [CrossRef]
135. Leathers, T.D. Pullulan. In *Biopolymers. Polysaccharides II: Polysaccharides from Eukaryotes*; Vandamme, E.J., De Baets, S., Steinbachel, A., Eds.; Wiley-VCH: Weincheim, Germany, 2002; Volume 6, pp. 1–35.
136. Kimoto, T.; Shibuya, T.; Shiobara, S. Safety studies of a novel starch, pullulan: Chronic toxicity in rats and bacterial mutagenicity. *Food Chem. Toxicol.* **1997**, *35*, 323–329. [CrossRef]
137. Prajapati, V.D.; Jani, G.K.; Khanda, S.M. Pullulan: An exopolysaccharide and its various applications. *Carbohydr. Polym.* **2013**, *95*, 540–549. [CrossRef]
138. Leathers, T.D. Biotechnological production and applications of pullulan. *Appl. Microbiol. Biotechnol.* **2003**, *62*, 468–473. [CrossRef]
139. Arnosti, C.; Repeta, D.J. Nuclear magnetic resonance spectroscopy of pullulan and isomaltose: Complete assignment of chemical shifts. *Starch* **1995**, *47*, 73–75. [CrossRef]
140. Petrov, P.T.; Shingel, K.I.; Scripko, A.D.; Tsarenkov, V.M. Biosynthesis of pullulan by *Aureobasidium pullulans* strain BMP-97. *Biotechnologia* **2022**, *1*, 36–48.
141. Kristo, E.; Biliaderis, C.G. Physical properties of starch nanocrystal-reinforced pullulan films. *Carbohydr. Polym.* **2007**, *68*, 146–158. [CrossRef]
142. Shingel, K.I. Current knowledge on biosynthesis, biological activity, and chemical modification of the exopolysaccharide, pullulan. *Carbohydr. Res.* **2004**, *339*, 447–460. [CrossRef] [PubMed]
143. Gniewosz, M.; Pobiega, K.; Kraśniewska, K.; Synowiec, A.; Chaberek, M.; Galus, S. Characterization and antifungal activity of pullulan edible films enriched with propolis extract for active packaging. *Foods* **2022**, *11*, 2319. [CrossRef] [PubMed]
144. FDA. *Agency Response Letter: GRAS Notice No. GRN 000099 [Pullulan]*; US Food and Drug Administration (US FDA), Center for Food Safety and Applied Nutrition (CFSAN), Office of Food Additive Safety: College Park, MD, USA, 2002.
145. European Parliament. *Regulation (EC) No 1333/2008 of the European Parliament and of the Council of 16 December 2008 on Food Additives. Annex II Community List of Food Additives Approved for Use in Foods and Conditions of Use*; European Parliament: Brussels, Belgium, 2008.
146. Grigoras, A.G. Drug delivery system using pullulan, a biocompatible polysaccharide produced by fungal fermentation of starch. *Environ. Chem. Lett.* **2019**, *17*, 1209–1223. [CrossRef]
147. Ganie, S.A.; Rather, L.J.; Li, Q. A review on anticancer applications of pullulan and pullulan derivative nanoparticles. *Carbohydr. Polym. Technol. Appl.* **2021**, *2*, 100115. [CrossRef]
148. Coltelli, M.-B.; Danti, S.; De Clerk, K.; Lazzeri, A.; Morganti, P. Pullulan for advanced sustainable body- and skin-contact applications. *J. Funct. Biomater.* **2020**, *11*, 20. [CrossRef]
149. Kraśniewska, K.; Ścibisz, I.; Gniewosz, M.; Mitek, M.; Pobiega, K.; Cendrowski, A. Effect of pullulan coating on postharvest quality and shelf-life of highbush blueberry (*Vaccinium corymbosum* L.). *Materials* **2017**, *10*, 965. [CrossRef]
150. Wu, S.; Chen, J. Using pullulan-based edible coatings to extend shelf-life of fresh-cut „Fuji" apples. *Int. J. Biol. Macromol.* **2013**, *55*, 254–257. [CrossRef]
151. Tzoumaki, M.V.; Biliaderis, C.G.; Vasilakakis, M. Impact of edible coatings and packaging on quality of white asparagus (*Asparagus officinalis*, L.) during cold storage. *Food Chem.* **2009**, *117*, 55–63. [CrossRef]
152. Kumar, N.; Neeraj, P.; Trajkovska-Petkovska, A.; Al-Hilifi, S.A.; Fawole, O.A. Effect of chitosan-pullulan composite edible coating functionalized with pomegranate peel extract on the shelf-life of mango (*Mangifera indica*). *Coatings* **2021**, *11*, 764. [CrossRef]

153. Gniewosz, M.; Synowiec, A. Antibacterial activity of pullulan films containing thymol. *Flavour Fragr. J.* **2011**, *26*, 389–395. [CrossRef]
154. Priyadarshi, R.; Riahi, Z.; Rhim, J.-W. Antioxidant pectin/pullulan edible coating incorporated with Vitis vinifera grape seed extract for extending the shelf-life of peanuts. *Postharvest Biol. Technol.* **2022**, *183*, 111740. [CrossRef]
155. Engineering ToolBox. Available online: https://www.engineeringtoolbox.com/polymer-properties-d_1222.html (accessed on 23 October 2022).
156. Mark, J.E. *Polymer Data Handbook*, 2nd ed.; Oxford University Press: Oxford, UK, 2009.
157. ChemSrc. Available online: https://www.chemsrc.com/en (accessed on 23 October 2022).
158. Farah, S.; Anderson, D.G.; Langer, R. Physical and mechanical properties of PLA, and their functions in widespread applications —A comprehensive review. *Adv. Drug Deliv. Rev.* **2016**, *107*, 367–392. [CrossRef] [PubMed]
159. Zubair, M.; Najeeb, F.; Mohamed, M.; Saeed, A.; Ramis, M.K.; Abdul Mujeeb, M.S.; Sarojini, B.K. Analysis of Chitosan Polymer Doped with Nano Al_2O_3 and Nano CuO. *Am. J. Polym Sci.* **2014**, *4*, 40–45.
160. Ioelovich, M. Crystallinity and Hydrophility of Chitin and Chitosan. *Res. Rev. J. Chem.* **2014**, 3, 7–14.
161. Ulery, B.D.; Nair, L.S.; Laurencin, C.T. Biomedical applications of biodegradable polymers. *J. Polym. Sci. Part B Polym. Phys.* **2011**, *49*, 832–864. [CrossRef]

polymers

MDPI

Article

Wood—Derived Polymers from Olefin—Functionalized Lignin and Ethyl Cellulose via Thiol–Ene Click Chemistry

Rongrong An [1], Chengguo Liu [2], Jun Wang [1,*] and Puyou Jia [2,*]

[1] School of Geographic and Biologic Information, Nanjing University of Posts and Telecommunications, Nanjing 210023, China

[2] Institute of Chemical Industry of Forest Products, Chinese Academy of Forestry, 16 Suojin North Road, Nanjing 210042, China

* Correspondence: wangj@njupt.edu.cn (J.W.); jiapuyou@icifp.cn (P.J.)

Abstract: Lignin and cellulose derivatives have vast potential to be applied in polymer materials. The preparation of cellulose and lignin derivatives through esterification modification is an important method to endow cellulose and lignin with good reactivity, processability and functionality. In this study, ethyl cellulose and lignin are modified via esterification to prepare olefin—functionalized ethyl cellulose and lignin, which are further used to prepare cellulose and lignin cross—linker polymers via thiol–ene click chemistry. The results show that the olefin group concentration in olefin—functionalized ethyl cellulose and lignin reached 2.8096 mmol/g and 3.7000 mmol/g. The tensile stress at break of the cellulose cross—linked polymers reached 23.59 MPa. The gradual enhancement in mechanical properties is positively correlated with the olefin group concentration. The existence of ester groups in the cross—linked polymers and degradation products makes them more thermally stable. In addition, the microstructure and pyrolysis gas composition are also investigated in this paper. This research is of vast significance to the chemical modification and practical application of lignin and cellulose.

Keywords: lignin; ethyl cellulose; cross—linked polymer; thiol–ene click chemistry

Citation: An, R.; Liu, C.; Wang, J.; Jia, P. Wood—Derived Polymers from Olefin—Functionalized Lignin and Ethyl Cellulose via Thiol–Ene Click Chemistry. *Polymers* **2023**, *15*, 1923. https://doi.org/10.3390/polym15081923

Academic Editors: Clara Delgado-Sánchez, Concepción Valencia-Barragán, Esperanza Cortés Triviño, Adrián Tenorio-Alfonso and Antonio M. Borrero-López

Received: 3 April 2023
Revised: 14 April 2023
Accepted: 14 April 2023
Published: 18 April 2023

1. Introduction

With the increasing depletion of fossil resources and the increasing environmental problems of "white pollution" and "microplastics" caused by the extensive use of non—degradable petroleum—based polymers, the use of bio—based raw materials, instead of petroleum—based compound raw materials, to prepare polymer materials has attracted widespread attention [1]. Cellulose has become an ideal raw material for bio—based polymer materials due to its rich sources, low price, excellent biodegradability, easy modification and many other advantages. Cellulose and its derivatives have been widely used in the fiber, paper, film, plastics, coatings and other industrial fields [2–4]. However, due to the existence of a large number of hydrogen bonds within and between the molecules of natural cellulose, the complexity of cellulose aggregation structure and high crystallinity, cellulose is insoluble in water and general organic solvents, in addition to not being able of being melted or processed as traditional plastics, which severely limits the application of cellulose materials. The chemical modification of cellulose includes oxidation, esterification, etherification and other grafting methods [5–7]. At present, the esterification derivatives of cellulose include cellulose acetate, cellulose propionate, cellulose butyrate and various cellulose—mixed esters, which are widely used in plastics, coatings, separation membranes, cigarette filters and other daily necessities [8–10].

Lignin is a rich natural resource. Compared with other biomass products, lignin has a relatively complex structure and contains a variety of functional groups [11]. It has broad research prospects for the development of appropriate methods for separating and extracting lignin, and then prepare functional composites. At present, lignin has been applied to the preparation of high−value materials, such as porous carbon materials, adsorption materials, capacitor electrode materials, graphene materials, surfactants and hydrogels, and has broad application prospects in the energy, medical, construction, agriculture and other fields [12–16]. The structural unit of lignin is similar to that of phenol. Preparing lignin−based phenolic resin by partially replacing phenol with lignin is the most feasible method [16–20]. Lignin and its derivatives can be used to synthesize bipolar plate materials for phenolic resin fuel cells, phenolic resin catalysts, phenolic resin foams, phenolic resin adhesives and other phenolic resin materials [21–23]. Since the 20th century, researchers have been exploring the preparation of lignin−based phenolic resin by replacing phenol with structurally modified lignin derivatives. After the 1990s, the preparation technology of lignin−based phenolic resin developed rapidly because many kinds of structurally modified lignin derivatives with flame retardancy were produced. However, due to the shortcomings of high pollution, high energy consumption, and a complex preparation process, lignin−based phenolic resin cannot be produced on a large scale.

As a simple C–S bonding reaction, "thiol–ene" reaction was discovered over 100 years ago [24]. This reaction has very attractive advantages: First, the C–S bonding reaction can be conducted under a variety of conditions. Secondly, various olefins can be used as suitable substrates, including activated and inactive polysubstituted olefins. Third, almost all mercaptans can be used for reactions, including highly functional substances. Finally, this reaction is very rapid, and the reaction conditions are mild, which can be conducted in the air environment [25–27]. The "thiol–ene "click reaction can be initiated by free radicals, initiated by ultraviolet light and free radicals using natural light, red light and redox system. The click reaction is fast and can be performed under normal temperature and pressure. If it is initiated by ultraviolet light, the reaction yield can reach more than 90% in a few seconds. As a means of constructing new materials, the "thiol–ene" click reaction has many advantages: olefin compounds are very rich, with high selectivity, and can synthesize a large number of compounds with various structures [28–31]. Recently, Jawerth reported that the ethanol−soluble fraction of Lignoboost Kraft lignin was selectively allylated using allyl chloride by means of a mild and industrially scalable procedure. The obtained modified lignin was then subsequently cross−linked to prepare thermosetting resin via thermally induced thiol–ene chemistry [29]. Cao et al. prepared thermosetting lignin−based polyurethane coatings with superior corrosion resistance and a high content of lignin by the polymerization of lignin−based polyol. Firstly, the phenolic hydroxyls of enzymatic hydrolysis lignin were, firstly, selectively converted to primary aliphatic hydroxyls by an allylation reaction. Subsequently, the thermal radical initiated thiol–ene click reaction was applied to efficiently prepare the lignin−based polyol [30]. Zeng et al. developed durable a superhydrophobic and oleophobic coating based on perfluorodecanethiol fluorosilicone polyurethane (PFDT−FSPU) and thiol−modified cellulose substrate. The cross−linked network structure was formed by the radical polymerization of double bonds in PFDT−FSPU when the polyurethane was irradiated with ultraviolet light, and it was anchored on the surface of the cotton fibers by click reaction between the thiol−modified cellulose substrate and PFDT−FSPU. The coated fabric showed excellent durability and can still maintain superhydrophobicity and oleophobicity even after 600 cycles of abrasion or 30 times of washing cycles or 168 h of accelerated aging test [31].

The chemical modification of ethyl cellulose and lignin to prepare olefin−functionalized lignin and ethyl cellulose can effectively increase reaction activity and enrich their application performance. The chemical modification of ethyl cellulose and lignin to construct new functional materials has become a major strategy to increase their added value. In this study, ethyl cellulose (EC) and lignin are modified via esterification to prepare olefin−functionalized ethyl cellulose and lignin, which are further used to prepare cellulose and lignin cross−linker polymers via thiol–ene click chemistry. The chemical structure of olefin−functionalized EC and lignin, as well as ethyl cellulose and lignin cross−linked polymers are characterized. In addition, the thermal stability, microstructure, mechanical property and pyrolysis gas composition are also investigated.

2. Experimental Procedure

2.1. Materials

Lignin (Hydroxyl content: 5.6 mmol/g), EC (Mw = 151,237, Mw unit = 454.5 g/mol, DP = 332, Hydroxyl content: 1.57–1.82 mmole/g), 1−Hydroxycyclohexyl phenyl ketone, undecylenic acid, 4−Dimethylaminopyridine (DMAP), trimethylacetic anhydride, dichloromethane (DCM), methanol, tetrahydrofuran (THF), ketone, azodiisobutyronitrile (AIBN), 3,6−Dioxa−1,8−octanedithiol and pentaerythritol tetra(3−mercaptopropionate) were obtained from commercial resources and used as received, unless otherwise mentioned.

2.2. Preparation of Olefin−Functionalized Ethyl Cellulose (OFE)

EC (3.0 g, 0.01975 mol), undecylenic acid (1.82 g, 0.009878 mol), DMAP (12 mg, 0.09878 mmol) and trimethylacetic anhydride (1.83 g, 0.009878 mol) were dissolved in 50 mL dry THF and stirred at 60 °C under a nitrogen atmosphere for 48 h. In order to remove the unreacted undecylenic acid, the purification process was conducted by repeating dissolution into dry tetrahydrofuran and precipitation from methanol. The solid product was dried under vacuum to obtain OFE and label as EC−1−0.5. When the mole ratios of EC, undecylenic acid, DMAP and trimethylacetic anhydride were 1:1:0.001:1 and 1:1.3:0.001:1, the obtained OFEs were labeled as EC−1−1 and EC−1−1.3, respectively. Scheme 1 shows the Synthesis of OFE.

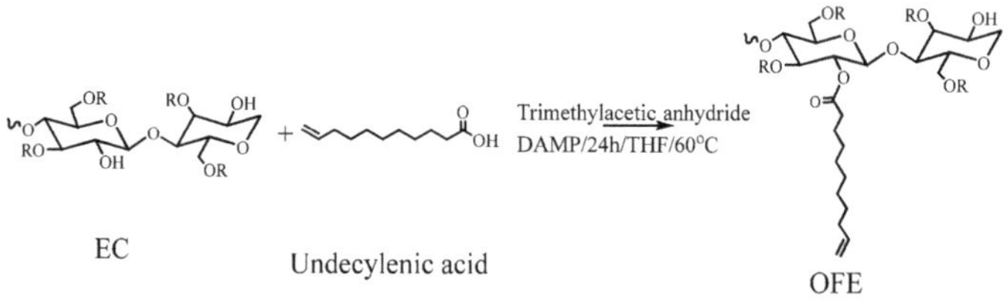

Scheme 1. Synthesis of OFE.

2.3. Preparation of Olefin−Functionalized Lignin (OFL)

Lignin (5.0 g, 28.75 mmol), undecylenic acid (2.645 g, 14.375 mmol), DMAP (17.53 mg, 0.14375 mmol) and trimethylacetic anhydride (2.677 g, 14.375 mmol) were dissolved in 50 mL dry THF. The mixture was stirred at 60 °C under a nitrogen atmosphere for 48 h. In order to remove the unreacted undecylenic acid, the purification process was conducted by repeating dissolution into dry tetrahydrofuran and precipitation from methanol. The solid product was dried under vacuum to obtain OFLs and labeled lignin−1−0.5. When the mole ratios of lignin, undecylenic acid, DMAP and trimethylacetic anhydride were 1:1:0.001:1 and 1:1.3:0.001:1, the obtained OFLs were labeled as lignin−1−1 and lignin−1−1.3, respectively. Scheme 2 shows the synthesis of OFL.

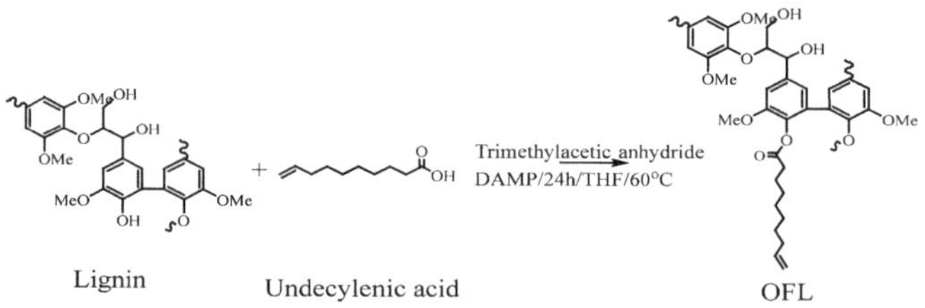

Scheme 2. Synthesis of OFL.

2.4. Determination of Olefin Group Concentration in OFE and OFL

The olefin group concentration in OFE and OFL was investigated with the internal standard method using 1,2,4,5−tetrachloro−3−nitrobenzene as the interior label.

OFE, OFL and 1,2,4,5−tetrachloro−3−nitrobenzene were dissolved in $CDCl_3$, and the ^1H NMR of the mixture was tested. The signals at 5.0 ppm and 5.82 ppm from the protons of the olefin group were also integrated. The signal at 7.75 ppm from the protons of 1,2,4,5−tetrachloro−3−nitrobenzene was 100. A detailed calculation is presented in [32].

2.5. Preparation of Wood−Derived Polymers through the Chemical Cross−Linking of OFE and OFL via Thiol–Ene Click Chemistry

OFL or OFL polymers, thiol cross−linkers and 1−Hydroxycyclohexyl phenyl ketone (5% of total mass) were dissolved in dry THF. The mixture was stirred for 30 min and ultrasound for 20 min at room temperature. Then, the mixture was poured into a polytetrafluoroethylene mold, dried under tin foil for 24 h and vacuumed for 24 h at room temperature; then, a UV light was used for 10 min (two 10 W UV lamp tubes). The reactants and obtained wood−derived polymers are shown in Table 1. A schematic diagram of the chemical structure of the wood−derived polymers is presented in Scheme 3.

Table 1. Reactants and the obtained wood−derived polymers.

Cross−linked Polymer	Reactants
EC−1−0.5−2SH	EC−1−0.5 and 3,6−Dioxa−1,8−octanedithiol, n(−CH=CH−) = n(−SH)
EC−1−1.0−2SH	EC−1−1.0 and 3,6−Dioxa−1,8−octanedithiol, n(−CH=CH−) = n(−SH)
EC−1−1.3−2SH	EC−1−1.3 and 3,6−Dioxa−1,8−octanedithiol, n(−CH=CH−) = n(−SH)
EC−1−0.5−4SH	EC−1−0.5 and pentaerythritol tetra(3−mercaptopropionate), n(−CH=CH−) = n(−SH)
EC−1−1.0−4SH	EC−1−1.0 and pentaerythritol tetra(3−mercaptopropionate), n(−CH=CH−) = n(−SH)
EC−1−1.3−4SH	EC−1−1.3 and pentaerythritol tetra(3−mercaptopropionate), n(−CH=CH−) = n(−SH)
EC−Lignin−2SH	EC−1−1, Lignin−1−1, and 3,6−Dioxa−1,8−octanedithiol, n(−CH=CH−)/n(−CH=CH−)/n(−SH) = 0.5:0.5:1
EC−Lignin−4SH	EC−1−1, lignin−1−1, and pentaerythritol tetra(3−mercaptopropionate), n(−CH=CH−)/n(−CH=CH−)/n(−SH) = 0.5:0.5:1
Lignin−1−1−2SH	Lignin−1−1 and 3,6−Dioxa−1,8−octanedithiol, n(−CH=CH−) = n(−SH)
Lignin−1−1−4SH	Lignin−1−1 and pentaerythritol tetra(3−mercaptopropionate), n(−CH=CH−) = n(−SH)

2.6. Characterization

Fourier transform infrared spectrometry (FTIR) spectra were obtained using a PerkinElmer spectrum 100 FTIR spectrometer. ^1H NMR spectra were recorded on a Bruker Avance III HD 300 spectrometer using CDCl3 as the solvent with tetramethylsilane (TMS) as the internal reference. The gel content (C_{gel}) was determined by Soxhlet extraction. First, approximately 0.5 g of polymer (recorded as m_0) was precisely weighed, extracted with acetone for 24 h, dried in a vacuum oven at 50 °C for 24 h and weighed again (recorded as m_1). The C_{gel} was calculated using $(m_1/m_0) \times 100\%$, and the three samples were tested for average. TG−FTIR spectra were analyzed by using a 409PC thermal analyzer coupled with a Nicolet IS10 FTIR instrument. The temperature was increased from 40 to 800 °C at a rate of 10 °C min^{-1} under a N_2 atmosphere. Tensile tests were conducted on an E43.104 Universal Testing Machine (MTS Instrument Crop., Shenzhen, China). Dog−bone shaped polymers with a length of 20 mm and a width of 5.0 mm were tested at room temperature with a crosshead speed of 20 mm min^{-1}. Five replicate samples were used to obtain an average value for each. The microstructure of the cross−linked polymer was determined using German Leica microscope DM750M.

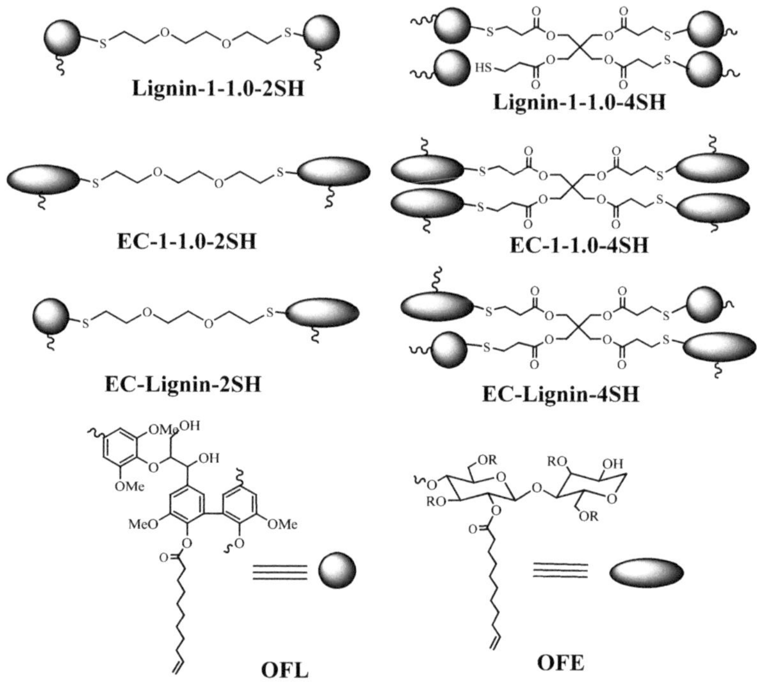

Scheme 3. Schematic diagram of the chemical structure of the wood−derived polymers through the chemical cross−linking of OFE and OFL via thiol–ene click chemistry.

3. Results and Discussion

The esterification products of lignin and EC with undecenoic acid were characterized with FT−IR and ^{1}H NMR, As seen in Figure 1, there are two strong proton signals at 5.0 ppm and 5.82 ppm in the ^{1}H NMR of undecenoic acid, which are attributed to the protons of olefin [33,34]. There are no proton signals at 5.0 ppm and 5.82 ppm in the ^{1}H NMR of EC. After esterification, proton signals at 5.0 ppm and 5.82 ppm gradually increased in the ^{1}H NMR of EC−1−0.5, EC−1−1.0 and EC−1−1.3. The results indicate that OFE products were obtained. As seen in Figure 2, the strong peak at 1725 cm^{-1} is attributed to the infrared absorption peak of the carbonyl group in the FT−IR of undecenoic acid [35–37]. There is no peak at 1725 cm^{-1} in FT−IR of EC. After esterification, infrared absorption peaks of the carbonyl group gradually increased in the FT−IR of EC−1−0.5, EC−1−1.0 and EC−1−1.3, which indicates that the carbonyl groups were formed after esterification. The peak at 1675 cm^{-1} in the FT−IR of EC−1−0.5, EC−1−1.0 and EC−1−1.3 is attributed to the olefin groups [35–37]. The formation of carbonyl groups and the appearance of olefin groups suggests that OFE products were obtained.

The ^{1}H NMR of OFL showed a similar characteristic absorption peak compared with that of OFE. As seen in Figure 3, there are no proton signals at 5.0 ppm and 5.82 ppm in the ^{1}H NMR of lignin. After esterification, proton signals gradually increased at 5.0 ppm and 5.82 ppm in the ^{1}H NMR of lignin−1−0.5, lignin−1−1.0 and lignin−1−1.3. The results indicate that OFL products were obtained. Similarly, the FT−IR of OFL showed a similar characteristic absorption peak compared that of with OFE. As seen from Figure 4, the strong peak at 1725 cm^{-1} is attributed to the infrared absorption peak of the carbonyl group in the FT−IR of undecenoic acid [33,34]. There is no peak at 1725 cm^{-1} in the FT−IR of lignin. After esterification, infrared absorption peaks of the carbonyl group gradually increased in the FT−IR of lignin−1−0.5, lignin−1−1.0 and lignin−1−1.3, which indicates that the carbonyl groups were formed after esterification. The peak at 1675 cm^{-1} in the FT−IR of lignin−1−0.5, lignin−1−1.0 and lignin−1−1.3 is attributed to olefin groups [35–37]. The formation of carbonyl groups and the appearance of olefin groups suggests that OFL products were obtained.

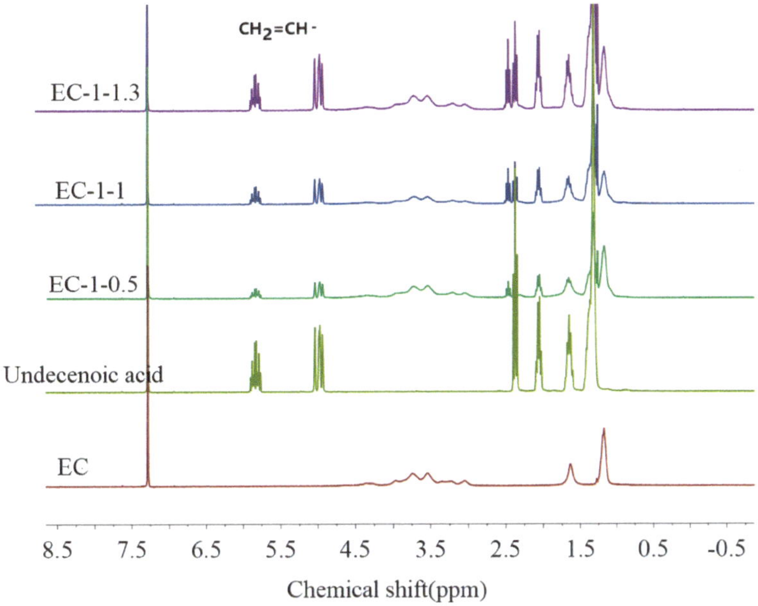

Figure 1. ^{1}H NMR of OFE, EC and undecenoic acid.

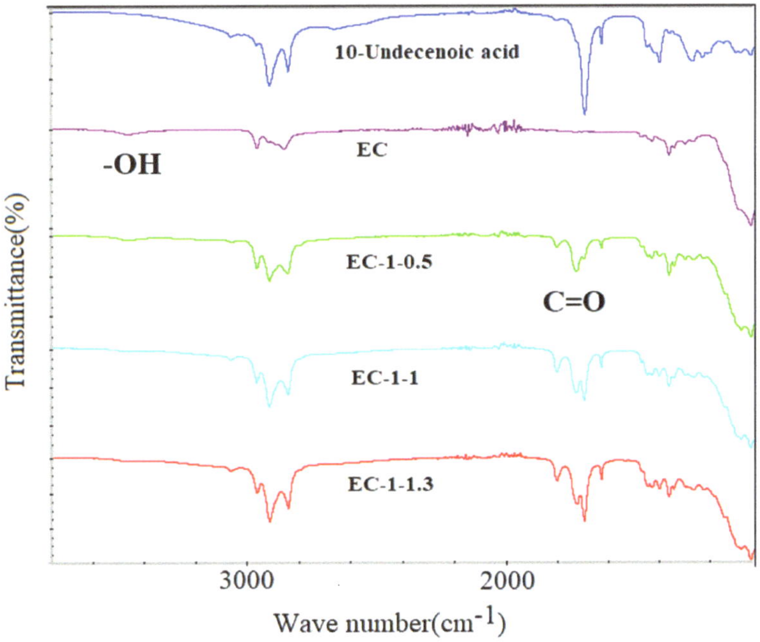

Figure 2. FT−IR of OFE, EC and undecenoic acid.

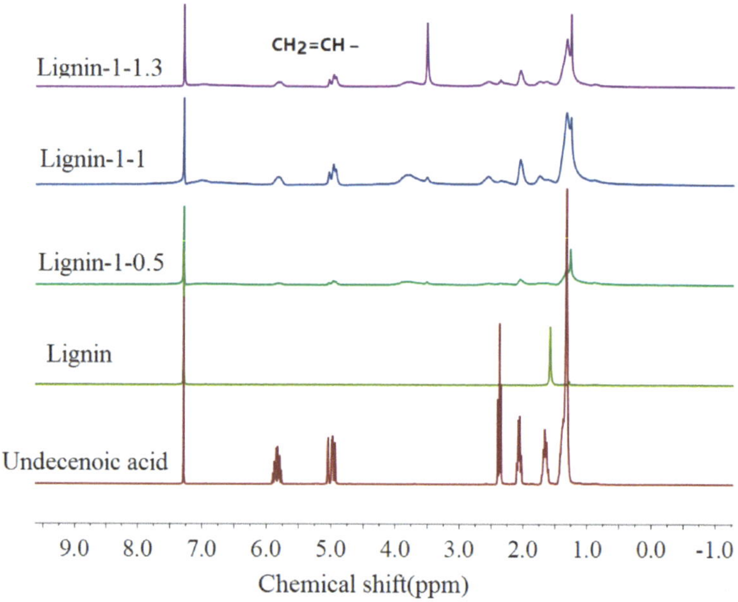

Figure 3. ^1H NMR of OFL, lignin and undecenoic acid.

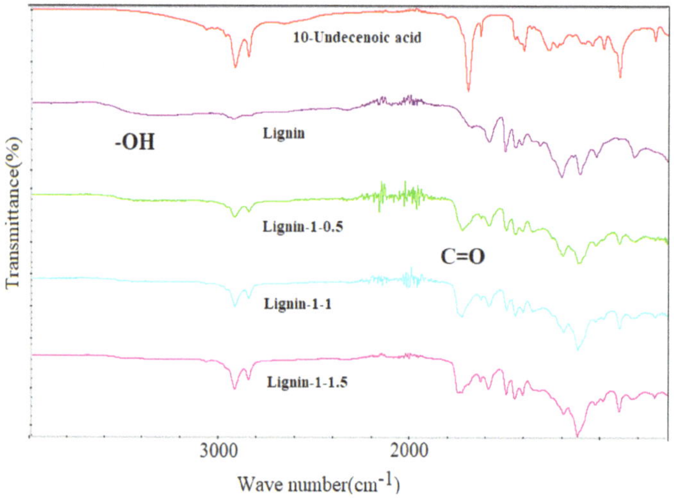

Figure 4. FT−IR of OFL, lignin and undecenoic acid.

The olefin group concentration in OFL and OFE was determined with the internal standard method using 2,3,5,6−Tetrachloro−3−nitrobenzene as the interior label. The ^1H NMR of OFL and OFE with the interior label was detected and the results are shown in Figures 5 and 6. The strong signal at 7.75 ppm in Figures 5 and 6 is attributed to the protons of 2,3,5,6−Tetrachloro−3−nitrobenzene [38,39]. The olefin group concentration in OFL and OFE was calculated according to the procedure presented in a recent study [32]. The results are shown in Table 2, revealing that the olefin group concentration in OFL increased from 1.8060 to 2.8096 mmol/g when the n(−COOH):n(−OH) increased from 0.5 to 1.3.

The olefin group concentration in OFE increased from 2.9200 to 3.7000 mmol/g when the n(−COOH):n(−OH) increased from 0.5 to 1.3.

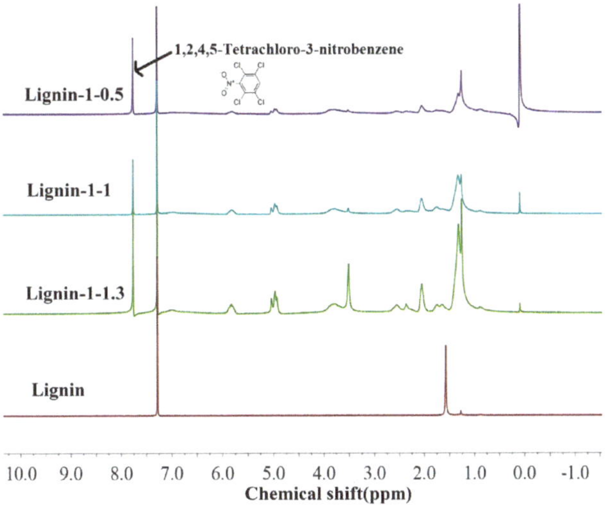

Figure 5. ^{1}H NMR of OFL with the interior label compared with lignin.

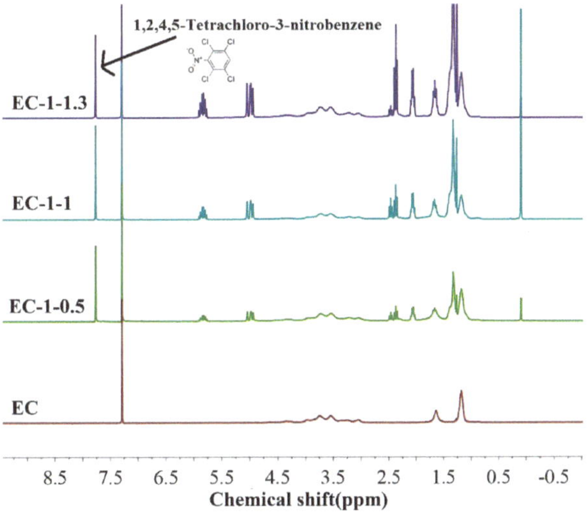

Figure 6. ^{1}H NMR of OFE with the interior label compared with EC.

Table 2. Olefin group concentration in OFL and OFE.

OFL and OFE Samples	Olefin Group Concentration in OFL and OFE (mmol/g)
Lignin−1−0.5	1.8060
Lignin−1−1	2.3818
Lignin−1−1.3	2.8096
EC−1−0.5	2.9200
EC−1−1	3.5873
EC−1−1.3	3.7000

The chemical structure of EC cross−linked polymers and lignin cross−linked polymers were investigated with FT−IR and ^{1}H NMR. As seen from Figure 7, the FT−IR of EC−1−1−2SH and EC−1−1−4SH was compared with EC−1−1. The peak at 1675 cm^{-1} in the FT−IR of EC−1−1 is attributed to the olefin groups [35–37]. After the thiol–ene click reaction, there is no peak at 1675 cm^{-1} in the FT−IR of EC−1−1−2SH and EC−1−1−24H. The chemical structure of the dissolved part for EC cross−linked polymers was investigated with ^{1}H NMR. As seen from Figure 8, the ^{1}H NMR of EC cross−linked polymer was compared with that of EC−1−1. The two strong proton signals at 5.0 ppm and 5.82 ppm in the ^{1}H NMR of EC−1−1 are attributed to the protons of olefin [33,34]. When the thiol–ene click reaction finished, there were no protons at 5.0 ppm and 5.82 ppm in the ^{1}H NMR of EC−1−1−2SH and EC−1−1−4SH, which indicates that there was a thiol–ene click reaction between OFL and the cross−linker (3,6−Dioxa−1,8−octanedithiol, pentaerythritol tetra(3−mercaptopropionate)), and that the EC cross−linked polymers were obtained.

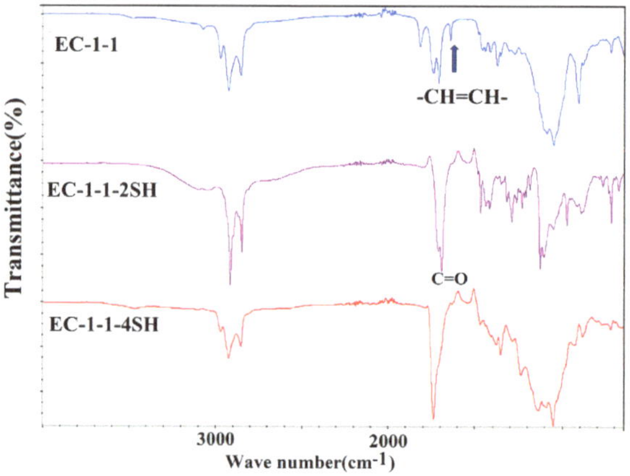

Figure 7. FT−IR of EC cross−linked polymers compared with that of EC−1−1.

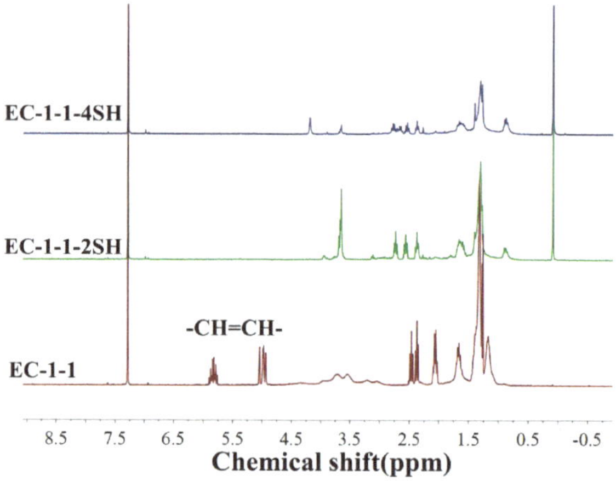

Figure 8. ^{1}H NMR of EC cross−linked polymer compared with that of EC−1−1.

Figure 9 shows the FT−IR of ligin−1−1−2SH and lignin−1−1−4SH compared with that of lignin−1−1. The peak at 1675 cm^{-1} in the FT−IR of ligin−1−1 is attributed to olefin groups [35–37]. When the thiol–ene click reaction finished, the peak at 1675 cm^{-1} in the FT−IR of ligin−1−1−2SH and ligin−1−1−4SH decreased obviously, which shows that there is still a small amount of the olefin group in the lignin cross−linked polymer.

Figure 10 shows the ¹H NMR of the chemical structure of the dissolved part for the lignin cross−linked polymers. The ¹H NMR of the lignin cross−linked polymer was compared with that of lignin−1−1. The two strong proton signals at 5.0 ppm and 5.82 ppm in the ¹H NMR of ligin−1−1 are attributed to the protons of olefin [34,34]. After the thiol–ene click reaction, the proton signals at 5.0 ppm and 5.82 ppm in the ¹H NMR of lignin−1−1−2SH and lignin−1−1−4SH appeared weak and almost disappeared, which indicates that the thiol–ene click reaction between OFL and cross−linker (3,6−Dioxa−1,8−octanedithiol, pentaerythritol tetra(3−mercaptopropionate)) is incomplete, because the precise proportion of olefin and sulfhydryl is difficult to control.

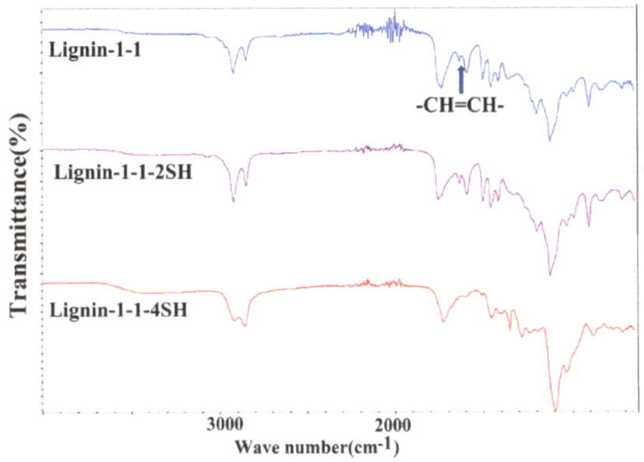

Figure 9. FT−IR of lignin cross−linked polymer compared with that of lignin.

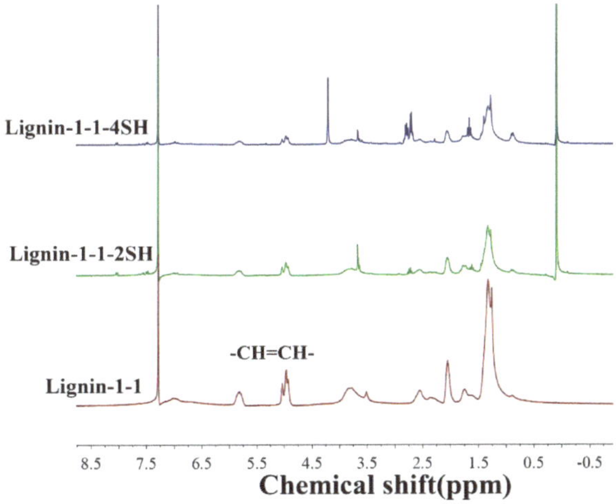

Figure 10. ¹H NMR of lignin cross−linked polymer compared with that of lignin.

Figure 11 shows the FT−IR of the EC–lignin cross−linked polymer compared with that of lignin−1−1, EC−1−1 and undecenoic acid. When the thiol–ene click reaction occurred, the peak at 1675 cm^{-1} in the FT−IR of EC−ligin−2SH and EC−ligin−4SH was weaker than those of lignin−1−1 and EC−1−1, which shows that there is still a small amount of the olefin group in the EC–lignin cross−linked polymer. Figure 12 shows the ¹H NMR of the chemical structure of the dissolved part for the EC–ligin cross−linked polymers, and the ¹H NMR of the lignin cross−linked polymers compared with that of lignin−1−1 and EC−1−1. The proton signals at 5.0 ppm and 5.82 ppm in the ¹H NMR of EC−ligin−2SH and EC−ligin−4SH were low, which was caused by the thiol–ene click reaction.

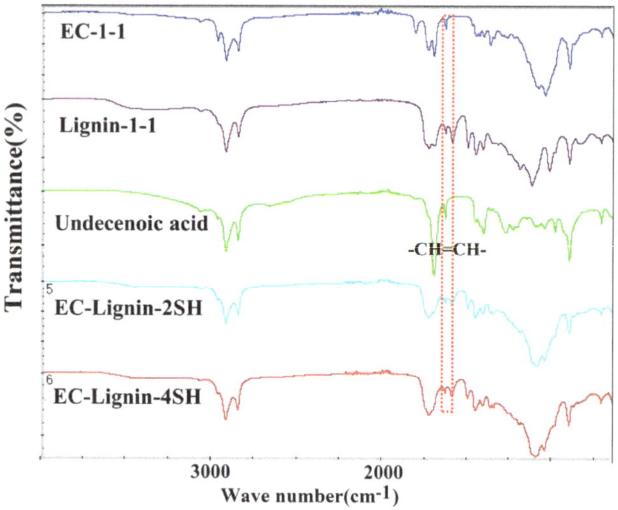

Figure 11. FT−IR of the EC–lignin cross−linked polymers compared with that of lignin−1−1, EC−1−1 and undecenoic acid.

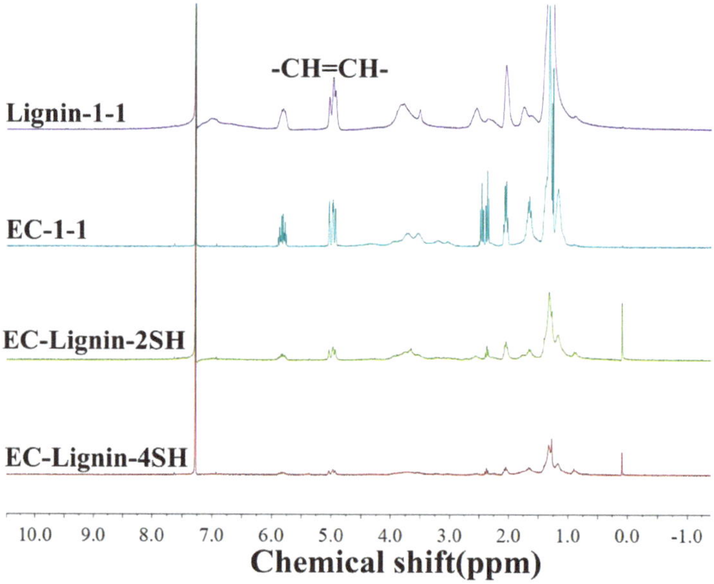

Figure 12. ^{1}H NMR of EC–lignin cross−linked polymer compared with those of lignin−1−1 and EC−1−1.

The tensile stress and strain at break of the EC and cross−linked polymers were investigated and the results are shown in Figures 13–16. As seen in Figure 13, the tensile stress and strain at break of EC were 146.8 MPa and 1.32%, respectively. For the cross−linked polymers, the tensile stress and strain at break increased when the olefin group concentration increased in OFE. When the olefin group concentration increased from 2.9200 to 3.7000 mmol/g, the tensile stress at break increased from 16.41 MPa to 23.59 MPa, as seen in Table 3, while the tensile strain at break firstly increased from 17.25% to 19.41% and then decreased to 18.12%. When the pentaerythritol tetra(3−mercaptopropionate) was used as cross–linker, as seen in Figure 15, the tensile stress at break increased from 6.86 MPa to 13.94 MPa, and the tensile strain increased from 14.85% to 22.38. EC–lignin–4SH showed excellent tensile properties compared to EC–lignin–2SH. The tensile stress at break for EC–lignin–4SH was 15.19 MPa, which is higher than that of EC–lignin–2SH at 13.72 MPa, and the tensile strain at break increased from 15.12% to 17.25%. The gradual enhancement in mechanical properties is positively correlated with the olefin group concentration. The mechanical properties of the lignin cross–linked polymers (lignin–1–1–2SH and lignin–1–1–4SH) were not tested because the polymers were in powder form. Compared with EC, the tensile strength of the cross–linked polymers decreased significantly, but the tensile strain increased sharply, which is caused that the flexible aliphatic hydrocarbon chain from undecylenic acid as the branched chains of EC and lignin increased the distance among the main chains, weakened the interaction force among the main chains and reduced the hydrogen bond interaction in the matrix of EC and lignin. The flexible aliphatic hydrocarbon chain from undecylenic acid contributed to plasticization by functioning as an internal plasticizer.

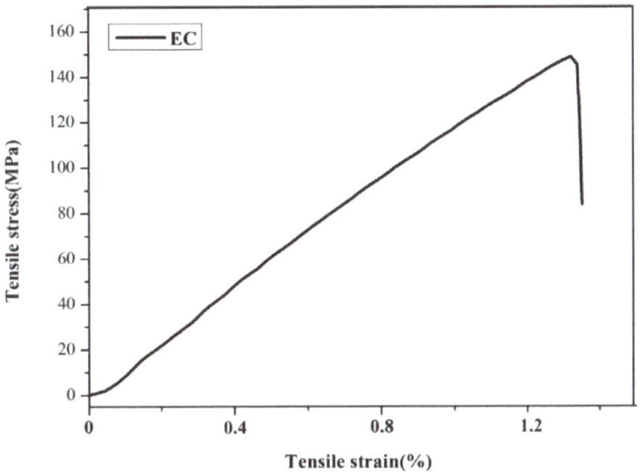

Figure 13. Tensile stress and strain of EC.

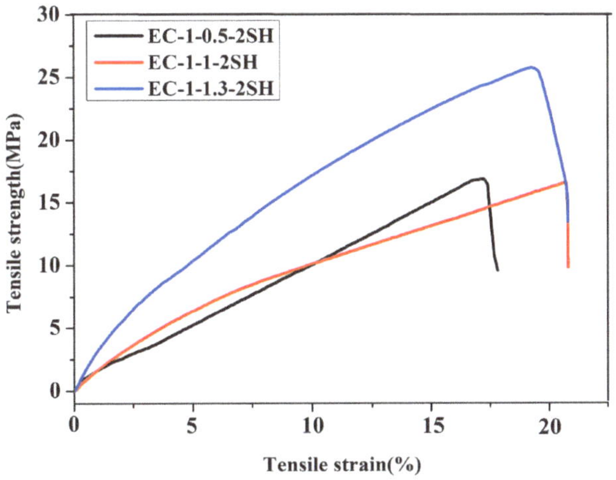

Figure 14. Tensile stress and strain of the EC cross–linked polymers using 3,6–Dioxa–1,8–octanedithiol as the cross–linker.

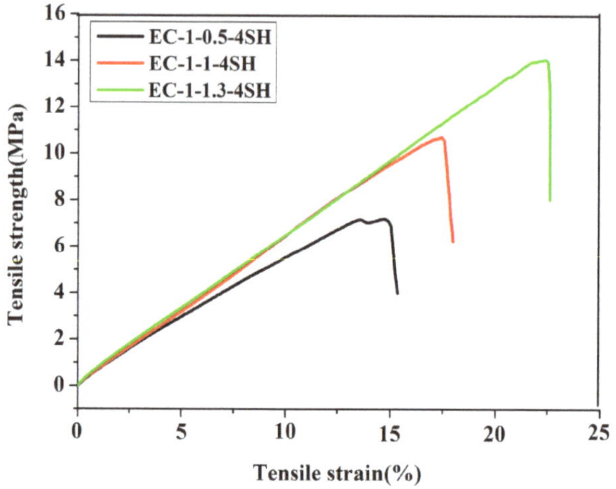

Figure 15. Tensile stress and strain of the EC cross−linked polymers using pentaerythritol tetra(3−mercaptopropionate) as the cross−linker.

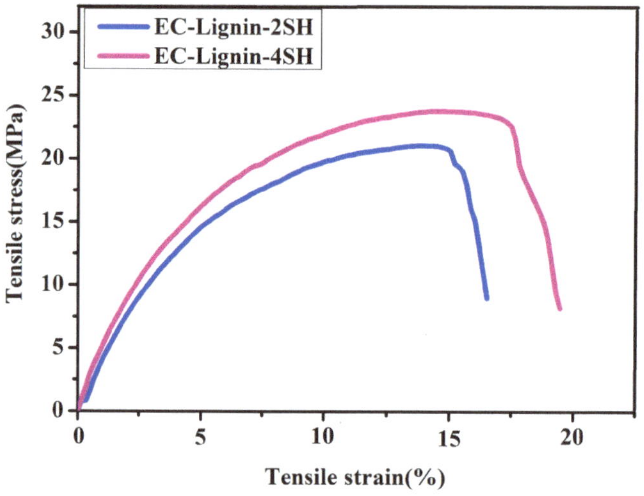

Figure 16. Tensile stress and strain of the EC–lignin cross−linked polymers.

Table 3. Tensile stress, tensile strain and C_{gel} of the cross−linked polymers.

Polymer Films	Tensile Stress (MPa)	Tensile Strain (%)	C_{gel} (%)
EC	146.89 ± 10.24	1.32 ± 0.02	–
EC−1−0.5−2SH	16.41 ± 1.22	17.25 ± 0.19	94.9 ± 0.8
EC−1−1.0−2SH	16.53 ± 1.10	19.41 ± 2.18	96.4 ± 0.4
EC−1−1.3−2SH	23.59 ± 3.01	18.12 ± 0.13	98.0 ± 0.4
EC−1−0.5−4SH	6.86 ± 0.21	14.85 ± 1.11	92.7 ± 0.6
EC−1−1.0−4SH	10.55 ± 0.89	16.92 ± 1.76	94.3 ± 0.7
EC−1−1.3−4SH	13.94 ± 1.00	22.38 ± 3.20	95.2 ± 0.9
EC−Lignin−2SH	13.72 ± 0.02	15.12 ± 1.19	94.8 ± 0.2
EC−Lignin−4SH	15.19 ± 1.03	17.25 ± 2.10	97.2 ± 0.5
Lignin−1−1−2SH	–	–	–
Lignin−1−1−4SH	–	–	–

The C_{gel} of the cross−linked polymers were detected and the results are shown in Table 3. The results show that C_{gel} is positively correlated to the tensile stress for the same type of polymers. For the EC−n−2SH cross−linked polymers, when the tensile stress increased from 16.41 MPa to 23.59 MPa, the C_{gel} increased from 94.9% to 98.0%. The other cross−linked polymers showed the same relationship between the gel content and the tensile strength.

The microstructure of the cross−linked polymers were investigated using a Leica DM750M optical microscope. As seen in Figure 17, the EC cross−linked polymers showed a uniform surface structure without large cracks or micropores, while the EC−lignin polymers showed an uneven surface structure with many large cracks and micropores, which may be due to solvent evaporation.

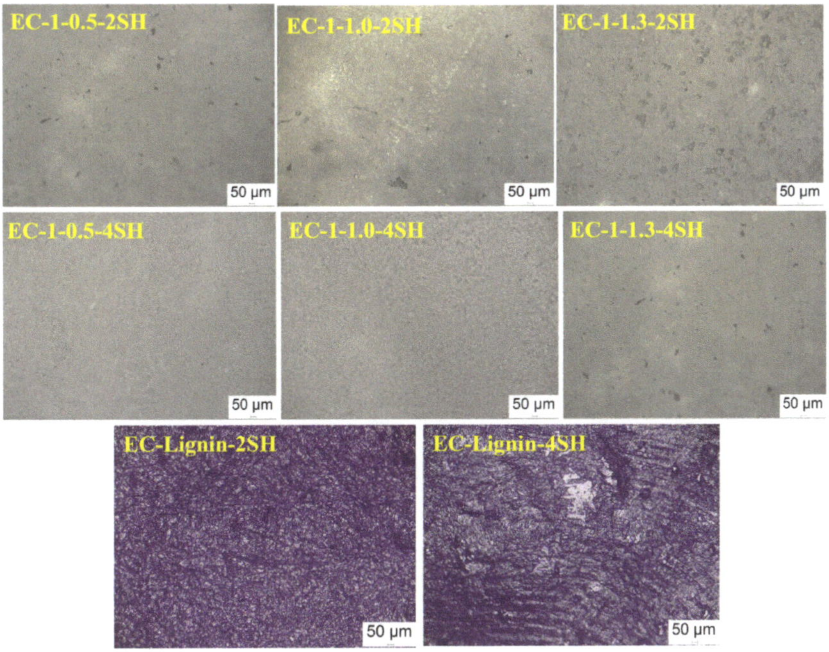

Figure 17. Microstructure of the cross−linked polymers.

Figure 18 shows the thermal stability of the EC and cross−linked polymers. Only one thermal degradation stage occurred at 360–370 °C for EC and all cross−linked polymers. It has been reported that, when cellulose−based polymers are degraded at high temperatures, their polymerization degree is reduced, the chemical composition also changes and the carbonyl groups increase [40,41]. When cellulose−based polymers are fully degraded, carbon monoxide, carbon dioxide, ethylene, water and carbon are produced [42,43]. Table 4 shows the thermal degradation temperature (T_d), the peak value of the thermal degradation temperature (T_P) and the char residue of the cross−linked polymers. The thermal stability of the cross−linked polymers (EC−1−1.0−2SH, EC−1−1.0−4SH and EC−1−1.3−4SH) was higher than that of EC. The T_d, T_P and char residue of the cross−linked polymers increased compared with those of EC. This is because OFE and the cross−linkers, such as 3,6−Dioxa−1,8−octanedithiol and pentaerythritol tetra(3−mercaptopropionate), contain thermostable ester groups, which makes the cross−linked polymers difficult degrade.

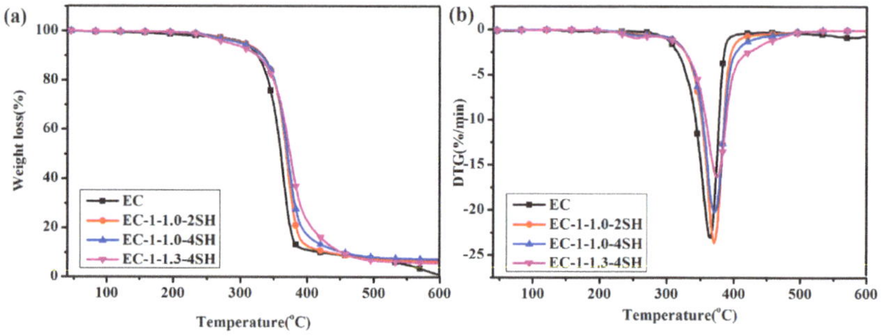

Figure 18. TGA (**a**) and DTG (**b**) curves of the cross−linked polymers.

Table 4. T_d, T_P and char residue of the cross−linked polymers.

Cross−linked Polymers	T_d (°C)	T_P (°C)	Char Residue (%)
EC	339.8	364.1	0.9
EC−1−1.0−2SH	348.2	370.2	6.6
EC−1−1.0−4SH	346.8	371.2	7.2
EC−1−1.3−4SH	344.0	374.1	5.7

In order to further investigate the composition of the thermal degradation products, TGA−FTIR was carried out. Figures 19 and 20 show the 3D and 2D FT−IR, respectively, of EC (a), EC−1−1.0−2SH (b), EC−1−1.0−4SH (c) and EC−1−1.3−4SH (d). The infrared characteristic absorption peak of the gas phase of the thermal degradation products can be clearly observed in Figure 20. The data were collected at the fastest decomposition temperature of 340 °C. The infrared characteristic absorption peaks at 3684, 2979, 2306 and 1747, 1391, 1057 cm^{-1} were attributed to H_2O, aliphatic hydrocarbon segments, CO_2 and degradation products containing ester groups, respectively [44–46]. The existence of ester groups in the cross−linked polymers and degradation products make them more thermally stable.

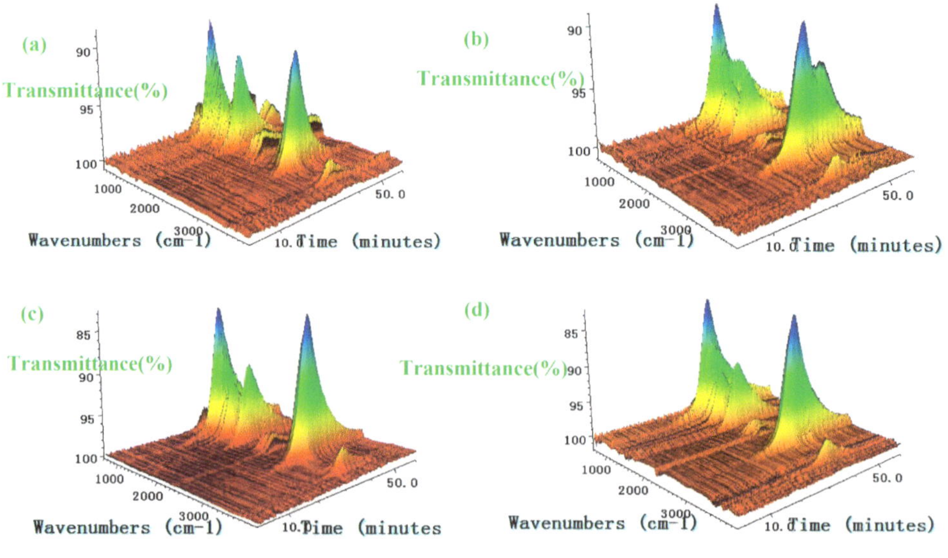

Figure 19. Three−dimensional FT−IR of EC (**a**), EC−1−1.0−2SH (**b**), EC−1−1.0 − 4SH (**c**) and EC−1−1.3−4SH (**d**).

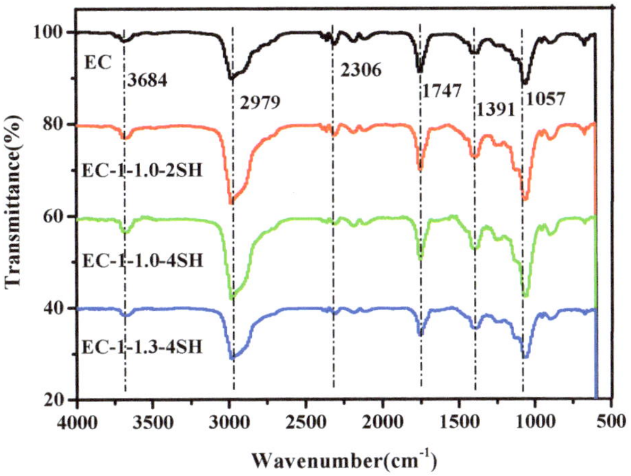

Figure 20. Two−dimensional FT−IR of EC, EC−1−1.0−2SH, EC−1−1.0−4SH and EC−1−1.3−4SH.

4. Conclusions

In this study, ethyl cellulose and lignin were modified via esterification to prepare olefin−functionalized ethyl cellulose and lignin, which were further used to prepare cellulose and lignin cross−link polymers via thiol–ene click chemistry. The results show that the olefin group concentration in the olefin−functionalized ethyl cellulose and lignin reached 2.8096 mmol/g and 3.7000 mmol/g, respectively. The tensile stress at break of the cellulose cross−linked polymers reached 23.59 MPa. The gradual enhancement in the mechanical properties was positively correlated with the olefin group concentration. Compared with EC, the tensile strength of the cross−linked polymers decreased significantly, but the tensile strain increased sharply. The flexible aliphatic hydrocarbon chain from undecylenic acid contributed to plasticization by functioning as an internal plasticizer. The changes in mechanical properties make the cross−linked polymers easier to process, expanding their application range. The EC–lignin polymers had an uneven surface structure and there were many large cracks and micropores due to the rigid benzene ring structure of lignin. The infrared characteristic absorption peaks at 3684, 2979, 2306 and 1747, 1391, 1057 cm^{-1} indicated that H_2O, aliphatic hydrocarbon segments, CO_2 and degradation products containing ester groups were released. The existence of ester groups in the cross−linked polymers and degradation products makes them more thermally stable. The obtained ethyl cellulose and lignin cross−link polymers are of great significance for practical applications and contribute towards the high−value−added utilization of lignin and cellulose.

Author Contributions: Conceptualization, P.J.; Writing—original draft, investigation, R.A.; Writing—review & editing, P.J. and J.W.; Supervision, C.L.; Funding acquisition, P.J. All authors have read and agreed to the published version of the manuscript.

Funding: This research was funded by Forestry Science and Technology Innovation and Extension Project of Jiangsu Province (LYKJ[2021]04).

Institutional Review Board Statement: Not applicable.

Informed Consent Statement: Not applicable.

Data Availability Statement: Not applicable.

Conflicts of Interest: The authors declare no conflict of interest. The funders had no role in the design of the study; in the collection, analyses, or interpretation of data; in the writing of the manuscript; or in the decision to publish the results.

References

1. Qin, M.; Chen, C.; Song, B.; Shen, M.; Cao, W.; Yang, H.; Zeng, G.; Gong, J. A review of biodegradable plastics to biodegradable microplastics: Another ecological threat to soil environments? *J. Clean. Prod.* **2021**, *312*, 127816. [CrossRef]
2. Shaghaleh, H.; Xu, X.; Wang, S. Current progress in production of biopolymeric materials based on cellulose, cellulose nanofibers, and cellulose derivatives. *RSC Adv.* **2018**, *8*, 825–842. [CrossRef] [PubMed]
3. Platnieks, O.; Gaidukovs, S.; Barkane, A.; Sereda, A.; Gaidukova, G.; Grase, L.; Thakur, V.K.; Filipova, I.; Fridrihsone, V.; Skute, M.; et al. Bio−Based Poly(butylene succinate)/Microcrystalline Cellulose/Nanofibrillated Cellulose−Based Sustainable Polymer Composites: Thermo−Mechanical and Biodegradation Studies. *Polymers* **2020**, *12*, 1472. [CrossRef] [PubMed]
4. Calvino, C.; Macke, N.; Kato, R.; Rowan, S.J. Development, processing and applications of bio−sourced cellulose nanocrystal composites. *Prog. Polym. Sci.* **2020**, *103*, 101221. Available online: https://www.sciencedirect.com/science/article/pii/S0079670020300149 (accessed on 10 April 2023). [CrossRef]
5. Li, C.; Wu, J.; Shi, H.; Xia, Z.; Sahoo, J.K.; Yeo, J.; Kaplan, D.L. Fiber-Based Biopolymer Processing as a Route toward Sustainability. *Adv. Mater.* **2022**, *34*, 2105196. [CrossRef]
6. Azimi, B.; Maleki, H.; Gigante, V.; Bagherzadeh, R.; Mezzetta, A.; Milazzo, M.; Guazzelli, L.; Cinelli, P.; Lazzeri, A.; Danti, S. Cellulose−based fiber spinning processes using ionic liquids. *Cellulose* **2022**, *29*, 3079–3129. [CrossRef]
7. Kumari, S.V.G.; Pakshirajan, K.; Pugazhenthi, G. Recent advances and future prospects of cellulose, starch, chitosan, polylactic acid and polyhydroxyalkanoates for sustainable food packaging applications. *Int. J. Biol. Macromol.* **2022**, *221*, 163–182. Available online: https://www.sciencedirect.com/science/article/pii/S0141813022019134 (accessed on 10 April 2023). [CrossRef]
8. Fox, S.C.; Li, B.; Xu, D.; Edgar, K.J. Regioselective Esterification and Etherification of Cellulose: A Review. *Biomacromolecules* **2011**, *12*, 1956–1972. [CrossRef]
9. Wang, Y.; Wang, X.; Xie, Y.; Zhang, K. Functional nanomaterials through esterification of cellulose: A review of chemistry and application. *Cellulose* **2018**, *25*, 3703–3731. [CrossRef]
10. Shokri, Z.; Seidi, F.; Saeb, M.R.; Jin, Y.; Li, C.; Xiao, H. Elucidating the impact of enzymatic modifications on the structure, properties, and applications of cellulose, chitosan, starch and their derivatives: A review. *Mater. Today Chem.* **2022**, *24*, 100780. Available online: https://www.sciencedirect.com/science/article/pii/S246851942200009X (accessed on 10 April 2023). [CrossRef]
11. Svinterikos, E.; Zuburtikudis, I.; Al−Marzouqi, M.H. Electrospun Lignin−Derived Carbon Micro− and Nanofibers: A Review on Precursors, Properties, and Applications. *ACS Sustain. Chem. Eng.* **2020**, *8*, 13868–13893. [CrossRef]
12. Lee, D.-W.; Jin, M.-H.; Park, J.-H.; Lee, Y.-J.; Choi, Y.-C. Flexible Synthetic Strategies for Lignin−Derived Hierarchically Porous Carbon Materials. *ACS Sustain. Chem. Eng.* **2018**, *6*, 10454–10462. [CrossRef]
13. Zhang, B.; Yang, D.; Qiu, X.; Qian, Y.; Yan, M.; Li, Q. Influences of aggregation behavior of lignin on the microstructure and adsorptive properties of lignin−derived porous carbons by potassium compound activation. *J. Ind. Eng. Chem.* **2020**, *82*, 220–227. Available online: https://www.sciencedirect.com/science/article/pii/S1226086X19305532 (accessed on 10 April 2023). [CrossRef]
14. Tian, Y.; Zhou, H. A novel nitrogen−doped porous carbon derived from black liquor for efficient removal of Cr(VI) and tetracycline: Comparison with lignin porous carbon. *J. Clean. Prod.* **2022**, *333*, 130106. Available online: https://www.sciencedirect.com/science/article/pii/S0959652621042724 (accessed on 10 April 2023). [CrossRef]
15. Xi, Y.; Yang, D.; Lou, H.; Gong, Y.; Yi, C.; Lyu, G.; Han, W.; Kong, F.; Qiu, X. Designing the effective microstructure of lignin−based porous carbon substrate to inhibit the capacity decline for SnO_2 anode. *Ind. Crop. Prod.* **2021**, *161*, 113179. Available online: https://www.sciencedirect.com/science/article/pii/S0926669020310967 (accessed on 10 April 2023). [CrossRef]
16. Gong, X.; Meng, Y.; Lu, J.; Tao, Y.; Cheng, Y.; Wang, H. A Review on Lignin-Based Phenolic Resin Adhesive. *Macromol. Chem. Phys.* **2022**, *223*, 2100434. [CrossRef]
17. Chen, Y.; Gong, X.; Yang, G.; Li, Q.; Zhou, N. Preparation and characterization of a nanolignin phenol formaldehyde resin by replacing phenol partially with lignin nanoparticles. *RSC Adv.* **2019**, *9*, 29255–29262. [CrossRef]
18. Huang, C.; Peng, Z.; Li, J.; Li, X.; Jiang, X.; Dong, Y. Unlocking the role of lignin for preparing the lignin−based wood adhesive: A review. *Ind. Crop. Prod.* **2022**, *187*, 115388. Available online: https://www.sciencedirect.com/science/article/pii/S0926669022008718 (accessed on 10 April 2023). [CrossRef]
19. Weng, S.; Li, Z.; Bo, C.; Song, F.; Xu, Y.; Hu, L.; Zhou, Y.; Jia, P. Design lignin doped with nitrogen and phosphorus for flame retardant phenolic foam materials. *React. Funct. Polym.* **2023**, *185*, 105535. Available online: https://www.sciencedirect.com/science/article/pii/S138151482300038X (accessed on 10 April 2023). [CrossRef]
20. Sarika, P.R.; Nancarrow, P.; Khansaheb, A.; Ibrahim, T. Bio−Based Alternatives to Phenol and Formaldehyde for the Production of Resins. *Polymers* **2020**, *12*, 2237. [CrossRef] [PubMed]
21. Asgher, M.; Qamar, S.A.; Bilal, M.; Iqbal, H.M.N. Bio−based active food packaging materials: Sustainable alternative to conventional petrochemical−based packaging materials. *Food Res. Int.* **2020**, *137*, 109625. Available online: https://www.sciencedirect.com/science/article/pii/S0963996920306505 (accessed on 10 April 2023). [CrossRef] [PubMed]
22. Yang, W.; Rallini, M.; Natali, M.; Kenny, J.; Ma, P.; Dong, W.; Torre, L.; Puglia, D. Preparation and properties of adhesives based on phenolic resin containing lignin micro and nanoparticles: A comparative study. *Mater. Des.* **2019**, *161*, 55–63. Available online: https://www.sciencedirect.com/science/article/pii/S0264127518308396 (accessed on 10 April 2023). [CrossRef]
23. Xu, Y.; Guo, L.; Zhang, H.; Zhai, H.; Ren, H. Research status, industrial application demand and prospects of phenolic resin. *RSC Adv.* **2019**, *9*, 28924–28935. [CrossRef]

24. Firdaus, M. Thiol−Ene (Click) Reactions as Efficient Tools for Terpene Modification. *Asian J. Org. Chem.* **2017**, *6*, 1702–1714. [CrossRef]
25. Firdaus, M.; Montero de Espinosa, L.; Meier, M.A.R. Terpene−Based Renewable Monomers and Polymers via Thiol–Ene Additions. *Macromolecules* **2011**, *44*, 7253–7262. [CrossRef]
26. Ahangarpour, M.; Kavianinia, I.; Harris, P.W.R.; Brimble, M.A. Photo−induced radical thiol-ene chemistry: A versatile toolbox for peptide−based drug design. *Chem. Soc. Rev.* **2021**, *50*, 898–944. [CrossRef] [PubMed]
27. Kazybayeva, D.S.; Irmukhametova, G.S.; Khutoryanskiy, V.V. Thiol−Ene "Click Reactions" as a Promising Approach to Polymer Materials. *Polym. Sci. Ser. B* **2022**, *64*, 1–16. [CrossRef]
28. Hoyle, C.E.; Bowman, C.N. Thiol−Ene Click Chemistry. *Angew. Chem. Int. Ed.* **2010**, *49*, 1540–1573. [CrossRef]
29. Jawerth, M.; Johansson, M.; Lundmark, S.; Gioia, C.; Lawoko, M. Renewable Thiol–Ene Thermosets Based on Refined and Selectively Allylated Industrial Lignin. *ACS Sustain. Chem. Eng.* **2017**, *5*, 10918–10925. [CrossRef]
30. Cao, Y.; Liu, Z.; Zheng, B.; Ou, R.; Fan, Q.; Li, L.; Guo, C.; Liu, T.; Wang, Q. Synthesis of lignin−based polyols via thiol−ene chemistry for high−performance polyurethane anticorrosive coating. *Compos. Part B Eng.* **2020**, *200*, 108295. [CrossRef]
31. Zeng, T.; Zhang, P.; Li, X.; Yin, Y.; Chen, K.; Wang, C. Facile fabrication of durable superhydrophobic and oleophobic surface on cellulose substrate via thiol−ene click modification. *Appl. Surf. Sci.* **2019**, *493*, 1004–1012. [CrossRef]
32. Yuan, L.; Zhang, Y.; Wang, Z.; Han, Y.; Tang, C. Plant Oil and Lignin−Derived Elastomers via Thermal Azide–Alkyne Cycloaddition Click Chemistry. *ACS Sustain. Chem. Eng.* **2018**, *7*, 2593–2601. [CrossRef]
33. Yuan, L.; Wang, Z.; Trenor, N.M.; Tang, C. Preparation and Applications of Polymers with Pendant Fatty Chains from Plant Oils. In *Sustainable Polymers from Biomass*; Wiley: Hoboken, NJ, USA, 2017; pp. 181–207. [CrossRef]
34. Yan, K.; Wang, J.; Wang, Z.; Yuan, L. Bio−based monomers for amide−containing sustainable polymers. *Chem. Commun.* **2023**, *59*, 382–400. [CrossRef] [PubMed]
35. Yasa, S.R.; Cheguru, S.; Krishnasamy, S.; Korlipara, P.V.; Rajak, A.K.; Penumarthy, V. Synthesis of 10−undecenoic acid based C22−dimer acid esters and their evaluation as potential lubricant basestocks. *Ind. Crop. Prod.* **2017**, *103*, 141–151. Available online: https://www.sciencedirect.com/science/article/pii/S092666901730225X (accessed on 10 April 2023). [CrossRef]
36. Kontham, V.; Ansari, K.R.; Padmaja, K.V. Tribological Properties of 10−Undecenoic Acid−Derived Schiff Base Lubricant Additives. *Arab. J. Sci. Eng.* **2021**, *46*, 5593–5603. [CrossRef]
37. Lluch, C.; Lligadas, G.; Ronda, J.C.; Galià, M.; Cadiz, V. "Click" Synthesis of Fatty Acid Derivatives as Fast−Degrading Polyanhydride Precursors. *Macromol. Rapid Commun.* **2011**, *32*, 1343–1351. [CrossRef]
38. Çiçek, S.S.; Girreser, U.; Zidorn, C. Quantification of the total amount of black cohosh cycloartanoids by integration of one specific 1 H NMR signal. *J. Pharm. Biomed. Anal.* **2018**, *155*, 109–115. Available online: https://www.sciencedirect.com/science/article/pii/S0731708518303376 (accessed on 10 April 2023). [CrossRef]
39. Crenshaw, M.D.; Tefft, M.E.; Buehler, S.S.; Brinkman, M.C.; Clark, P.I.; Gordon, S.M. Determination of nicotine, glycerol, propylene glycol and water in electronic cigarette fluids using quantitative ^{1} H NMR. *Magn. Reson. Chem.* **2016**, *54*, 901–904. [CrossRef]
40. Puls, J.; Wilson, S.A.; Hölter, D. Degradation of Cellulose Acetate−Based Materials: A Review. *J. Polym. Environ.* **2011**, *19*, 152–165. [CrossRef]
41. Lucena, M.d.C.C.; de Alencar, A.E.V.; Mazzeto, S.E.; Soares, S.d.A. The effect of additives on the thermal degradation of cellulose acetate. *Polym. Degrad. Stab.* **2003**, *80*, 149–155. Available online: https://www.sciencedirect.com/science/article/pii/S0141391002003968 (accessed on 10 April 2023). [CrossRef]
42. Mülhaupt, R. Green Polymer Chemistry and Bio−based Plastics: Dreams and Reality. *Macromol. Chem. Phys.* **2013**, *214*, 159–174. [CrossRef]
43. Williams, P.T.; Onwudili, J. Subcritical and Supercritical Water Gasification of Cellulose, Starch, Glucose, and Biomass Waste. *Energy Fuels* **2006**, *20*, 1259–1265. [CrossRef]
44. Jia, P.; Zhang, M.; Liu, C.; Hu, L.; Feng, G.; Bo, C.; Zhou, Y. Effect of chlorinated phosphate ester based on castor oil on thermal degradation of poly (vinyl chloride) blends and its flame retardant mechanism as secondary plasticizer. *RSC Adv.* **2015**, *5*, 41169–41178. [CrossRef]
45. Jia, P.; Hu, L.; Zhang, M.; Zhou, Y.-H. TG−FTIR and TG−MS analysis applied to study the flame retardancy of PVC–castor oil−based chlorinated phosphate ester blends. *J. Therm. Anal. Calorim.* **2016**, *124*, 1331–1339. [CrossRef]
46. Jia, P.; Zhang, M.; Hu, L.; Liu, C.; Feng, G.; Yang, X.; Bo, C.; Zhou, Y. Development of a vegetable oil based plasticizer for preparing flame retardant poly(vinyl chloride) materials. *RSC Adv.* **2015**, *5*, 76392–76400. [CrossRef]

polymers

MDPI

Article

Influence of Enzymatically Hydrophobized Hemp Protein on Morphology and Mechanical Properties of Bio-Based Polyurethane and Epoxy Foams

Guillem Ferreres [1], Sílvia Pérez-Rafael [1], Angela Gala Morena [1], Tzanko Tzanov [1] and Liudmyla Gryshchuk [2,*]

1 Grup de Biotecnologia Molecular i Industrial, Universitat Politècnica de Catalunya, Edifici Gaia, TR14, Rambla Sant Nebridi, 22, 08222 Terrassa, Spain; guillem.ferreres@upc.edu (G.F.); silvia.perez.rafael@upc.edu (S.P.-R.); angela.gala.morena@upc.edu (A.G.M.); tzanko.tzanov@upc.edu (T.T.)

2 Leibniz-Institut für Verbundwerkstoffe GmbH, Erwin-Schrödinger-Straße 58, 67663 Kaiserslautern, Germany

* Correspondence: liudmyla.gryshchuk@ivw.uni-kl.de

Citation: Ferreres, G.; Pérez-Rafael, S.; Morena, A.G.; Tzanov, T.; Gryshchuk, L. Influence of Enzymatically Hydrophobized Hemp Protein on Morphology and Mechanical Properties of Bio-Based Polyurethane and Epoxy Foams. *Polymers* **2023**, *15*, 3608. https://doi.org/10.3390/polym15173608

Academic Editors: Antonio M. Borrero-López, Concepción Valencia-Barragán, Esperanza Cortés Triviño, Adrián Tenorio-Alfonso and Clara Delgado-Sánchez

Received: 7 August 2023
Revised: 26 August 2023
Accepted: 29 August 2023
Published: 31 August 2023

Abstract: Biomass fillers offer the possibility to modify the mechanical properties of foams, increasing their cost-effectiveness and reducing their carbon footprint. In this study, bio-based PU (soft, open cells for the automotive sector) and epoxy (EP, hard, closed cells for construction applications) composite foams were prepared by adding pristine and laccase-mediated lauryl gallate-hydrophobized hemp protein particles as filler (HP and HHP, respectively). The fillers were able to modify the density, the mechanical properties and the morphology of the PU and EP foams. The addition of HP filler increases the density of PU foams up to 100% and significantly increases the σ values by 40% and Emod values. On the other hand, the inclusion of the HHP as filler in PU foams mostly results in reduced density, by almost 30%, and reduced σ values in comparison with reference and HP-filled foams. Independently from filler concentration and type, the biomass increased the Emod values for all foams relative to the reference. In the case of the EP foams, the tests were only conducted for the foams filled with HHP due to the poor compatibility of HP with the EP matrix. HHP decreased the density, compressive strength and Emod values of the composites. For both foams, the fillers increased the size of the cells, while reducing the amount of open cells of PU foams and the amount of closed cells for EP foams. Finally, both types of foams filled with HHP reduced the moisture uptake by 80 and 45%, respectively, indicating the successful hydrophobization of the composites.

Keywords: polyurethane and epoxy composite foams; hemp protein; laccase-assisted hydrophobization; bio-fillers; mechanical properties

1. Introduction

Polyurethane foams (PUFs) are extensively utilized across various industries due to their versatility, lightweight nature, and exceptional thermal insulation properties. Despite the domination of polyurethane foams on the foam market [1], the development of other foam types, for example, polystyrene (PS), poly(vinyl chloride) (PVC), polyethylene (PE), polypropylene (PP) or poly(methyl methacrylate) [1], phenolic [2–4], and (bio)-epoxy foams [5–9], is spreading continuously. Although each type of foam presents its own advantages, density, thermal stability, and mechanical properties are key parameters for its ultimate application. One possible way to improve the density as well as mechanical properties of PU and epoxy foams is the introduction of filler(s) [10,11]. Moreover, filler can reduce costs while maintaining acceptable performance levels, especially if the filler components are generally cheaper than the base materials [12–15].

A promising filler material that has gained attention for use in PUFs is hemp biomass. Hemp is a fast-growing plant that can be cultivated without excessive water or chemical supplementation, making it cost-effective relative to other crops [16]. It is considered a renewable resource, aligning with sustainability goals and reducing reliance on non-renewable materials [17]. Additionally, the strong and stiff fibers or particles derived from hemp can enhance the foam's tensile strength, flexural strength, and impact resistance, thereby increasing the foam's overall performance [18,19].

However, hemp and other types of biomass fillers do present some challenges. Hemp moisture adsorption capacity can negatively affect the long-term dimensional stability and mechanical properties of foams (including polyurethane), as well as the reaction kinetics and the foam expansion during the production of the material [20–23]. Moreover, biomass fillers may face issues related to degradation, durability, and dispersion within the polyurethane matrix [24–26]. They can also contribute to increased flammability and hinder the flame retardant properties of the foam, requiring additional additives or treatments to comply with fire safety regulations [27].

To address these concerns, hydrophobized hemp biomass fillers offer several improvements. The incorporation of hydrophobic fillers has the potential to enhance flame retardant effectiveness, improve compatibility with the matrix, promote better dispersion, and result in greater consistency of performance. Moreover, the application of this modified biomass could contribute to increasing the mechanical properties of the foam [28–30].

Herein, we modified hemp protein by laccase-catalyzed oxidative grafting of lauryl gallate (LG)—a phenolic compound with an alkaline chain. Grafting was carried out by adapting a previously reported method to produce hydrophobized cellulose and wool [31–33]. In this process, biomass was pre-activated enzymatically using acetosyringone. This mediator facilitates the laccase-assisted oxidation of chemical groups that would otherwise be inaccessible to the enzyme. By implementing this approach, we were able to achieve the LG grafting onto the hemp biomass in a waterborne reaction without the need for hazardous reagents used in other chemical hydrophobization reactions, such as periodate, or harsh conditions, such as combustion methods. The hydrophobized biomass was used as filler for polyurethane and epoxy foams, and its influence on the foam properties was investigated.

2. Materials and Methods

2.1. Materials and Reagents

Polycarbonatediols Cardyon® LC 05 (made using Covestro's CO_2-technology integrating up to 20 percent CO_2 into polyol, OH n = 53.5 KOH/g) and ETERNACOLL UT-200 (OH n = 56 KOH/g) were supplied by Covestro (Leverkusen, Germany) and UBE Corporation Europe (Castellón de la Plana, Spain), respectively. Poly(propylene glycol) 4000, Polyethylene glycol 600, Aspartic acid, Formic acid, Dibutyltin dilaurate, Tween 80, Poly(methylhydrosiloxane), lauryl gallate (LG) and carboxymethylcellulose sodium salt were purchased from Sigma-Aldrich (St. Louis, MO, USA). Exolit® OP 560 (co-reactive flame retardant with an OH value of 450 mg KOH/g) was supplied by Clariant (Muttenz, BL, Switzerland). Ortegol 500 was supplied by Evonik (Essen, Germany). LED-103 (reactive, acid blocked catalyst, OH n = 2405 KOH/g) and Niax silicone L-6164 were supplied by Momentive Performance Materials Inc. (Antwerp, Belgium) Iso 133/6 poly(4,4′-Diphenylmethandiisocyanat) with 32% of NCO groups and Ongronat CO5700—PMDI/PPG-prepolymer with 8.5% NCO content were supplied by BASF (Ludwigshafen am Rhein, LU, Germany) and BorsodChem (Kazincbarcika, Hungary), respectively.

Bio-based epoxy resin SR Greenpoxy 56 and amino-hardener SZ 8525 (from Sicomin Epoxy systems) were purchased from Time Out Composite oHG, Bornheim-Sechtem, Germany. Epoxidized Cardanol Cardolite® NC-513 and CNSL Novolac resin NX-4001 were supplied by Cardolite Corporation. Novozymes (Bagsværd, Denmark) supplied fungal laccase Novozym 51003 from *Myceliophthora thermophile* (EC1.10.3.2). 3′,5′-dimethoxy-4′-hydroxyacetophenone (acetosyringone) was obtained from ACROS Organics (Geel, Belgium). Hemp protein residues obtained from the oil-pressing process of hempseeds were kindly provided by Kroppenstedter Olmühle (Kroppenstedt, Germany). All reagents for foam preparation were used without any additional purification.

2.2. Hydrophobization of Hemp Protein

The hydrophobization of hemp protein powder was accomplished by an enzymatic, laccase-mediated functionalization with LG in a bioreactor Labfors 5 (Infors HT, Bottmingen, Switzerland), following a previously described protocol, with some modifications [23]. First, 7.5 mM of acetosyringone was dissolved in 2 L of 50 mM sodium acetate buffer at pH 5.5. Then, 10 mg/mL of biomass was added to the mixture. Upon complete dissolution of the reagents, the hemp protein was pre-activated using laccase (13 U/mL) for 1 h at 50 °C. Subsequently, LG solution containing 40 vol.% ethanol was introduced into the mixture to initiate the grafting process on the biomass to a final concentration of 6 mM of LG and 20 vol.% of ethanol. After 2 h of reaction, the modified hemp powder was separated by centrifugation at 10,000× g for 30 min in order to eliminate the unreacted compounds. The resulting pellet was frozen at −80 °C and subsequently freeze-dried to obtain the final functionalized hemp protein product.

2.3. Polyurethane Foam Preparation

Bio-based PU foams were prepared using a one-step method. At first, a solution consisting of a blend of polyols Cardyon® LC 05, Eternacoll UT-200, Poly(propylene glycol) 4000, and Polyethylene glycol 600 was prepared. To the obtained solution, aspartic and formic acids were added as blocking agents; LED-103 and Dibutyltin dilaurate were added as blowing and gelling catalysts, respectively. Niax silicone L-6164 and Ortegol 500 were added as cell-openers, and Tween 80 used as a bio-based co-surfactant. A water solution of 2.5 wt% carboxymethylcellulose was used as chemical blowing agent and bio-based co-surfactant. Exolit® OP 560 and Poly(methylhydrosiloxane) were added as co-reactive flame retardant and blowing agent, respectively. The resulting mixture (denoted as Component A) was mixed for 20 min at 2000 rpm for homogenization.

In the case of preparation of filled foam, the filler was added in appropriate concentration to Component A before homogenization. More specifically, the obtained fillers were added at the following concentrations: 0.25; 0.5; 1, 1.5; 2, 2.5 and 3 wt%. For foams with filler amount ≥3.5 wt%, post-reaction shrinking of more than 4% was observed. That is why these foams were not considered for further testing. Moreover, in general, for HHP-filled foams, we observed lower shrinking in comparison with HP-filled foams.

Next, an appropriate amount ([NCO]/[OH] = 1.05) of Component B, a blend of Iso 133/6 and Ongronat CO5700, was added to Component A, and their combination was stirred with mechanical stirring for 20 s at 2000 rpm. Immediately afterwards, the resultant mixture was transferred into an open cylindrical mold, allowing free rising at room temperature. For the sake of brevity, the produced composite foams were named as PU_Name_y, where "Name" is the abbreviation of corresponding filler and "y" refers to wt% of the filler added to the PU matrix. For example, the "HP" in the sample name PU_HP_1 referred to hemp protein as filler, and the "1" indicated the wt% of HP added in the PU matrix, while PU_HHP_y was used for foams with hydrophobized hemp protein, respectively. Unfilled PU foam named PU_Ref was used as reference. Table 1 reports the amounts of reagents used in PU formulation for 100 g of total foam.

2.4. Epoxy Foam Preparation

Bio-based epoxy foams were prepared using a one-step method. At first, an appropriate amount of SR GreenPoxy 56 as base resin, CNSL Novolac resin NX-4001 as co-resin and Cardolite® NC-513 as co-reactive diluent were thoroughly mixed together at 50–55 °C for 10 min with a propeller mixer at 2000 rpm. After that, the appropriate amount of hardener SZ 8525 was added to the resin blend and mixed at 2000 rpm for 2 min. In the last step, Poly(methylhydrosiloxane) as blowing agent was added, and all components were mixed for 1 min at 2000 rpm. Afterwards, the resultant reactive mixture was immediately transferred into an open Al mold for free-rise, kept at RT for 1 h, and then post-cured at 70 °C for 4 h.

Table 1. Amounts of reagents used in PU formulations for 100 g of total foam.

Component	Amount, g
Cardyon® LC 05	16.37
Poly(propylene glycol)4000	4.01
Eternacoll UT-200	3.04
Polyethylene glycol 600	1.96
Water + CMC_2.5%	2.50
Aspartic acid	1.02
Exolit® OP 560	5.73
Dibutyltin dilaurate	1.60
Formic acid	1.39
LED-103	0.05
Tween 80	1.02
Niax silicone L-6164	1.02
Ortegol 500	1.23
Poly(methylhydrosiloxane)	0.82
iso 133/6	19.24
Ongronat CO5700	39.00
	100.00

In the case of filled foam preparation, the appropriate amount of filler (0.25; 0.5; 1, 1.5; 2, 2.5 and 3 wt%) was added in resin blend before homogenization. The produced composite foams were named as Epoxy_Name_y, where "Name" is the abbreviation of the corresponding filler, and "y" refers to the wt% of filler added to the epoxy matrix. For example, in a sample named Epoxy_HHP_1, "HHP" refers to the filler, and "1" indicates the wt% of HHP added in the epoxy matrix. Unfilled epoxy foam named as Epoxy_Ref was used as reference. For composite epoxy foams, only HHP was used as filler because of the significantly worse dispersibility of HP in resin. Table 2 reports the amounts of reagents used in epoxy formulations for 100 g of total foam.

Table 2. Amounts of reagents used in EP formulations for 100 g of total foam.

Component	Amount, g
SR GreenPoxy 56	72.46
Cardolite® NX-4001	3.62
Cardolite® NC-513	3.62
SZ 8525	18.12
Poly(methylhydrosiloxane)	2.18
	100.00

2.5. Characterization of the Hydrophobized Hemp Protein

To evaluate the enzymatic grafting modification, FTIR analysis of the biomass was recorded over the 4000–650 cm^{-1} range, performing 64 scans with a PerkinElmer Spectrum 100 (PerkinElmer, Waltham, MA, USA). The baseline was corrected, and the spectra were normalized using the PerkinElmer Spectrum software v.1.0, with the maximum absorbance intensity value serving as the reference. The hydrophobicity of the biomass was determined using the sessile drop method. A layer of hemp powder was applied to a glass support, and then a 2 μL water droplet was casted onto the biomass. Subsequently, the contact angle of the drop was measured using a Drop Shape Analyzer (Krüss, Hamburg, Germany).

2.6. Polyurethane and Epoxy Foam Characterization

2.6.1. Characterization of Density and Mechanical Properties

The density of PU and epoxy foams was determined at 23 °C with 50% relative humidity [34]. The density value reported is the average value of 10 specimens with size 30 mm × 30 mm × 30 mm (length × width × thickness).

Mechanical compressive strength of PU/epoxy foams was determined according to [35] and carried out through a Zwick 1445 Retroline machine (ZwickRoell GmbH and Co. KG, Berlin, Germany). The following parameters were used for measurement: initial load 0.5 N, E-modulus velocity 10 mm/min, testing velocity 10%/min, maximal deformation 70%. Compressive strength at 10% and 40% strain and according to values of compressive modulus were performed. Then, 10 specimens were tested, and an average value was taken along with the standard deviation.

2.6.2. Determination of Moisture Uptake

Hydrophobicity of filled PU and epoxy foams was determined by moisture uptake test in humidity camera at 23 °C and relative humidity of 90%. Foam samples before testing were dried at 40 °C to constant weight and then placed in the humidity camera. At intervals of 24 h, the samples were weighed to control weight increases due to water absorption. The experiment was considered fully completed if the last three weight measurements showed a weight with a maximal difference of 0.00001 g.

3. Results and Discussion

3.1. Characterization of the Hydrophobized Hemp Protein

The laccase-assisted method to graft lauryl gallate onto hemp protein yielded hydrophobized biomass. This modification was evaluated through FTIR and contact angle. The spectrum of the hydrophobized biomass showed an increase in the signals at ~2919 and ~2857 cm^{-1} compared to the unmodified sample, corresponding to the C-H stretching absorption of the lauryl gallate. Furthermore, the signal associated with the C=C-C of the aromatic ring at ~1619 cm^{-1} also increased in the modified hemp spectrum. Additionally, the LG moieties caused the appearance of signals in the regions of ~1200 and ~700 cm^{-1} due to C-O and C-H bonds, respectively (Figure 1A) [33]. To assess the hydrophobicity of the modified material, the contact angle was measured and compared with the pristine biomass. The contact angle of the hemp after treatment increased from 96.4° to 124.2° (Figure 1B), thereby confirming the successful hydrophobization of the material. LG is an ester of gallic acid and dodecanol; the reaction with laccase couples the phenolic groups with the ones present in the hemp protein, exposing the long hydrocarbon chain. The non-polar nature of these chains repels the water, conferring to the biomass hydrophobic properties.

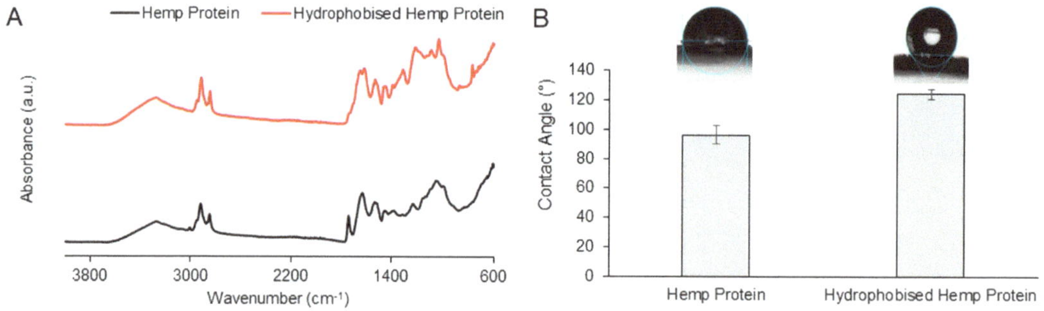

Figure 1. FTIR spectra of hemp protein (black) and hydrophobized hemp protein (red) (**A**) and contact angle of the biomass before and after the hydrophobization (**B**).

In order to test the influence of hydrophobic modification on the morphology of hemp protein particles, the biomass was analyzed by SEM. As is visible from the SEM images, the hydrophobization had a significant influence on the surface structure of the hemp protein particles (Figure 2). The HP SEM images displayed a smooth surface, while the HHP presented attached on the surface nano-structures due to the modification with LG. Similar surface morphology changes have been previously described in lignocellulosic biomass modified with this compound [36].

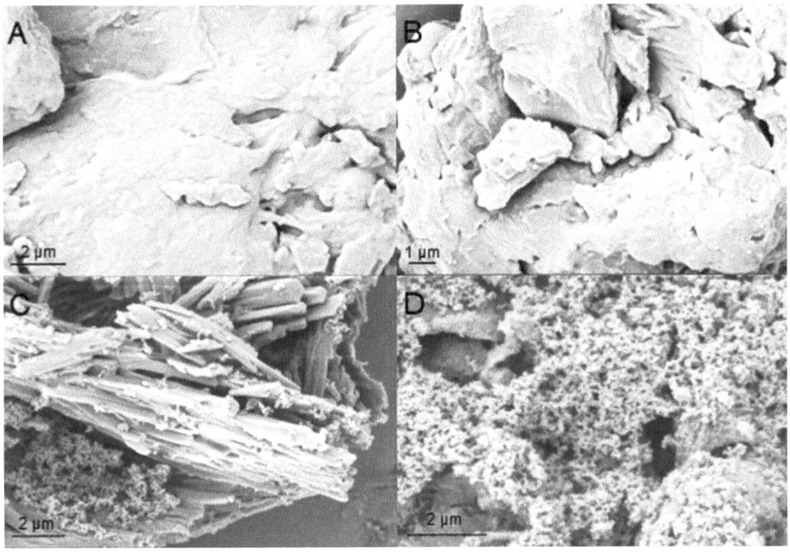

Figure 2. SEM images of pristine (**A**,**B**) and hydrophobized (**C**,**D**) hemp protein under 20,000× and 30,000× magnification, respectively.

3.2. Polyurethane Foam Characterization

In order to test the influence of hemp protein hydrophobization on the polyurethane foams' mechanical properties, determination of the filled foam density and the assessment of compressive strength values were carried out. Before foam preparation, fillers were dispersed in Component A (detailed description in Section 2.3). For foams filled with hemp protein, the density increased relative to the reference foam. Modified hemp protein-filled PU foams presented lower density than the reference foams and the HP-filled ones (Figure 3). It must be noted that hydrophobized filler demonstrated significantly better dispersibility in polyol blend.

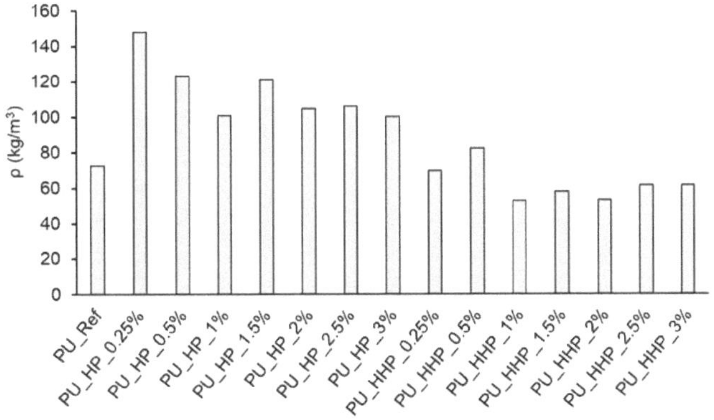

Figure 3. Density of bio-based polyurethane foams filled with pristine (HP) and hydrophobized hemp protein (HHP) (max. standard deviation 4.16%).

Testing of mechanical properties demonstrated a significant increase in σ40% values for HP-filled PU foams; however, the σ10% values did not differ significantly. Only the foams with 3 wt% of filler displayed an increase in both σ10% and σ40% values in comparison with the reference in ca. 100%. PU foams filled with hydrophobized hemp protein exhibited the highest σ10% and σ40% values at 1 wt% filler content, surpassing the values of the reference foams. In general, addition of HHP as filler in PU foams mostly resulted in decreased density and compressive strength in comparison with reference and the HP-filled foams (Figure 4A). Values of Emod increased for foams with both filler types independent of the filler concentration. HP-filled foams with 3 wt% of filler presented the highest value (increasing 6.5-fold in comparison with the reference). The foams filled with HHP, with the exception of the foam with a filler content of 3 wt%, presented higher Emod values than those of the HP-filled ones, which indicates better resistance to deformation by external forces (Figure 4B).

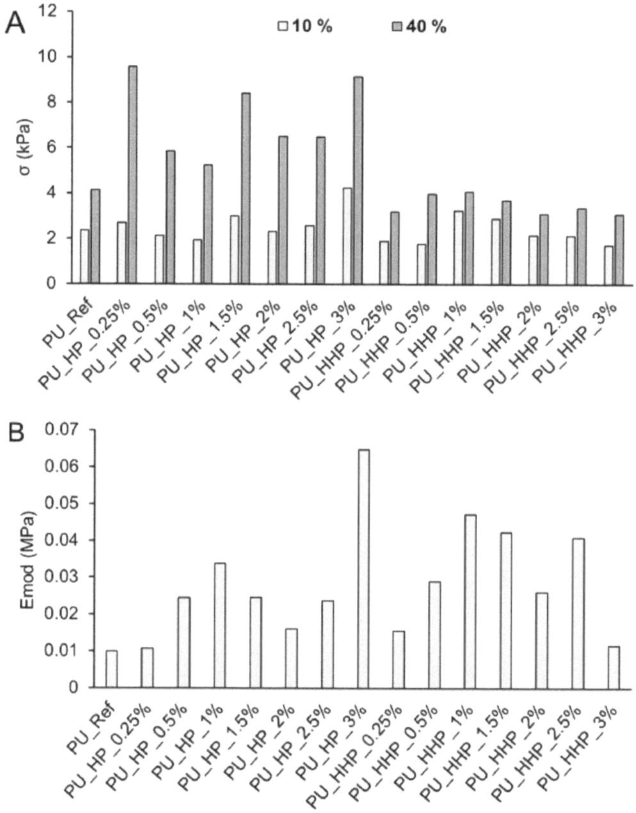

Figure 4. Compressive strength values at 10% and 40% deformation (**A**) and Emod values (**B**) for reference polyurethane foam and foam filled with pristine and hydrophobized hemp protein (max. standard deviation 6.85% and 9.01% for σ and Emod values, respectively).

Such significant differences in the mechanical properties of foams filled with hydrophobized and non-hydrophobized hemp protein can be explained by the different morphology of the foams. The higher hydrophilicity of the pristine fillers reduced their dispersibility in the polyurethane matrix, producing bigger agglomerates. Moreover, at higher concentrations of the pristine HP (for example, 3 wt%), the agglomerates were embedded in the matrix (Figure 5A, areas marked with red arrows). For the foam filled with the same concentration of HHP, there was an obviously better dispersion of the filler in the polyurethane matrix, presenting agglomerates of much smaller size (Figure 5B). At the same time, bigger cells were formed in the HP-filled foams (Figure 5C,E) relative to the HHP-filled ones (Figure 5D,F). A similar trend in open-cell flexible PU foams—an increase in cell size with increasing hydrophilicity of the filler—was reported by Sung et al. [30]. This effect can be explained by improved compatibility of the hydrophobic or, as in our case, hydrophobized biomass with the matrix, which leads to an increase in interfacial adhesion. Furthermore, the addition of both fillers in the PU foams increased the foams' cell size in comparison with the non-filled foams and decreased the amount of open cells (Figure 5G,H). Because the morphology of the foams has a direct influence on the mechanical properties, analyzing the data from the mechanical tests and SEM studies, it can be concluded that the introduction of fillers increases the cell size and σ and Emod values.

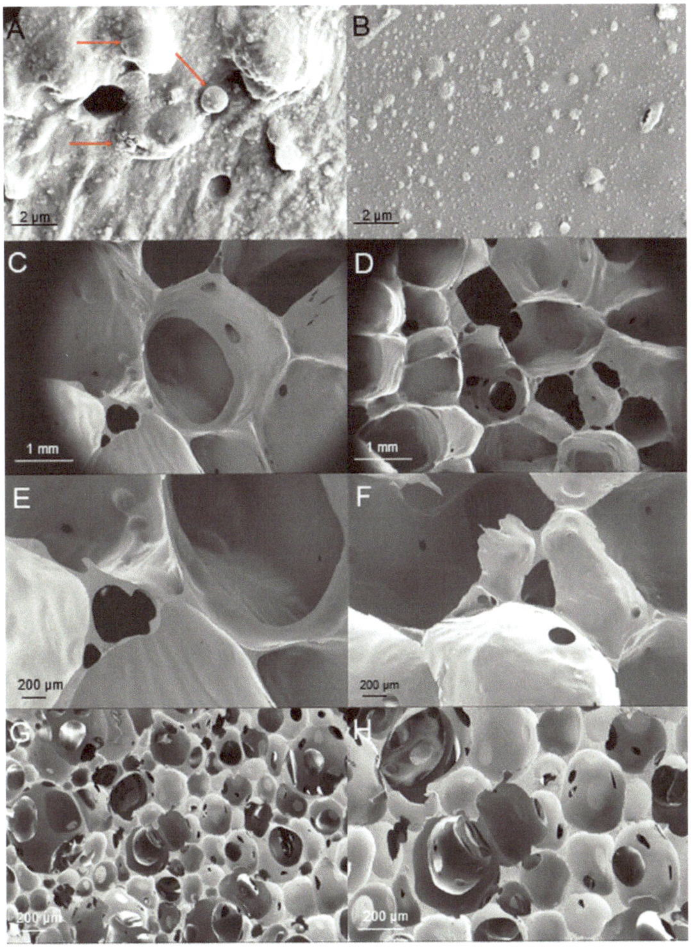

Figure 5. SEM images of polyurethane foams filled with 3 wt% of pristine (**A**,**C**,**E**) and 3 wt% of hydrophobized (**B**,**D**,**F**) hemp protein under 50×, 100× and 20,000× magnification, respectively. (**G**,**H**) SEM images of the reference polyurethane foam (100× and 200× magnification).

However, the presence of agglomerates may cause a reduction in these properties at higher reinforcement content. A similar tendency has been reported for soft PU composite foams filled with SiO_2 [37].

3.3. Epoxy Foam Characterization

In the case of epoxy foams, only HHP fillers were used in the formulation due to the poor dispersibility of HP in the resin. The addition of the HHP fillers into the epoxy mixture (hard foams with closed cells) reduced the density of the foams in comparison with the reference. The tendency is similar to that of polyurethane foams filled with HHP, as the foam filled with 1 wt% HHP was the one with the lowest density (Figure 6).

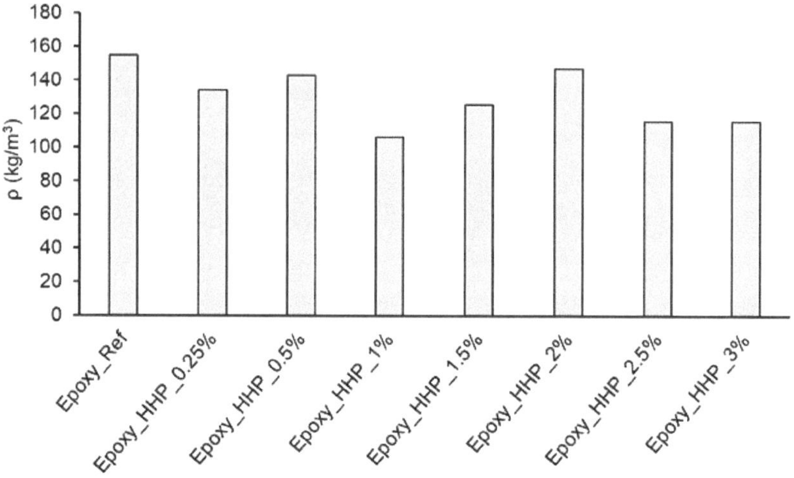

Figure 6. Density of reference epoxy foam and foam filled with hydrophobized hemp protein (max. standard deviation 4.98%).

The compressive strength test showed that the inclusion of HHP in epoxy foam decreased both σ10% and σ40% values with increasing filler amount (Figure 7A). The same tendency was observed for Emod values (Figure 7B). The epoxy foam filled with 1 wt% of HHP presented the lowest density, the lowest values of compressive strength at 10% and 40% deformation as well as Emod.

The SEM analysis demonstrated that the fillers increased the size of the cells and decreased the number of closed cells (Figure 7C,D). This tendency has been previously described; the density and mechanical properties of rigid polyurethane foams filled with precipitated silica were decreased with increased filler loading due to cell damage [38].

3.4. Moisture Uptake of the Foams

Finally, the moisture uptake of both types of foams was measured. The addition of pristine hemp protein to polyurethane foam increased the hydrophobicity insignificantly. At the maximal tested amount (3 wt%) of the HP-filler, moisture uptake decreased by 18% in comparison with reference. At the same time, the addition of the hydrophobized hemp protein resulted in decreasing the moisture uptake by 80% in comparison with the reference. The same tendency was observed for the epoxy foams filled with the HHP. Although the non-filled epoxy foam was strongly hydrophobic, increasing the hydrophobized filler amount resulted in decreasing the moisture uptake by 45%, showing the feasibility of the hydrophobization of hemp protein (Table 3).

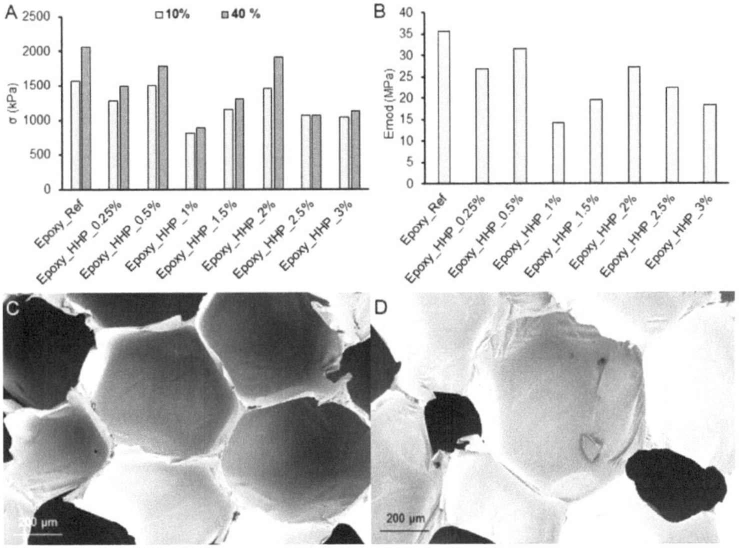

Figure 7. Compressive strength values at 10% and 40% deformation (**A**) and Emod values (**B**) for reference and epoxy foam filled with hydrophobized hemp protein (max. standard deviation 6.85% and 7.54% for σ and Emod values, respectively); SEM images of the reference (**C**) and filled (1 wt% of HHP) epoxy foam (**D**) under 200× magnification.

Table 3. Moisture uptake for reference and filled PU and epoxy foams.

Sample	Moisture Uptake,%	Sample	Moisture Uptake,%	Sample	Moisture Uptake,%
PU_Ref	5.06 ± 0.21			Epoxy_Ref	1.23 ± 0.27
PU_HP_0.25%	4.99 ± 0.19	PU_HHP_0.25%	4.04 ± 0.50	Epoxy_HHP_0.25%	1.10 ± 0.58
PU_HP_0.5%	4.97 ± 0.19	PU_HHP_0.5%	3.20 ± 0.44	Epoxy_HHP_0.5%	0.95 ± 0.19
PU_HP_1%	4.91 ± 0.21	PU_HHP_1%	2.81 ± 0.34	Epoxy_HHP_1%	0.90 ± 0.42
PU_HP_1.5%	4.90 ± 0.14	PU_HHP_1.5%	2.12 ± 0.31	Epoxy_HHP_1.5%	0.85 ± 0.30
PU_HP_2%	4.84 ± 0.23	PU_HHP_2%	1.86 ± 0.60	Epoxy_HHP_2%	0.82 ± 0.50
PU_HP_2.5%	4.23 ± 0.16	PU_HHP_2.5%	1.28 ± 0.38	Epoxy_HHP_2.5%	0.79 ± 0.23
PU_HP_3%	4.18 ± 0.30	PU_HHP_3%	1.04 ± 0.29	Epoxy_HHP_3%	0.68 ± 0.14

4. Conclusions

Hydrophobized hemp protein was successfully obtained through a laccase-mediated modification process. To the best of our knowledge, HHP was used for the first time as a filler for PU and EP foams. The introduction of HHP into the formulation of either PU or EP foams resulted in decreasing the foams' density, which is an advantage for the application of such materials in lightweight construction and thermal insulation areas. For the PU foams, the inclusion of HPP in their formulation increased the Emod, potentially reducing the deformation of the composite. In the case of EP foams, HPP reduced the compressive strength and the Emod. Incorporation of the HPP as filler led to a decrease in the number of open cells in PU foams. As a result, an increase in the Emod of PU foams was observed. For EP foams, the introduction of HHP filler reduced the number of closed cells, which led to a decrease in Emod; from another point of view, such foams could be suitable for better sound absorption. Furthermore, both foams presented reduced moisture uptake when filled with modified biomass, demonstrating the effective hydrophobization of the composites. Finally, differences in the influence of the hydrophobized hemp protein on the morphology and, as result, mechanical properties of open- and closed-cell foams pose new questions and challenges. Such different tendencies with increasing (for PU foams) and decreasing (for EP foams) amounts of closed cells and increasing cell size with the addition of filler, as well as significant changes in density and mechanical properties, especially Emod values, requires further research, which is already planned by the authors: (i) synthesis and investigation of nanoparticles of HHP; (ii) investigation of dispersing methods for selection of the method most suitable for agglomerate-free nano-composite foam preparation; (iii) investigation of the influence of bio-based nano-fillers on the density and mechanical properties of composite foams.

Author Contributions: Conceptualization, L.G. and T.T.; funding acquisition, L.G. and T.T; formal analysis, L.G. and S.P.-R.; foam preparation, investigation of mechanical properties, moisture uptake tests, SEM investigations, L.G.; filler preparation, S.P.-R. and A.G.M., IR spectroscopy and contact angle determination, A.G.M., G.F. and S.P.-R.; writing—original draft preparation, L.G., S.P.-R. and G.F.; writing—review and editing, L.G. and T.T. All authors have read and agreed to the published version of the manuscript.

Funding: This research was funded by the EU (Horizon 2020 project "An Open Innovation Test Bed for Nano-Enabled Bio-Based PUR Foams and Composites" (BIOMAT), Grant agreement No.953270, and Horizon Europe project "New Bio-Based and Sustainable Raw Materials Enabling Circular Values Chains of High Performance Lightweight Biocomposites" (r-LIGHTBIOCOM), Grant agreement No. 101091691).

Institutional Review Board Statement: Not applicable.

Data Availability Statement: The data presented in this study are available on request from the corresponding author.

Conflicts of Interest: The authors declare no conflict of interest.

References

1. Coste, G.; Negrell, C.; Caillol, S. From gas release to foam synthesis, the second breath of blowing agents. *Eur. Polym. J.* **2020**, *140*, 110029. [CrossRef]
2. Kim, B.G.; Gil Lee, D. Development of microwave foaming method for phenolic insulation foams. *J. Mater. Process. Technol.* **2008**, *201*, 716–719. [CrossRef]
3. Basso, M.; Giovando, S.; Pizzi, A.; Celzard, A.; Fierro, V. Tannin/furanic foams without blowing agents and formaldehyde. *Ind. Crop. Prod.* **2013**, *49*, 17–22. [CrossRef]
4. Celzard, A.; Zhao, W.; Pizzi, A.; Fierro, V. Mechanical properties of tannin-based rigid foams undergoing compression. *Mater. Sci. Eng. A* **2010**, *527*, 4438–4446. [CrossRef]
5. Altuna, F.I.; Ruseckaite, R.A.; Stefani, P.M. Biobased Thermosetting Epoxy Foams: Mechanical and Thermal Characterization. *ACS Sustain. Chem. Eng.* **2015**, *3*, 1406–1411. [CrossRef]
6. Ito, A.; Semba, T.; Taki, K.; Ohshima, M. Effect of the molecular weight between crosslinks of thermally cured epoxy resins on the CO_2-bubble nucleation in a batch physical foaming process. *J. Appl. Polym. Sci.* **2014**, *131*, 40407. [CrossRef]
7. Omonov, T.S.; Curtis, J.M. Biobased epoxy resin from canola oil. *J. Appl. Polym. Sci.* **2014**, *131*, 40142. [CrossRef]
8. Chen, K.; Tian, C.; Lu, A.; Zhou, Q.; Jia, X.; Wang, J. Effect of SiO_2 on rheology, morphology, thermal, and mechanical properties of high thermal stable epoxy foam. *J. Appl. Polym. Sci.* **2013**, *131*, 40068. [CrossRef]
9. Jalalian, M.; Jiang, Q.; Bismarck, A. Air Templated Macroporous Epoxy Foams with Silica Particles as Property-Defining Additive. *ACS Appl. Polym. Mater.* **2019**, *1*, 335–343. [CrossRef]
10. Choe, H.; Lee, J.H.; Kim, J.H. Polyurethane composite foams including CaCO3 fillers for enhanced sound absorption and compression properties. *Compos. Sci. Technol.* **2020**, *194*, 108153. [CrossRef]
11. Wu, G.; Xie, P.; Yang, H.; Dang, K.; Xu, Y.; Sain, M.; Turng, L.-S.; Yang, W. A review of thermoplastic polymer foams for functional applications. *J. Mater. Sci.* **2021**, *56*, 11579–11604. [CrossRef]
12. Zhang, L.; Roy, S.; Chen, Y.; Chua, E.K.; See, K.Y.; Hu, X.; Liu, M. Mussel-Inspired Polydopamine Coated Hollow Carbon Microspheres, a Novel Versatile Filler for Fabrication of High Performance Syntactic Foams. *ACS Appl. Mater. Interfaces* **2014**, *6*, 18644–18652. [CrossRef]
13. Nofar, M. Effects of nano-/micro-sized additives and the corresponding induced crystallinity on the extrusion foaming behavior of PLA using supercritical CO_2. *Mater. Des.* **2016**, *101*, 24–34. [CrossRef]
14. Chandni, T.; Anand, K. Utilization of recycled waste as filler in foam concrete. *J. Build. Eng.* **2018**, *19*, 154–160. [CrossRef]

15. Żukowska, W.; Kosmela, P.; Wojtasz, P.; Szczepański, M.; Piasecki, A.; Barczewski, R.; Barczewski, M.; Hejna, A. Comprehensive Enhancement of Prepolymer-Based Flexible Polyurethane Foams' Performance by Introduction of Cost-Effective Waste-Based Ground Tire Rubber Particles. *Materials* **2022**, *15*, 5728. [CrossRef]
16. Schumacher, A.G.D.; Pequito, S.; Pazour, J. Industrial hemp fiber: A sustainable and economical alternative to cotton. *J. Clean. Prod.* **2020**, *268*, 122180. [CrossRef]
17. Ahmed, A.T.M.F.; Islam, Z.; Mahmud, S.; Sarker, E.; Islam, R. Hemp as a potential raw material toward a sustainable world: A review. *Heliyon* **2022**, *8*, e08753. [CrossRef]
18. Członka, S.; Strąkowska, A.; KAIRYTĖ, A. The Impact of Hemp Shives Impregnated with Selected Plant Oils on Mechanical, Thermal, and Insulating Properties of Polyurethane Composite Foams. *Materials* **2020**, *13*, 4709. [CrossRef]
19. Osabohien, E.; Egboh, S.H.O. Utilization of bowstring hemp fiber as a filler in natural rubber compounds. *J. Appl. Polym. Sci.* **2008**, *107*, 210–214. [CrossRef]
20. Yeh, S.-K.; Hsieh, C.-C.; Chang, H.-C.; Yen, C.C.; Chang, Y.-C. Synergistic effect of coupling agents and fiber treatments on mechanical properties and moisture absorption of polypropylene–rice husk composites and their foam. *Compos. Part A Appl. Sci. Manuf.* **2015**, *68*, 313–322. [CrossRef]
21. Marcovich, N.E.; Aranguren, M.I.; Reboredo, M.M. Dependence of the mechanical properties of woodflour-polymer composites on the moisture content. *J. Appl. Polym. Sci.* **1998**, *68*, 2069–2076. [CrossRef]
22. Amran, U.A.; Zakaria, S.; Chia, C.H.; Roslan, R.; Jaafar, S.N.S.; Salleh, K.M. Polyols and rigid polyurethane foams derived from liquefied lignocellulosic and cellulosic biomass. *Cellulose* **2019**, *26*, 3231–3246. [CrossRef]
23. Spada, J.C.; Jasper, A.; Tessaro, I.C. Biodegradable Cassava Starch Based Foams Using Rice Husk Waste as Macro Filler. *Waste Biomass-Valorization* **2020**, *11*, 4315–4325. [CrossRef]
24. Zhang, J.; Hori, N.; Takemura, A. Effect of natural biomass fillers on the stability, degradability, and elasticity of crop straws liquefied polyols-based polyurethane foams. *J. Appl. Polym. Sci.* **2023**, *140*, e53324. [CrossRef]
25. Jonjaroen, V.; Ummartyotin, S.; Chittapun, S. Algal cellulose as a reinforcement in rigid polyurethane foam. *Algal Res.* **2020**, *51*, 102057. [CrossRef]
26. Li, H.; Liang, Y.; Li, P.; He, C. Conversion of biomass lignin to high-value polyurethane: A review. *J. Bioresour. Bioprod.* **2020**, *5*, 163–179. [CrossRef]
27. Branda, F.; Malucelli, G.; Durante, M.; Piccolo, A.; Mazzei, P.; Costantini, A.; Silvestri, B.; Pennetta, M.; Bifulco, A. Silica Treatments: A Fire Retardant Strategy for Hemp Fabric/Epoxy Composites. *Polymers* **2016**, *8*, 313. [CrossRef]
28. Lei, Z.; Zheng, P.; Niu, L.; Yang, Y.; Shen, J.; Zhang, W.; Wang, C. Ultralight, robustly compressible and super-hydrophobic biomass-decorated carbonaceous melamine sponge for oil/water separation with high oil retention. *Appl. Surf. Sci.* **2019**, *489*, 922–929. [CrossRef]
29. Hwang, U.; Lee, B.; Oh, B.; Shin, H.S.; Lee, S.S.; Kang, S.G.; Kim, D.; Park, J.; Shin, S.; Suhr, J.; et al. Hydrophobic lignin/polyurethane composite foam: An eco-friendly and easily reusable oil sorbent. *Eur. Polym. J.* **2022**, *165*, 110971. [CrossRef]
30. Sung, G.; Kim, J.H. Influence of filler surface characteristics on morphological, physical, acoustic properties of polyurethane composite foams filled with inorganic fillers. *Compos. Sci. Technol.* **2017**, *146*, 147–154. [CrossRef]
31. Cusola, O.; Valls, C.; Vidal, T.; Tzanov, T.; Roncero, M.B. Electrochemical Insights on the Hydrophobicity of Cellulose Substrates Imparted by Enzymatically Oxidized Gallates with Increasing Alkyl Chain Length. *ACS Appl. Mater. Interfaces* **2015**, *7*, 13834–13841. [CrossRef] [PubMed]
32. Hossain, K.G.; González, M.D.; Monmany, J.M.D.; Tzanov, T. Effects of alkyl chain lengths of gallates upon enzymatic wool functionalisation. *J. Mol. Catal. B Enzym.* **2010**, *67*, 231–235. [CrossRef]
33. Garcia-Ubasart, J.; Colom, J.F.; Vila, C.; Hernández, N.G.; Roncero, M.B.; Vidal, T. A new procedure for the hydrophobization of cellulose fibre using laccase and a hydrophobic phenolic compound. *Bioresour. Technol.* **2012**, *112*, 341–344. [CrossRef]
34. *ASTM D1622-03*; Standard Test Method for Apparent Density of Rigid Cellular Plastics, Book of Standards Volume: 08.01. ASTM International: West Conshohocken, PA, USA, 2010.
35. ISO 844:2021. Rigid Cellular Plastics—Determination of Compression Properties, Technical Committee: ISO/TC 61/SC 10 Cellular plastics. Available online: https://www.iso.org/standard/73560.html (accessed on 1 August 2023).
36. Peng, D.; Li, W.; Liang, X.; Zheng, L.; Guo, X. Enzymatic preparation of hydrophobic biomass with one-pot synthesis and the oil removal performance. *J. Environ. Sci.* **2023**, *124*, 105–116. [CrossRef] [PubMed]
37. Moghim, M.H.; Keshavarz, M.; Zebarjad, S.M. Effect of SiO_2 nanoparticles on compression behavior of flexible polyure-thane foam. *Polym. Bull.* **2019**, *76*, 227–239. [CrossRef]
38. Thirumal, M.; Khastgir, D.; Singha, N.K.; Manjunath, B.; Naik, Y. Mechanical, Morphological and Thermal Properties of Rigid Polyurethane Foam: Effect of the Fillers. *Cell. Polym.* **2007**, *26*, 245–259. [CrossRef]

polymers

Article

Effects of Phenolics on the Physicochemical and Structural Properties of Collagen Hydrogel

Sadia Munir [1], Wei Yue [1], Jinling Li [1], Xiaoyue Yu [1], Tianhao Ying [1], Ru Liu [1], Juan You [1], Shanbai Xiong [1] and Yang Hu [1,2,*]

1 College of Food Science and Technology, Huazhong Agricultural University, Wuhan 430070, China; bajwa.uos@gmail.com (S.M.); xiaoyueyu@webmail.hzau.edu.cn (X.Y.); yingtianhao@webmail.hzau.edu.cn (T.Y.); liuru@mail.hzau.edu.cn (R.L.); juanyou@mail.hzau.edu.cn (J.Y.); xiongsb@mail.hzau.edu.cn (S.X.)
2 Bioactive Peptide Technology Hubei Engineering Research Center, Jingzhou 434000, China
* Correspondence: huyang@mail.hzau.edu.cn; Tel./Fax: +86-27-8728-8375

Abstract: In the current era, the treatment of collagen hydrogels with natural phenolics for the improvement in physicochemical properties has been the subject of considerable attention. The present research aimed to fabricate collagen hydrogels cross-linked with gallic acid (GA) and ellagic acid (EA) at different concentrations depending on the collagen dry weight. The structural, enzymatic, thermal, morphological, and physical properties of the native collagen hydrogels were compared with those of the GA/EA cross-linked hydrogels. XRD and FTIR spectroscopic analyses confirmed the structural stability and reliability of the collagen after treatment with either GA or EA. The cross-linking also significantly contributed to the improvement in the storage modulus, of 435 Pa for 100% GA cross-linked hydrogels. The thermal stability was improved, as the highest residual weight of 43.8% was obtained for the hydrogels cross-linked with 50% GA in comparison with all the other hydrogels. The hydrogels immersed in 30%, 50%, and 100% concentrations of GA also showed improved swelling behavior and porosity, and the highest resistance to type 1 collagenase (76.56%), was obtained for 50% GA cross-linked collagen hydrogels. Moreover, GA 100% and EA 100% obtained the highest denaturation temperatures (Td) of 74.96 °C and 75.78 °C, respectively. In addition, SEM analysis was also carried out to check the surface morphology of the pristine collagen hydrogels and the cross-linked collagen hydrogels. The result showed that the hydrogels cross-linked with GA/EA were denser and more compact. However, the improved physicochemical properties were probably due to the formation of hydrogen bonds between the phenolic hydroxyl groups of GA and EA and the nitrogen atoms of the collagen backbone. The presence of inter- and intramolecular cross-links between collagen and GA or EA components and an increased density of intermolecular bonds suggest potential hydrogen bonding or hydrophobic interactions. Overall, the present study paves the way for further investigations in the field by providing valuable insights into the GA/EA interaction with collagen molecules.

Keywords: collagen; gallic acid; ellagic acid; hydrogels; storage modules

Citation: Munir, S.; Yue, W.; Li, J.; Yu, X.; Ying, T.; Liu, R.; You, J.; Xiong, S.; Hu, Y. Effects of Phenolics on the Physicochemical and Structural Properties of Collagen Hydrogel. *Polymers* **2023**, *15*, 4647. https://doi.org/10.3390/polym15244647

Academic Editors: Antonio M. Borrero-López, Concepción Valencia-Barragán, Esperanza Cortés Triviño, Adrián Tenorio-Alfonso and Clara Delgado-Sánchez

Received: 9 August 2023
Revised: 30 November 2023
Accepted: 2 December 2023
Published: 8 December 2023

1. Introduction

In the present era, hydrogels are an attractive class of materials. Hydrogels have been known for several years as appealing scaffolds because have they highly expanded and interconnected structural networks, which give them the ability to encapsulate the bioactive compounds and effectively transfer mass [1]. Recently, more and more researchers have paid attention to natural polymer hydrogels due to their high safety level, low immunogenicity, good biocompatibility, biodegradability, hydrophilic nature, and abundant availability [1–3], including proteins. Collagen is abundantly found in human tissues and is particularly abundant in load-bearing structural systems such as bones, skin, lungs, and tendons. In addition, collagen can also improve cellular adherence and promote extracellular matrix production in proliferating cells [4]. Furthermore, collagen molecules have many reactive groups that can be used to modify the collagen [5] and also function in numerous organic natural body functions such as tissue regeneration, healing, control of tissue-related diseases, cellular response, and structuring [6]. In addition, collagen suffers from the limitations of relatively weak mechanical properties, chemical stability, and resistance to enzymatic degradation [5], although it is widely used as an ingredient to improve the consistency, elasticity, and stability of food products.

Plant-based phenolic composites are considered to be the most vital bioactive composites. These compounds contain a number of hydroxyl groups and have diverse biological functions such as structural support, pigmentation, chemical defense, and radiation prevention [7,8]. Gallic acid (GA) is also a phenolic compound of plant origin and is known as (3,4,5-trihydroxybenzoic acid). Moreover, GA has been shown to have a variety of properties in biomedical studies, including anti-allergic, anti-fungal, anti-inflammatory, anti-cancer, anti-viral, anti-mutagenic, and anti-carcinogenic properties [7,9,10]. In addition, from a medical point of view, gallic acid plays an important role in the protective mechanism against reactive oxygen species and free radicals. It breaks the free radical chains through hydroxyl groups [6,11]. Furthermore, GA was reported by Thanyacharoen et al. to be a bioactive and stable agent in chitosan/PVA-based hydrogels [12]. It was also reported by Jiang et al. [13] that GA can increase the release efficiency of chitin-based hydrogels.

Alternatively, ellagic acid (EA), a representative of flavonoids found in a variety of fruits such as pomegranate, pecans, and berries, has received extensive attention due to its numerous antioxidant, cytotoxic, radical-scavenging, anti-viral, anti-inflammatory, anti-carcinogenic, and anti-apoptotic properties [14,15]. Ellagic acid is a dilactone of hexahydrooxydiphenic acid. It is usually produced by plants through the hydrolysis of tannins such as ellagitannins [16]. It contains four hydroxyl groups and these hydroxyl groups can increase the antioxidative action of lipid peroxidation to protect the cell from oxidative destruction [14]. Therefore, EA has hydrophilic characteristics due to its structural appearance, which includes a planar biphenyl, and a lipophilic component connected by two lactone rings, and four hydroxyl groups. These hydroxyl groups combine with the lactone groups to form a hydrophilic unit [17]. The hydrophilic region of the EA molecule plays an important role in its biological activity. Notably, this is due to the presence of both hydrogen-bonding acceptor (lactone) and donor (-OH) sites. In particular, the phenolic hydroxyl groups in EA can be separated under physiological conditions [18,19]. In addition, it has been reported by Huang et al. that EA can cross-link with PEG-based hydrogels and reduce the viability of human oral cancer cells [20].

The primary purpose of this study was to prepare collagen hydrogels with improved physicochemical properties via the cross-linking of natural phenolic compounds. Gallic acid or ellagic acid were added as potential cross-linkers. In particular, the influences of phenolic compounds on the physicochemical properties, such as the thermal, structural, enzymatic, and morphological properties of collagen-based hydrogels, were examined. In previous studies, gallic acid and ellagic acid have never been prepared in collagen-based hydrogels and compared to each other. In addition, XRD, FTIR, water retention, enzyme degradation, porosity, and swelling ratio were investigated. Finally, the SEM examination of the cross-linked hydrogels was also carried out. This study could contribute to providing a new vision in the biomaterials or biomedical industries based on the cross-linking of collagen hydrogels with natural phenolic compounds such as gallic acid or ellagic acid.

2. Material and Methods

2.1. Materials

Collagen extraction was performed using fresh grass carp (*Ctenopharyngodon idella*) obtained from a local slaughterhouse. Following the procedure described by Zhu et al. [21], the extraction method used was a combination of acid and pepsin extraction. The pepsin enzyme used in the extraction process was purchased from Bio-Sharp Company. Gallic acid and ellagic acid used in the study were purchased from Shanghai Aladdin Bio-Chem Technology Co., Ltd., Shanghai, China. Lyophilized type I collagenase isolated from *Clostridium histolyticum* (freeze-dried powder, ≥125 CDU/mg solid) was purchased from Sigma, St. Louis, MO, USA. All other reagents used in the experiments were of high analytical grade.

2.2. Production of Collagen

Using a modified approach, collagen was isolated from the skin of freshly bought grass carp *(Ctenopharyngodon idella)* at 4 °C [21]. To summarize, the skin was washed, sliced into small pieces, and soaked for 72 h in 0.01 M NaOH (1:20, *w*/*v*). To remove non-collagenous constituents, the solution was changed after every 8 h. The skin pieces were then immersed in 10% (*w*/*w*) isopropanol for 24 h to remove fat, and then neutralized with double-distilled water. Based on the dry weight of the skin, the pieces were combined with 0.5 M acetic acid (1:50, *w*/*v*) and 2% pepsin (1:3000; Sigma, USA). The mixture was stirred at 4 °C for 2 days. To separate the pepsin-soluble collagen solution, it was filtered and centrifuged (10,000 rpm, 10 min). For purification, 1.5 M $(NH_4)_2SO_4$ was added overnight, followed by 0.5 M acetic acid. The resulting purified collagen was resolubilized in 0.5 M acetic acid and then dialyzed with 0.04 M disodium dihydrogen pyrophosphate solution for 2 days, 0.02 M disodium dihydrogen pyrophosphate solution for 3 days, and finally with double-distilled water for 2 days. The collagen solution was lyophilized and kept in a dark and dry place for future studies. This ensured the high quality of the collagen from the grass carp skin.

2.3. Fabrication of Cross-Linked Collagen Hydrogels

Pure collagen hydrogels were prepared in accordance with the method of Zhu et al. [21], with slight modifications as illustrated in Figure 1. Freeze-dried collagen was reconstituted by dissolving the collagen in 0.5 M acetic acid. The ratio was (10 mg:1 mL) 10 parts of collagen to 1 part acetic acid, and the process was performed at 4 °C. The solution of collagen and acetic acid was subjected to gentle stirring until the complete dissolution of the collagen was achieved. The pH of the collagen solution was adjusted to a neutral range between pH 7.0 and 7.5 and it was adjusted using either 2 M NaOH or 2 M acetic acid with continuous monitoring and necessary adjustments to ensure the desired pH. Next, 5 mL of the neutralized collagen solution was added to individual 24-well plates. These plates were then incubated at 37 °C for 4 h. This incubation time allowed the collagen to modify as a gel. The hydrogels were stored at 4 °C overnight after the initial incubation. Further gelation and overall stabilization of the collagen hydrogels were promoted by this additional step.

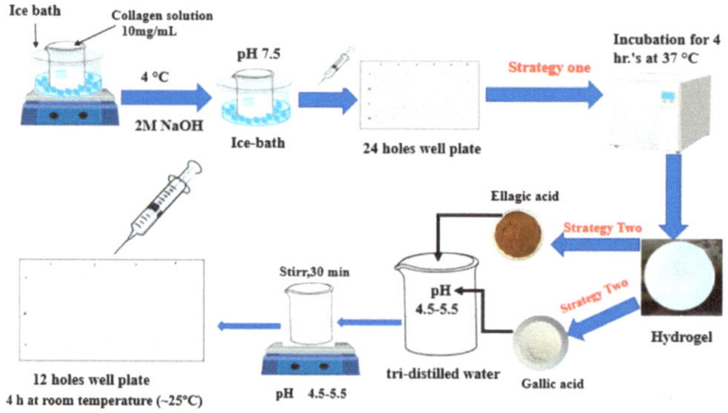

Figure 1. Schematic diagram demonstrating the overall strategy of the current work.

For the preparation of cross-linked collagen hydrogels, GA and EA were dissolved individually in tri-distilled water at different concentrations (0%, 1%, 5%, 10%, 30%, 50%,

100%, and 200% w/w based on the dry weight of the collagen hydrogels) at a pH range of 4.5–5.5. The solutions containing GA and EA were gently shaken at room temperature for approximately 30 min. The collagen hydrogels were then immersed in these prepared solutions. The immersion was performed at room temperature, approximately 25 °C, for a period of 4 h. The cross-linked collagen hydrogels were carefully washed with distilled water to remove any unbound GA and EA after the 4 h immersion. Finally, the gels were stored at 4 °C for further processing.

2.4. Collagen Hydrogel Characterization

2.4.1. Dehydration of Hydrogels

Prior to lyophilization, all hydrogels were placed at −80 °C for complete freezing. Dehydration of the frozen hydrogels was performed in a lyophilizer for a minimum of 24 h to remove all the moisture. Lyophilized hydrogels were subjected to thermogravimetric curve, DSC, SEM, porosity, water-holding capacity, swelling behavior, FTIR, and XRD characterization to verify the water-holding capacity of the network in the structure after lyophilization.

2.4.2. Determination of Porosity

The determination of the porosity of the hydrogel was in accordance with Zhu et al. [21]. First, a 25 mL beaker was filled with ethanol, and its weight was m_1. The lyophilized sample (m_0) was then immersed into the beaker, which was then ultrasonically degassed to permeate the hydrogel with ethanol, and the total weight was calculated as m_2. The ethanol was then carefully scraped from the hydrogel's surface, and the weight of the left half was measured as m_3. Following that, the hydrogel's porosity was then estimated using the following formula:

$$\text{Porosity (\%)} = (m_2 - m_0 - m_3)/(m_1 - m_3) \times 100 \quad (1)$$

2.4.3. Swelling Ratio and Water-Holding Capability

The swelling ratio was quantified according to Zhu et al. [21] at 25 °C by soaking the weighed (W_0) freeze-dried cylindrical specimens in phosphate-buffered saline (0.1 M PBS, pH 7.4). After soaking for 1, 3, 5, 8, 12, 20, 50, and 90 min, the swollen hydrogels were removed. After the removal of excess water, the hydrogels were immediately weighed (W_t). The following formula can be used to calculate the swelling ratio (SR):

$$SR = (W_t - W_0)/W_0 \quad (2)$$

The following method was used to determine the water-holding capacity (WHC). Briefly, the freshly prepared samples (equilibrium swollen hydrogel in water) were first weighed after the water had been wiped off from the surface with filter paper ($W_{swollen}$). Subsequently, the samples were then lyophilized and weighed as $W_{lyophilized}$. The following formula can be used to calculate the percentage of WHC:

$$\text{WHC (\%)} = (W_{swollen} - W_{lyophilized})/W_{swollen} \times 100 \quad (3)$$

2.4.4. Thermogravimetric Analysis (TGA) of Collagen Hydrogels

Thermogravimetric (TG) curves of collagen hydrogels were obtained using a TGA-2050 instrument (Mettler-Toledo). The instrument was operated under a nitrogen atmosphere and heated between 30–550 °C at a heating rate of 10 °C/min. The rehydrated hydrogels were used for the TGA test with a sample size of approximately 5 mg.

2.4.5. Thermal Transition Analysis of Collagen Hydrogels

The thermal transition temperature of the collagen hydrogels was studied using differential scanning calorimetry (DSC) (DSC 200PC, Netzsch, Germany. The specimens (5 mg) were precisely weighed and placed in sealed aluminum dishes. At a heating rate of 5 °C/min over a temperature range of 25–110 °C, the aluminum pans were scanned under a nitrogen atmosphere. The empty, sealed aluminum pan, which was sealed, was used as the reference point. The temperature of the endothermic peak was reported as the temperature of denaturation (Td).

2.4.6. Analysis by X-Ray Diffraction (XRD)

The X-ray diffraction structures of the collagen hydrogels were analyzed using CuKa radiation from a rotating anode generator operated at 40 kV and 40 mA in the 2θ range (4–60 °C) with a mono-single filter at a scan rate of 10 °C min^{-1} (D8 Advance, Bruker, Germany).

2.4.7. Fourier Transform Infrared Spectroscopy (FTIR) Analysis

On a germanium, single crystal FTIR spectrophotometer, the infrared spectra of all lyophilized gel powders were obtained from tablets containing (0.8–1 mg) collagen hydrogel in approximately 100 mg potassium bromide (KBr). All freeze-dried gel powders' IR spectra were obtained from tablets containing (0.8–1 mg) collagen hydrogel in 100 mg potassium bromide (KBr). The tablets were placed on the single-reflection germanium crystal cell using an FTIR spectrophotometer. Signals from 4000 to 400 cm^{-1} were acquired for 64 scans at a data acquisition rate of 4 cm^{-1} per point and compared to a background spectrum collected from a clean, empty cell.

2.4.8. Measurements of Dynamic Rheology

The hydrogels (diameter = 20 mm, thickness = 5 mm) prepared from self-assembled neutral collagen solutions (10 mg/mL) and incubated for 2 h were subjected to a dynamic time sweep for 60 min (37 °C) at the different frequencies (1–10 Hz) to monitor the gelling behavior. Furthermore, the dynamic temperature was also set at 37 °C. During the test, a rheometer (AR2000ex, TA, Woodland, CA, USA), with a parallel stainless-steel plate (diameter = 40 mm, gap = 1 mm) was used to determine the storage and loss modulus values. The frequency sweep was adjusted to 0.01–10 Hz. A deformation of 2% was selected for all samples. The temperature was controlled by using a Peltier temperature controller. A solvent trap was used to prevent water loss from the samples during the measurement. The value of the tangent δ (tan δ) was also calculated as the G''/G' ratio, which reflects the thermal energy loss.

2.4.9. Enzymatic Analysis of Stability

The in vitro enzymatic degradation of collagen-based hydrogels was carried out with the use of type I collagenase that was derived from Clostridium. Lyophilized hydrogels were first swollen by immersion in phosphate-buffered saline (PBS) at a pH of 7.0 until complete swelling was achieved. Each hydrogel was then placed in 1 mL enzyme hydrolysate containing type I collagenase (200U, Sigma) and 0.01 M $CaCl_2$ and incubated at 37 °C for 24 h. 0.2 mL of 0.25 M EDTA was added to the hydrogel mixture followed by cooling in an ice bath to complete the degradation process. The mixture was then centrifuged at 5000× *g* for 10 min at 4 °C. A quantity of 2 mL of the resulting supernatant was collected. Quantities of 1 mL of chloramine T and perchloric acid were added to the supernatant. Each was allowed to stand for 20 min and 5 min, respectively. Then, 1 mL of paradimethylaminobenzaldehyde (DMAB) was added. The mixture was incubated in a water bath at 60 °C for 20 min. Ultraviolet spectroscopy at 560 nm was used to quantify the hydroxyproline content. The conversion coefficient between collagen and hydroxyproline in aquatic animals was calculated to assess the degree of hydrogel biodegradation. Specifically, in the absence of GA or EA, the percentage of hydroxyproline released from the collagen-based hydrogel was compared to that of fully degraded collagen. For reference purposes, untreated collagen was used as a control.

2.4.10. Scanning Electron Microscopy (SEM)

Collagen hydrogels cross-linked with GA or EA for morphological characterization were prepared as described by Liu et al. [22] with slight modifications. Briefly, freshly prepared hydrogels were cut into fragments (~2 × 2 mm) and then immersed in 2.5% glutaraldehyde in 0.2 M phosphate (pH 7.2) overnight and dehydrated in graded ethanol solution with a series of concentrations (30%, 50%, 70%, 80%, 95%, and 100%). Subsequently, the samples were then treated with isoamyl acetate for a period of 15 min. The hydrogels were then freeze-dried with the use of a lyophilizer. After that, the hydrogels were coated with a layer of gold. An ultrahigh-resolution field emission scanning electron microscope was used to observe the microstructure, and SEM images were observed using a JSM-5610 SEM (JEOL, Tokyo, Japan), with 20 kV acceleration voltage. The magnification was 3 K, 8 K, and 15 K times.

2.4.11. Statistical Evaluation

Prism and Origin software versions 8.5 (SAS Institute Inc., Cary, NC, USA) were used for data analysis. To detect statistically significant differences, Duncan's multiple range test was performed. The level of significance was fixed at $p < 0.05$. By finding significant differences across experimental groups, this statistical technique provides compelling evidence for the reported results.

3. Results and Discussion

3.1. Porosity Measurements

The porosity, an important factor for collagen hydrogels, is illustrated in Figure 2A. In general, higher porosities have been studied, particularly for collagen-based hydrogels. These were thought to be valuable for supporting nutrition and cell metabolism during cell proliferation and adhesion in tissues [23]. The control hydrogels exhibited a lower porosity level of ~87.51%, and a similar tendency was observed in GA 1%, EA 1%, GA 200%, and EA 100% (~88.96%, 88.21%, 89.83%, and 89.51%, respectively). Notably a concentration-dependent pattern was developed in the porosity of the collagen cross-linked hydrogel, which increased with the increasing concentrations of GA but decreased with the GA 200%, which might be due to the higher concentration of GA. Conversely, the porosity of EA cross-linked hydrogels presented an increase with increasing EA concentrations up to 30%, followed by a decrease with higher EA concentrations. GA 50% exhibited the highest porosity (98.24%) among all hydrogels containing GA and EA at all concentrations. Gallic acid exhibited significantly superior porosity than ellagic acid in the cross-linking of collagen hydrogels according to the findings of the present study. The distinction in porosity can be attributed to the differences in chemical structures and the capacity of GA and EA to bind with collagen. At higher concentrations, the porosity of GA/EA cross-linked hydrogels decreased as GA and EA bound to collagen through hydrogen and hydroxyl bonds [24]. In particular, higher concentrations of phenolic acid appeared to disrupt the self-assembly of collagen, resulting in heterogeneous pore topologies with lower porosity at 200% of GA and 50%, 100%, and 200% of EA. Remarkably, the porosity of collagen hydrogels cross-linked with GA/EA demonstrated dynamic behavior. It was characterized by an initial increase followed by a subsequent reduction in the porosity. The initial porosity rise can be attributed to the introduction of cross-linking agents into the hydrogels [25]. GA or EA facilitated the formation of interconnected pores within the hydrogel matrix. The subsequent reduction in porosity with an increasing concentration of GA/EA cross-linked hydrogels decreased because of matrix enhancement, potentially driven by stronger intermolecular interactions and structural rearrangements within the hydrogel network. Prior studies have also reported that higher porosity is correlated with improved cell viability, tissue formation, wound healing, cell proliferation, and cell permeation [21,25].

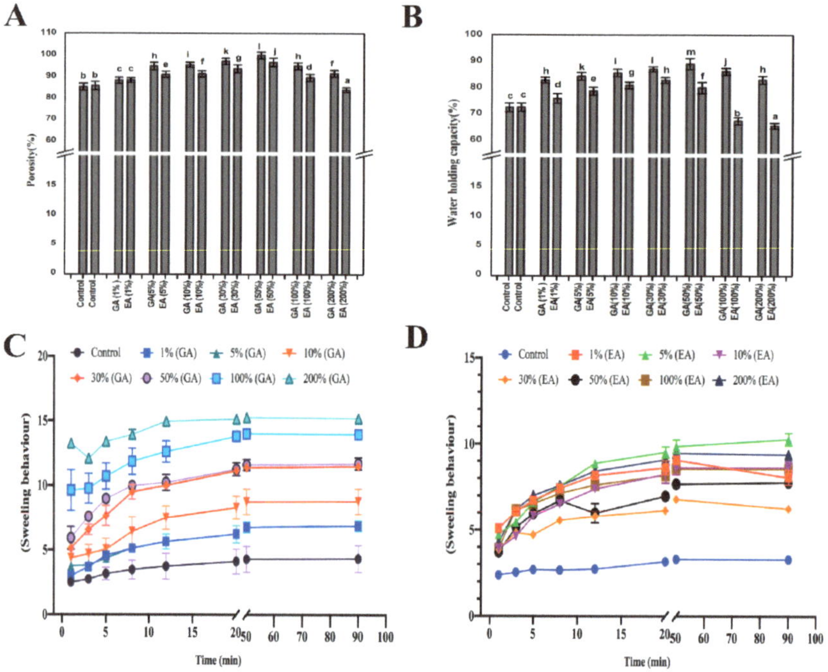

Figure 2. Evaluation of physical properties of pristine and GA/EA cross-linked hydrogels. (**A**) Porosity. (**B**) Water-holding capacity (WHC). (**C**,**D**) Swelling behavior of pure, GA, and EA hydrogels with different concentrations of gallic and ellagic acid according to the dry weight of collagen. Different letters (a, b, c, etc.) above concentrations indicate significant ($p < 0.05$) differences.

3.2. Water-Holding Capability

Numerous critical elements have contributed to the observed phenomenon in the context of water-holding capacity. Hence, immersion of GA and EA within the collagen matrix boosted the hydrogel's hydrophilicity, aggregating both water in absorption and water retention [26,27]. Furthermore, the water-holding capacity (WHC) of GA/EA cross-linked hydrogels presented concentration-dependent descriptions, as shown in Figure 2B. There was an increase in WHC at lower concentrations of GA or EA cross-linked hydrogels. A number of assorted factors support these phenomena. Furthermore, the immersion of GA and EA inside the hydrogel structure may have provided additional water molecule binding sites, contributing to an increase in WHC in cross-linked collagen hydrogels [28]. However, when the concentration of GA/EA in cross-linked hydrogels exceeded a predefined limit, WHC decreased. The reduction was due to the increasing density and compactness of the hydrogel matrix at higher cross-linking concentrations. Increased concentrations of GA or EA result in an extraordinary production of binding sites within the collagen network, resulting in a denser matrix that inhibits water entry and retention. When the WHC of GA and EA hydrogels at equal concentrations was compared, it was clear that GA hydrogels had better WHC compared to EA. The difference can be attributed to the different chemical structures and cross-linking abilities of GA and EA. Gallic acid has a higher tendency for water molecule binding and a more dynamic interaction with collagen, resulting in increased water absorption and retention [29]. On the other hand, the WHC of collagen hydrogels cross-linked with EA increased with increasing EA content up to a concentration of 30%. Then, the WHC gradually decreased. Moreover, the highest WHC was found in collagen hydrogels cross-linked with GA, especially at a 50% GA concentration (89.07%), whereas the lowest WHC was found in collagen hydrogels cross-linked with EA at a

200% EA concentration (64.92%). These results highlight the significant influence of GA and EA cross-linking on WHC, which may have implications for a variety of biomaterial applications. However, Lin et al. [30] conducted research that also supported the idea that improved WHC in cross-linked collagen hydrogels is beneficial for biomedical engineering in biomaterials.

3.3. Swelling Property

The swelling property is also known as a vital property of hydrogels and is usually related to the moisture transfer across the hydrogels or the retention of moisture within the system from the environment. The swelling ratio (SR) was primarily affected by external solution variables such as charge number and ionic strength, as well as polymer features such as network flexibility, the presence of hydrophilic functional groups, and the degree of cross-linking density [31]. The effects of cross-linking on the swelling properties of GA/EA cross-linked hydrogels are shown inFigure 2C,D. The swelling ratio in the neutral solution (native collagen) was lower than that in the acidic solution (GA/EA cross-linked hydrogels). All hydrogels presented swelling within the first 20 min after immersion in the solution. The swelling ratio improved marginally from 8 to 50 min, then remained nearly constant from 50 to 90 min, indicating that the swelling equilibrium had been reached. The GA/EA addition enhanced the swelling ratio of GA/EA cross-linked collagen hydrogels. Hydrogels quickly absorb aqueous solutions in acidic circumstances, and the hydroxyl and amino groups in the hydrogels become highly protonated [32]. However, the immersion of EA promoted the swelling ratio when the EA was less than 100% based on the dry weight of collagen. The formation of hydrogen bonds between the phenolic hydroxyl groups of GA and EA and the nitrogen atoms of the collagen backbone is most likely responsible for the enhanced swelling ratio. On the other hand, the swelling ratio of EA cross-linked hydrogels with higher EA concentrations (100% and 200%) was lower than that of pure collagen hydrogel. However, this was due to the significant self-polymerization of EA, which inhibits the extension of collagen fibrils in PBS solution to achieve a reduced swelling ratio [33,34]. However, the swelling ratio was remarkably increased (about 3–15% ($p < 0.05$)) with the rise of GA/EA concentrations and time duration during the period from 1 to 50 min and then remained constant from 50 min to 90 min. The molecular chain was stretched with charge repulsion and the capacity for swelling increased. Collagen hydrogels are sensitive to pH, indicating that ionic groups in hydrogels play an important role in absorbing water in the gel, as stated by Wang et al. [32]. This effect suggested that the GA/EA cross-linked hydrogels were suitable for application in acidic environments, where they could make good use of their absorptive property.

3.4. Thermogravimetric Analysis (TGA) of Collagen Hydrogels

Figure 3A,B, and Table 1 depict the thermal stability analysis of collagen hydrogels cross-linked with GA and EA. The thermal degradation characteristics of the hydrogels exhibited variability contingent upon the origin and concentration of phenolic compounds within GA or EA. The quantity of hydroxyl groups present in phenolic compounds is known to exert a substantial influence on the interaction between proteins and phenolic compounds [35,36]. The presence of carboxylic and hydroxyl groups in phenolic compounds could result in intra- and intermolecular interactions such as hydrogen, ionic, covalent, and non-covalent bonds, which change the chemical connotation of proteins and phenolic chemicals [35,37]. The interactions contributed to the improvement in thermal stability. In the thermogravimetric (TG) analysis, weight loss was observed in the initial stage ($\Delta w1 = 6.5\%$, 99.8 °C) for all hydrogel samples. The weight loss during the specific temperature range was attributed to the release of two types of water states, namely, free and bound water, absorbed by the hydrogels [38]. Furthermore, the hydrogels treated with GA/EA lost less weight than the control group, implying that the presence of GA/EA reduced the water content of the collagen hydrogels due to the greater hydrophobicity of the phenolic compounds.

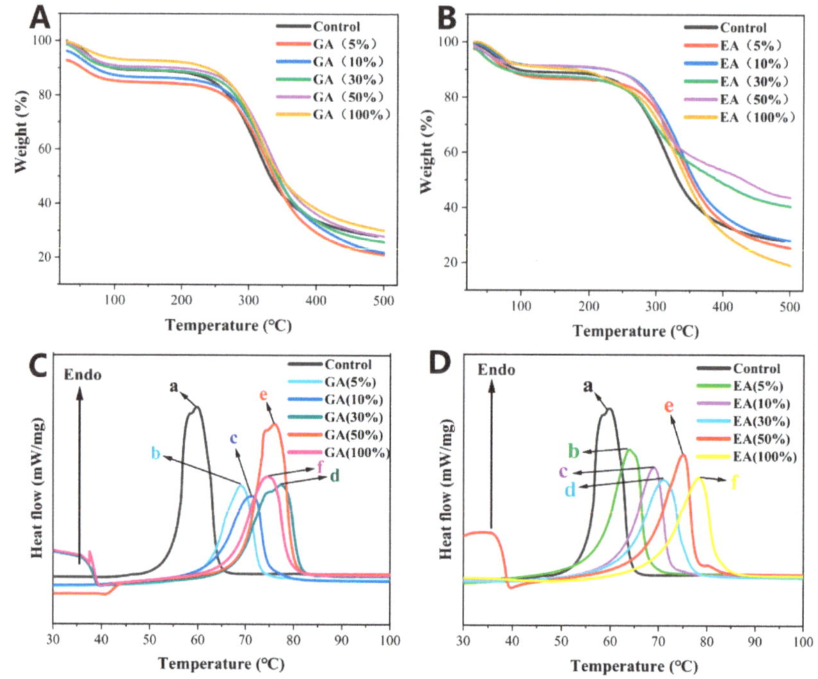

Figure 3. (**A**) TGA curves of GA cross-linked collagen hydrogels. (**B**) TGA curves of EA cross-linked collagen hydrogels. (**C**) DSC curves of GA cross-linked collagen hydrogels. (**D**) DSC curves of EA cross-linked collagen hydrogels. Different letters (a, b, c, etc.) above the concentrations indicate significant ($p < 0.05$) differences.

Table 1. Thermal degradation temperature (Td, °C) and weight loss (Δw, %) of collagen hydrogels immersed with GA and EA with different concentrations according to the dry weight of collagen. Δ_1, Δ_2, Δ_3, and Δ_4 represent the first, second, third, and fourth stages of weight loss of hydrogels during the heating scan, respectively.

Samples	(*w*/*w* of Protein)	Δ1		Δ2		Δ3		Δ4		Residue (%)
		Td1	Δw1	Td2	Δw2	Td3	Δw3	Td4	Δw4	
	Control	96.4	10.1	228.3	3.0	366.7	47.6	489.3	11.1	28.2
GA	5%	85.3	10.8	247.3	5.1	392.6	43.1	539.6	11.3	28.7
	10%	87.6	7.5	257.9	6.2	397.4	37.7	538.2	16.1	32.5
	30%	91.1	9.4	252.2	5.9	397.7	25.7	540.4	18.5	40.5
	50%	99.8	7.2	253.5	5.7	388.5	35.2	541.2	8.1	43.8
	100%	99.8	6.5	246.5	5.3	400.1	51.7	543.5	19.1	17.4
EA	5%	99.3	11.5	255.1	4.7	380.8	45.6	540.3	14.3	23.9
	10%	91.4	6.5	251.3	5.3	406.1	50.4	540.1	10.7	27.1
	30%	77.5	11	234.1	2.6	415.8	40.2	541.3	7.1	39.1
	50%	89.7	7.3	232.5	3.3	330.1	27.9	539.8	18.6	42.9
	100%	84.3	7.9	256.1	9.7	405.4	52.4	540.6	13.9	16.1

The second stage of weight loss was observed in all hydrogels at a temperature range of 228.3–257.9 °C, with a weight loss percentage (Δw2) ranging from 2.6% to 9.7%. The weight loss at the second stage was typically associated with the release of structurally bound water and low molecular weight proteins, specifically collagen [38]. In conclusion, the thermal stability analysis reveals that the thermal degradation behavior of collagen hydrogels was influenced by the presence of phenolic compounds from GA/EA cross-linking. The interactions between proteins and phenolic compounds, driven by the hydroxyl and carboxylic groups, contributed to the increased thermal stability observed in the GA/EA cross-linked collagen hydrogels. The weight loss observed during TG analysis corresponds to the release of water states and low molecular weight proteins present in the collagen hydrogels [39]. The observed temperature range in the second stage of weight loss was found to be higher than the decomposition temperature of phenolic compounds [40]. The present study suggested that the cross-linking between collagen and GA/EA in the hydrogels occurs through hydrogen bonding. The lower weight loss (Δw2) observed in the GA/EA-treated collagen hydrogels compared to the control further supports the strong cross-linking between the protein (collagen) and phenolic compounds (GA/EA) via hydrogen or covalent bonds [37]. The decomposition temperatures (Td1 and Td2) of collagen protein-based hydrogels containing GA and EA were reduced compared to those of the control. On the other hand, the decomposition temperature changed with the different concentrations of GA/EA. The results indicated that hydrogels immersed in GA or EA solutions exhibited higher heat resistance than those of the control group.

During the third stage, which was characterized by Δw3 ranging from 25.7% to 52.4% and Td3 between 330.1 °C and 415.8 °C, the weight loss was attributed to the degradation of larger cross-linked proteins within the collagen hydrogels. Additionally, a fourth stage of weight loss (Δw4 = 7.1–19.1%) was observed in the temperature range of 489.3 °C to 541.3 °C. The weight loss was predominantly associated with the degradation of high-temperature stable components. The immersion of collagen hydrogels in GA or EA solutions at different concentrations significantly influenced the thermal stability of the hydrogels ($p < 0.05$), primarily through strong interactions between the proteins and phenolic compounds, and particularly through covalent cross-linking [37]. Notably, hydrogels immersed in 30% and 50% GA/EA concentrations, based on the dry weight of collagen, exhibited enhanced thermal stability as compared to all other hydrogels.

3.5. Differential Scanning Calorimetry (DSC) Measurements

DSC measurements of GA/EA cross-linked hydrogels are presented in Figure 3C,D. The collagen hydrogels immersed in GA or EA phenolic acids showed an improved Td value compared to the original hydrogels, as described in Table 2. When the collagen-based hydrogels were heated, the helix–coil conversion took place, and, as a result, the helix disappeared and was progressively separated into three randomly coiled peptide α-chains [41]. Typically, the thermal transformation of collagen has been proposed to be the collapse of the triple helical structure of collagen into random coils, with Td being the major endothermic peak [21,41]. The process of triple helical structure modification of GA/EA cross-linked collagen hydrogels was represented by a typical endothermic peak in the range of 56.78 °C to 75.78 °C. It has been found that when the structural integrity of collagen is improved, the endothermic peak is shifted to a lower temperature, resulting in a significantly lower Td than that of collagen with structural integrity [4,21,42]. The denaturation temperature (Td) increased to approximately 19 °C with the increasing concentration of GA or EA from 1 to 100% according to the dry weight of collagen. Moreover, GA 100% and EA 100% obtained the highest denaturation temperatures (Td) of 74.96 °C and 75.78 °C, respectively. However, the higher denaturation temperature (Td) observed in hydrogels cross-linked with GA or EA as compared to that of native collagen could be attributed to numerous factors. First, the presence of these cross-linking agents increases the thermal stability of the collagen hydrogels, as evidenced by the increase in Td with increasing concentrations of GA or EA. A possible explanation for the increase in thermal stability may be an increase in the accessibility of active sites in the collagen molecules [29]. The presence of GA or EA might be facilitated by the collagen interaction with cross-linkers, and the bonding between collagen and GA/EA could increase the resistance of the network to thermal denaturation and contribute to a more stable network structure [43]. The hydrophobic interfaces, mainly maintained by glycine residues and H-bonds formed between GA or EA depositions, play an important role in the stabilization of collagen molecules. In addition, the other possible reason could be that the GA/EA reactive sites could react with the amino group (NH_2) of the collagen side chains [44]. In short, it was also noted that the thermal transition temperature represents the energy required for the destruction of the amide bonds, hydrogen bonds, and van der Waals forces that maintain the collagen triple helical structure. It should be noted that the increase in denaturation temperature is promising for maintaining the structural reliability of collagen, which is fundamental for the assembly of collagen-based hydrogels. Indeed, the increased hydrophobicity of GA or EA molecules, as well as their ability to form peptide bonds, can be important in their incorporation into certain regions of collagen fibrils [43]. This inclusion helps to stabilize the collagen scaffold structure. Because GA and EA are hydrophobic, they can interact with hydrophobic areas inside collagen fibrils. Hydrophobic interactions could happen between nonpolar areas of GA or EA and specific amino acid residues within collagen molecules, including proline or hydrophobic clusters. The interaction between hydrophobic binding improves GA or EA affinity and binding to collagen [45].

Table 2. Thermal denaturation (Td, °C) of collagen hydrogels immersed with GA and EA with different concentrations according to the dry weight of collagen.

Treatments	Sample	Thermal Denaturation Temperature (Td) °C
GA	Control	56.78
	5%	63.35
	10%	65.88
	30%	68.93
	50%	71.19
	100%	74.96
EA	5%	62.15
	10%	66.13
	30%	67.84
	50%	70.51
	100%	75.78

3.6. Analysis by X-Ray Diffraction (XRD)

The structural stability of collagen molecules after modification with GA or EA was evaluated using X-ray diffraction (XRD). Figure 4A,B presented the XRD spectra of GA/EA cross-linked collagen hydrogels. The XRD spectra revealed a diffuse scattering pattern with a small peak observed at about 8°, indicating the presence of intermolecular packing gaps between the molecular chains. An earlier study by Hu et al. [46], suggested that the observation was compatible with the Schmitt model's description of the assembly of collagen molecules into fibrils [46]. Furthermore, a broader peak in the 20° to 25° range was detected, corresponding to the widespread scattering of collagen fiber synthesis [45]. The peak pattern in the XRD spectra of the GA/EA cross-linked hydrogels was identical to that of the control collagen hydrogels. The closeness implies that the collagen's structural reliability was intact after treatment with GA or EA solutions. According to the XRD examination, the cross-linking of collagen with GA or EA had no significant effect on the overall structural properties of the collagen fibers. However, a previous study suggested that the presence of GA or EA in collagen hydrogels represented the intermolecular binding sites and that the assembly of collagen molecules into fibrils remained constant [47]. Therefore, the findings supported the assumption that GA/EA cross-linking does not alter collagen's structural stability, making it an appropriate modification approach for retaining collagen's inherent structure for biological applications [46,48].

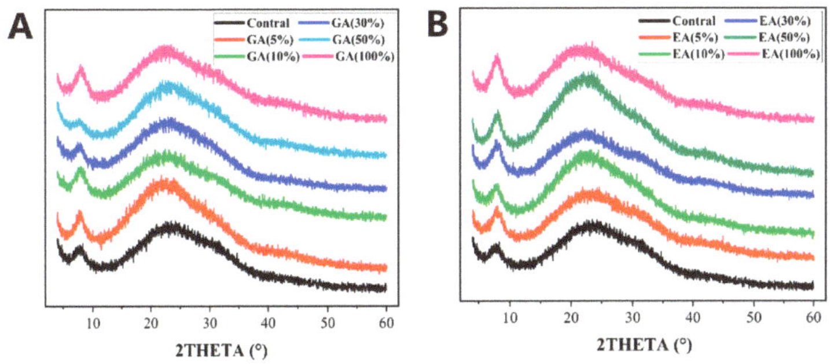

Figure 4. (**A**) XRD curves of GA cross-linked and pristine hydrogels with different concentrations of gallic acid according to the dry weight of collagen. (**B**) XRD curves of EA cross-linked and pristine hydrogels with different concentrations of ellagic acid according to the dry weight of collagen.

3.7. Fourier Transform Infrared Spectroscopy (FTIR) Analysis

Fourier transform infrared spectroscopy (FTIR) was used to describe the chemical structure of collagen and collagen GA or EA cross-linked hydrogels. Figure 5A,B explain the amide bands of collagen fibers with or without GA or EA cross-linking. The amide A band, observed in the wave number range of 3430–3470 cm^{-1}, was a characteristic feature of collagen cross-linked hydrogels. The amide A band was associated with hydrogen bonding interactions and results from the unfolding vibration of N-H bonds. Moreover, it provided valuable evidence about the network of hydrogen bonds within the structure of collagen. In addition, another band was detected in the wave number range of 3230–3290 cm^{-1}, and was known as the amide B band. The asymmetric stretching of the CH_2 groups within the collagen molecules was responsible for the amide B band. The conformational arrangement and structural properties of the collagen hydrogels can be determined from the presence and characteristics of the band. In addition, a band at 3140–3180 cm^{-1} was also observed in GA or EA cross-linked collagen fibers. However, the amide A and amide B bands vanished in pristine collagen hydrogels. In addition to the classic amide A and amide B bands, a novel peak in the wave number range of 1700–1730 cm^{-1} was identified in the spectra of collagen hydrogels cross-linked with GA or EA, but it was not present in the spectra of pristine collagen hydrogels. The unfolding vibrations of C=O (carbonyl) groups of chemical intermediates produced during GA or EA cross-linking are primarily responsible for this new peak. The emergence of the peak demonstrated that C=O groups participate in the cross-linking event, resulting in the production of more stable C-OH (hydroxyl) groups [45]. The unfolding vibrations of the C=O groups can provide insight into the structural changes and chemical modifications that occur within the collagen matrix throughout the cross-linking process. The inclusion of GA or EA could be the reason for these changes and could increase the stability and structural integrity of collagen-based hydrogels. Amides (A, B) promote consistency in the triple helical structure of collagen [49]. Amide (I) was located between 1620–1700 cm^{-1} and consisted of three main components: a band at 1650 cm^{-1} correlated with the α-helix/random coil confirmation, a band developed at 1620–1640 cm^{-1} corresponding to the β-sheet conformation, and a band appearing at 1660–1670 cm^{-1} corresponding to the β-turn [48]. In general, amide (I) was associated with the secondary structure of the protein and represents the trembling of amide carbonyls along with the polypeptide backbone and the native triple helix, which were transformed into C- and N- telopeptides in the collagen [49]. In addition, an amide (II) band of collagen fibers was found between 1550–1570 cm^{-1} and attributed to CH_2 bending vibration. The C-N unfolding and N-H bending vibrations of the amidic bond, and the wiggle vibrations of the glycine backbone and the CH_2 group of the proline side chain, are responsible for the complex amidic peaks [21]. On the other hand, FTIR shows strong binding as the concentration of GA or EA increased from 1 to 200%, except for GA 1% and EA 100% based on the dry weight of collagen, which might be due to the difference in their structure and binding ability at low and higher concentrations. The FTIR results indicated that GA or EA cross-linking has a significant effect on the structural properties and could be suitable for future experiments compared to the native collagen hydrogels, as the obtained results showed that clearer (amide A, B) and amide (I, II, III) bands at 1410–1480 cm^{-1}, corresponding to the unfolding vibrations of CH_2 and C=O, were altered in comparison to the native collagen hydrogels [50].

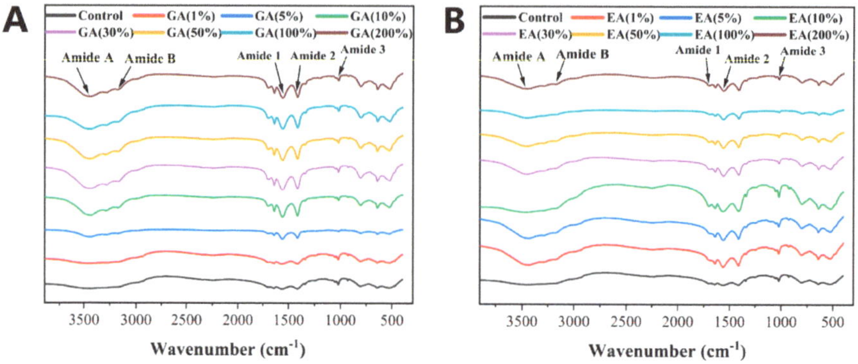

Figure 5. (**A**,**B**) FTIR spectra of GA or EA cross-linked or pristine collagen hydrogels with different concentrations of GA/EA according to the dry weight of collagen.

3.8. Dynamic Rheological Measurements

The storage modulus (G′) represents the elastic response of the hydrogels, and it was an indication of the ability to store and recover energy upon deformation. Moreover, the storage modulus (G′) also indicates the stiffness and rigidity of the hydrogel network. In contrast, the loss modulus (G″) reflects the viscous behavior and energy dissipation within the gel. The loss modulus (G″) was also known to measure the hydrogel's ability to flow and deform under applied stress. The relationship between G′ and G″ aids in the understanding of the dominant behavior, i.e., whether the hydrogel was more elastic or more viscous. The loss factor, often expressed as the phase angle tangent (tan δ), was calculated as the ratio of G″ to G′. However, tangent (tan δ) provided valuable information about the relative contribution of the elasticity and the viscosity within the hydrogel. On the other hand, a low value of tan δ mostly indicates the elastic response where the storage modulus leads toward the loss modulus. Conversely, a higher value of tan δ indicates more viscous behavior; however, the loss modulus dominates. The frequency dependence curves of storage modulus G', loss modulus G'', and tan δ of all COL-GA/EA hydrogels are presented in Figure 6A–F. The results obtained in the present study demonstrated that both the loss modulus (G″) and the storage modulus (G′) exhibit an increasing trend with increasing sweep frequencies up to 100% of GA or EA concentration based on the dry weight of collagen. However, both moduli decreased at 200% of GA or EA concentration. It was noteworthy that the storage modulus of elasticity (G′) was significantly higher than the loss modulus of elasticity (G′) at a constant frequency. The results indicated an expansion in the flexibility and a reduction in the mobility of the hydrogels after cross-linking with GA or EA [51,52]. The loss tangent (tan δ), which was the ratio of G″ to G′, serves as a measure of the transition from liquid-like to solid-like behavior. A value of tan δ close to 1 indicates a distinct elastomeric behavior [21]. In our study, all hydrogels exhibited tan δ values less than 1, signifying a superiority of solid-like gel properties. Moreover, tan δ showed a weak frequency dependence, suggesting the presence of stable gel-like properties in the GA/EA cross-linked hydrogels. In addition, tan δ showed a similar trend concerning the concentration of GA or EA, and was highest at a 100% concentration, and then decreased to the lowest value at a concentration of 200%, based on the dry weight of collagen. These findings suggested that the hydrogels maintain inter- and intramolecular cross-linked bonds between collagen and GA or EA components. There was evidence of improvement in the density of intermolecular bonds in collagen, which could be due to hydrogen bonding or hydrophobic interactions.

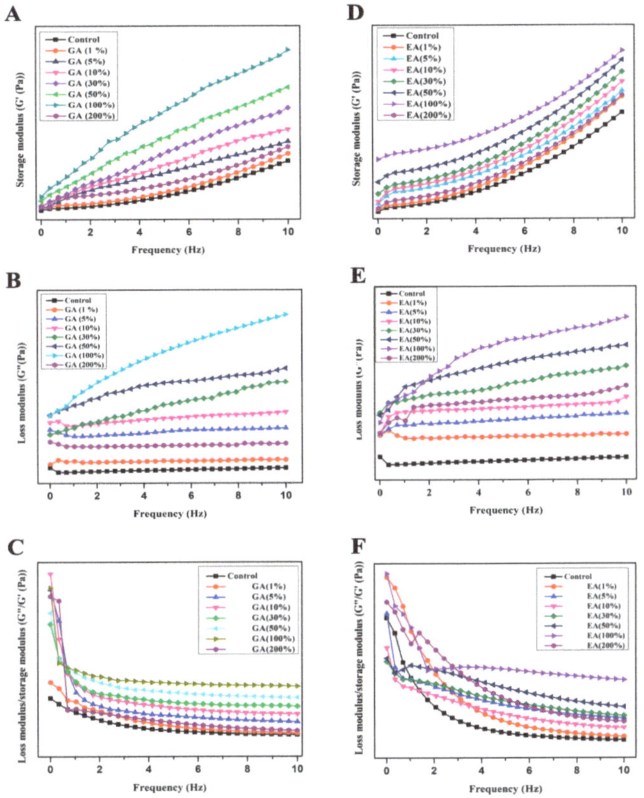

Figure 6. Effects of gallic acid and ellagic acid according to the dry weight of collagen with different concentrations on the dynamic viscoelasticity of GA/EA cross-linked and pure collagen hydrogels. (**A**) Storage modulus of gallic acid (G′). (**B**) Loss modulus of gallic acid (G″). (**C**) Loss factor of gallic acid (tan δ) (**D**) Storage modulus of ellagic acid (G′). (**E**) Loss modulus of ellagic acid (G″) (**F**) Loss factor of ellagic acid (tan δ).

3.9. Enzymatic Degradation of Hydrogels

In the present study, the amount of hydroxyproline after enzymatic degradation was used to evaluate the enzymatic inhibition ability of pristine hydrogels and GA/EA cross-linked collagen hydrogels. Figure 7 shows the percentage residue after enzymatic degradation of pristine and cross-linked hydrogels after immersion in collagenase solution at 37 °C for 24 h. As the concentration of GA or EA in the collagen hydrogels was increased, the percentage of the residual amount of collagen was also increased. In addition, pristine hydrogels have significantly ($p < 0.05$) lower % residues as compared to GA/EA cross-linked hydrogels. The findings indicated that the resistance to collagenase of GA or EA cross-linked collagen hydrogels was significantly ($p < 0.05$) higher compared to that of the native collagen hydrogels. On the other hand, hydrogels cross-linked with GA have notably better resistance compared to EA, particularly at GA 30% to GA 100%; for example, GA 5–100% achieved (28.25%, 42.09%, 56.20%, 62.41%, and 76.56%, respectively), and GA 200% achieved 30.50%. However, the EA percent (%) residue was enhanced from EA 5% to

EA 30%, (9.08–47.66%), and then decreased with the increasing concentration of EA; for example, EA 50% reached 34.98, EA 100% reached 32.66, and EA 200% reached 29.86 percent residue. Furthermore, the results showed that the pristine collagen hydrogels have lower resistance to the collagenase enzyme compared to cross-linked hydrogels. Additionally, phenolic acid cross-linked hydrogels were primarily dependent on the formation of a covalent bond between the polymer and the cross-linking agent, e.g., via the Michael or Schiff reaction, and, as a result, such hydrogels had resistance to enzymatic degradation. Moreover, according to Bam et al. [43] GA is a mixture of -OH and -COOH groups and was described as a potential cross-linker by increasing collagen stability, as indicated by greater resistance to collagenase activity. These functional groups can react with functional groups found in collagen, such as the -NH_2 and -OH groups. In general, the breaking of covalent bonds between intermolecular interactions, such as the van der Waals force and the hydrogen bond, requires more energy and has been associated with a lower degree of degradation [29].

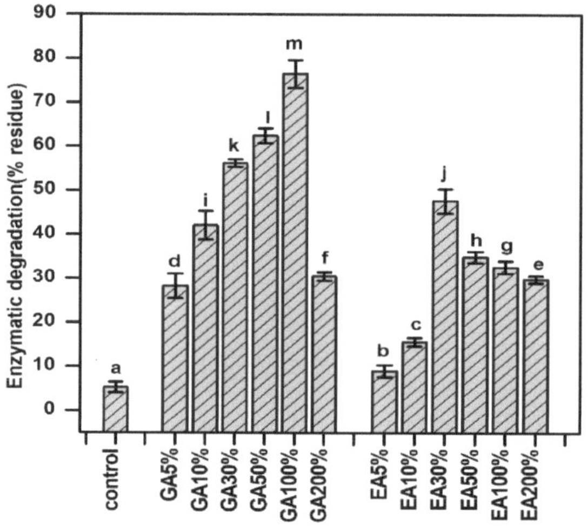

Figure 7. Enzymatic degradation of GA & EA cross-linked collagen hydrogels with different concentrations of gallic and ellagic acid (C) control, and 5%, 10%, 30%, 50%, 100%, and 200% were different concentrations according to the dry weight of collagen. Different letters (a, b, c, etc.) above the concentrations indicate significant ($p < 0.05$) differences.

3.10. Scanning Electron Microscopy (SEM)

The influence of GA or EA on the microstructure of collagen-based hydrogels was studied using scanning electron microscopy (SEM). Figure 8A–G demonstrates the SEM images and the surface areas of the hydrogels. SEM analysis revealed the presence of a typical tough assembly in both collagen and collagen-cross-linked hydrogels, which was attributed to the association of collagen molecules within the hydrogel matrix. In accordance with the results of XRD, FTIR, and physicochemical parameter analysis, the microstructure of the hydrogels showed changes depending on the presence of GA or EA and the concentration used. It was observed that hydrogels with a relatively rougher surface, showing an accumulation of material on the surface, were obtained at higher concentrations of GA or EA. As reported in previous studies [49,53], the increased irregularity of the hydrogel surface can be attributed to enhanced covalent and non-covalent interactions between collagen proteins and phenolic compounds [43]. The use of GA or EA to cross-link collagen molecules provides additional bonding opportunities, both between collagen molecules and within individual collagen molecules [33]. As a result, the structural properties of the hydrogel are densified and compacted. The modification in structural properties has been linked to the formation of intermolecular bonds between collagen molecules, resulting in a network of interconnected fibrils within the hydrogel matrix [34]. The hydrogels exhibited significant changes in microstructure as the concentration of GA or EA increased from 30% to 100% (as shown in Figure 8). In particular, the presence of fibrils within the hydrogels became more pronounced, and the number of fibrils increased at higher concentrations of either GA or EA. The hydrogels' fibrillar structure led to their increased water-retaining capacity. The increased number of fibrils increased the surface area of the hydrogel and improved its ability to retain water molecules, resulting in higher swelling values when compared to pristine hydrogels. The creation of fibrils can be linked to an increase in the density of intermolecular interactions between collagen and GA or EA, which promotes the tangling and aggregation of collagen molecules, resulting in a more interconnected and porous structure. These intermolecular associations include covalent bonds, hydrogen bonds, and numerous non-covalent interactions. The increasing predominance of these intermolecular linkages causes the collagen molecules to become entangled and intertwined, resulting in a denser structure [53]. Additionally, intermolecular associations included covalent bonds, hydrogen bonds, and numerous non-covalent interactions. The increasing predominance of these intermolecular linkages causes the collagen molecules to become entangled and intertwined, resulting in a denser structure [54]. The formation of intramolecular associations inhibited the mobility of collagen molecules and the resulting hydrogels had a more compact and dense microstructure.

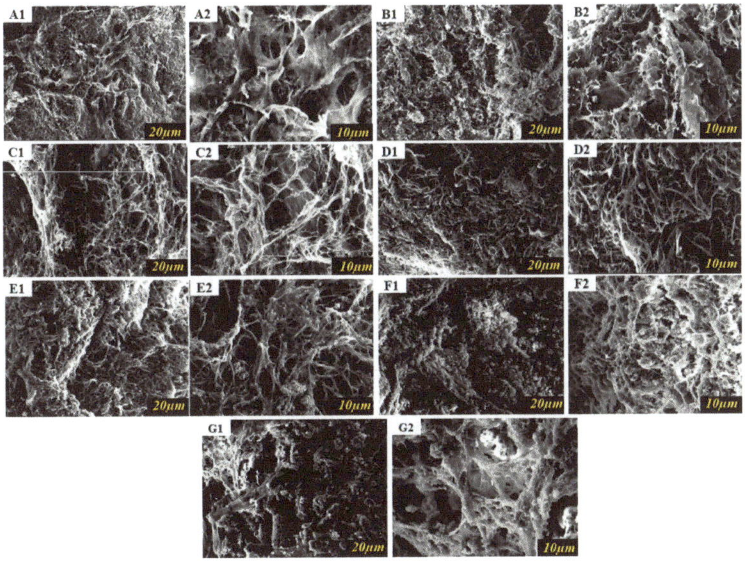

Figure 8. Morphological images of collagen hydrogels visualized by scanning electron microscopy. (**A1**,**A2**) SEM images of the pristine collagen; (**B1**,**B2**) SEM images of GA 30%; (**C1**,**C2**) SEM images of GA 50%; (**D1**,**D2**) SEM images of GA 100%; (**E1**,**E2**) SEM images of EA 30%; (**F1**,**F2**) SEM images of EA 50%; (**G1**,**G2**) SEM images of EA 100% crosslinked collagen hydrogels.

4. Conclusions

This study successfully presented fabricated collagen-based hydrogels cross-linked with GA or EA and demonstrated remarkable physicochemical, thermal, structural, and morphological properties. Cross-linking collagen-based hydrogels with GA and EA resulted in considerable improvements in a range of physicochemical parameters. These enhancements covered all the properties, such as the storage modulus, loss modulus, resistance to enzymatic degradation by type I collagenase, heat stability, porosity, swelling response, and water retention. The GA/EA cross-linking method was crucial in retaining the structural integrity of native collagen, as evidenced by XRD and FTIR spectroscopic investigations. Furthermore, SEM analysis demonstrated that the surface morphology of the GA/EA cross-linked hydrogels was improved when compared to the native hydrogels. Collectively, the results indicated that the incorporation of GA and EA into collagen hydrogels resulted in significant improvements in their enzymatic, thermal, and structural properties. These improvements were attributed to the formation of inter- and intramolecular cross-links between the collagen and GA/EA components. The cross-linked structure resulted in improved functional properties of the hydrogels by providing increased stability and resistance to degradation. The results of the present study indicated that the obtained GA/EA cross-linked collagen hydrogels are important for the design of collagen biomaterials for different biomedical applications and provide a promising approach to improve the properties and performance of collagen hydrogels for future studies.

Author Contributions: S.M.: Design the experiments, conducted the formal experiments, collected the data, carried out the statistical analysis, and drafted the manuscript. W.Y.: After peer review, revised the manuscript. J.L.: Designed and visualized the methodology. X.Y.: Programmed the software. T.Y.: Revised the figures after peer review. Y.H.: Designed the experiments. Participated in all aspects of the study and revised the manuscript. R.L. and J.Y.: Provided critical assistance in polishing the manuscript. S.X.: Supervised the work and solicited financial support. All authors have read and agreed to the published version of the manuscript.

Funding: This research received financial support from Hubei Provincial Natural Science Foundation (No. 2022CFB162), Huazhong Agricultural University-Agricultural Genomics Institute of Shenzhen Cooperation Fund (NO. SZYJY2021009), and China Agriculture Research System (NO. CARS-45-28).

Institutional Review Board Statement: Not applicable.

Data Availability Statement: The data presented in the current study is available on request from the corresponding author.

Acknowledgments: This study received financial support from Hubei Provincial Natural Science Foundation (No. 2022CFB162), Huazhong Agricultural University-Agricultural Genomics Institute of Shenzhen Cooperation Fund (NO. SZYJY2021009), and China Agriculture Research System (NO. CARS-45-28).

Conflicts of Interest: The authors declare no conflict of interest.

References

1. Tian, Z.; Duan, L.; Wu, L.; Shen, L.; Li, G. Rheological Properties of Glutaraldehyde-Crosslinked Collagen Solutions Analyzed Quantitatively Using Mechanical Models. *Mater. Sci. Eng. C* **2016**, *63*, 10–17. [CrossRef]
2. Deng, C.; Liu, Y.; Li, J.; Yadav, M.P.; Yin, L. Diverse Rheological Properties, Mechanical Characteristics and Microstructures of Corn Fiber Gum/Soy Protein Isolate Hydrogels Prepared by Laccase and Heat Treatment. *Food Hydrocoll.* **2018**, *76*, 113–122. [CrossRef]
3. Skopinska-Wisniewska, J.; Kuderko, J.; Bajek, A.; Maj, M.; Sionkowska, A.; Ziegler-Borowska, M. Collagen/Elastin Hydrogels Cross-Linked by Squaric Acid. *Mater. Sci. Eng. C* **2016**, *60*, 100–108. [CrossRef] [PubMed]
4. Yu, X.; Tang, C.; Xiong, S.; Yuan, Q.; Gu, Z.; Li, Z.; Hu, Y. Modification of Collagen for Biomedical Applications: A Review of Physical and Chemical Methods. *Curr. Org. Chem.* **2016**, *20*, 1797–1812. [CrossRef]
5. Myllyharju, J.; Kivirikko, K.I. Collagens and Collagen-Related Diseases. *Ann. Med.* **2001**, *33*, 7–21. [CrossRef] [PubMed]
6. Rajan, V.K.; Muraleedharan, K. A Computational Investigation on the Structure, Global Parameters and Antioxidant Capacity of a Polyphenol, Gallic Acid. *Food Chem.* **2017**, *220*, 93–99. [CrossRef] [PubMed]
7. Omotoso, G.O.; Gbadamosi, I.T.; Olajide, O.J.; Dada-Habeeb, S.O.; Arogundade, T.T.; Yawson, E.O. Moringa Oleifera Phytochemicals Protect the Brain against Experimental Nicotine-Induced Neurobehavioral Disturbances and Cerebellar Degeneration. *Pathophysiology* **2018**, *25*, 57–62. [CrossRef] [PubMed]
8. El-Lakkany, N.M.; El-Maadawy, W.H.; Seif el-Din, S.H.; Saleh, S.; Safar, M.M.; Ezzat, S.M.; Mohamed, S.H.; Botros, S.S.; Demerdash, Z.; Hammam, O.A. Antifibrotic Effects of Gallic Acid on Hepatic Stellate Cells: In Vitro and in Vivo Mechanistic Study. *J. Tradit. Complement. Med.* **2018**, *9*, 45–53. [CrossRef]
9. Gong, W.; Wang, R.; Huang, H.; Hou, Y.; Wang, X.; He, W.; Gong, X.; Hu, J. Construction of Double Network Hydrogels Using Agarose and Gallic Acid with Antibacterial and Anti-Inflammatory Properties for Wound Healing. *Int. J. Biol. Macromol.* **2023**, *227*, 698–710. [CrossRef]

10. Yu, S.-H.; Mi, F.-L.; Pang, J.-C.; Jiang, S.-C.; Kuo, T.-H.; Wu, S.-J.; Shyu, S.-S. Preparation and Characterization of Radical and pH-Responsive Chitosan–Gallic Acid Conjugate Drug Carriers. *Carbohydr. Polym.* **2011**, *84*, 794–802. [CrossRef]
11. Kang, B.; Vales, T.; Cho, B.-K.; Kim, J.-K.; Kim, H.-J. Development of Gallic Acid-Modified Hydrogels Using Interpenetrating Chitosan Network and Evaluation of Their Antioxidant Activity. *Molecules* **2017**, *22*, 1976. [CrossRef] [PubMed]
12. Thanyacharoen, T.; Chuysinuan, P.; Techasakul, S.; Nooeaid, P.; Ummartyotin, S. Development of a Gallic Acid-Loaded Chitosan and Polyvinyl Alcohol Hydrogel Composite: Release Characteristics and Antioxidant Activity. *Int. J. Biol. Macromol.* **2018**, *107*, 363–370. [CrossRef] [PubMed]
13. Jiang, H.; Kobayashi, T. Ultrasound Stimulated Release of Gallic Acid from Chitin Hydrogel Matrix. *Mater. Sci. Eng. C* **2017**, *75*, 478–486. [CrossRef] [PubMed]
14. Al-Obaidi, M.M.J.; Al-Bayaty, F.H.; Al Batran, R.; Hassandarvish, P.; Rouhollahi, E. Protective Effect of Ellagic Acid on Healing Alveolar Bone after Tooth Extraction in Rat—A Histological and Immunohistochemical Study. *Arch. Oral Biol.* **2014**, *59*, 987–999. [CrossRef] [PubMed]
15. Ruan, J.; Yang, Y.; Yang, F.; Wan, K.; Fan, D.; Wang, D. Novel Oral Administrated Ellagic Acid Nanoparticles for Enhancing Oral Bioavailability and Anti-Inflammatory Efficacy. *J. Drug Deliv. Sci. Technol.* **2018**, *46*, 215–222. [CrossRef]
16. Guvvala, P.R.; Ravindra, J.P.; Selvaraju, S.; Arangasamy, A.; Venkata, K.M. Ellagic and Ferulic Acids Protect Arsenic-Induced Male Reproductive Toxicity via Regulating Nfe2l2, Ppargc1a and StAR Expressions in Testis. *Toxicology* **2019**, *413*, 1–12. [CrossRef] [PubMed]
17. Tokutomi, H.; Takeda, T.; Hoshino, N.; Akutagawa, T. Molecular Structure of the Photo-Oxidation Product of Ellagic Acid in Solution. *ACS Omega* **2018**, *3*, 11179–11183. [CrossRef]
18. Zuccari, G.; Baldassari, S.; Ailuno, G.; Turrini, F.; Alfei, S.; Caviglioli, G. Formulation Strategies to Improve Oral Bioavailability of Ellagic Acid. *Appl. Sci.* **2020**, *10*, 3353. [CrossRef]
19. Sharifi-Rad, J.; Quispe, C.; Castillo, C.M.S.; Caroca, R.; Lazo-Vélez, M.A.; Antonyak, H.; Polishchuk, A.; Lysiuk, R.; Oliinyk, P.; De Masi, L.; et al. Ellagic Acid: A Review on Its Natural Sources, Chemical Stability, and Therapeutic Potential. *Oxid. Med. Cell. Longev.* **2022**, *2022*, 3848084. [CrossRef]
20. Huang, Z.; Delparastan, P.; Burch, P.; Cheng, J.; Cao, Y.; Messersmith, P.B. Injectable Dynamic Covalent Hydrogels of Boronic Acid Polymers Cross-Linked by Bioactive Plant-Derived Polyphenols. *Biomater. Sci.* **2018**, *6*, 2487–2495. [CrossRef]
21. Zhu, S.; Gu, Z.; Xiong, S.; An, Y.; Liu, Y.; Yin, T.; You, J.; Hu, Y. Fabrication of a Novel Bio-Inspired Collagen–Polydopamine Hydrogel and Insights into the Formation Mechanism for Biomedical Applications. *RSC Adv.* **2016**, *6*, 66180–66190. [CrossRef]
22. Liu, L.; Wen, H.; Rao, Z.; Zhu, C.; Liu, M.; Min, L.; Fan, L.; Tao, S. Preparation and Characterization of Chitosan—Collagen Peptide/Oxidized Konjac Glucomannan Hydrogel. *Int. J. Biol. Macromol.* **2018**, *108*, 376–382. [CrossRef] [PubMed]
23. Loh, Q.L.; Choong, C. Three-Dimensional Scaffolds for Tissue Engineering Applications: Role of Porosity and Pore Size. *Tissue Eng. Part B Rev.* **2013**, *19*, 485–502. [CrossRef] [PubMed]
24. Foudazi, R.; Zowada, R.; Manas-Zloczower, I.; Feke, D.L. Porous Hydrogels: Present Challenges and Future Opportunities. *Langmuir* **2023**, *39*, 2092–2111. [CrossRef] [PubMed]
25. Lu, Z.; Gao, J.; He, Q.; Wu, J.; Liang, D.; Yang, H.; Chen, R. Enhanced Antibacterial and Wound Healing Activities of Microporous Chitosan-Ag/ZnO Composite Dressing. *Carbohydr. Polym.* **2016**, *156*, 460–469. [CrossRef] [PubMed]
26. Das, S.; Baker, A.B. Biomaterials and Nanotherapeutics for Enhancing Skin Wound Healing. *Front. Bioeng. Biotechnol.* **2016**, *4*, 82. [CrossRef] [PubMed]
27. Kaczmarek, B.; Mazur, O. Collagen-Based Materials Modified by Phenolic Acids—A Review. *Materials* **2020**, *13*, 3641. [CrossRef]
28. Chuang, C.-H.; Lin, R.-Z.; Melero-Martin, J.M.; Chen, Y.-C. Comparison of Covalently and Physically Cross-Linked Collagen Hydrogels on Mediating Vascular Network Formation for Engineering Adipose Tissue. *Artif. Cells Nanomed. Biotechnol.* **2018**, *46*, S434–S447. [CrossRef]
29. Sapuła, P.; Bialik-Wąs, K.; Malarz, K. Are Natural Compounds a Promising Alternative to Synthetic Cross-Linking Agents in the Preparation of Hydrogels? *Pharmaceutics* **2023**, *15*, 253. [CrossRef]
30. Lin, H.; Dan, W.; Dan, N. The Water State in Crosslinked Poly(Vinyl Alcohol)–Collagen Hydrogel and Its Swelling Behavior. *J. Appl. Polym. Sci.* **2012**, *123*, 2753–2761. [CrossRef]
31. Caló, E.; Khutoryanskiy, V.V. Biomedical Applications of Hydrogels: A Review of Patents and Commercial Products. *Eur. Polym. J.* **2015**, *65*, 252–267. [CrossRef]
32. Wang, M.; Li, J.; Li, W.; Du, Z.; Qin, S. Preparation and Characterization of Novel Poly (Vinyl Alcohol)/Collagen Double-Network Hydrogels. *Int. J. Biol. Macromol.* **2018**, *118*, 41–48. [CrossRef] [PubMed]
33. Yu, C.; Naeem, A.; Liu, Y.; Guan, Y. Ellagic Acid Inclusion Complex-Loaded Hydrogels as an Efficient Controlled Release System: Design, Fabrication and In Vitro Evaluation. *J. Funct. Biomater.* **2023**, *14*, 278. [CrossRef] [PubMed]
34. Zhang, T.; Guo, L.; Li, R.; Shao, J.; Lu, L.; Yang, P.; Zhao, A.; Liu, Y. Ellagic Acid–Cyclodextrin Inclusion Complex-Loaded Thiol–Ene Hydrogel with Antioxidant, Antibacterial, and Anti-Inflammatory Properties for Wound Healing. *ACS Appl. Mater. Interfaces* **2023**, *15*, 4959–4972. [CrossRef]
35. Buitimea-Cantúa, N.E.; Gutiérrez-Uribe, J.A.; Serna-Saldívar, S.O. Phenolic–Protein Interactions: Effects on Food Properties and Health Benefits. *J. Med. Food* **2017**, *21*, 188–198. [CrossRef]
36. Tian, X.; Wang, Y.; Duan, S.; Hao, Y.; Zhao, K.; Li, Y.; Dai, R.; Wang, W. Evaluation of a Novel Nano-Size Collagenous Matrix Film Cross-Linked with Gallotannins Catalyzed by Laccase. *Food Chem.* **2021**, *351*, 129335. [CrossRef]

37. Quan, T.H.; Benjakul, S.; Sae-leaw, T.; Balange, A.K.; Maqsood, S. Protein–Polyphenol Conjugates: Antioxidant Property, Functionalities, and Their Applications. *Trends Food Sci. Technol.* **2019**, *91*, 507–517. [CrossRef]
38. Yan, M.; An, X.; Duan, S.; Jiang, Z.; Liu, X.; Zhao, X.; Li, Y. A Comparative Study on Cross-Linking of Fibrillar Gel Prepared by Tilapia Collagen and Hyaluronic Acid with EDC/NHS and Genipin. *Int. J. Biol. Macromol.* **2022**, *213*, 639–650. [CrossRef]
39. Durga, R.; Jimenez, N.; Ramanathan, S.; Suraneni, P.; Pestle, W.J. Use of Thermogravimetric Analysis to Estimate Collagen and Hydroxyapatite Contents in Archaeological Bone. *J. Archaeol. Sci.* **2022**, *145*, 105644. [CrossRef]
40. Boles, J.S.; Crerar, D.A.; Grissom, G.; Key, T.C. Aqueous Thermal Degradation of Gallic Acid. *Geochim. Cosmochim. Acta* **1988**, *52*, 341–344. [CrossRef]
41. Zeugolis, D.I.; Paul, G.R.; Attenburrow, G. Cross-Linking of Extruded Collagen Fibers—A Biomimetic Three-Dimensional Scaffold for Tissue Engineering Applications. *J. Biomed. Mater. Res. Part A* **2009**, *89A*, 895–908. [CrossRef] [PubMed]
42. Zhang, M.; Li, J.; Ding, C.; Liu, W.; Li, G. The Rheological and Structural Properties of Fish Collagen Cross-Linked by N-Hydroxysuccinimide Activated Adipic Acid. *Food Hydrocoll.* **2013**, *30*, 504–511. [CrossRef]
43. Bam, P.; Bhatta, A.; Krishnamoorthy, G. Design of Biostable Scaffold Based on Collagen Crosslinked by Dialdehyde Chitosan with Presence of Gallic Acid. *Int. J. Biol. Macromol.* **2019**, *130*, 836–844. [CrossRef] [PubMed]
44. Deming, T.J. Synthesis of Side-Chain Modified Polypeptides. *Chem. Rev.* **2016**, *116*, 786–808. [CrossRef] [PubMed]
45. Yu, C.; Chen, X.; Zhu, W.; Li, L.; Peng, M.; Zhong, Y.; Naeem, A.; Zang, Z.; Guan, Y. Synthesis of Gallic Acid-Loaded Chitosan-Grafted-2-Acrylamido-2-Methylpropane Sulfonic Acid Hydrogels for Oral Controlled Drug Delivery: In Vitro Biodegradation, Antioxidant, and Antibacterial Effects. *Gels* **2022**, *8*, 806. [CrossRef] [PubMed]
46. Hu, Y.; Dan, W.; Xiong, S.; Kang, Y.; Dhinakar, A.; Wu, J.; Gu, Z. Development of Collagen/Polydopamine Complexed Matrix as Mechanically Enhanced and Highly Biocompatible Semi-Natural Tissue Engineering Scaffold. *Acta Biomater.* **2017**, *47*, 135–148. [CrossRef] [PubMed]
47. Xiao, Y.; Wang, C.; Zhou, J.; Wu, J.; Lin, W. Modular Design of Vegetable Polyphenols Enables Covalent Bonding with Collagen for Eco-Leather. *Ind. Crops Prod.* **2023**, *204*, 117394. [CrossRef]
48. Andonegi, M.; Heras, K.L.; Santos-Vizcaíno, E.; Igartua, M.; Hernandez, R.M.; de la Caba, K.; Guerrero, P. Structure-Properties Relationship of Chitosan/Collagen Films with Potential for Biomedical Applications. *Carbohydr. Polym.* **2020**, *237*, 116159. [CrossRef]
49. Yu, X.; Yuan, Q.; Yang, M.; Liu, R.; Zhu, S.; Li, J.; Zhang, W.; You, J.; Xiong, S.; Hu, Y. Development of Biocompatible and Antibacterial Collagen Hydrogels via Dialdehyde Polysaccharide Modification and Tetracycline Hydrochloride Loading. *Macromol. Mater. Eng.* **2019**, *304*, 1800755. [CrossRef]
50. Wu, X.; Liu, Y.; Liu, A.; Wang, W. Improved Thermal-Stability and Mechanical Properties of Type I Collagen by Crosslinking with Casein, Keratin and Soy Protein Isolate Using Transglutaminase. *Int. J. Biol. Macromol.* **2017**, *98*, 292–301. [CrossRef]
51. Xie, M.; Hu, B.; Yan, Y.; Zhou, L.; Ou, S.; Zeng, X. Rheological Properties of Gallic Acid-Grafted-Chitosans with Different Substitution Degrees. *LWT* **2016**, *74*, 472–479. [CrossRef]
52. Choi, I.; Lee, S.E.; Chang, Y.; Lacroix, M.; Han, J. Effect of Oxidized Phenolic Compounds on Cross-Linking and Properties of Biodegradable Active Packaging Film Composed of Turmeric and Gelatin. *LWT* **2018**, *93*, 427–433. [CrossRef]
53. Jayachandran, B.; Parvin, T.N.; Alam, M.M.; Chanda, K.; MM, B. Insights on Chemical Crosslinking Strategies for Proteins. *Molecules* **2022**, *27*, 8124. [CrossRef] [PubMed]
54. Amirrah, I.N.; Lokanathan, Y.; Zulkiflee, I.; Wee, M.F.M.R.; Motta, A.; Fauzi, M.B. A Comprehensive Review on Collagen Type I Development of Biomaterials for Tissue Engineering: From Biosynthesis to Bioscaffold. *Biomedicines* **2022**, *10*, 2307. [CrossRef]

polymers

MDPI

Article

Cellulose in Secondary Xylem of Cactaceae: Crystalline Composition and Anatomical Distribution

Agustín Maceda [1,*], Marcos Soto-Hernández [2] and Teresa Terrazas [1,*]

1 Instituto de Biología, Universidad Nacional Autónoma de México, Mexico City 04510, Mexico
2 Programa de Botánica, Colegio de Postgraduados, Montecillo 56230, Mexico
* Correspondence: biologoagustin@hotmail.com (A.M.); tterrazas@ib.unam.mx (T.T.); Tel.: +5255-5622-9116 (A.M. & T.T.)

Abstract: Cellulose is the main polymer that gives strength to the cell wall and is located in the primary and secondary cell walls of plants. In Cactaceae, there are no studies on the composition of cellulose. The objective of this work was to analyze the crystallinity composition and anatomical distribution of cellulose in Cactaceae vascular tissue. Twenty-five species of Cactaceae were collected, dried, and milled. Cellulose was purified and analyzed with Fourier transform infrared spectroscopy, the crystallinity indexes were calculated, and statistical analyzes were performed. Stem sections were fixed, cut, and stained with safranin O/fast green, for observation with epifluorescence microscopy. The crystalline cellulose ratios had statistical differences between *Echinocereus pectinatus* and *Coryphantha pallida*. All cacti species presented a higher proportion of crystalline cellulose. The fluorescence emission of the cellulose was red in color and distributed in the primary wall of non-fibrous species; while in the fibrous species, the distribution was in the pits. The high percentages of crystalline cellulose may be related to its distribution in the non-lignified parenchyma and primary walls of tracheary elements with helical or annular thickenings of non-fibrous species, possibly offering structural rigidity and forming part of the defense system against pathogens.

Keywords: crystalline cellulose; FTIR; fluorescence microscopy; crystallinity indexes; Cactaceae

Citation: Maceda, A.; Soto-Hernández, M.; Terrazas, T. Cellulose in Secondary Xylem of Cactaceae: Crystalline Composition and Anatomical Distribution. *Polymers* **2022**, *14*, 4840. https://doi.org/10.3390/polym14224840

Academic Editor: Antonio M. Borrero-López

Received: 18 October 2022
Accepted: 8 November 2022
Published: 10 November 2022

1. Introduction

Plant cell walls give rigidity, delimit the cell space, and function as a physical barrier against pathogens [1]. The main polymers of the cell wall are cellulose, hemicelluloses, and lignin, in addition to other compounds such as pectins, structural proteins, and glycoproteins [2]. The cell wall is divided in two: the primary wall with abundant cellulose microfibers, some hemicelluloses, xyloglucans, and pectins [3]. The primary wall develops during cell growth and maintains some elasticity during the initial stage. The secondary wall has three layers: S1, S2, and S3. A secondary wall develops once cell growth ends, mainly lignin accumulates, and cellulose occurs in a lesser quantity than other polysaccharides such as hemicelluloses [4].

Cellulose is the main component of the cell wall [1] and is a homopolysaccharide composed of repeated glucose residues linked by β-(1→4) bonds that generate long and rigid microfibrils [5] made of 18 cellulose polymers [6]. The presence of cellulose microfibrils in the cell wall confers mechanical and enzymatic resistance [7], structural rigidity in the primary wall [2], and forms the scaffolding for binding with pectins and hemicelluloses [5]. Two types of cellulose are characterized by their orientation and packaging. The cellulose formed by microfibrils is also called crystalline cellulose [8], which is packaged because the microfibrils are linked by hydrogen β-(1→4)-linked D-glucose units, making it more compact, rigid, and ordered [9]. The cellulose matrix that has no order (amorphous) forms a network where the hemicelluloses, pectins, lignin, and phenolic compounds are inserted and joined [3,10].

The importance of studying the crystalline composition of cellulose is due to its use in the paper industry [8] and for the production of biofuels [11]. In addition, Cactaceae species are considered second generation plants for biofuel production because, despite their slow growth, they resist drought conditions and high temperatures, and are not an essential part of human consumption [12]. The species with the highest amount of amorphous cellulose [13] and the lowest presence of lignin or lignin, and with the highest accumulation of syringyl monomers [14], are those that have potential for use in the previously mentioned industries. On the other hand, cellulose studies also focus on anatomical-structural analysis, to identify cell wall interactions with biotic and abiotic factors [15]. At this point, different authors have analyzed the stress effects of the environment on cellulose accumulation [16] and the interaction between cellulose and pathogens [17].

In succulent species, studies are scarce and most have focused on the genus *Agave*, a member of the Asparagaceae family [18]. In the Cactaceae family, few studies have identified the chemical composition of lignin [19–22]; furthermore, the crystallinity of cellulose was only analyzed in *Opuntia ficus-indica*, due to the potential use of its cellulosic compounds in the biofuel and paper industry [23]. However, of the other species of the Opuntioideae, Pereskioideae, and Cactoideae subfamilies, there have been no studies on the composition of cellulose and its distribution in the secondary xylem (wood), which would allow us to understand the structure and functioning of cellulose in the cell wall of the parenchyma, water conductive, and supporting cells. Therefore, this study aimed to analyze the crystalline composition of cellulose and its distribution in wood. The hypothesis was that there would be variation in the crystalline composition of cellulose among the different groups of cacti, due to the type of wood present.

2. Materials and Methods

2.1. Crystalline Analysis of Cellulose

To analyze the crystalline cellulose of Cactaceae species, adult and healthy plants of representative species of Cactaceae (Table 1) were collected. The vascular cylinder of all species was isolated and dried for 72 h. After that, 2 g of each species was weighed, dried, and ground. Subsequently, to obtain free-extractive wood, successive extractions were applied with ethanol: benzene (1:2), ethanol 96%, and hot water at 90 °C, according to the method proposed by Reyes-Rivera and Terrazas [24].

For the free-extractive wood, cellulose purification was performed using the procedure of Maceda et al. [20] based on the Kürschner-Hoffer method. Whereby, 0.5 g of extractive-free wood was weighed and added to 12.5 mL of HNO3/ethanol (1:4 v/v), kept in a reflux system for 1 h, allowed to cool to room temperature, and the sample was decanted. Subsequently, 12.5 mL of HNO_3/ethanol was added and the process of reflux, cooling, and decantation was repeated two more times. In the last process, 12.5 mL of 1% aqueous KOH solution was added, kept under reflux for an additional 30 min, and finally filtered through a fine-pore Büchner filter. The residue (cellulose) was dried at 60 °C for 12 h.

Obtainment of the crystalline and amorphous cellulose proportion was performed with attenuated total reflection Fourier transform infrared spectroscopy (ATR-FTIR), because the covalent and non-covalent interactions of cellulose could be identified, and the crystallinity of cellulose could be measured [25]. A small amount of dry cellulose from each species was taken, readings in the spectrum range of 4000–650 cm^{-1} (30 scans with a resolution of 4 cm^{-1}, 15 s per sample) of each sample were made in an FTIR Spectrometer (Agilent Cary 630 FTIR), and a baseline correction was made with MicroLab PC software (Agilent Technologies, Santa Clara, CA, USA). The raw spectra were converted from transmittance to absorbance, and the average of three spectra was obtained using the Resolution Pro FTIR Software program (Agilent Technologies, Santa Clara, CA, USA).

The total crystallinity index (TCI) proposed by Nelson and O'Connor [26], or also called the proportion of crystallinity [13], was calculated from the ratio between the intensity absorption peaks 1370 cm^{-1} and 2900 cm^{-1} [27]. The lateral order index (LOI) [26,27], or the second proportion of crystallinity [13], was calculated from the ratio between the intensity absorption peaks 1430 cm^{-1} and 893 cm^{-1}. Hydrogen bonding intensity (HBI) was calculated from the ratio between 3350 cm^{-1} and 1315 cm^{-1} [28].

The data obtained in triplicate for TCI, LOI, and HBI for each species were analyzed with non-parametric statistics, due to the data having no normality with the Kolmogorov-Smirnov and Shapiro–Wilk tests, even with the square root and log transformations. A non-parametric Kruskal–Wallis test was used to determine if there were differences among the species, followed by Dunn's post hoc test.

Table 1. Morpho-anatomical characteristics of the 25 species of Cactaceae.

Subfamily	Species	Collection Number	Wood Type	Stem
Cactoideae	*Coryphantha clavata* (Scheidw.) Backeb.	BV2535	Non-fibrous	Cylindrical
	C. cornifera (DC.) Lem.	BV2534	Non-fibrous	Cylindrical
	C. delaetiana (Quehl) A. Berger	BV2542	Non-fibrous	Globose
	C. delicata L. Bremer	SA1927	Non-fibrous	Cylindrical
	C. hintoriorum Dicht & A. Lüthy	BV2539	Non-fibrous	Cylindrical
	C. macromeris (Engelmann) Lemaire	BV2600	Non-fibrous	Globose
	C. pallida Britton & Rose	SA860	Non-fibrous	Globose
	C. pseudoechinus Boed.	BV2543	Non-fibrous	Cylindrical
	C. ramillosa Cutak	HSM3775	Non-fibrous	Globose
	C. retusa Britton & Rose	SG55	Non-fibrous	Globose
	Echinocereus cinerascens (DC.) Lem. subsp. *tulensis*	SA1744	Non-fibrous	Cylindrical
	E. dasyacanthus Engelm.	SA2077	Non-fibrous	Cylindrical
	E. pectinatus (Scheidw.) Engelm.	SA1918	Non-fibrous	Cylindrical
	E. pentalophus (DC.) Lem.	SA1740	Non-fibrous	Cylindrical
	Mammillaria carnea Zucc. Ex Pfeiff.	DA241	Non-fibrous	Cylindrical
	M. dixathocentron Backeb. Ex Mottram	CPNL133	Non-fibrous	Cylindrical
	M. magnifica Buchenau	UG1411	Non-fibrous	Columnar
	M. mystax Mart.	DA238	Non-fibrous	Cylindrical
	Neolloydia conoidea (DC.) Britton & Rose	BV2595	Non-fibrous	Cylindrical
Opuntioideae	*Cylindropuntia imbricata* (Haw.) F. M. Knuth	TT990	Fibrous	Tree
	C. kleiniae (DC.) F. M. Knuth	TT1000	Fibrous	Shrub
	C. leptocaulis (DC.) F. M. Knuth	TT994	Fibrous	Shrub
	Opuntia stenopetala Lem.	TT997	Fibrous	Shrub
	O. stricta (Haw.) Haw.	TT998	Fibrous	Shrub
Pereskioideae	*Leuenbergeria lychnidiflora* (DC.) Lodé	TT967	Fibrous	Tree

The vouchers were deposited in the National Herbarium of Mexico (MEXU). Initial collectors were BV, Balbina Vázquez; SA, Salvador Arias; HSM, Hernando Sánchez-Mejorada; UG, Ulises Guzmán; DA, David Aquino; CPNL, Carmen P. Novoa; TT, Teresa Terrazas. SA, verified species identification.

2.2. *Epifluorescence Microscopy*

To analyze the anatomical distribution of cellulose in Cactaceae species, epifluorescent microscopy was used. Epifluorescence is a technique used to identify the distribution of structural components such as lignin, cellulose, phenolic compounds, starch, and proteins [29]. In the particular case of epifluorescence, it allows the observation of thin sections of plant samples, where the tissues are anisotropic [30]. The method used to make the observations was based on the results obtained by De Micco and Aronne [31] and Maceda et al. [21,22] with epifluorescence and safranin O/fast green staining (SF). However, to confirm that the fluorescence emission by safranin/fast green staining corresponded to the fluorescence of lignified tissues and cellulose, samples of *Ferocactus latispinus* (HAW.) Britton and Rose were used as a model for comparison with two other typical stains for fluorescence, such as acridine orange–Congo red (AO), in addition to calcofluor (CA), based on the method proposed by Nakaba et al. [32].

Representative samples of wood from the base of the stem of the Cactaceae species were obtained, fixed with a solution of formalin–acetic acid–ethanol (10:5:85) [33], and washed, before dehydrating the samples [30]. The non-fibrous species samples were embedded with paraffin and cut with a rotatory microtome, and the fibrous species were cut with a sliding microtome [20]. All the samples were stained with safranin O/fast green and mounted in synthetic resin according to Loza-Cornejo and Terrazas [34].

Due to the different structural components of the cell wall, presenting autofluorescence and fluorescence emission [29], three different excitation and emission bands were used at the same time [21]. This process allowed obtaining images with a "true color" [29]. Therefore, a wide-field fluorescence microscope (Zeiss Axio Imager Z2) with Apotome 2.0 (Zeiss Apotome.2), an AxioCam MRc 5 (Zeiss), and a microscope metal halide fluorescence light source (Zeiss HXP 120) were used. The multicolor images were obtained with a triple excitation/emission bandwidth: DAPI (blue) with an excitation of 365 nm and emission of 445/50 nm, FITC (green) with an excitation of 470/40 nm and emission of 525/50 nm, and TRITC (red) with an excitation of 546/12 nm and emission of 575–640 nm. Each sample was exposed to fluorescence light with a low power for a maximum of one minute, as proposed by Baldacci-Cresp et al. [35], to avoid overexposure of the samples and cause their photobleaching. Images were obtained with the Zen Blue 2.5 lite (Zeiss, Jena, Germany) program, and brightness adjustments were made to the entire image.

3. Results

3.1. Cellulose Composition

Figure 1 illustrates the cellulose spectra and Table 2 shows the allocation of the main bands. The absence or presence of small bands at 1269 cm^{-1} corresponding to lignin and hemicelluloses; 1595 cm^{-1}, 1512 cm^{-1}, and 1463 cm^{-1} assigned to lignin; and 1735 cm^{-1} corresponding to xylenes and hemicelluloses reflected the effectiveness of the extraction and cellulose purification. The presence of 1640 cm^{-1} reflected the O-H vibration of absorbed water. The parameters for crystallinity, TCI, LOI, and HBI, had statistically significant differences ($p < 0.05$) with a non-parametric Kruskal–Wallis analysis (Table 3); and using Dunn's post hoc test, the species that were statistically different were identified (Table 4 and the website on Data Availability Statement contain the tables for each variable and the comparison between species).

The TCI showed that all species had values above one, because cacti have a higher proportion of crystalline cellulose than amorphous cellulose (Table 4). *E. pectinatus* had the highest TCI values and had significant differences with *C. pallida* and *C. clavata*, which presented the lowest values (Table 4). The lateral order index (LOI) reflected the degree of order in cellulose and the presence of crystalline cellulose II or amorphous cellulose; all species had similar values, except for *E. pectinatus, C. pallida,* and *C. ramillosa,* because the last two presented significant differences from the first (Table 4). In HBI, which relates to the crystal system and the bound water, the significant differences were between *C. pallida* with the highest values, and *E. pectinatus*, which presented the lowest HBI of the cacti (Table 4). The crystallinity indexes TCI and HBI were high in all Cactaceae species, but had lower values in the LOI index, because the cacti species had a lesser order (disorder) in the crystalline structure, but a higher proportion of crystalline cellulose.

Table 2. Assignment of FTIR absorption bands for the cellulose of Cactaceae species.

Wavenumber (cm^{-1})	Assignments
3000–3600	OH stretching [27]
2900	CH stretching [13,27]
1430	CH_2 symmetric bending (crystalline and amorphous cellulose) [13,26,27]
1372	C-H and C-O bending vibration bonds [27]
1336	C-O-H in-plane bending (amorphous cellulose) [27]
1315	CH_2 wagging vibration (crystalline cellulose) [26]
1163	C-O-C asymmetrical stretching [36]
893	Out-of-plane asymmetrical stretching of cellulose ring [13]
670	C-O-H out-of-plane stretching [37]

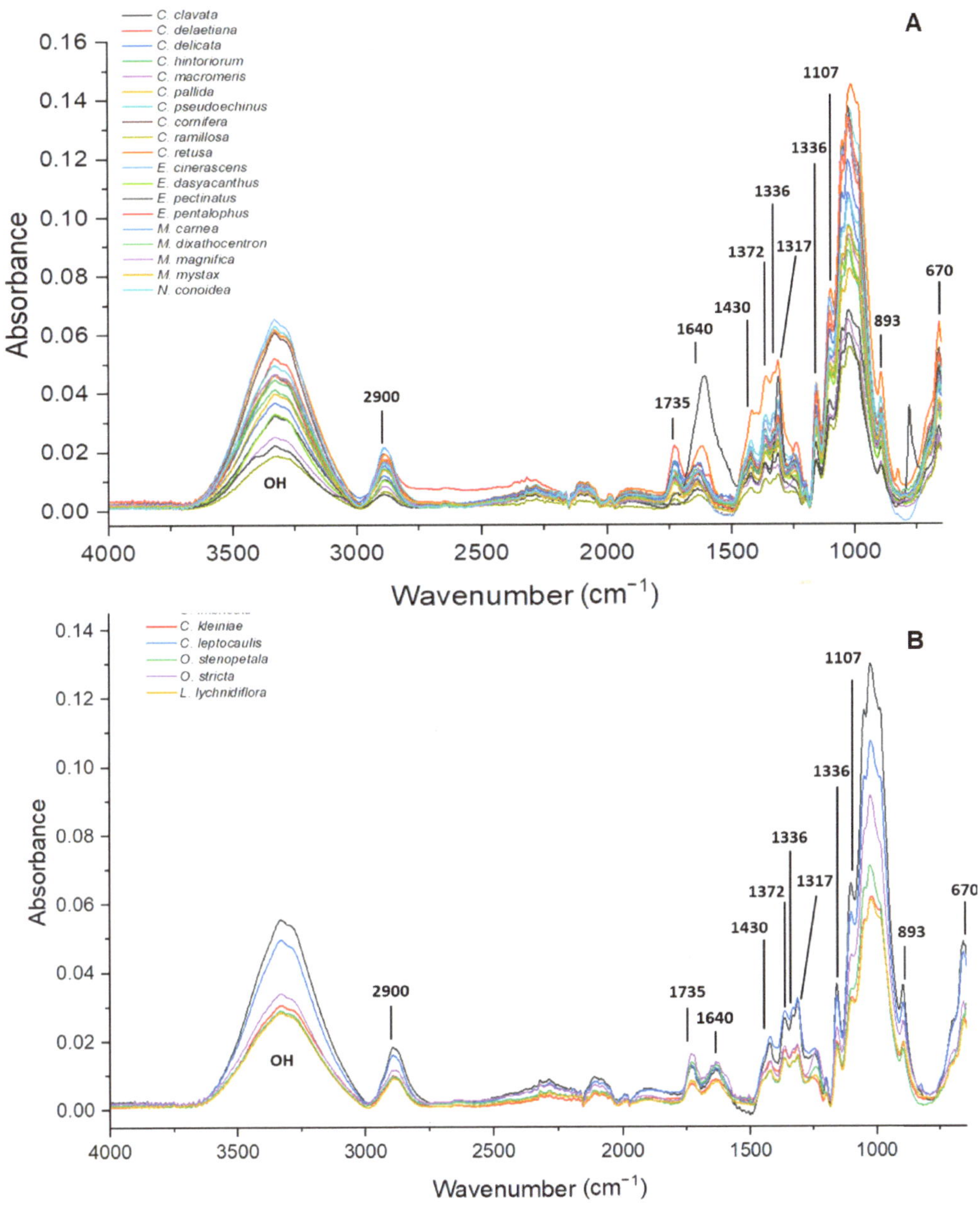

Figure 1. FTIR spectra for obtaining cellulose crystalline indexes for the 25 Cactaceae species. (**A**) Non-fibrous species. (**B**) Fibrous species.

Table 3. Kruskal–Wallis analysis for the crystalline indexes and hydrogen bond index.

Crystalline Indexes	χ-Square	Df	Significance
TCI	65.81053	24	9.25×10^{-6}
LOI	52.00702	24	7.81×10^{-4}
HBI	61.89333	24	3.44×10^{-5}

Table 4. Crystallinity indexes with standard deviations, to determine the crystalline cellulose of the 25 Cactaceae species.

Species	TCI (A1370/A2900)	LOI (A1430/A893)	HBI (A3400/A1315)
C. pallida	1.118 ± 0.019 a	0.503 ± 0.007 a	1.365 ± 0.032 a
C. clavata	1.126 ± 0.012 ab	0.507 ± 0.014 ab	1.301 ± 0.048 ab
C. delaetiana	1.263 ± 0.020 abcd	0.515 ± 0.020 ab	1.148 ± 0.018 ab
C. delicata	1.227 ± 0.021 abcd	0.523 ± 0.010 ab	1.231 ± 0.033 ab
C. hintoriorum	1.273 ± 0.022 abcd	0.561 ± 0.009 ab	1.131 ± 0.024 ab
C. macromeris	1.291 ± 0.023 abcd	0.565 ± 0.009 ab	1.131 ± 0.024 ab
C. pseudoechinus	1.197 ± 0.030 abcd	0.540 ± 0.009 ab	1.287 ± 0.030 ab
E. cinerascens	1.198 ± 0.026 abcd	0.538 ± 0.011 ab	1.339 ± 0.037 ab
E. dasyacanthus	1.234 ± 0.039 abcd	0.535 ± 0.014 ab	1.202 ± 0.034 ab
E. pentalophus	1.165 ± 0.018 abcd	0.540 ± 0.012 ab	1.270 ± 0.028 ab
M. dixathocentron	1.243 ± 0.021 abcd	0.513 ± 0.012 ab	1.279 ± 0.026 ab
M. magnifica	1.219 ± 0.025 abcd	0.532 ± 0.012 ab	1.289 ± 0.025 ab
M. mystax	1.179 ± 0.028 abcd	0.511 ± 0.013 ab	1.252 ± 0.017 ab
N. conoidea	1.212 ± 0.015 abcd	0.513 ± 0.012 ab	1.280 ± 0.030 ab
C. imbricata	1.141 ± 0.021 abcd	0.526 ± 0.013 ab	1.277 ± 0.021 ab
C. kleiniae	1.192 ± 0.015 abcd	0.528 ± 0.006 ab	1.293 ± 0.034 ab
C. leptocaulis	1.242 ± 0.021 abcd	0.567 ± 0.010 ab	1.250 ± 0.019 ab
O. stenopetala	1.299 ± 0.028 abcd	0.535 ± 0.011 ab	1.185 ± 0.042 ab
O. stricta	1.287 ± 0.023 abcd	0.541 ± 0.012 ab	1.139 ± 0.035 ab
L. lychnidiflora	1.202 ± 0.034 abcd	0.510 ± 0.010 ab	1.270 ± 0.037 ab
M. carnea	1.323 ± 0.024 abcd	0.540 ± 0.009 ab	1.097 ± 0.028 ab
C. ramillosa	1.362 ± 0.027 abcd	0.472 ± 0.011 a	1.085 ± 0.036 ab
C. retusa	1.486 ± 0.039 bcd	0.569 ± 0.017 ab	1.030 ± 0.019 ab
E. pectinatus	1.606 ± 0.042 cd	0.647 ± 0.015 b	0.516 ± 0.007 b

Different letters in each column indicate significant differences ($p < 0.05$). Mean ± standard deviation (SD).

3.2. Staining Methods for Fluorescence

When comparing the bright-field images with the fluorescence images, it was observed that in the bright-field images, the blue tones corresponded to the non-lignified tissue, while the red tones corresponded to the lignified tissue. In the fluorescence images, the green tones reflected the lignified tissue and the red the non-lignified (Figure 2A,B). In the SF, AO, and CA stains (Figure 2) it was observed that the secondary walls of the vessel elements (v) and wide-band tracheids (wbt) showed fluorescence emission in green to bluish tones in all three types of staining. In non-lignified tissues, such as the parenchyma or the primary walls of the vascular tissue, in the SF and AO stains, the cellulose fluoresced in red tones, while, for CA, the tones were bluish to greenish (Figure 2D).

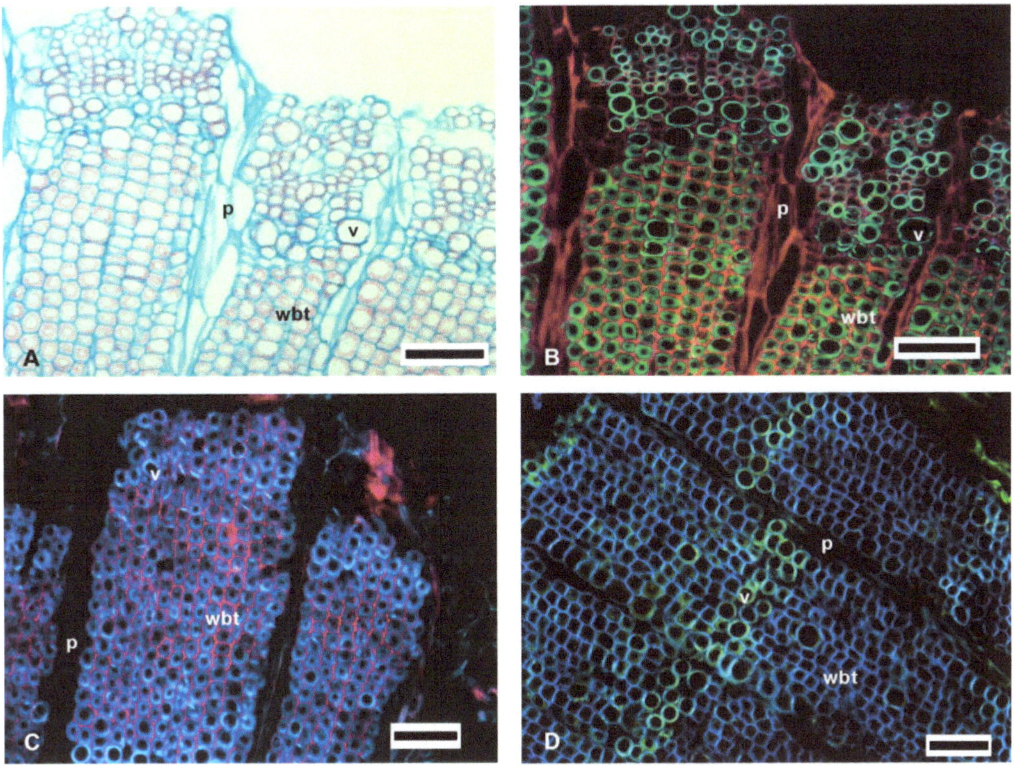

Figure 2. Images of the vascular tissue of *F. latispinus*. (**A**) Bright-field image of SF. (**B**) Fluorescence image of SF. (**C**) Fluorescence image of AO. (**D**) Fluorescence image of CA. p = parenchyma, v = vessel, wbt =wide-band tracheids. Scale: 100 μm.

3.3. Cellulose Distribution in Cells of the Secondary Xylem

Figures 3–6 show representative species of fibrous (Pereskioideae and Opuntioideae) and non-fibrous (Cactoideae) wood. In the bright-field microscopy images, the lignified cell walls were observed in red tones, with the non-lignified in blue tones. With fluorescence microscopy, the fibrous species reflected bluish-green to yellow tones; while in the non-fibrous species, the tones of the lignified walls were greenish. The presence of cellulose and other components of the non-lignified walls reflected reddish tones in all species. *Cylindropuntia leptocaulis* (Figure 3A,B) had starch within some parenchyma (p) cells, which were cyan-colored. The distribution of cellulose was different between the fibrous species and non-fibrous species.

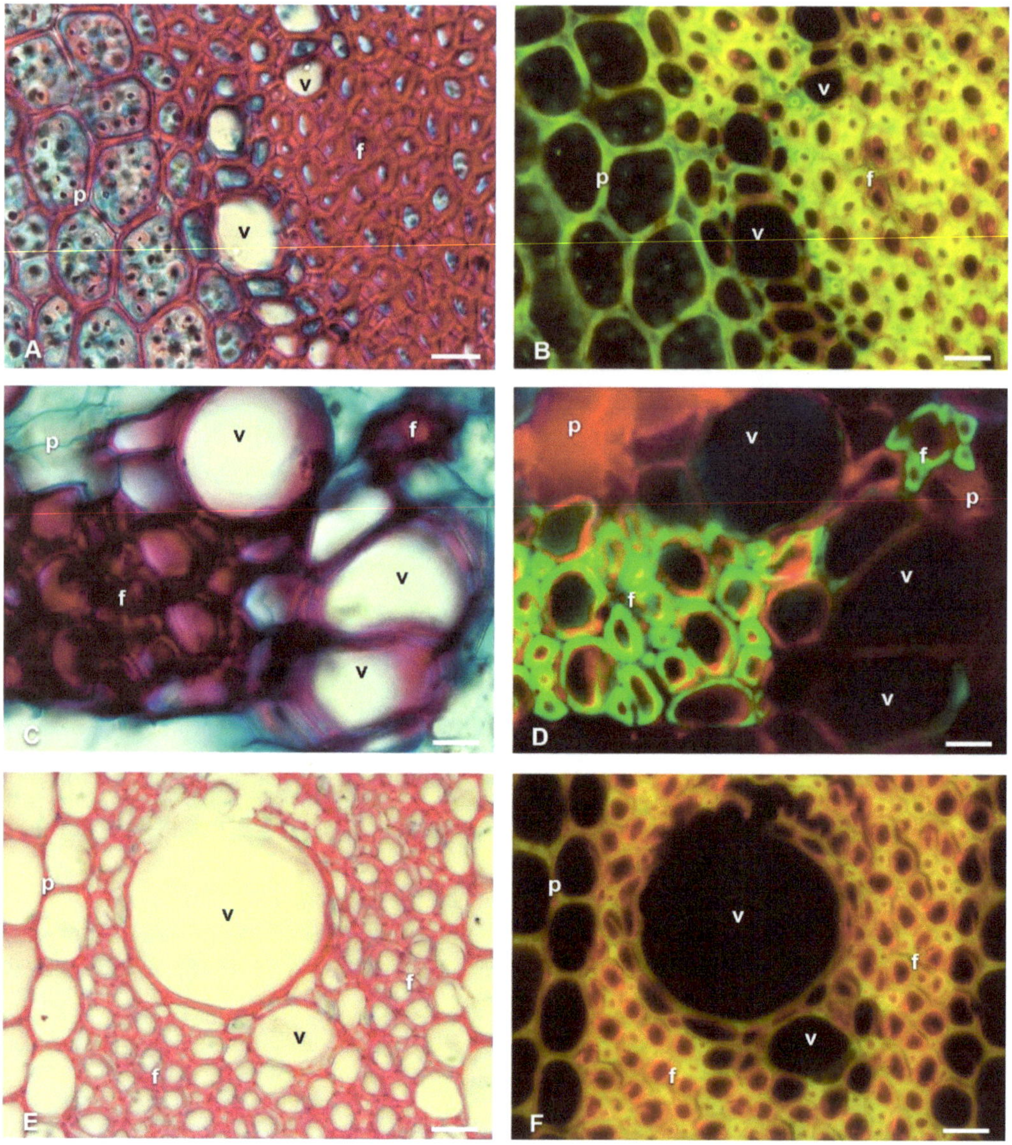

Figure 3. Fluorescence emission of transverse sections of fibrous wood species of Cactaceae. (**A**,**B**) *Cylindropuntia leptocaulis*. (**C**,**D**) *Opuntia stenopetala*. (**E**,**F**) *Leuenbergeria lychnidiflora*. f = fibers, p = parenchyma, v = vessel. Scale: 20 μm.

In longitudinal sections of *Cylindropuntia*, *Opuntia*, and *L. lychnidiflora*, cellulose was detected in the pits of the VEs (Figure 4A,B,D). In addition, the presence of cellulose was mainly in the simple pits of the fibers (f) of *Cylindropuntia* and *L. lychnidiflora*. (Figure 4A,B,D). The unlignified parenchyma of *O. stenopetala* in longitudinal sections showed primary walls with the presence of cellulose (Figure 4C); whereas, in the lignified parenchyma of *Cylindropuntia* and *Leuenbergeria*, lignin was mainly accumulated in the cell wall, and cellulose was exclusively in the pits (Figure 4A,B,D).

In non-fibrous species (Figure 5), cellulose fluorescence emission was detected in the primary wall of p, wbt, and v; unlike in fibrous species, where fluorescence emission was not detected in the primary wall but in the S3 layer of the secondary wall adjacent to the lumen. Furthermore, the non-fibrous species showed secondary walls as helical and annular thickenings in v and wbt, so the primary wall was visible in the longitudinal sections (Figure 6). In all non-fibrous species, the p was abundant and unlignified, in addition to having a greater accumulation of cellulose (Figures 5B,D,F,H and 6B).

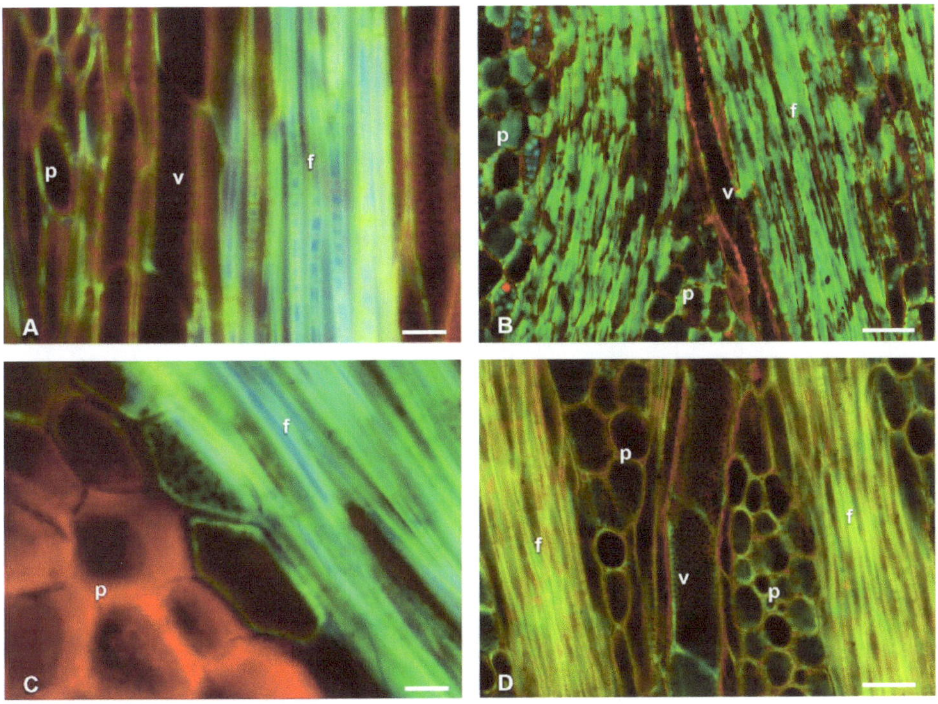

Figure 4. Fluorescence emission of longitudinal sections of fibrous wood species of Cactaceae. (**A**) *Cylindropuntia kleiniae*. (**B**) *Cylindropuntia leptocaulis*. (**C**) *Opuntia stenopetala*. (**D**) *Leuenbergeria lychnidiflora*. f = fibers, p = parenchyma, v = vessel. Scale: 20 µm: (**A**,**C**); 50 µm: (**B**,**D**).

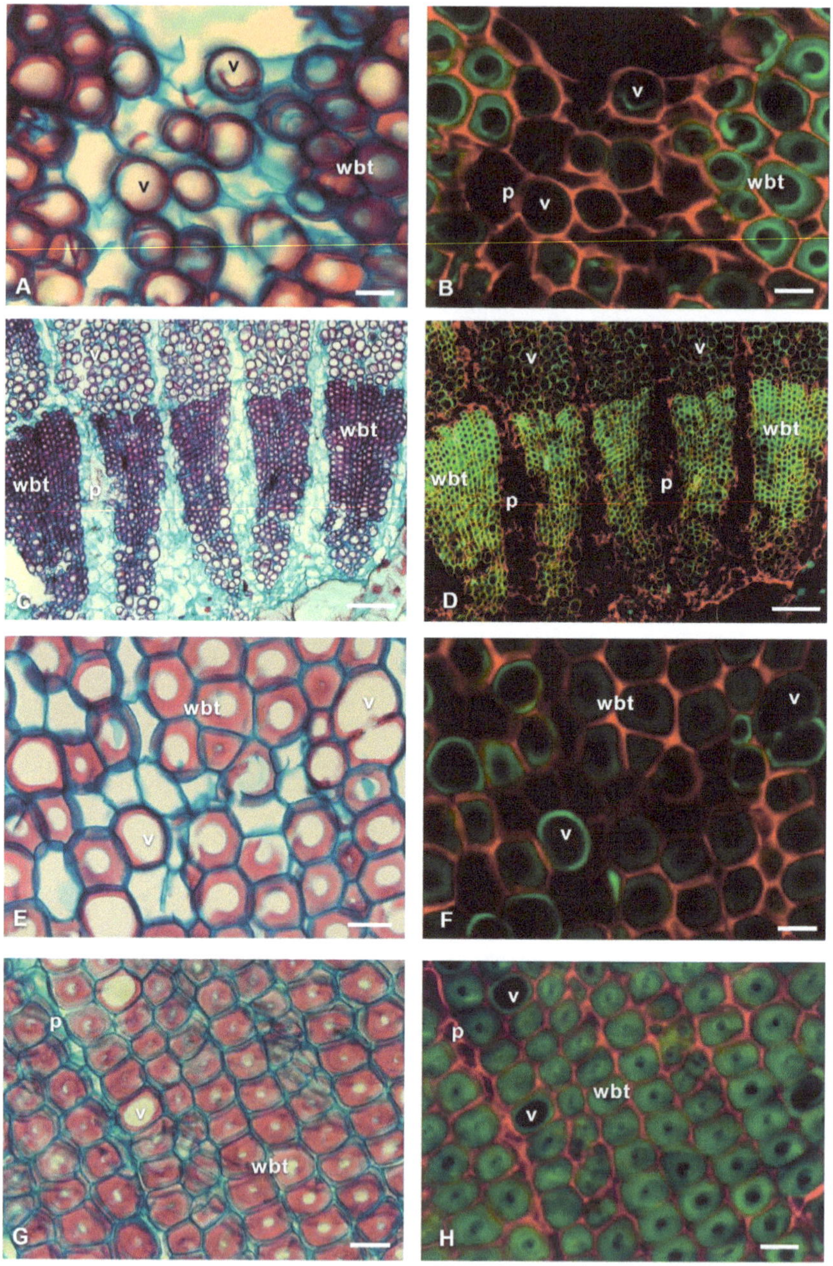

Figure 5. Fluorescence emission of transverse sections of non-fibrous wood species of Cactaceae. (**A**,**B**) *Coryphantha clavata*. (**C**,**D**) *Echinocereus cinerascens*. (**E**,**F**) *Mammillaria dixathocentron*. (**G**,**H**) *Neolloydia conoidea*. p = parenchyma, v = vessel, wbt = wide-band tracheids. Scale: 20 µm: (**A**,**B**,**E**,**H**); 200 µm: (**C**,**D**).

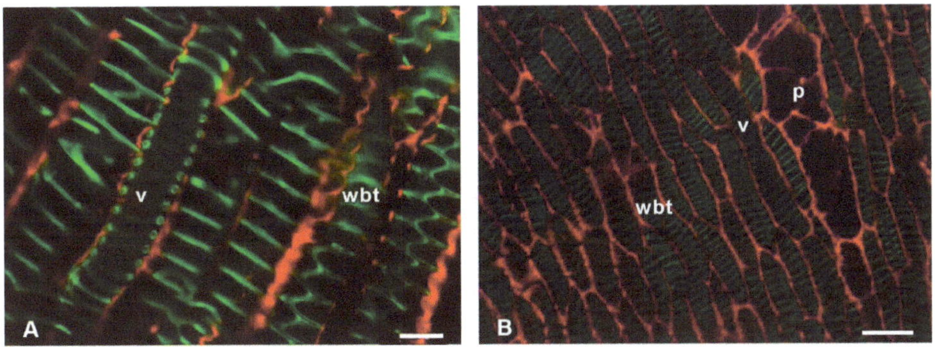

Figure 6. Fluorescence emission of longitudinal sections of non-fibrous wood species of Cactaceae. (**A**) *Carpathia* delaetiana. (**B**) *Neolloydia conoidea*. p = parenchyma, v = vessel, wbt = wide-band tracheids. Scale: 20 µm: (**A**,**B**).

4. Discussion

All Cactaceae species had a higher cellulose crystalline proportion, and the distribution of cellulose was in unlignified parenchyma, v, and wbt; all with secondary walls as helical and annular thickenings.

4.1. Crystalline Indexes

The TCI index was based on the proportion between the intensity peaks of 1370 cm^{-1} and 2900 cm^{-1}, according to Nelson and O'Conner [26]. The TCI is proportional to the degree of crystallinity of cellulose; therefore, the higher the ratio, the higher the percentage of crystalline cellulose, as reported by various authors for different species, fibers, and materials [27,28,38]. In the case of the LOI index, the peak of 1430 cm^{-1} corresponds to the presence of crystalline cellulose I, while the peak of 893 cm^{-1} reflects the presence of crystalline cellulose II and amorphous cellulose [39]. LOI reflects the ordered regions perpendicular to the chain direction, which is influenced by the chemical extraction and purification of cellulose [40]. The low values of LOI for Cactaceae species could have been the result of the presence of crystalline cellulose II, which influences the peak 893 cm^{-1} [41], or the effect of the temperature and concentration of the NaOH solution during the purification of cellulose [39]. In the case of HBI, these values are useful for interpreting qualitative changes in cellulose crystallinity; the lower values indicate the presence of crystalline cellulose, and if the values are higher, this could indicate the presence of cellulose II or amorphous cellulose. However, these values could represent the amount of bound water in the fiber structure and the presence of extractives, hemicelluloses, and lignin that increase HBI values [27].

The results observed in Table 4 reflect that all cacti species had high proportions of crystalline cellulose, because they had high values of TCI and LOI, and similar values in HBI. The species *E. pectinatus*, *C. retusa*, and *C. ramillosa* (except in the LOI value) had the highest crystalline proportions, due to having the highest values of TCI and LOI, and lower values of HBI, which corresponds with the fibers and materials with high crystalline cellulose proportions reported by Colom and Carrillo [42], Carrillo et al. [43], Široký et al. [39], and Poletto et al. [27].

Comparing with the reports in the literature for other cacti species, the high proportions of crystallinity cellulose were similar to most angiosperms reported by Agarwal et al. [44] and to the bark of the cactus *Cereus forbesii*, which had a percentage of 82% crystalline cellulose [45]. *Opuntia ficus-indica* is a species with varying crystalline cellulose percentages in different parts of the plant: cladodes 27% [23] and 79% [46], spines 34–70% [47,48], vascular tissue 22–28% [49], fruit epidermis 38% [50], and 60% in seeds [51]. Maceda et al. [46] reported percentages of crystalline cellulose of 76% and 74%, respectively, for *Opuntia streptacantha* and *O. robusta*. These values are very similar to those reported in this study for non-fibrous and fibrous species. Therefore, high proportions or percentages of crystalline cellulose is a constant in different Cactaceae species, probably due to the presence of non-lignified parenchyma or non-lignified primary cell walls, as observed in the anatomical distribution. Further studies comparing the purity of cellulose using the Kürschner-Hoffer method with other methods, such as the Seifert [52] method, would allow corroborating the proportions of crystalline cellulose in the Cactaceae species analyzed.

4.2. Staining Methods for Fluorescence

With the three staining methods (SF, AO, and CA; Figure 2), the distribution of the lignified tissue from the non-lignified was identified. The tones in the fluorescence emission for lignin in SF were similar to what was observed in AO and CA, and to what was reported by other authors [22,29,30]. The use of SF for bright-field and fluorescence is advantageous over AO and CA, because images can only be taken using fluorescence microscopy. On the other hand, the AO and CA techniques are semi-permanent, so over time the fluorescence is lost, when the dyes become diluted in the mounting medium used (phosphate buffer:glycerol 1:1 v/v) [30]; while for SF, a permanent mounting medium is used, so slides can be stored without loss of fluorescence.

Finally, the difference in lignin tones was visible in SF and CA, such as v, whose intensity and fluorescence tonalities were different from wbt; while for AO, the tonalities were similar in v and wbt. For the distribution of cellulose in the non-lignified parenchyma and in the vascular tissue, SF was clearly observed in red tones; while in AO and CA, the fluorescence emission of the parenchyma was obscured by the fluorescence of the vascular tissue. Therefore, SF was an efficient method for determining the distribution of lignified tissues and cellulose, as was previously reported in the literature [21,22].

4.3. Cellulose Distribution and Crystalline Composition

The tonalities observed in the fluorescence emission for cellulose and lignin correspond to those reported in other similar studies with a safranin O/fast green staining technique and three excitation bands [21,22,29–31]. This technique allowed the detection of differences in the distribution of both structural compounds (cellulose and lignin), between non-fibrous and fibrous species. The safranin O dye allows the detection of lignin autofluorescence and fluorescence in analyzed tissues [30,31], with tonalities of blue to green [29,35]. In the case of celluloses and hemicelluloses that do not have autofluorescence, such as lignin [29], with safranin O/fast green staining, the fluorescence emission of cellulose can be detected at 570–620 nm and its tonalities were reddish, similarly to the results of Maceda et al. [21,22]. However, further studies with immunohistochemistry or specific fluorophores could help confirm the distribution of cellulose [2,29].

When comparing the results with those reported by Maceda et al. [21] for the primary xylem, cellulose accumulated in the primary walls and lignin in the secondary ones of the helical and annular thickenings of the protoxylem and metaxylem of fibrous species, similarly to what was observed in non-fibrous adult plants. Only in the metaxylem of *Leuenbergeria lychnidiflora* was there a decrease in the accumulation of cellulose in the primary walls and a greater accumulation in the intervascular pits, as obtained in the secondary xylem of the fibrous species.

The presence of high proportions of crystalline cellulose in all studied cacti, mainly in the non-fibrous species (*E. pectinatus*, *C. retusa*, and *C. ramillosa*), could be related to the abundant unlignified parenchyma and the lower accumulation of lignin in the xylem, as seen in Figures 4 and 5B. Even when including fibrous species of *Cylindropuntia* and *Opuntia*, there was a high proportion of crystalline cellulose, similarly to non-fibrous species, possibly because the xylem had a greater accumulation of lignin in the cell walls of the v, f, and p [20]. This high proportion of lignin may function as a physical [53] and chemical barrier against the attack of pathogens [54]. Zhao and Dixon [7] and Bacete and Hamann [55] mentioned that the cell wall is a dynamic barrier in conditions of abiotic stress.

Therefore, in some cells, the presence of increased lignin accumulation can be observed [56]; while in others, similarly to in the gelatinous layer (G), cellulose is mainly accumulated [8].

In the cells with non-lignified primary cell walls, the accumulation of crystalline cellulose packed by hydrogen chain β-(1→4)-linked D-glucose units [9] makes cellulose more hydrophobic than amorphous cellulose [8], conferring structural support [9] and causing a decrease in the efficiency of cellulase enzymes [57], by not presenting sites for binding with enzymes and making it difficult to degrade [58]. In contrast, amorphous cellulose is slightly hydrophilic [55], susceptible, and degrades rapidly, due to the action of cellulase enzymes [59] and pH changes from pathogens [60].

Infection with some pathogenic fungi occurs when the hyphae invade the roots and subsequently the vascular tissue [16]. In the fibrous species of Cactaceae, the accumulation of lignin works as a physical barrier in Vs [61]. However, in non-fibrous species with little accumulation of lignin in the helical or annular thickenings, the presence of crystalline cellulose in the primary wall of the tracheary elements and the unlignified parenchyma function as a physical barrier, to prevent the spread of the fungus by reducing the effectiveness of cellulase enzymes [55]. For the species analyzed in this work, it has been reported that they have higher percentages of cellulose than lignin in the vascular tissue [20]; thus, possibly, the presence of crystalline cellulose reinforces and protects the vessels with helical and annular secondary thickenings. In subsequent studies using transmission electron microscopy techniques [62], the presence of crystalline cellulose in the primary wall of the tracheary elements could be analyzed and characterized, which will support and confirm this assertion.

The presence of high proportions of crystalline cellulose in fiber species could be due to the succulence of their stems and their distribution in humid environments, such as for *Leuenbergeria lychnidiflora*, [63] or in the extreme conditions of arid and semi-arid climates with seasons of high humidity. The resistance of plants to stressful conditions is energetically expensive, in addition to the constitutive expression of defense mechanisms, such as the accumulation of callose [1,5], pectins [55], or secondary metabolites [64]. This is not always the best strategy against the colonization of pathogens or diseases, because it can restrict physiological processes and have negative impacts on the plant, such as a reduced seed production and biomass [14]. Therefore, the presence of primary physical barriers, such as lignin [65,66] and crystalline cellulose [9] that inhibit the spread of pathogens, could decrease the expression of the defensive systems (callose, pectins, secondary metabolites) and be energetically expensive [15]. The heterogeneity in the composition of cell walls between species reflects the diversity of defensive mechanisms against the degrading enzymes that pathogens have developed to break down plant cell walls, such as the numerous cell wall-degrading enzymes (CWDEs), polygalacturonases, and xylanases [67].

In these Cactaceae species, as mentioned previously, cellulose may work to confer structural support to the primary wall, without losing flexibility [68], and thus maintaining the cell structure during periods of water stress and rain [69]. Furthermore, the increased amount of crystalline cellulose could function as a defense against pathogens [57], by providing resistance to degradation by glycosyl hydrolase enzymes [9] and enzymes produced by pathogenic fungi [17]. Further analyses in a larger number of dimorphic and non-fibrous cacti, together with other families of succulent plants, may confirm that the presence of crystalline cellulose is due to the presence of non-lignified parenchyma, as was observed in the species analyzed in this work.

In addition, the crystalline cellulose can be used for the production of microfibrillated cellulose nanofibers that could be applied in the production of medical equipment [70,71] or in paper recycling [72]. Amorphous cellulose could be enzymatically degraded for the production of biofuels [23,73]; thus, cacti species have potential for future use [74]. However, it is essential to analyze the profitability in terms of cultivation and plant growth.

5. Conclusions

The non-fibrous species presented a high proportion of crystalline cellulose, reflected in their TCI, LOI, and HBI proportions. The distribution of the cellulose was in the primary cell wall of the tracheary elements and the unlignified parenchyma. In fibrous species, the distribution was in the cell wall near the lumen and the simple and alternate pits of vessel elements and fibers. The high proportion of crystalline cellulose could be related to resistance to pathogens, due to the presence of a non-lignified primary cell wall in all the cacti species.

Author Contributions: Conceptualization, A.M. and T.T.; Formal analysis, A.M., M.S.-H. and T.T.; Funding acquisition, T.T.; Investigation, A.M. and T.T.; Methodology, A.M., M.S.-H. and T.T.; Project administration, T.T.; Resources, M.S.-H.; Writing—original draft, A.M.; Writing—review and editing, M.S.-H. and T.T. All authors have read and agreed to the published version of the manuscript.

Funding: Funding was provided by DGAPA-UNAM postdoctoral fellowship to AM (document number: CJ IC/CTIC I5OO7I2O2I) and by the DGAPA-PAPIIT: UNAM grants IN209012 and IN210115 to TT.

Institutional Review Board Statement: Not applicable.

Informed Consent Statement: Not applicable.

Data Availability Statement: Raw data are available in the Figshare repository: https://doi.org/10.6084/m9.figshare.20264472.v1, https://figshare.com/articles/dataset/Data_from_Cellulose_in_secondary_xylem_of_Cactaceae_crystalline_composition_and_anatomical_distribution/20264472 (accessed on 11 June 2022).

Acknowledgments: Authors thank Pedro Mercado Ruaro from Laboratorio de Morfo-Anatomía y Citogenética (LANABIO, UNAM) for allowing us to use the fluorescence microscope; Rubén San Miguel Chávez for allowing us to use the FTIR, and Julio César Montero-Rojas for artwork.

Conflicts of Interest: The authors declare no conflict of interest.

References

1. Srivastava, V.; McKee, L.S.; Bulone, V. Plant Cell Walls. In *eLS.*; John Wiley & Sons, Ltd.: Chichester, UK, 2017; pp. 1–17.
2. Bidhendi, A.J.; Chebli, Y.; Geitmann, A. Fluorescence visualization of cellulose and pectin in the primary plant cell wall. *J. Microsc.* **2020**, *278*, 164–181. [CrossRef] [PubMed]
3. Kuki, H.; Yokoyama, R.; Kuroha, T.; Nishitani, K. Xyloglucan is not essential for the formation and integrity of the cellulose network in the primary cell wall regenerated from *Arabidopsis* protoplasts. *Plants* **2020**, *9*, 629. [CrossRef] [PubMed]
4. Harris, P.J. Primary and secondary plant cell walls: A comparative overview. *N. Zeal. J. For. Sci.* **2006**, *36*, 36–53.
5. Lampugnani, E.R.; Khan, G.A.; Somssich, M.; Persson, S. Building a plant cell wall at a glance. *J. Cell Sci.* **2018**, *131*, jcs207373. [CrossRef]
6. Kubicki, J.D.; Yang, H.; Sawada, D.; O'Neill, H.; Oehme, D.; Cosgrove, D. The shape of native plant cellulose microfibrils. *Sci. Rep.* **2018**, *8*, 13983. [CrossRef]
7. Zhao, Q.; Dixon, R.A. Altering the cell wall and its impact on plant disease: From forage to bioenergy. *Annu. Rev. Phytopathol.* **2014**, *52*, 69–91. [CrossRef]
8. Festucci-Buselli, R.A.; Otoni, W.C.; Joshi, C.P. Structure, organization, and functions of cellulose synthase complexes in higher plants. *Brazilian J. Plant Physiol.* **2007**, *19*, 1–13. [CrossRef]
9. Malinovsky, F.G.; Fangel, J.U.; Willats, W.G.T. The role of the cell wall in plant immunity. *Front. Plant Sci.* **2014**, *5*, 178. [CrossRef]
10. Mnich, E.; Bjarnholt, N.; Eudes, A.; Harholt, J.; Holland, C.; Jørgensen, B.; Larsen, F.H.; Liu, M.; Manat, R.; Meyer, A.S.; et al. Phenolic cross-links: Building and de-constructing the plant cell wall. *Nat. Prod. Rep.* **2020**. [CrossRef]
11. Carere, C.R.; Sparling, R.; Cicek, N.; Levin, D.B. Third generation biofuels via direct cellulose fermentation. *Int. J. Mol. Sci.* **2008**, *9*, 1342–1360. [CrossRef]
12. Sims, R.E.H.; Mabee, W.; Saddler, J.N.; Taylor, M. An overview of second generation biofuel technologies. *Bioresour. Technol.* **2010**, *101*, 1570–1580. [CrossRef] [PubMed]
13. Ciolacu, D.; Ciolacu, F.; Popa, V.I. Amorphous cellulose-structure and characterization. *Cellul. Chem. Technol.* **2011**, *45*, 13–21.
14. Alves, A.; Simoes, R.; Stackpole, D.J.; Vaillancourt, R.E.; Potts, B.M.; Schwanninger, M.; Rodrigues, J. Determination of the syringyl/guaiacyl ratio of *Eucalyptus globulus* wood lignin by near infrared-based partial least squares regression models using analytical pyrolysis as the reference method. *J. Near Infrared Spectrosc.* **2011**, *19*, 343–348. [CrossRef]
15. Bacete, L.; Mélida, H.; Miedes, E.; Molina, A. Plant cell wall-mediated immunity: Cell wall changes trigger disease resistance responses. *Plant J.* **2018**, *93*, 614–636. [CrossRef]

16. Le Gall, H.; Philippe, F.; Domon, J.-M.; Gillet, F.; Pelloux, J.; Rayon, C. Cell wall metabolism in response to abiotic stress. *Plants* **2015**, *4*, 112–166. [CrossRef]
17. Kesten, C.; Gámez-Arjona, F.M.; Menna, A.; Scholl, S.; Dora, S.; Huerta, A.I.; Huang, H.; Tintor, N.; Kinoshita, T.; Rep, M.; et al. Pathogen-induced pH changes regulate the growth-defense balance in plants. *EMBO J.* **2019**, *38*, e101822. [CrossRef]
18. Hidalgo-Reyes, M.; Caballero-Caballero, M.; HernáNdez-Gómez, L.H.; Urriolagoitia-Calderón, G. Chemical and morphological characterization of *Agave angustifolia* bagasse fibers. *Bot. Sci.* **2015**, *93*, 807–817. [CrossRef]
19. Reyes-Rivera, J.; Canché-Escamilla, G.; Soto-Hernández, M.; Terrazas, T. Wood chemical composition in species of Cactaceae the relationship between lignification and stem morphology. *PLoS ONE* **2015**, *10*, e0123919. [CrossRef]
20. Maceda, A.; Soto-Hernández, M.; Peña-Valdivia, C.B.; Terrazas, T. Chemical composition of cacti wood and comparison with the wood of other taxonomic groups. *Chem. Biodivers.* **2018**, *15*, e1700574. [CrossRef]
21. Maceda, A.; Soto-Hernández, M.; Peña-Valdivia, C.B.; Trejo, C.; Terrazas, T. Differences in the structural chemical composition of the primary xylem of Cactaceae: A topochemical perspective. *Front. Plant Sci.* **2019**, *10*, 1497. [CrossRef]
22. Maceda, A.; Reyes-Rivera, J.; Soto-Hernández, M.; Terrazas, T. Distribution and chemical composition of lignin in secondary xylem of Cactaceae. *Chem. Biodivers.* **2021**, *18*, e2100431. [CrossRef] [PubMed]
23. Yang, L.; Lu, M.; Carl, S.; Mayer, J.A.; Cushman, J.C.; Tian, E.; Lin, H. Biomass characterization of *Agave* and *Opuntia* as potential biofuel feedstocks. *Biomass Bioenergy* **2015**, *76*, 43–53. [CrossRef]
24. Reyes-Rivera, J.; Terrazas, T. Lignin analysis by HPLC and FTIR. *Methods Mol. Biol.* **2017**, *1544*, 193–211. [CrossRef]
25. Kruer-Zerhusen, N.; Cantero-Tubilla, B.; Wilson, D.B. Characterization of cellulose crystallinity after enzymatic treatment using Fourier transform infrared spectroscopy (FTIR). *Cellulose* **2018**, *25*, 37–48. [CrossRef]
26. Nelson, M.L.; O'Connor, R.T. Relation of certain infrared bands to cellulose crystallinity and crystal latticed type. Part I. Spectra of lattice types I, II, III and of amorphous cellulose. *J. Appl. Polym. Sci.* **1964**, *8*, 1311–1324. [CrossRef]
27. Poletto, M.; Ornaghi, H.L.; Zattera, A.J. Native cellulose: Structure, characterization and thermal properties. *Materials* **2014**, *7*, 6105–6119. [CrossRef]
28. Cichosz, S.; Masek, A. IR study on cellulose with the varied moisture contents: Insight into the supramolecular structure. *Materials* **2020**, *13*, 4573. [CrossRef]
29. Donaldson, L. Autofluorescence in plants. *Molecules* **2020**, *25*, 2393. [CrossRef]
30. Kitin, P.; Nakaba, S.; Hunt, C.G.; Lim, S.; Funada, R. Direct fluorescence imaging of lignocellulosic and suberized cell walls in roots and stems. *AoB Plants* **2020**, *12*, plaa032. [CrossRef]
31. De Micco, V.; Aronne, G. Combined histochemistry and autofluorescence for identifying lignin distribution in cell walls. *Biotech. Histochem.* **2007**, *82*, 209–216. [CrossRef]
32. Nakaba, S.; Kitin, P.; Yamagishi, Y.; Begum, S.; Kudo, K.; Nugroho, W.D.; Funada, R. Three-dimensional imaging of cambium and secondary xylem cells by confocal laser scanning microscopy. In *Plant Microtechniques and Protocols*; Springer: Cham, Switzerland, 2015; pp. 431–465. [CrossRef]
33. Ruzin, S.E. *Plant Microtechnique and Microscopy*; Oxford University Press: Oxford, UK, 1999; ISBN 0195089561.
34. Loza-Cornejo, S.; Terrazas, T. Anatomía del tallo y de la raíz de dos especies de *Wilcoxia* Britton & Rose (Cactaceae) del noreste de México. *Bot. Sci.* **1996**, *59*, 13–23. [CrossRef]
35. Baldacci-Cresp, F.; Spriet, C.; Twyffels, L.; Blervacq, A.; Neutelings, G.; Baucher, M.; Hawkins, S. A rapid and quantitative safranin-based fluorescent microscopy method to evaluate cell wall lignification. *Plant J.* **2020**, *102*, 1074–1089. [CrossRef] [PubMed]
36. Lionetto, F.; Del Sole, R.; Cannoletta, D.; Vasapollo, G.; Maffezzoli, A. Monitoring wood degradation during weathering by cellulose crystallinity. *Materials* **2012**, *5*, 1910–1922. [CrossRef]
37. Liu, Y.; Kim, H.-J. Fourier Transform Infrared Spectroscopy (FT-IR) and simple algorithm analysis for rapid and non-destructive assessment of developmental cotton fibers. *Sensors* **2017**, *17*, 1469. [CrossRef] [PubMed]
38. Poletto, M.; Pistor, V.; Santana, R.M.C.; Zattera, A.J. Materials produced from plant biomass: Part II: Evaluation of crystallinity and degradation kinetics of cellulose. *Mater. Res.* **2012**, *15*, 421–427. [CrossRef]
39. Široký, J.; Blackburn, R.S.; Bechtold, T.; Taylor, J.; White, P. Attenuated total reflectance Fourier-transform Infrared spectroscopy analysis of crystallinity changes in lyocell following continuous treatment with sodium hydroxide. *Cellulose* **2010**, *17*, 103–115. [CrossRef]
40. Hofmann, D.; Fink, H.P.; Philipp, B. Lateral crystallite size and lattice distortions in cellulose II samples of different origin. *Polymer* **1989**, *30*, 237–241. [CrossRef]
41. Kljun, A.; Benians, T.A.S.; Goubet, F.; Meulewaeter, F.; Knox, J.P.; Blackburn, R.S. Comparative analysis of crystallinity changes in cellulose I polymers using ATR-FTIR, X-ray diffraction, and carbohydrate-binding module probes. *Biomacromolecules* **2011**, *12*, 4121–4126. [CrossRef]
42. Colom, X.; Carrillo, F. Crystallinity changes in lyocell and viscose-type fibres by caustic treatment. *Eur. Polym. J.* **2002**, *38*, 2225–2230. [CrossRef]
43. Carrillo, F.; Colom, X.; Suñol, J.J.; Saurina, J. Structural FTIR analysis and thermal characterisation of lyocell and viscose-type fibres. *Eur. Polym. J.* **2004**, *40*, 2229–2234. [CrossRef]

44. Agarwal, U.P.; Reiner, R.R.; Ralph, S.A. Estimation of cellulose crystallinity of lignocelluloses using near-IR FT-Raman spectroscopy and comparison of the Raman and Segal-WAXS methods. *J. Agric. Food Chem.* **2013**, *61*, 103–113. [CrossRef] [PubMed]
45. Orrabalis, C.; Rodríguez, D.; Pampillo, L.G.; Londoño-Calderón, C.; Trinidad, M.; Martínez-García, R. Characterization of nanocellulose obtained from *Cereus forbesii* (a South American cactus). *Mater. Res.* **2019**, *22*, 20190243. [CrossRef]
46. Maceda, A.; Soto-Hernández, M.; Peña-Valdivia, C.B.; Trejo, C.; Terrazas, T. Characterization of lignocellulose of *Opuntia* (Cactaceae) species using FTIR spectroscopy: Possible candidates for renewable raw material. *Biomass Convers. Biorefinery* **2020**, *12*, 5165–5174. [CrossRef]
47. Vignon, M.R.; Heux, L.; Malainine, M.-E.; Mahrouz, M. Arabinan-cellulose composite in *Opuntia ficus-indica* prickly pear spines. *Carbohydr. Res.* **2004**, *339*, 123–131. [CrossRef] [PubMed]
48. Marin-Bustamante, M.Q.; Chanona-Pérez, J.J.; Güemes-Vera, N.; Cásarez-Santiago, R.; Pereaflores, M.J.; Arzate-Vázquez, I.; Calderón-Domínguez, G. Production and characterization of cellulose nanoparticles from nopal waste by means of high impact milling. *Procedia Eng.* **2017**, *200*, 428–433. [CrossRef]
49. Greco, A.; Maffezzoli, A. Rotational molding of biodegradable composites obtained with PLA reinforced by the wooden backbone of *Opuntia ficus-indica* cladodes. *J. Appl. Polym. Sci.* **2015**, *132*, 42447. [CrossRef]
50. Habibi, Y.; Mahrouz, M.; Vignon, M.R. Microfibrillated cellulose from the peel of prickly pear fruits. *Food Chem.* **2009**, *115*, 423–429. [CrossRef]
51. Habibi, Y.; Heux, L.; Mahrouz, M.; Vignon, M.R. Morphological and structural study of seed pericarp of *Opuntia ficus-indica* prickly pear fruits. *Carbohydr. Polym.* **2008**, *72*, 102–112. [CrossRef]
52. Tribulová, T.; Kačík, F.; Evtuguin, D.V.; Čabalová, I.; Ďurkovič, J. The effects of transition metal sulfates on cellulose crystallinity during accelerated ageing of silver fir wood. *Cellulose* **2019**, *26*, 2625–2638. [CrossRef]
53. Hamann, T. Plant cell wall integrity maintenance as an essential component of biotic stress response mechanisms. *Front. Plant Sci.* **2012**, *3*, 77. [CrossRef]
54. Ďurkovič, J.; Kačík, F.; Olčák, D.; Kučerová, V.; Krajňáková, J. Host responses and metabolic profiles of wood components in dutch elm hybrids with a contrasting tolerance to dutch elm disease. *Ann. Bot.* **2014**, *114*, 47–59. [CrossRef] [PubMed]
55. Bacete, L.; Hamann, T. The role of mechanoperception in plant cell wall integrity maintenance. *Plants* **2020**, *9*, 574. [CrossRef] [PubMed]
56. Polo, C.C.; Pereira, L.; Mazzafera, P.; Flores-Borges, D.N.A.; Mayer, J.L.S.; Guizar-Sicairos, M.; Holler, M.; Barsi-Andreeta, M.; Westfahl, H.; Meneau, F. Correlations between lignin content and structural robustness in plants revealed by X-ray ptychography. *Sci. Rep.* **2020**, *10*, 6023. [CrossRef] [PubMed]
57. Thomas, L.H.; Trevor Forsyth, V.; Šturcová, A.; Kennedy, C.J.; May, R.P.; Altaner, C.M.; Apperley, D.C.; Wess, T.J.; Jarvis, M.C. Structure of cellulose microfibrils in primary cell walls from collenchyma. *Plant Physiol.* **2013**, *161*, 465–476. [CrossRef] [PubMed]
58. Rytioja, J.; Hildén, K.; Yuzon, J.; Hatakka, A.; de Vries, R.P.; Mäkelä, M.R. Plant-polysaccharide-degrading enzymes from Basidiomycetes. *Microbiol. Mol. Biol. Rev.* **2014**, *78*, 614–649. [CrossRef]
59. Suchy, M.; Linder, M.B.; Tammelin, T.; Campbell, J.M.; Vuorinen, T.; Kontturi, E. Quantitative assessment of the enzymatic degradation of amorphous cellulose by using a quartz crystal microbalance with dissipation monitoring. *Langmuir* **2011**, *27*, 8819–8828. [CrossRef]
60. Kubicek, C.P.; Starr, T.L.; Glass, N.L. Plant cell wall–degrading enzymes and their secretion in plant-pathogenic fungi. *Annu. Rev. Phytopathol.* **2014**, *52*, 427–451. [CrossRef]
61. Liu, Q.; Luo, L.; Zheng, L. Lignins: Biosynthesis and biological functions in plants. *Int. J. Mol. Sci.* **2018**, *19*, 335. [CrossRef]
62. Ruel, K.; Nishiyama, Y.; Joseleau, J.P. Crystalline and amorphous cellulose in the secondary walls of *Arabidopsis*. *Plant Sci.* **2012**, *193*, 48–61. [CrossRef]
63. Edwards, E.J.; Donoghue, M.J. *Pereskia* and the origin of the cactus life-form. *Am. Nat.* **2006**, *167*, 777–793. [CrossRef]
64. Meraj, T.A.; Fu, J.; Raza, M.A.; Zhu, C.; Shen, Q.; Xu, D.; Wang, Q. Transcriptional factors regulate plant stress responses through mediating secondary metabolism. *Genes* **2020**, *11*, 346. [CrossRef] [PubMed]
65. Miedes, E.; Vanholme, R.; Boerjan, W.; Molina, A. The role of the secondary cell wall in plant resistance to pathogens. *Front. Plant Sci.* **2014**, *5*, 358. [CrossRef]
66. Maceda, A.; Soto-Hernández, M.; Peña-Valdivia, C.; Trejo, C.; Terrazas, T. Lignina: Composición, síntesis y evolución. *Madera Bosques* **2021**, *27*, e2722137. [CrossRef]
67. Annis, S.L.; Goodwin, P.H. Production and regulation of polygalacturonase isozymes in Canadian isolates of *Leptosphaeria maculans* differing in virulence. *Can. J. Plant Pathol.* **1997**, *19*, 358–365. [CrossRef]
68. Moura, J.C.M.S.; Bonine, C.A.V.; de Oliveira Fernandes Viana, J.; Dornelas, M.C.; Mazzafera, P. Abiotic and biotic stresses and changes in the lignin content and composition in plants. *J. Integr. Plant Biol.* **2010**, *52*, 360–376. [CrossRef] [PubMed]
69. Garrett, T.Y.; Huynh, C.-V.; North, G.B. Root contraction helps protect the "living rock" cactus *Ariocarpus fissuratus* from lethal high temperatures when growing in rocky soil. *Am. J. Bot.* **2010**, *97*, 1951–1960. [CrossRef]
70. Dlouhá, J.; Suryanegara, L.; Yano, H. The role of cellulose nanofibres in supercritical foaming of polylactic acid and their effect on the foam morphology. *Soft Matter* **2012**, *8*, 8704–8713. [CrossRef]
71. Wegner, T.H.; Jones, P.E. Advancing cellulose-based nanotechnology. *Cellulose* **2006**, *13*, 115–118. [CrossRef]
72. Robles, N.F.; Saucedo, A.R.; Delgado, E.; Sanjuán, R.; Turrado, J. Effect of cellulose microfibers on paper with high contents of recycled fiber. *Rev. Mex. Ciencias For.* **2014**, *5*, 70–78.

73. Zheng, Y.; Pan, Z.; Zhang, R. Overview of biomass pretreatment for cellulosic ethanol production. *Int. J. Agric. Biol. Eng.* **2009**, *2*, 51–68. [CrossRef]
74. Khamis, G.; Papenbrock, J. Newly established drought-tolerant plants as renewable primary products as source of bioenergy. *Emirates J. Food Agric.* **2014**, *26*, 1067–1080. [CrossRef]

polymers

MDPI

Article

Construction and Characterization of Fitting Equations for a New Wheat Straw Pulping Method

Xiaoli Liang [1,2,†], Shan Wei [2,†], Yanpeng Xu [1,2], Liang Yin [3], Ruiming Wang [1,2], Piwu Li [1,2] and Kaiquan Liu [1,2,*]

[1] State Key Laboratory of Biobased Material and Green Papermaking (LBMP), Qilu University of Technology (Shandong Academy of Sciences), Jinan 250353, China; 10431211135@stu.qlu.edu.cn (X.L.); 10431211110@stu.qlu.edu.cn (Y.X.); wrm@qlu.edu.cn (R.W.); piwuli@qlu.edu.cn (P.L.)
[2] Key Laboratory of Shandong Microbial Engineering, College of Bioengineering, Qilu University of Technology (Shandong Academy of Sciences), Jinan 250353, China; 1043118282@stu.qlu.edu.cn
[3] Gansu Engineering Technology Research Center for Microalgae, Hexi University, Zhangye 734000, China; yinl03@163.com
* Correspondence: liukq@qlu.edu.cn
† These authors contributed equally to this work.

Abstract: The pretreatment of pulp with enzymes has been extensively studied in the laboratory. However, due to cost constraints, the application of enzymes in the pulp and paper industry is very limited. In this paper, an environment-friendly and efficient pulping method is proposed as an alternative to traditional pulping and papermaking methods. This new method overcomes the low efficiency and extreme pollution problems associated with traditional pulping methods. In addition, fitting equations for the new pulping method are constructed using data on enzyme treatments, which reflect the effect of enzymes and enable the realization of real-time control of the pulping process. The experimental results show that the efficiency of the pulping and papermaking process can be improved using biological enzymes, and the separation of cellulose can be facilitated using mixed enzymes, which have a better effect than single enzymes.

Keywords: wheat straw; lignin; Biopulping; xylanase; pectinase

Citation: Liang, X.; Wei, S.; Xu, Y.; Yin, L.; Wang, R.; Li, P.; Liu, K. Construction and Characterization of Fitting Equations for a New Wheat Straw Pulping Method. *Polymers* **2023**, *15*, 4637. https://doi.org/10.3390/polym15244637

Academic Editors: Antonio M. Borrero-López, Concepción Valencia-Barragán, Esperanza Cortés Triviño, Adrián Tenorio-Alfonso and Clara Delgado-Sánchez

Received: 19 October 2023
Revised: 2 December 2023
Accepted: 6 December 2023
Published: 7 December 2023

1. Introduction

According to relevant statistics, the total output of paper and paperboard in the global pulp and paper industry in 2022 is about 417 million tons, of which the total output of the United States is about 65.95 million tons, and the total output of China is about 124.32 million tons [1]. As the world's largest developing country, China's total volume of paper products ranks first in the world, but the per capita output of paper products is much lower than that of developed countries. According to statistics, in 2022, China's pulp, paper and paper products industry achieved a total output of 283.91 million tons of pulp, paper, cardboard and paper products, with an average annual growth of 1.32%. Among them, the output of paper and paperboard was 124.25 million tons, an increase of 2.64% over the previous year. Pulp output was 85.87 million tons, an increase of 5.01% over the previous year. The output of paper products was 73.79 million tons, 4.65% less than the previous year [1]. In 2022, there are about 2500 paper and cardboard production enterprises in the country, and the national production of paper and cardboard is 124.25 million tons, an increase of 2.64% over the previous year. Over the previous year, consumption was 124.03 million tons, an increase of −1.94%, with a per capita annual consumption of 87.84 kg (1.412 billion people). According to the survey data of the China Paper Association, the total pulp production in 2022 will be 85.87 million tons, an increase of 5.01% over the previous year [2]. This includes 21.15 million tons of wood pulp, an increase of 16.92%; 59.14 million tons of waste pulp, an increase of 1.72% over the previous year; non-wood pulp was 5.58 million tons, an increase of 0.72% over the previous year [3]. In 2022, there will be 4727 paper products production enterprises above designated size in China, with a production capacity of 73.79 million tons, an increase of −4.65% over the previous year. The consumption was 68.97 million tons, an increase of −5.89% over the previous year. The import volume was 160,000 tons and the export volume was 4.98 million tons. From 2013 to 2022, the average annual growth rate of paper product production is 3.69%, and the average annual growth rate of consumption is 3.45% [2].

Although the demand for paper and paperboards has remained high in recent years, the shortage of raw materials has limited the development of the paper industry. The production capacity of pulp, particularly wood pulp, is insufficient to compensate for the high dependence on raw material imports. As a gramineous plant, wheat straw is an important source of biomass, which is grown globally. It is one of the three largest cereal crops in the world [4,5]. As agricultural waste, wheat straw employed in pulping not only reduces environmental pollution but also maximize straw resources [6]. Wheatgrass is an abundant crop by-product in China. With the continuous development of agricultural production science and technology and the improvement of wheat yields, the annual harvest of rice-wheatgrass has also increased [7]. Wheatgrass papermaking is an important aspect to promote the development of China's papermaking industry. Of China's non-wood pulp products, rice straw pulp and bamboo pulp make up the majority. In 2022, the domestic wheat straw pulp output was 1.5 million tons, an increase of 0.72% over the previous year, accounting for 26.9% of non-wood pulp output. The output of bamboo pulp was 2.46 million tons, accounting for 44.1% [2,8].

The chemical components of common deciduous wood were 58.61% cellulose, 22.71% pentosan, 17.04% lignin, 0.52% ash and 1.74% benzene-alcohol extract [8]. The chemical composition and content of wheat straw stem were: cellulose 47.09%, pentosan 32.28%, lignin 10.23%, ash 8.94% and benzene-alcohol extract 5.36% [9]. The average fiber length of wheatgrass was 1.32 mm, which was larger than that of broadleaf wood (1.03 mm) and smaller than that of coniferous wood (3.40 mm) [9]. Wheat straw has high cellulose content, short and fine fibers, low lignin molecular weight and wood ratio, and contains a large number of phenolic hydroxyl and ether bonds. It has strong lyophilic ability under alkaline environments, and can dissolve lignin at low temperatures to reduce energy waste [7].

Pulping mainly refers to the production process of using certain means, such as chemical reagents, papermaking machinery, or a combination of the two, to break down wood or other plant fiber raw materials, so that the fibers are dissociated and become unbleached color paste or bleached pulp [8]. The industry's history dates back to ancient times, but its modernization began in the mid-19th century. After entering the 20th century, the pulp and paper industry began to widely use chemical pulping, mechanical pulping and chemical mechanical pulping [10]. With the continuous development of the pulp and paper industry, since entering the 21st century, the global pulp and paper industry has also begun to develop in the direction of green environmental protection, promoting the recycling of waste paper and green pulping technology to reduce the harm to the environment [9]. Traditional pulping methods include alkaline pulping and kraft pulping. To remove lignin and separate cellulose, these processes must be conducted under high-temperature conditions, requiring special equipment and long processing times [11]. Presently, sodium hydroxide pretreatment is the most widely used pulping method. This is because OH^- can act on the ether and ester bond of lignin to separate lignin and hemicellulose [12–14]. The solubility of lignin in different environments depends on different precursors or combinations of precursors [15]. Therefore, the enzyme activity should be considered when using chemical reagents to treat wheat straw.

The existing problems in the paper industry also include high energy consumption and environmental pressure. The chemical treatment of wheat straw will produce many harmful substances, including chlorophenol, dioxin, furan, fatty acid, resin acid and chlorolignin compounds. It has been confirmed that these chlorinated hydrocarbon organic pollutants are mainly produced by changes to the chemical structure of lignin during cooking and bleaching. These substances are harmful to the environment. Dioxins are easily produced when bleaching pulp with chlorine. These substances are detrimental to the environment [16]. Eventually, they will destroy the self-healing ability of the environment and cause irreversible harm.

Wheat straw pretreatment can reduce the obstinacy of the cell wall and increase the accessibility of enzymes to the carbohydrates in the cell wall [17]. Hydrothermal pretreatment changes the structure and molecular weight of lignin, which is mainly because of the breaking of chemical bonds and the formation of new carbon–carbon double bonds [18]. The hydrothermal pretreatment of lignin at high temperatures can exert a high-intensity inhibition effect on the enzyme, which is mainly because of the nonproduction adsorption and enzyme inactivation [17].

Studies have shown that the tensile index and energy absorption of the xylanase-treated wheat straw do not increase; however, its tear index decreases significantly [19]. Xylanase enhances the bleaching process and has a positive impact on pulp, paper and wastewater treatment, aiming to reduce the use of bleaching chemicals in the pulp refining process [6,20]. Xylanase can be employed in the pulping process, and its application is mainly facilitated by the existence of refractory lignin. The refractory property of lignin residues is partly attributed to the existence of xylan in hardwood kraft pulp, which is not easy to degrade and separate from the fiber [21,22]. The precipitated xylan forms a barrier on the fiber surface, which prevents the residual lignin from diffusing from the fiber wall. Moreover, hemicellulose combines with other fibrous substances and pectin through noncovalent and covalent bonds, indicating that the application of xylanase in the pulping process can improve the chemical extraction efficiency of lignin. Pectinase-treated pulp fiber has the characteristics of long fiber length, small fine fiber length, and high flexible fiber content, which are conducive for subsequent pulping.

In this paper, we propose a new strategy for improving the environmental impact of the pulping and papermaking processes. We attempt to produce pulp in a neutral environment. Through this approach, the production cost of the enzyme can be significantly reduced and the conditions for obtaining high-quality and low-cost paper can be achieved. In addition, pulping in a neutral environment reduces environmental pollution. In order to improve the pulping efficiency of the factory, we consider combining a certain index in the pulp after pulping with the degree of beater, exploring the law of the change of the index and the degree of beater, and describing the change of the index and the degree of beater through the fitting equation, so as to achieve the online control of biomechanical pulping [23,24].

2. Materials and Methods

2.1. Materials

The wheat straw used in this study was obtained from wheat fields in five different regions of China (Dezhou City, Shandong Province; Linfen City, Shanxi Province; Huaihua City, Anhui Province; Suqian City, Jiangsu Province; Puyang City, Henan Province). After the composition analysis of wheat straw, the wheat straw from Dezhou City, Shandong Province was used for the follow-up experiment. Before pulping, the raw materials were cleaned, cut and sampled with a length of approximately 1 cm, followed by drying under the sun.

The xylanase and pectinase used in the experiments were provided by Shandong Longkete Enzyme Preparation Co., Ltd. (Yishui, China). All the other reagents used were analytically pure.

2.2. Pretreatment

The dried wheat straw and water in the ratio of 1:8 were placed in an 80 °C water bath for 2 h for swelling treatment, and enzymes were added simultaneously. After 2 h, a 1.5% potassium hydroxide (KOH) solution was added. The mixture solution was left to stand for 1 h, after which the pH was adjusted to neutral with phosphoric acid.

2.3. Enzymolysis

Under neutral conditions, enzyme treatment was conducted at different temperatures (50–90 °C) for 1 h, after which the treated samples were ground using a refiner with revolutions of 3000 rpm. After refining, the enzyme was inactivated by placing the pulp in a water bath at 100 °C for 10 min, after which the samples were taken for analysis.

2.4. Analysis Method

2.4.1. Composition Analysis of Wheat Straw

The contents of cellulose, hemicellulose and lignin in the wheat straw samples were determined according to the standard laboratory analysis procedure tp-510-42618 of the Renewable Energy Laboratory (NREL) [25]. The percentages of cellulose, hemicellulose and lignin in the samples were determined relative to the dry basis. The content of pectin was determined by carbazole colorimetry. The wheat straw was dried to a constant weight at (105 ± 2) °C, after which its water content was calculated. The wheat straw was placed in a muffle furnace and burned at (550 ± 10) °C for 2 h to determine its ash content (NREL/TP-510-42622) [26]. The protein content of the wheat straw was measured by the Kjeldahl method, according to NREL/TP-510-42625, and the nitrogen factor was 5.70 [27]. The fat content was determined by alkaline hydrolysis.

2.4.2. Determination and Definition of the Enzyme Activity

The activities of xylanase and pectinase were determined using the standard 3,5-dinitrosalicylic acid method [28]. The enzyme activity in the enzyme treatment is defined as the amount of enzyme required to generate 1 μmol of the substrate in 1 min.

2.4.3. Determination of the Main Detection Indexes

The reducing sugar content was determined by Fehling's Reagent Titration method [29]. Three parallel determinations were carried out in each group. The content of soluble solids was measured by drying at 105 °C, and each experiment was carried out three times in parallel. The pulp's Schober beating degree was measured using the IMT-DJD02 beating degree tester (Dongguan international material tester precision instrument Co., LTD, Dongguan, China); the data were kept to two decimal places, and each group of experiments was conducted three times in parallel.

2.4.4. Establishment of the Fitting Equation

A Plackett–Burman test with Design-Expert 12.0 software was used to screen out three significant influencing factors, and the evaluation index was selected as the beating degree. The Box–Behnken response surface method was used to optimize the pulping process. MATLAB R2017a software was used to nonlinear fit the change of reducing sugar content and soluble solid content in the pulp after pulping and the change of beating degree.

2.4.5. Scanning Electron Microscopy (SEM) Analysis

Using the Thermo Verios XHR SEM model scanning electron microscope (Thermo Fisher Scientific (China) Co., Ltd., Shanghai, China). The sample was freeze-dried to remove moisture. A conductive adhesive tape was glued to the sample table, on which the freeze-dried samples were dispersed. Due to the poor conductivity of the sample to be observed, it was necessary to spray gold.

The morphology of the enzyme treated wheat straw was analyzed by SEM. The secondary electron resolution is 3.5 nm; acceleration voltage 200 V–30 KV; and magnification 20–200,000×.

2.4.6. Fourier Transform Infrared Analysis (FT-IR)

Using the BRUKER TENSOR Fourier infrared spectrometer (Bruker Corporation, Billerica, MA, USA). The samples were ground after freeze-drying and analyzed using a

FT-IR. The wavelength absorption range was 400–4000 cm^{-1}, the resolution was 0.5 cm^{-1}, and the signal-to-noise ratio was 4000:1.

2.4.7. The X-ray Diffraction (XRD) Sample Preparation and Crystallinity Analysis

Using the Rigaku SmartLab SE X-ray diffractometer (Rigaku Corporation, Tokyo, Japan), cellulose of wheat straw was isolated following the method described in other previous research. The freeze-dried material is ground into fine particles and subsequently put in the sample tank. Diffraction patterns were obtained using an X-ray powder diffractometer, and a copper target was used; Cu Kα generates X-rays under the conditions of an acceleration voltage of 40 kV and a current of 40 mA. The scanning angle range was 5–50°. After scanning, the crystallinity of the sample was determined.

The calculation formula for the crystallinity index is as follows:

$$CrI\ (\%) = \frac{I_{200} - I_{\text{am}}}{I_{200}} \times 100\%$$

where I_{200} is the intensity of the crystalline peak at the maximum 2θ value between 22° and 23° for cellulose I (between 18° and 22° for cellulose II), and I_{am} is the intensity at the minimum 2θ between 18° and 19° for cellulose I (between 13° and 15° for cellulose II) [30–32].

3. Results and Discussion

3.1. Components of Wheat Straw

The main components of wheat straw from five different sources were quantitatively determined, and the results are shown in Table 1. The results showed that the cellulose content of wheat straw in Dezhou City of Shandong Province and Huaihua City of Anhui Province were 47.09% and 47.15%, respectively. The lignin content and pectin content of the former were 2.52% and 0.51% lower than the latter, respectively. Therefore, the wheat straw from Dezhou City, Shandong Province, was selected for the experiment. Table 1 shows a comparison of the compositions of the wheat straws obtained from different producing areas.

Table 1. Comparison of the compositions of the wheat straws obtained from different producing areas.

Composition	Dezhou City, Shandong Province	Linfen City, Shanxi Province	Huaihua City, Anhui Province	Suqian City, Jiangsu Province	Puyang City, Henan Province
Cellulose (dry basis)	47.09%	34.47%	47.15%	36.06%	41.90%
Hemicellulose (dry basis)	32.28%	36.43%	35.06%	35.09%	31.67%
Lignin (dry basis)	10.23%	7.72%	12.75%	7.94%	8.51%
Pectin (dry basis)	1.45%	2.52%	1.96%	2.48%	1.95%
Ash (dry basis)	8.94%	4.89%	4.18%	5.47%	8.51%
Water content	9.11%	8.04%	9.01%	6.81%	7.17%
Protein	2.01%	2.86%	1.27%	2.51%	3.09%
crude fat	0.40%	0.20%	0.40%	0.20%	0.10%

3.2. Enzyme Activity Determination at Different Temperatures and pH Values

The actual enzyme activities of xylanase and pectinase were measured at 40–90 °C and pH 5.0–10.0, separately. Figure 1 shows that when the temperature was fixed, the actual enzyme activity of pectinase was relatively low at pH 7.0 and 8.0. A possible reason is that the types of enzymes commercially sourced are different; thus, they show different pH preferences. When the temperature is fixed, the enzyme activity of xylanase decreases as the pH increases. The optimum temperature of xylanase and pectinase is 50 °C.

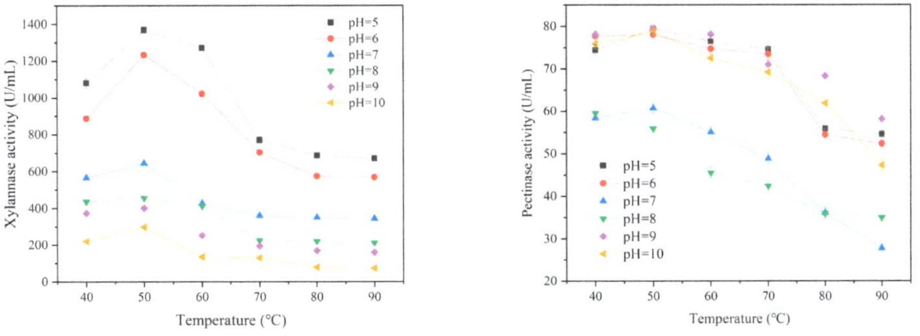

(a) Xylanase (b) Pectinase

Figure 1. The activities of xylanase and pectinase at different temperatures (40–90 °C) and pH (5–10).

3.3. *Effect of Enzyme Treatment on the Reducing Sugar and Soluble Solid Contents*

Figure 2 shows that the optimal temperature for xylanase to act in the pulping process is approximately 50–60 °C. The reducing sugar content was the highest at 60 °C, reaching 3.69 mg/mL, whereas the soluble solid content was the highest at 50 °C, which was 1.46%. The reducing sugar content increased notably in the range of 40–60 °C, whereas it exhibited a downward trend in the range of 60–90 °C. The soluble solid content increased from 40 °C to 50 °C and subsequently decreased from 50 °C to 90 °C. The optimal temperature of pectinase in the pulping process was 50 °C, at which point the reducing sugar content was the highest, reaching 1.52 mg/mL. The second highest content was observed at 60 °C. To facilitate the effect of pectinase in the pulping process, 70 °C was employed, at which point the soluble solid content was the highest (2.84%). However, it should be noted that the soluble solid content increased with the temperature from 40 °C to 50 °C. When the temperature increased to 60 °C, the soluble solid content decreased. The soluble solid content increased between 60 °C and 80 °C. A possible reason for this phenomenon is that the soluble solids were affected by the superposition of the optimal temperature and the enzyme. After 60 °C, the effect of the temperature on the soluble solids was greater than that of the enzyme.

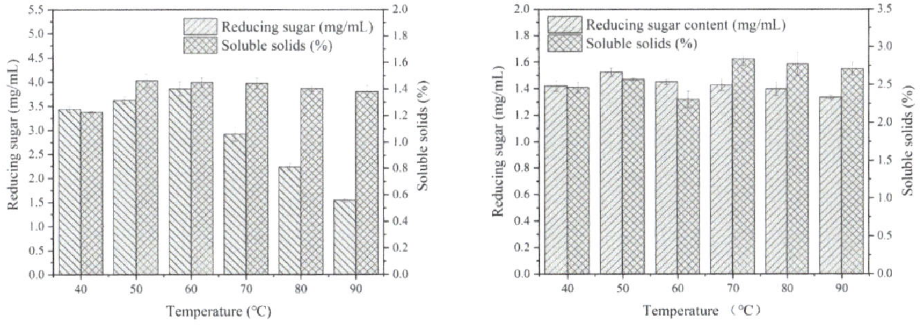

(a) Xylanase (b) Pectinase

Figure 2. The reducing sugar and soluble solid contents in different enzyme treatments.

3.4. *Box–Behnken Response Surface Methodology*

This method evaluates the three significant factors in the Plackett–Burman experiment while keeping the other factors at their central level. The range and level of the investigated variables are shown in Table 2a. The Box–Behnken design has three factors and three levels, including four replications of the center point, which are used to fit the second-order response surface. A polynomial quadratic equation is obtained to determine the influence of each variable on the response. The design and results of the trial scheme are shown in Table 2b, and the analysis of variance is shown in Table 2c.

Table 2. (a) Factors and levels of the Box–Behnken response surface experiment. (b) Box–Behnken response surface methodology test scheme design. (c) Analysis of variance table.

a

Factor	**Level**		
	−1	**0**	**1**
KOH dosage (%)	1.5	2.0	2.5
Liquid solid ratio	7	8	9
revolutions (r)	2500	3000	3500

b

Serial Number	KOH Dosage (%)	Liquid Solid Ratio	Revolutions (r)	Beating Degree (°SR)
1	2.0	8	3000	37.75
2	2.0	8	3000	37.33
3	2.5	7	3000	34.00
4	1.5	7	3000	46.00
5	1.5	8	2500	47.33
6	2.0	7	3500	36.50
7	1.5	9	3000	39.75
8	2.0	8	3000	37.50
9	2.0	7	2500	44.33
10	2.0	9	2500	38.75
11	2.0	8	3000	37.50
12	2.5	9	3000	28.33
13	2.0	8	3000	38.00
14	2.5	8	2500	34.25
15	1.5	8	3500	39.33
16	2.0	9	3500	30.75
17	2.5	8	3500	26.25

c

Source	Sum of Squares	df	Mean Square	F Value	*p* Value Prob > F	
Model	503.42	9	55.94	313.47	<0.0001	significant
KOH dosage (%) (*A*)	307.27	1	307.27	1722	<0.0001	
Liquid solid ratio (*B*)	69.03	1	69.03	386.86	<0.0001	
revolutions (r) (*C*)	124.66	1	124.66	698.63	<0.0001	
AB	0.0841	1	0.0841	0.4713	0.5145	
AC	0	1	0	0	1	
BC	0.0441	1	0.0441	0.2471	0.6343	
A^2	2.22	1	2.22	12.42	0.0097	
B^2	0.0706	1	0.0706	0.3957	0.5493	
C^2	0.0425	1	0.0425	0.2383	0.6403	
Residual	1.25	7	0.1784			
Lack of fit	0.9749	3	0.325	4.74	0.0834	not significant
Pure error	0.2741	4	0.0685			
Cor total	504.67	16				

The binomial fitting equation obtained by statistical analysis is as follows:

$$\begin{aligned} Y = \ & 37.62 - 6.20 \times A - 2.94 \times B - 3.95 \times C + 0.1450 \times AB \\ & -0.1050 \times BC - 0.7255 \times A^2 + 0.1295 \times B^2 - 0.1005 \times C^2 \end{aligned}$$

The p value of the fitting equation is less than 0.05, which indicates that the binomial model has reached a significant degree, the fitting of the model is good and the response value can be detected.

3.5. Establishment of the Fitting Equation

(1) Establishment of the fitting equation between the beating degree and reducing sugar content.

Matlab R2017a was used to fit the reducing sugar content and beating degree, and the fitting equation of the Fourier transform distribution was obtained (Formula (1)). The values and confidence limits of each constant at 95% confidence are shown in Table S1.

$$\begin{aligned} f(x) = \ & 822.8 + 1553 * \cos(x*5.464) + 916.2 * \sin(x*5.464) \\ & +584.3 * \cos(2*x*5.464) + 2019 * \sin(2*x*5.464) \\ & -1162 * \cos(3*x*5.464) + 1696 * \sin(3*x*5.464) \\ & -1613 * \cos(4*x*5.464) + 51.8 * \sin(4*x*5.464) \\ & -621.1 * \cos(5*x*5.464) - 762 * \sin(5*x*5.464) \\ & +102 * \cos(6*x*5.464) - 417.6 * \sin(6*x*5.464) \\ & +93.34 * \cos(7*x*5.464) - 51.8 * \sin(7*x*5.464) \end{aligned} \quad (1)$$

(2) Establishment of the fitting equation between the beating degree and the soluble solids.

Matlab R2017a was used to fit the values of the soluble solid content and beating degree, and the fitting equation of the Fourier transform distribution was obtained (Formula (2)). The values and confidence limits of each constant at 95% confidence are shown in Table S2.

$$\begin{aligned} f(x) = \ & -3387 + 1894 * \cos(x*8.213) + 6240 * \sin(x*8.213) \\ & +4649 * \cos(2*x*8.213) - 3242 * \sin(2*x*8.213) \\ & -3667 * \cos(3*x*8.213) - 2506 * \sin(3*x*8.213) \\ & -452 * \cos(4*x*8.213) + 2940 * \sin(4*x*8.213) \\ & +1461 * \cos(5*x*8.213) - 581.9 * \sin(5*x*8.213) \\ & -501.9 * \cos(6*x*8.213) - 327.9 * \sin(6*x*8.213) \\ & +24.22 * \cos(7*x*8.213) + 129.3 * \sin(7*x*8.213) \end{aligned} \quad (2)$$

The goodness of fit of the obtained fitting equation was tested, and the results are shown in Table S3.

When the reducing sugar content was 2.2 mg/mL, the beating degree predicted according to Formula (1) was approximately 35.05°SR, and the error from the actual measured value was 1.26%; at this time, the soluble solid content was 1.64%. The beating degree predicted according to Formula (2) was approximately 35.57°SR, and the error from the actual value was 0.20%. Briefly, from the comparison of the goodness of fit of the above two equations, it was found that the fit equations of the degree of beating and soluble solid content are better than those of the degree of beating and reducing sugar content. Therefore, in the actual production process, one can choose to use the fitting equations of the beating degree and soluble solid content for the online control of the beating process.

3.6. Scanning Electron Microscopy Analysis of the Pulping Effect

Compared with the blank control without enzyme treatment, the use of xylanase increases the curvature of the fiber, and a small amount of fiber is separated; consequently, many fine fibers appear on the surface edge of the fiber. Compared with the xylanase-treated wheat straw case, the change in the fiber surface of the wheat straw treated with pectinase is noticeable, and more flakes of fiber are detached and attached to the whole fiber surface (as shown in Figure S1).

3.7. Infrared Analysis of the Pulping Effect

In the study of pulp properties, FT-IR is mainly used for the qualitative analysis of raw fiber, mainly through the analysis of the characteristic functional groups of fibrous substances to preliminarily determine the structural changes in the fibrous substances. Figure S2 shows a strong absorption band at 3408 cm^{-1}, corresponding to the O–H bond stretching vibration of the untreated wheat straw, which indicates that the wheat straw fiber swells easily at this time. In addition, the absorption band at 1055 cm^{-1} may be assigned to the stretching vibration of the alkoxy group in the acetyl moiety and that of the carbon–oxygen double bond. After chemical and mechanical treatments, the absorption band intensities at 3408 and 1055 cm^{-1} decreased considerably, indicating that the structure of the hemicellulose was destroyed to some extent. The decrease in the absorption band intensities at 1512 and 1245 cm^{-1} also indicated that the structure of the lignin was destroyed considerably after treatment.

After the wheat straw was treated with different enzymes, FT-IR analysis was carried out on the pulp samples. The absorption band at 3418 cm^{-1}, corresponding to the –OH stretching vibration in the pulp samples treated with xylanase and pectinase, was the strongest, indicating that more phenolic hydroxyl groups were produced and the hydration degree of the pulp was high. The formation of hydrogen bonds in cellulose reduces the hygroscopicity of fiber and paper. The strength of paper depends on the strength of the fiber itself and the bonding strength between the fibers. The beating process refines the fibers and exposes more hydroxyl groups on the surface. When the fibers are pulped in the paper machine and the paper is dried, hydrogen bonds are formed between the fibers, and the binding force is increased, resulting in a certain paper strength [33]. The absorption bands of the blank control at 1460 and 1423 cm^{-1} were significantly different from those after the enzyme treatment, indicating that enzyme treatment could change the benzene ring structure of lignin. The band at 1046 cm^{-1} corresponds to the symmetric stretching vibration of the C–O–C glycosidic bond. Compared with the blank control, the increase of absorption band intensity indicated the increase of cellulose content. Other functional groups and characteristic substances with notable structural changes are listed in Table 3.

Table 3. FT-IR spectrum analysis after enzyme treatment.

Wavemunber (cm^{-1})	Corresponding Structure
3348–3408	O–H stretching vibration
2895–2902	C–H stretching vibration, CH_3, CH_2
1595–1597	Stretching vibration of benzene ring (lignin)
1460	CH_2 deformation vibration, Carbon skeleton vibration of benzene ring
1423	CH_2 shear vibration, CH_2 bending vibration (lignin), Benzene ring vibration
1365–1371	C–H bending vibration
1327–1228	C–O–C stretching vibration (lignin phenol ether bond), Syringyl, Condensation guaiacol
1232–1234	Acetyl and hydroxyl vibration, Syringa type C=O stretching vibration
1034–1056	C–O–C glucoside bond symmetric stretching vibration

3.8. X-ray Diffraction Analysis of the Pulping Effect

Compared with that of the untreated wheat straw, the peak intensities of the crystallization zone of the mechanically and chemically treated wheat straw decreased, and the crystallinity index decreased from 71.98% to 49.04%. According to Figure S3, the corresponding crystallinity indexes of the pulp treated by the enzymes are shown in Table 4. By comparison, it was found that the crystallinity of cellulose increased by varying degrees depending on the treatment, and the same conclusion has been reported elsewhere [34]. The crystallinity index of wheat straw increased by 1.19% after pectinase treatment, which was mainly because pectinase degraded the free poly galacturonic acid in pulp, reduced the anion waste in the system, reduced the proportion of amorphous area and increased the crystallization zone. The increase in the crystallinity of the pulp will increase its physical strength considerably. Notably, the effect of xylanase was the most significant. The crystallinity index of the wheat straw was the lowest after xylanase treatment: only 0.49% higher. The addition of xylanase degraded a part of the hemicellulose, which reduced the noncrystallized zone (mainly in the form of a noncrystalline zone) in the pulp, consequently increasing the crystallinity. Raw material and final pulp effect are shown in Figure S4.

Table 4. Comparison of the cellulose crystallinity index under different treatment conditions.

	Blank Control	Xylanase	Pectinase	Xylanase + Pectinase
Crystallinity index (*CrI*%)	49.04	49.53	50.23	51.09

4. Conclusions

In response to the problems of low yield and serious pollution in chemical pulping, we propose a biomechanical pulping method that combines xylanase and pectinase treatment under normal temperature and pressure conditions, which can reduce the dosage of drugs and the generation of pulping wastewater. The application of biological enzymes promotes the dissociation of cellulose, hemicellulose, lignin and other components in wheat straw, reduces the difficulty and improves the efficiency of pulping. At the same time, by establishing fitting equations for beating degree, reducing sugar and soluble solids, real-time monitoring of reducing sugar, soluble solids and other indicators can be achieved to achieve online control of important indicators such as beating degree in pulping, which is conducive to improving the automation level of straw pulping. In addition, some methods were used to characterize the pulping effect of the composite enzyme, and the results proved that the addition of complex enzymes had a significant positive influence on the pulping effect.

Supplementary Materials: The following supporting information can be downloaded at: https://www.mdpi.com/article/10.3390/polym15244637/s1, Table S1. Values and confidence limits of the parameters in Formula 1 with 95% confidence; Table S2. Values and confidence limits of the parameters in Formula 2 with 95% confidence; Table S3. Goodness of fit test of the fitting equation; Figure S1. Scanning electron microscopy images of the blank control, xylanase, and pectinase treated wheat straw (from left to right); Figure S2. Fourier transform infrared image after the enzyme treatment; Figure S3. X-ray diffraction patterns of the pulp prepared by enzyme pretreatment; Figure S4. Starting material and final pulping effect.

Author Contributions: K.L. conceived and designed the experiments. X.L., S.W. and Y.X. performed the experiments. K.L., X.L., S.W., L.Y., P.L. and R.W. analyzed the data. K.L., X.L. and S.W. drafted the manuscript. All authors have read and agreed to the published version of the manuscript.

Funding: This work was supported by the National Key Research and Development Program of China (No. 2019YFC1905902); and the Foundation (No. 202008) of Qilu University of Technology of Cultivating Subject for Biology and Biochemistry; the foundation (No. 2022GH026) of International Technology Cooperation Project from Qilu University of Technology (Shandong Academy of Sciences); the foundation (No. FWL2021065) of Shandong Provincial Key Laboratory of Biophysics; the foundation (No. 21YF5FA129 and No. 22YF7FG188) of Gansu Provincial Key R & D Program-Social Development; the funders had no role in study design, data collection and analysis, decision to publish, or preparation of the manuscript.

Institutional Review Board Statement: This article does not contain any studies with human participants or animals performed by any of the authors.

Data Availability Statement: Data are contained within the article and Supplementary Materials.

Conflicts of Interest: The authors declare no conflict of interest.

References

1. Nations, Food and Agriculture Organization of the United. Pulp and Paper Capacities, Survey 2019–2024/Capacités de la Pâte et du Papier, Enquête 2019–2024/Capacidades de Pulpa y Papel, Estudio 2019–2024. Available online: https://www.fao.org/home/en (accessed on 1 December 2023).
2. China Paper Association. China Pulp and Paper Industry: Annual Report 2022. *China Pulp Pap. Ind.* **2023**, *44*, 21–30+26.
3. Liu, Z.; Wang, H.; Hui, L. Pulping and Papermaking of Non-Wood Fibers. *Pulp Pap. Process.* **2018**, *1*, 4–31.
4. Alemdar, A.; Sain, M. Isolation and characterization of nanofibers from agricultural residues: Wheat straw and soy hulls. *Bioresour. Technol.* **2008**, *99*, 1664–1671. [CrossRef] [PubMed]
5. Liu, Q.; Lu, Y.; Aguedo, M.; Jacquet, N.; Ouyang, C.; He, W.; Yan, C.; Bai, W.; Guo, R.; Goffin, D.; et al. Isolation of High-Purity Cellulose Nanofibers from Wheat Straw through the Combined Environmentally Friendly Methods of Steam Explosion, Microwave-Assisted Hydrolysis, and Microfluidization. *ACS Sustain. Chem. Eng.* **2017**, *5*, 6183–6191. [CrossRef]
6. Ge, P.J.; Zhao, J.; Xu, J.; Qu, Y.B.; You, J.X. Changes of chemical composition and elements of wheat straw in the process of xylanase-alkali mechanical pulping. *J. Cell. Sci. Technol.* **2005**, 1–7. [CrossRef]
7. Kuang, S.J. Discussion on some problems of wheat straw pulp and paper making. *China Pulp Pap.* **1992**, *6*, 4–12.
8. Chen, X.; Wang, Z.J.; Wang, J. Study on fiber morphology and chemical composition of four common broadleaf wood. *Hunan Papermak. Process Technol.* **2009**, *1*, 5–7.
9. Sain, M.; Panthapulakkal, S. Bioprocess preparation of wheat straw fibers and their characterization. *Ind. Crops Prod.* **2006**, *23*, 1–8. [CrossRef]
10. Khristova, P.; Kordsachia, O.; Patt, R.; Karar, I.; Khider, T. Environmentally friendly pulping and bleaching of bagasse. *Ind. Crops Prod.* **2006**, *23*, 131–139. [CrossRef]
11. Hideno, A. Short-time alkaline peroxide pretreatment for rapid pulping and efficient enzymatic hydrolysis of rice straw. *Bioresour. Technol.* **2017**, *230*, 140–142. [CrossRef]
12. Brinchi, L.; Cotana, F.; Fortunati, E.; Kenny, J.M. Production of nanocrystalline cellulose from lignocellulosic biomass: Technology and applications. *Carbohydr. Polym.* **2013**, *94*, 154–169. [CrossRef]
13. Kim, J.; Sunagawa, M.; Kobayashi, S.; Shin, T.; Takayama, C. Developmental localization of calcitonin gene-related peptide in dorsal sensory axons and ventral motor neurons of mouse cervical spinal cord. *Neurosci. Res.* **2016**, *105*, 42–48. [CrossRef]
14. Yue, Y.; Han, J.; Han, G.; Zhang, Q.; French, A.D.; Wu, Q. Characterization of cellulose I/II hybrid fibers isolated from energycane bagasse during the delignification process: Morphology, crystallinity and percentage estimation. *Carbohydr. Polym.* **2015**, *133*, 438–447. [CrossRef]
15. Grabber, J.H. How Do Lignin Composition, Structure, and Cross-Linking Affect Degradability? A Review of Cell Wall Model Studies. *Crop Sci.* **2005**, *45*, 820–831. [CrossRef]
16. Singh, A.K.; Chandra, R. Pollutants released from the pulp paper industry: Aquatic toxicity and their health hazards. *Aquat. Toxicol.* **2019**, *211*, 202–216. [CrossRef] [PubMed]
17. Kellock, M.; Maaheimo, H.; Marjamaa, K.; Rahikainen, J.; Zhang, H.; Holopainen-Mantila, U.; Ralph, J.; Tamminen, T.; Felby, C.; Kruus, K. Effect of hydrothermal pretreatment severity on lignin inhibition in enzymatic hydrolysis. *Bioresour. Technol.* **2019**, *280*, 303–312. [CrossRef] [PubMed]
18. Donohoe, B.S.; Decker, S.R.; Tucker, M.P.; Himmel, M.E.; Vinzant, T.B. Visualizing lignin coalescence and migration through maize cell walls following thermochemical pretreatment. *Biotechnol. Bioeng.* **2008**, *101*, 913–925. [CrossRef] [PubMed]
19. Przybysz Buzała, K.; Kalinowska, H.; Borkowski, J.; Przybysz, P. Effect of xylanases on refining process and kraft pulp properties. *Cellulose* **2017**, *25*, 1319–1328. [CrossRef]
20. Dukare, A.; Sharma, K.; Kautkar, S.; Dhakane-Lad, J.; Yadav, R.; Nadanathangam, V.; Saxena, S. Microbial xylanase aided biobleaching effect on multiple components of lignocelluloses biomass based pulp and paper: A review. *Nord. Pulp Pap. Res. J.* **2023**, *38*, 459–480. [CrossRef]
21. Gil, N.; Gil, C.; Amaral, M.E.; Costa, A.P.; Duarte, A.P. Use of enzymes to improve the refining of a bleached Eucalyptus globulus kraft pulp. *Biochem. Eng. J.* **2009**, *46*, 89–95. [CrossRef]
22. Singh, G.; Kaur, S.; Khatri, M.; Arya, S.K. Biobleaching for pulp and paper industry in India: Emerging enzyme technology. *Biocatal. Agric. Biotechnol.* **2019**, *17*, 558–565. [CrossRef]
23. Tofani, G.; Cornet, I.; Tavernier, S. Multiple linear regression to predict the brightness of waste fbres mixtures before bleaching. *Chem. Pap.* **2022**, *76*, 4351–4365. [CrossRef]
24. Adamopoulos, S.; Karageorgos, A.; Rapti, E.; Birbilis, D. Predicting the properties of corrugated base papers using multiple linear regression and artificial neural networks. *Drew. Pract. Nauk. Doniesienia Komun.* **2016**, *59*, 161–172.
25. Sluiter, A.; Hames, B.; Ruiz, R.; Scarlata, C.; Sluiter, J.; Templeton, D. Determination of structural carbohydrates and lignin in biomass national renewable. *Energy Lab.* **2011**, *10*, 1–15.
26. Sluiter, A.; Hames, B.; Ruiz, R.O.; Scarlata, C.; Sluiter, J.; Templeton, D. Determination of Ash in Biomass. *Natl. Renew. Energy Lab.* **2008**, 1–6.
27. Hames, B.; Scarlata, C.; Sluiter, A. *Determination of Protein Content in Biomass*; National Renewable Energy Laboratory: Golden, CO, USA, 2008.

28. *GB/T 5513-2019*; Grain and Oil Testing—Determination of Reducing Sugars and Non-Reducing Sugars in Grain. State Administration for Market Regulation; National Standardization Administration: Beijing, China, 2019. Available online: https://kns.cnki.net/kcms2/article/abstract?v=z-q19lQZUWEiNhWAfCSeAn5Fx0O7HB21hvvas22byeprisLplIuwl4nhmFgi8OKOQVJxVNtNlRYqc_j9hiCDzakGfh8CnamSWTzzgkdcZWGsbO650bJn_ji-OA5cJJfBPEstdnwpL9c=&uniplatform=NZKPT&language=CHS (accessed on 1 December 2023).
29. *GB/T 23874-2009*; Determination of Xylanase Activity of Feed Additive—Spectropho-Tometric Method. General Administration of Quality Supervision, Inspection and Quarantine of the People's Republic of China; Standardization Administration of China: Beijing, China, 2009. Available online: https://kns.cnki.net/kcms2/article/abstract?v=z-q19lQZUWHhwGZb99SgUJgD3IMnTAym3TsV2-Hy0SRJMYojb3dC4MjETzgCn--o8MfFYgbc3ccH246_vLtAOFzrugMmN0W-wogOj-e--levO4BS6dCdZ_AqRFWsG5dk&uniplatform=NZKPT&language=CHS (accessed on 1 December 2023).
30. Segal, L.; Creely, J.J.; Martin, A.E.; Conrad, C.M. An Empirical Method for Estimating the Degree of Crystallinity of Native Cellulose Using the X-Ray Diffractometer. *Text. Res. J.* **1959**, *29*, 786–794. [CrossRef]
31. Blanco, Á.; Negro, C.; Díaz, L.L.; Saarimaa, V.; Sundberg, A.; Holmbom, B.R. Influence of Thermostable Lipase Treatment of Thermomechanical Pulp (TMP) on Extractives and Paper Properties. *Appita Technol. Innov. Manuf. Environ.* **2009**, *62*, 113–117.
32. Roncero, M.B.; Colom, J.F.; Vidal, T. Cellulose protection during ozone treatments of oxygen delignified. *Carbohydr. Polym.* **2003**, *51*, 243–254. [CrossRef]
33. Duan, C.; Wang, X.; Zhang, Y.; Xu, Y.; Ni, Y. Fractionation and cellulase treatment for enhancing the properties of kraft-based dissolving pulp. *Bioresour. Technol.* **2017**, *224*, 439–444. [CrossRef] [PubMed]
34. Dutta, P.D.; Neog, B.; Goswami, T. Xylanase enzyme production from Bacillus australimaris P5 for prebleaching of bamboo (*Bambusa tulda*) pulp. *Mater. Chem. Phys.* **2020**, *243*, 122227. [CrossRef]

polymers

MDPI

Article

Processing, Characterization and Disintegration Properties of Biopolymers Based on Mater-Bi® and Ellagic Acid/Chitosan Coating

Carolina Villegas, Sara Martínez, Alejandra Torres, Adrián Rojas, Rocío Araya and Abel Guarda et al.

Center for Packaging Innovation (LABEN), Center for the Development of Nanoscience and Nanotechnology (CEDENNA), Technology Faculty, University of Santiago de Chile (USACH), Santiago 9170201, Chile
* Correspondence: carolina.villegasv@usach.cl (C.V.); alejandra.torresm@usach.cl (A.T.)

Abstract: Among the most promising synthetic biopolymers to replace conventional plastics in numerous applications is MaterBi® (MB), a commercial biodegradable polymer based on modified starch and synthetic polymers. Actually, MB has important commercial applications as it shows interesting mechanical properties, thermal stability, processability and biodegradability. On the other hand, research has also focused on the incorporation of natural, efficient and low-cost active compounds into various materials with the aim of incorporating antimicrobial and/or antioxidant capacities into matrix polymers to extend the shelf life of foods. Among these is ellagic acid (EA), a polyphenolic compound abundant in some fruits, nuts and seeds, but also in agroforestry and industrial residues, which seems to be a promising biomolecule with interesting biological activities, including antioxidant activity, antibacterial activity and UV-barrier properties. The objective of this research is to develop a film based on commercial biopolymer Mater-Bi® (MB) EF51L, incorporating active coating from chitosan with a natural active compound (EA) at two concentrations (2.5 and 5 wt.%). The formulations obtained complete characterization and were carried out in order to evaluate whether the incorporation of the coating significantly affects thermal, mechanical, structural, water-vapor barrier and disintegration properties. From the results, FTIR analysis yielded identification, through characteristic peaks, that the type of MB used is constituted by three polymers, namely PLA, TPS and PBAT. With respect to the mechanical properties, the values of tensile modulus and tensile strength of the MB-CHI film were between 15 and 23% lower than the values obtained for the MB film. The addition of 2.5 wt.% EA to the CHI layer did not generate changes in the mechanical properties of the system, whereas a 5 wt.% increase in ellagic acid improved the mechanical properties of the CHI film through the addition of natural phenolic compounds at high concentrations. Finally, the disintegration process was mainly affected by the PBAT biopolymer, causing the material to not disintegrate within the times indicated by ISO 20200.

Keywords: Mater-Bi®; ellagic acid; chitosan; coating

Citation: Carolina Villegas, Sara Martínez, Alejandra Torres, Adrián Rojas, Rocío Araya and Abel Guarda et al. Processing, Characterization and Disintegration Properties of Biopolymers Based on Mater-Bi® and Ellagic Acid/Chitosan Coating. *Polymers* **2023**, *15*, 1548. https://doi.org/10.3390/polym15061548

Academic Editors: Antonio M. Borrero-López, Concepción Valencia-Barragán, Esperanza Cortés Triviño, Adrián Tenorio-Alfonso and Clara Delgado-Sánchez

Received: 1 February 2023
Revised: 16 March 2023
Accepted: 17 March 2023
Published: 21 March 2023

1. Introduction

Nowadays, the main concerns in the food industry are oriented towards two different directions. The first one is related to food safety, preservation and quality assurance, and while second is the reduction in plastic packaging waste. One way to address these problems is the development of compostable and/or biodegradable materials to reduce environmental contamination without losing safe food [1]. Due to this, the development of biopolymers provides a pathway to accomplishing a sustainable environment by reducing dependency on non-renewable fossil fuel raw materials [2].

Currently, the most important polymers on the market are divided into three subgroups as follows: polymers based on renewable resources (starch and cellulose); biodegradable polymers based on biodegradable monomers (vegetable oils and lactic acid); and

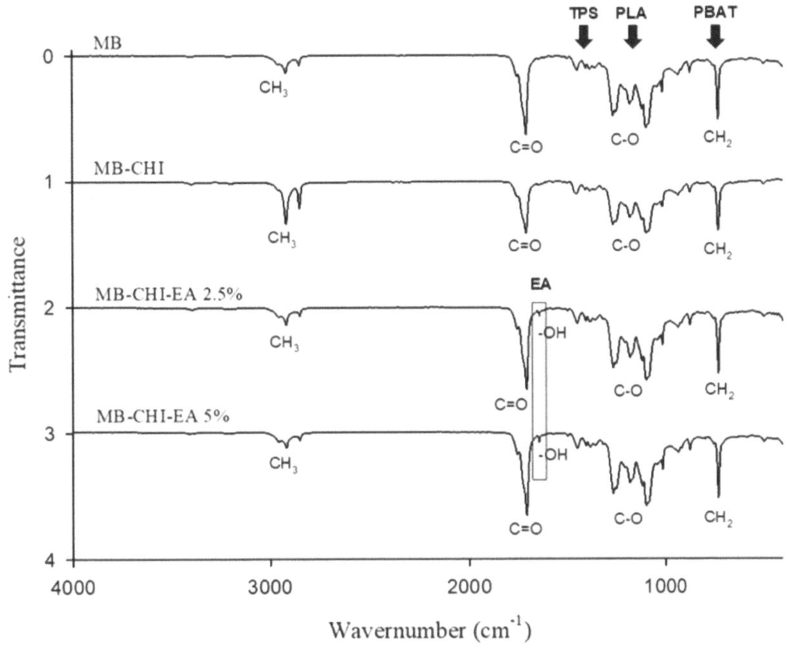

Figure 2. IR spectra of the biopolymers and active coating biopolymers.

In addition, it is possible observe an increase associated with the bands at 3004–2830 cm^{-1} between the MB and MB-CHI biopolymers. This behavior could be due to the overlapping of νCH vibrations in the CH_3, CH_2 and $=CH_2$ groups coming from biopolymers and chitosan coating [1] that incorporated active coating which causes the same band to decrease due to low affinity of EA in the hydrophilic coating.

The EA spectrum shows three bands at 3562 cm^{-1}, 3468 cm^{-1} and 3149 cm^{-1}, corresponding to the axial stretch of OH. The bands at 1656 cm^{-1}, 1600 cm^{-1}, 1710 cm^{-1} and 1157 cm^{-1} are associated with the aromatic C=C stretch, C–C stretch, C=O stretch and C–O–C, respectively. The band at 1348 cm^{-1} corresponds to the hydrogen bonds (HO–H). Finally, the band at 758 cm^{-1} indicates the presence of the phenyl ring [42].

3.2. Differential Calorimetry Scanning (DSC)

DSC analysis was performed to determine the effect of active coating on the thermal transitions of MB; the results are shown in Figure 3, and in Figure 3a, it is possible recognize three endothermic peaks. The first peak at ~61 °C is related the glass transition (Tg) of PLA [43], the second endothermic peak at ~119 °C corresponds to the melting of PBAT (Tm_1), which is found between 115 and 125 °C [44], and the last peak ~160 °C (Tm_2) is related to the melting of PLA [39]. With respect to the TPS, Surendren and coworkers indicated that the endothermic peaks between 165 °C and 180 °C could belong to the melting temperature (Tm) of thermoplastic starch (TPS) [45], while other research indicates that TPS has a wide melting range between 160 and 380 °C, which depends on the botanical origin and proportion of the starch the amylose/amylopectin ratio of the starch structure, in addition to the plasticizer used and the plasticizing conditions [46]. The behavior of Tg, Tm_1, Tm_2 and Tc is similar to that reported by Bianchi and Morreale (2023). They studied the thermal properties of PLA/PBAT, where it was possible to observe the first- and second-order temperatures associated with each polymer. With respect to PLA, two temperatures were identified, namely the glass transition temperature (60 °C) and the melting temperature (155 °C); meanwhile, the crystallization (78 °C) and melting (117 °C) temperatures correspond to PBAT [47]. These results are in accordance with the structural properties.

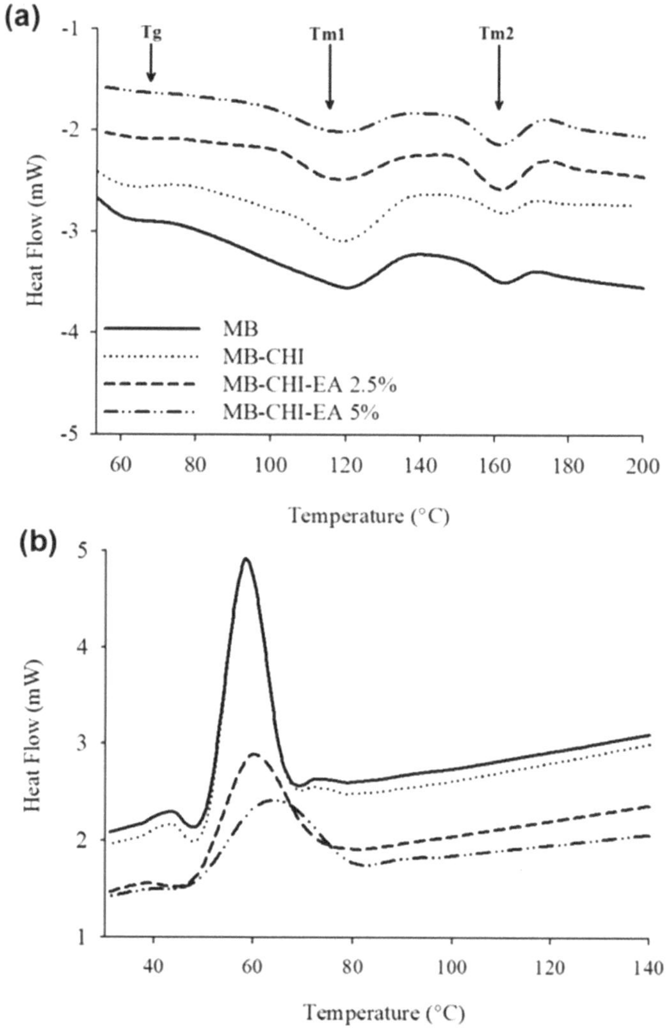

Figure 3. DSC thermograms during the (**a**) first heating scan and (**b**) cooling scan for all samples.

The incorporation of CHI-EA active coating caused a slight decrease in Tg, Tm_1 and Tm_2 due to the plasticizing effect. On the other hand, in Figure 3b, it is possible to observe a decrease in the crystallization peak, which could be due to the hydrogen bonding of the carbonyl group of the biopolymers and the OH and NH_2 groups in chitosan; these interactions occur in the amorphous regions, suppressing the extent of crystallization [22].

3.3. Thermogravimetric Analysis (TGA)

TGA is a very important technique used to evaluate the thermal stability of commercial biopolymers and active coating biopolymers. TGA and the first derivative (DTG) thermograms of different samples are illustrated in Figure 4. When evaluating the weight loss of MB, three mass losses are observed around 324, 336 and 400 °C, which describe the degradation of different biopolymers of the commercial sample.

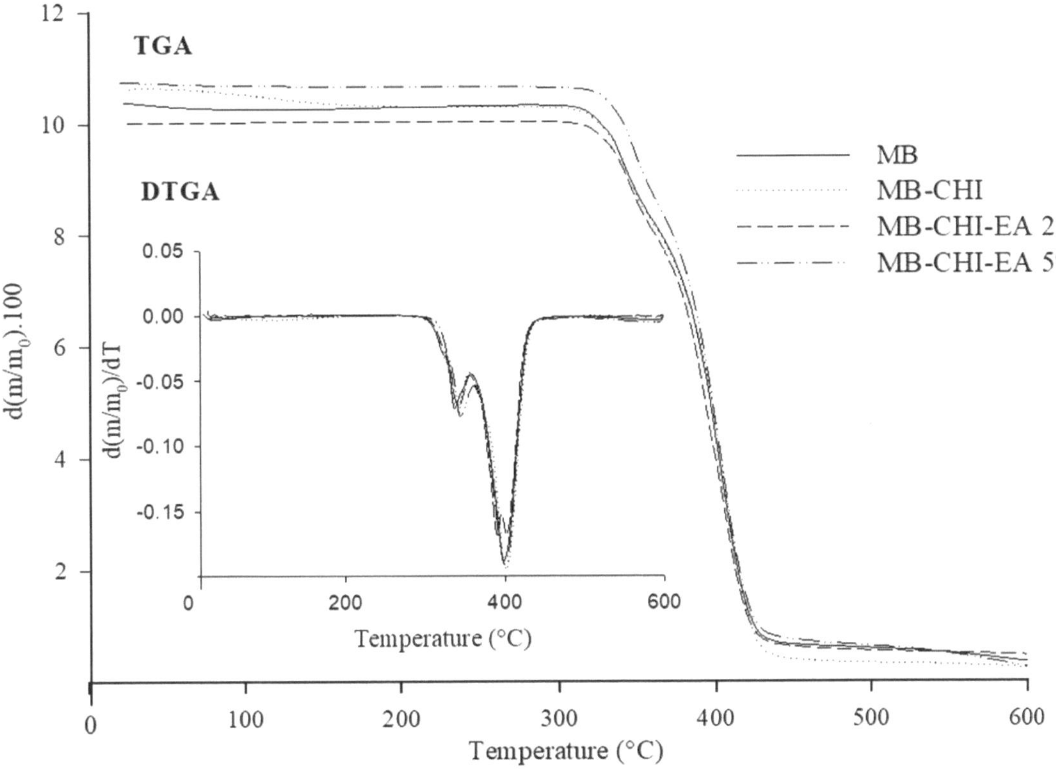

Figure 4. TGA and DTGA curves obtained for biopolymers and active coating biopolymers.

Different researchers have been able to establish degradation temperature ranges between 50 and 320 °C for thermoplastic starch (TPS), which can be correlated with the elimination of water, glycerol and other low-molecular-weight compounds [48–50]. The temperatures range from 30 to 200 °C, which represents the evaporation of water adsorbed by starch and the plasticizers used, together with the evaporation of low-molecular-weight compounds [51] at ~320 °C, which represents the degradation of amylose and amylopectin [46].

It has been reported that the thermal degradation temperature of PLA ranges from 304 to 380°C [43,51]. Meanwhile, PBAT has a degradation temperature range between 390 and 410 °C [52,53]. Based on the literature, the first degradation inflection (324 °C) is attributed to TPS degradation, while the second (336 °C) corresponds to PLA decomposition and the third inflection (400 °C) corresponds to PBAT.

On the other hand, if the films with and without coating are compared, the latter show early minor weight loss (~100 °C), which is attributed to the desorption of moisture as hydrogen-bound water and acetic acid are present in the coating [54]. Furthermore, in films with coating, another inflection was observed close to 300 °C. Similar behavior was described by Vilela and coworkers (2017), who identified that chitosan presented a weight loss temperature at about 290 °C associated with the degradation of the CHI skeleton [22]. Meanwhile, Bonilla and coworkers associated this inflection to the degradation of the polymer structure, including the dehydration of saccharide rings and the decomposition of acetylated and deacetylated units at 298.3 °C [55]. The results point to the fact that the thermal stability of MB is not affected for CHI/EA active coating because of the presence of EA, due to most probably due to low loadings.

3.4. Mechanical Properties and Water-Vapor Permeability

The effect of the addition of a chitosan (CHI) coating layer, with different concentrations (2.5 and 5%) of ellagic acid (EA), on the mechanical properties of films based on Mater-Bi® (MB) as a polymeric substrate was analyzed following the method detailed in Section 2.4.3.

Table 1 shows the tensile modulus (TM), tensile strength (TS) and elongation at break (EB) of the MB film, the MB biopolymer coated with chitosan (MB-CHI) and the MB biopolymer coated with CHI layers loaded with EA at 2.5% (MB-CHI-EA 2.5%) and 5% (MB-CHI-EA 5%). The MB monolayer film presented TM, TS and EB values consistent with the literature data [14]. TM and TS values for the MB-CHI film, composed of an MB substrate layer coated with a low-molecular-weight (MW) chitosan layer, were 15 and 23% lower than the values obtained for the MB film, respectively.

Table 1. Mechanical properties for the different samples developed.

Samples	Thickness (μm)	Tensile Modulus—TM (MPa)	Tensile Strength—TS (MPa)	Elongation of Break—EB (%)
MB	109 ± 7 [a]	286 ± 20 [c]	31 ± 3 [b]	670 ± 79 [b]
MB-CHI	161 ± 5 [b]	244 ± 15 [a]	24 ± 2 [a]	568 ± 54 [a]
MB-CHI-EA 2.5%	171 ± 15 [c]	335 ± 21 [b]	25 ± 2 [a]	565 ± 67 [a]
MB-CHI-EA 5%	183 ± 20 [c]	267 ± 22 [d]	30 ± 3 [b]	722 ± 94 [b]

Mean value ($n = 10$) ± SD. Parameters in columns denoted with the same letters ([a–d]) do not differ statistically at the level of confidence (0.05).

The effect of a CHI coating layer added over a polymer substrate using a solvent-based process on the mechanical properties of the resulting bilayer film depends on the balance between the plasticizing effect of the organic solvent on the polymer substrate and the reinforcing effect of the added CHI layer. Particularly, the polymer substrate reinforcement degree, achieved by adding a CHI layer is influenced by, among other factors, the molecular weight (MW) of CHI. Zhang and coworkers (2022) reported that the mechanical properties of CHI films, including TM and TS, significantly increased as the MW of CHI increased from 30 to 200 kDa [56]. In this context, effective increases in the mechanical properties of different biopolymeric substrates have been reported using CHI layers with medium and high MW, instead of using CHI layers with low MW. On the other hand, Fiore and coworkers (2021) reported a significant increase in TM and TS of poly acid(lactic) (PLA) films coated with a layer of CHI of medium MW [57]. In other work, such as that reported by Tanpichai and coworkers (2022), an increase in TM and TS of cellulose-based paper due to its coating with a layer of CHI of high MW was reported. The authors attributed this enhancement to the filling of voids between cellulose fibers with CHI and to the formation of additional hydrogen bonds between the anionic charges of cellulose and the cationic charges of CHI [58].

In this context, the decrease in TM and TS in the MB-CHI film, with respect to the monolayer MB film, could be related to the plasticizing effect of the CHI solution on the polymer structure and to the decrease in polymer substrate crystallinity. Both phenomena were reported in Section 3.2 and seem to be dominant over the reinforcing effect of adding a layer of CHI of low MW. On the other hand, it seems that CHI coating limits the extension capacity of the resulting MB-CHI film, decreasing EB from 670% to 568%. This behavior has been previously reported for other biopolymeric systems using a CHI layer as coating [57].

On the other hand, the addition of EA at 2.5% to CHI coating did not generate changes in the mechanical properties of the bilayer system (MB-CHI-EA 2.5%), with respect to the system without EA (MB-CHI). This fact agrees with the results reported by Vilela (2017), who reported production by solvent casting CHI films with EA up to 5% [22]. The opposite behavior was found in our study by adding EA at 5 % to the CHI layer. Particularly, TM and TS values for the MB-CHI-EA 5.0% film were higher than the values obtained for the MB-CHI system. The improvement in the mechanical properties of CHI films via the addition of naturally occurring phenolic compounds at high concentrations has previously been reported. Siripatrawan and Harte reported an increase in tensile modulus from 23.66 to 27.55 by adding green tea extract (GTE) up to 20%. The authors attributed this fact to the interaction between the CH matrix and phenolic compounds of GTE. Finally, the increase in EB in the bilayer system due to the addition of EA at 5% could be related to the plasticizing effect on the polymer structure [59].

Table 2 shows the water-vapor permeability for different developed samples. From these results, it can be observed that the water-vapor permeability of the MB film remained unalterable after the addition of CHI and the activation of CHI-EA coating. These results agree with the results reported for other biopolymers coated with CHI, such as cellulose-based paper [58], as well as with the results reported for biopolymers coated with a CHI layer loaded with naturally occurring phenolic compounds, such as PLA film coated with CHI, incorporated with rosemary essential oil at 2% [57].

Table 2. Water-vapor permeability for the different samples developed.

Samples	WVP (g/m^2/día)
MB	$7.4 \times 10^{-14} \pm 5 \times 10^{-15\,a}$
MB-CHI	$4.9 \times 10^{-14} \pm 3 \times 10^{-15\,b}$
MB-CHI-EA 2.5%	$8.20 \times 10^{-14} \pm 1,8 \times 10^{-15\,a}$
MB-CHI-EA 5%	$7.6 \times 10^{-14} \pm 7 \times 10^{-15\,a}$

Mean value (n = 10) ± SD. Parameters in columns denoted with the same letters (^{a,b}) do not differ statistically at the level of confidence (0.05).

3.5. Disintegration under Composting Conditions

3.5.1. Visual Appearance of Films during Composting

The visual inspection carried out of MB, MB-CHI and MB-CHI-EA (2.5 and 5%) after times of disintegration under composting conditions are shown in Figure 5a. From the results, it is possible to observe the slow rate of disintegration of the samples under composting conditions. After only 1 day of incubation, most of the formulations changed their color and became opaquer, losing transparency, which provoked a change in the refraction index of the materials as a result of water absorption and/or the presence of low-molecular-weight compounds formed by an enzymatic attack on the glycosidic bonds of the starch component of the Mater-Bi matrix [37].

After just 7 days, the samples presented signs of erosion, and aspect was greatly changed due to contact with the organic waste matrix presenting organic matter deposition on the surface. At day 50, some samples just started to break. The samples with the incorporation of CHI and CHI-EA (2.5 and 5%) that were still recoverable at 90 days presented slow degradation compared to other biopolymers.

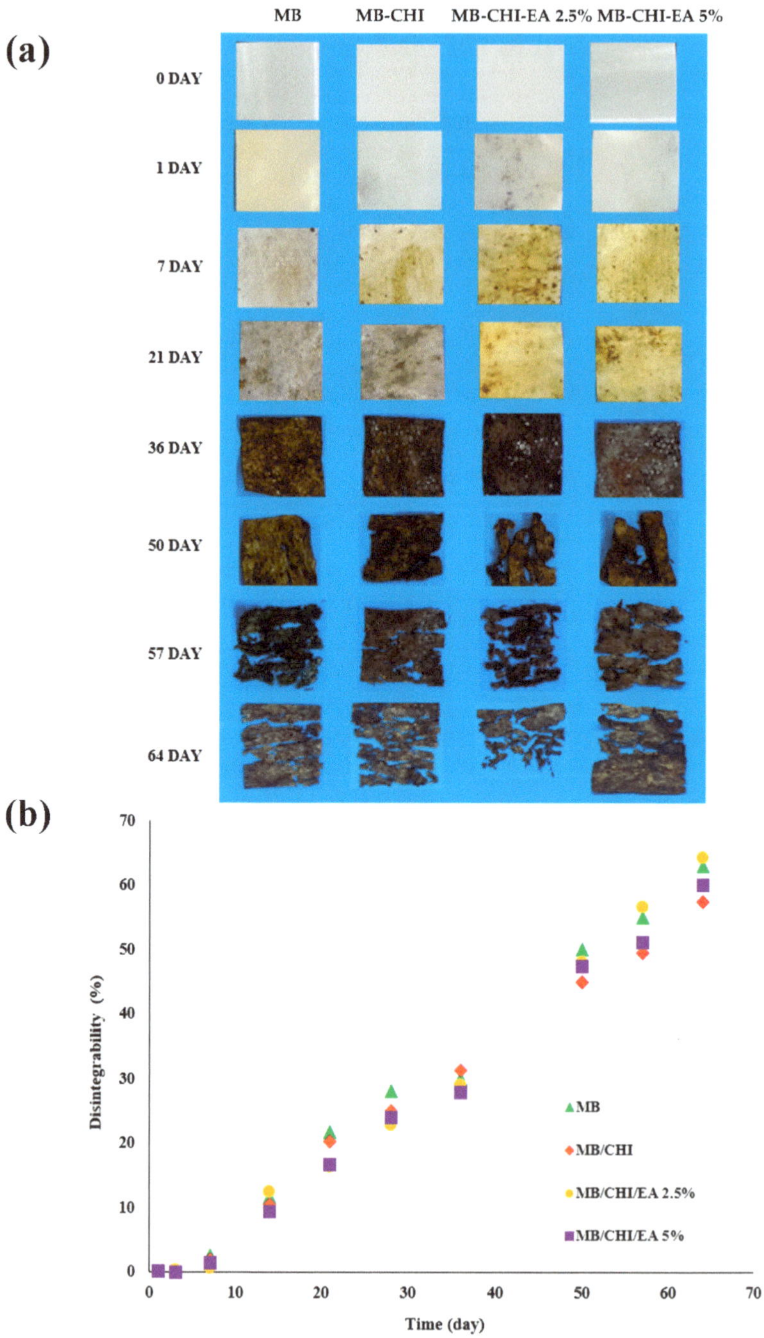

Figure 5. (**a**) Visual appearance of developed materials at different times under composting conditions; (**b**) film disintegration degree under composting conditions as a time function.

3.5.2. Disintegration Degree of Films during Composting

Visual observations were confirmed by calculating the disintegration degree in terms of mass loss as a function of incubation time using the Boltzmann function to correlate the sigmoidal behavior of the mass loss during disintegrability in the composting process, which is also presented in Figure 5b. From the image, it can be observed that all the samples reached a degree of disintegration of appropriately 70% with respect to their weight loss; however, these values do not comply with current legislation for biodegradable materials, which indicates that after a maximum time of 12 weeks, the samples must reach a degree of disintegration equal to or greater than 90% [35].

As discussed in Section 3.1, this commercial biopolymer is made up of PLA, TPS and PBAT. Regarding PLA, different research indicates that these biopolymers present a higher rate of disintegration when compared with the other biopolymers that make up MB. Villegas and coworkers (2019) studied the disintegration of PL. In their study, on day 7, small pieces of the analyzed polymers were collected due to physical and/or chemical degradation of the polymer, which occurred because it lost flexibility [46]. Due to this, PLA hydrolysis begins in the amorphous region of the polymer structure. Additionally, it has been reported that the intensity of the –C=O band increases with the composting time due to hydrolytic degradation, resulting in an increase in the number of carboxylic end groups in the polymer chains [36].

The degradation of TPS starts with non-enzymatic hydrolysis, which leads to a significant molar weight reduction, followed by enzyme action from the microorganisms present in the compost medium throughout the bulk of the polymeric matrix [60], as well as the biodegradation of starch-based polymers is a result of an enzymatic attack at the glucosidic linkages of the long-chain sugar units, leading to their breakdown into oligosaccharides, disaccharides and monosaccharides that are readily accessible to enzymatic attacks [61]. Sessini and coworkers (2019) studied thermal degradation and disintegrability under composting conditions of melt-processed blends based on ethylene-vinyl acetate and thermoplastic starch, such as EVA/TPS, as well as their nanocomposites, reinforced with natural bentonite. The results of the disintegration test showed that EVA/TPS blends and their nanocomposites presented positive interactions, which delay the disintegration of the TPS matrix in compost, thus improving TPS stability reaching 100% disintegrability in less than 60 days [60]. Meanwhile, blending biodegradable polymers, such as TPS, with non-biodegradable polymers, such as EVA, leads to an increase in compostable polymer percentage in partially degradable materials, giving a possible solution for the end-life of these materials after their use.

It is reported that the degradation of PBAT under composting conditions is considerably slower than that of PLA [62]. It is important to note that neat PLA almost degrades at 16 days, while a much longer incubation time is required to degrade PLA/PBAT blends and PBAT [63]. Recently, an investigation indicated that PBAT is a synthetic aromatic-aliphatic co-polyester with a molecular structure more complex than starch [64]. As such, it takes a longer period of time to be completely assimilated by microorganisms and transformed into stable products, and it is much more likely to subjected to process conditions, such as insufficient moistening content [65]. It has also been reported that PBAT disintegration is much slower and can take up to 12 weeks to disintegrate by 40% [66], although other authors have reported that PBAT takes 230 days to reach just 35% disintegration under the same composting conditions in this study (ISO 20200) because its structure must be hydrolyzed before microorganisms consume it as a source of nutrients [67]. On the other hand, Xie and coworkers (2023) developed large-size reed-reinforced PBAT composites with different filler degrees, and the properties and biodegradation behavior of the composites were investigated. With respect to the enzymatic degradation tests, it has been shown that the degradation rates of the composites were all greater than those of PBAT, and they all conformed to the surface erosion degradation mode. Furthermore, the main site of an enzymatic attack on PBAT during degradation is the ester bond, which breaks and increases the hydroxyl and carboxyl groups [68].

Similar results to this investigation were obtained by Aldas and coworkers (2021). They blended Mater-Bi® NF866 with different additives to produce bio-based and compostable films for food packaging or agricultural mulch films. Disintegration was carried out in a test under composting conditions in which Mater-Bi® reached 28% disintegrability over the 180 days of the composting test [16].

These results indicate that the disintegration of MB depends on the polymers that constitute it, but in this case, it stands out that TPS and PLA are easily biodegradable polymers, mainly due to the fact that starch-based materials are more available to microorganisms, improving the degree of disintegration; meanwhile, the high content of PBAT in the matrix explains the low disintegration degree of the material.

4. Conclusions

The effect of the addition of a chitosan (CHI) coating layer with different concentrations (2.5 and 5 wt.%) of ellagic acid (EA) on a commercial Mater-Bi® (MB) biopolymer was studied. From the results regarding structural (FTIR) and thermal properties (DSC and TGA), it was possible to identify different temperatures and peaks associated with the three following biopolymers: PLA, TPS and PBAT.

With respect to the incorporation of CHI-EA, active coating caused a slight decrease in Tg, Tm_1 and Tm_2 due to the plasticizing effect; moreover, the addition of EA at 2.5 wt.% to the CHI layer did not generate changes in the mechanical properties of the system, while an increase in ellagic acid 5 wt.% improved the mechanical properties of CHI films by the addition of naturally occurring phenolic compounds at high concentrations. Regarding mechanical properties, the tensile modulus and tensile strength values of the MB-CHI film, composed of an MB substrate layer coated with a low-molecular-weight chitosan layer, were 15 and 23% lower, respectively, than the values obtained for the MB film. The effect of a CHI coating layer added over a polymer substrate via a solvent-based process on the mechanical properties of the resulting film depends on the balance between the plasticizing effect of the organic solvent on the polymer substrate and the reinforcing effect of the added CHI layer. Finally, during the disintegration process, all the samples reached a degree of disintegration of appropriately 70% with respect to their weight loss; however, these values do not comply with current legislation for biodegradable materials, which indicates that after a maximum time of 12 weeks, the samples must reach a degree of disintegration equal to or greater than 90%. PBAT is principally responsible for this slow degradation because it takes a longer period of time to be completely assimilated by microorganisms and transformed into stable products.

Author Contributions: Conceptualization, C.V. and S.M.; methodology, C.V. and S.M.; software, S.M. and R.A.; validation, C.V., A.T. and A.R.; formal analysis, C.V., S.M. and R.A.; investigation, C.V., A.T. and A.R.; resources, C.V. and A.T.; data curation, C.V. and A.T.; writing—original draft preparation, C.V. and A.R.; writing—review and editing, C.V., A.T. and A.R.; visualization, S.M.; supervision, C.V. and A.T.; project administration, C.V. and A.T.; funding acquisition, M.J.G. and A.G. All authors have read and agreed to the published version of the manuscript.

Funding: This research was supported by the National Fund for Scientific and Technological Development (FONDECYT) Postdoctoral 3210434 (ANID-CHILE), Scientific and Technological Research Direction (DICYT-USACH) and the Basal Financing Program for Scientific and Technological Centers of Excellence CEDENNA (grant number AFB220001).

Institutional Review Board Statement: Not applicable.

Data Availability Statement: Data sharing not applicable.

Acknowledgments: C. Villegas gratefully acknowledges the Scientific and Technological Development (FONDECYT) Postdoctoral grant number 3210434 (ANID-CHILE) and FACTEC scholarship 2020 of the Technology Faculty, USACH. A. Torres gratefully acknowledges Project 082271TM_DAS, University of Santiago de Chile.

Conflicts of Interest: The authors declare no conflict of interest.

References

1. Stoleru, E.; Vasile, C.; Irimia, A.; Brebu, M. Towards a Bioactive Food Packaging: Poly(Lactic Acid) Surface Functionalized by Chitosan Coating Embedding Clove and Argan Oils. *Molecules* **2021**, *26*, 4500. [CrossRef] [PubMed]
2. Subash, A.; Naebe, M.; Wang, X.; Kandasubramanian, B. Biopolymer—A sustainable and efficacious material system for effluent removal. *J. Hazard. Mater.* **2022**, *443*, 130168. [CrossRef] [PubMed]
3. Ibrahim, S.; Riahi, O.; Said, S.B.; Sabri, M.F.; Rozali, S. Biopolymers from Crop Plants. In *Reference Module in Materials Science and Materials Engineering*; Elsevier: Amsterdam, The Netherlands, 2019.
4. Meereboer, K.W.; Misra, M.; Mohanty, A.K. Review of recent advances in the biodegradability of polyhydroxyalkanoate (PHA) bioplastics and their composites. *Green Chem.* **2020**, *22*, 5519–5558. [CrossRef]
5. Demiriz, B.; Kars, G.; Yücel, M.; Eroğlu, İ.; Gündüz, U. Hydrogen and poly-β-hydroxybutyric acid production at various acetate concentrations using Rhodobac-ter capsulatus DSM 1710. *Int. J. Hydrogen Energy* **2019**, *44*, 17269–17277. [CrossRef]
6. Eslami, H.; Grady, M.; Mekonnen, T.H. Biobased and compostable trilayer thermoplastic films based on poly (3-hydroxybutyrate-co-3-hydroxyvalerate) (PHBV) and thermoplastic starch (TPS). *Int. J. Biol. Macromol.* **2022**, *220*, 385–394. [CrossRef]
7. Palai, B.; Biswal, M.; Mohanty, S.; Nayak, S.K. In situ reactive compatibilization of polylactic acid (PLA) and thermoplastic starch (TPS) blends; synthesis and evaluation of extrusion blown films thereof. *Ind. Crop. Prod.* **2019**, *141*, 111748. [CrossRef]
8. Andrade, M.; Barbosa, C.; Cerqueira, M.; Azevedo, A.G.; Barros, C.; Machado, A.; Coelho, A.; Furtado, R.; Belo, C.; Saraiva, M.; et al. PLA films loaded with green tea and rosemary polyphenolic extracts as an active packaging for almond and beef. *Food Packag. Shelf Life* **2023**, *36*, 101041. [CrossRef]
9. Phothisarattana, D.; Wongphan, P.; Promhuad, K.; Promsorn, J.; Harnkarnsujarit, N. Blown film extrusion of PBAT/TPS/ZnO nanocomposites for shelf-life extension of meat packaging. *Colloids Surf. B* **2022**, *214*, 112472. [CrossRef]
10. Zhai, X.; Li, M.; Zhang, R.; Wang, W.; Hou, H. Extrusion-blown starch/PBAT biodegradable active films incorporated with high retentions of tea polyphenols and the release kinetics into food simulants. *Int. J. Biol. Macromol.* **2023**, *227*, 851–862. [CrossRef]
11. Shen, L.; Haufe, J.; Patel, M. Product overview and market projection of emerging bio-based plastics PRO-BIP 2009. *Rep. Eur. Polysacch. Netw. Excell. EPNOE Eur. Bioplastics* **2009**, *243*, 1–245.
12. Scaffaro, R.; Maio, A.; Lopresti, F. Physical properties of green composites based on poly-lactic acid or Mater-Bi®filled with Posidonia Oceanica leaves. *Compos. Part A Appl. Sci. Manuf.* **2018**, *112*, 315–327. [CrossRef]
13. Aldas, M.; Rayón, E.; López-Martínez, J.; Arrieta, M.P. A deeper microscopic study of the interaction between gum rosin derivatives and a Mater-Bi type bioplastic. *Polymers* **2020**, *12*, 226. [CrossRef] [PubMed]
14. Scaffaro, R.; Sutera, F.; Botta, L. Biopolymeric bilayer films produced by co-extrusion film blowing. *Polym. Test* **2018**, *65*, 35–43. [CrossRef]
15. Setti, C.; Suarato, G.; Perotto, G.; Athanassiou, A.; Bayer, I.S. Investigation of in vitro hydrophilic and hydrophobic dual drug release from polymeric films produced by sodium alginate-MaterBi® drying emulsions. *Eur. J. Pharm. Biopharm.* **2018**, *130*, 71–82. [CrossRef] [PubMed]
16. Aldas, M.; Pavon, C.; Ferri, J.M.; Arrieta, M.P.; López-Martínez, J. Films Based on Mater-Bi® Compatibilized with Pine Resin Derivatives: Optical, Barrier, and Disintegration Properties. *Polymers* **2021**, *13*, 1506. [CrossRef] [PubMed]
17. Martelli, G.; Giacomini, D. Antibacterial and antioxidant activities for natural and synthetic dual-active compounds. *Eur. J. Med. Chem* **2018**, *158*, 91–105. [CrossRef]
18. de Oliveira, C.A.; Darío, M.F. *Cosméticos Bioactivos. En Manual de Ecomateriales;* Martínez, L.M.T., Kharissova, O.V., Kharisov, B.I., Eds.; Springer International Publishing: Cham, Switzerland, 2017; pp. 1–23.
19. Zhang, M.; Zhang, Y.; Cai, Z.F. Selective determination of ellagic acid in aqueous solution using blue-green emissive copper nanoclusters. *Spectrochim. Acta Part A* **2023**, *295*, 122597. [CrossRef]
20. Nyamba, I.; Lechanteur, A.; Semdé, R.; Evrard, B. Physical formulation ap-proaches for improving aqueous solubility and bioavailability of ellagic acid: A review. *Eur. J. Pharm. Biopharm.* **2021**, *159*, 198–210. [CrossRef]
21. Venkatasubramanian, A.; Thiyagaraj, A.; Subbiah, V.; Solairaja, S.; Arumugam, S.; Ra-maligam, S.; Venkatabalasubramanian, S. Ameliorative role of ellagic acid against acute liver steatosis in adult zebrafish experimental model. *Comp. Biochem. Physiol.* **2021**, *247*, 109061. [CrossRef]
22. Vilela, C.; Pinto, R.J.; Coelho, J.; Domingues, M.R.; Daina, S.; Sadocco, P.; Santos, S.A.O.; Freire, C.S. Bioactive chitosan/ellagic acid films with UV-light protection for active food packaging. *Food Hydrocoll.* **2017**, *73*, 120–128. [CrossRef]
23. Tavares, W.S.; Tavares-Júnior, A.G.; Otero-Espinar, F.J.; Martín-Pastor, M.; Sousa, F.F. Design of ellagic acid-loaded chitosan/zein films for wound bandaging. *J. Drug Deliv. Sci. Technol.* **2020**, *59*, 101903. [CrossRef]
24. Zhu, W.; Chen, J.; Dong, Q.; Luan, D.; Tao, N.; Deng, S.; Wang, L.; Hao, Y.; Li, L. Development of organic-inorganic hybrid antimicrobial materials by mechanical force and application for active packaging. *Food Packag. Shelf Life* **2023**, *37*, 101060.
25. Sandhya, A.; Suchart, S.; Jyotishkumar, P. Essential oils as antimicrobial agents in biopolymer-based food packaging—A comprehensive review. *Food Biosci.* **2020**, *38*, 100785.
26. Santagata, G.; Mallardo, S.; Fasulo, G.; Lavermicocca, P.; Valerio, F.; Di Biase, M.; Di Stasio, M.; Malinconico, M.; Volpe, M.G. Pectin-honey coating as novel dehydrating bioactive agent for cut fruit: Enhancement of the functional properties of coated dried fruits. *Food Chem.* **2018**, *258*, 104–110. [CrossRef] [PubMed]

27. Valerio, F.; Volpe, M.G.; Santagata, G.; Boscaino, F.; Barbarisi, C.; Di Biase, M.; Bavaro, A.R.; Lonigro, S.L.; Lavermicocca, P. The viability of probiotic Lactobacillus paracasei IMPC2.1 coating on apple slices during dehydration and simulated gastro-intestinal digestion. *Food Biosci.* **2020**, *34*, 100533. [CrossRef]
28. Moeini, A.; Germann, N.; Malinconico, M.; Santagata, G. Formulation of secondary compounds as additives of biopolymer-based food packaging: A review. *Trends Food Sci. Technol.* **2021**, *114*, 342–354. [CrossRef]
29. Riseh, R.S.; Vatankhah, M.; Hassanisaadi, M.; Kennedy, J.F. Chitosan-based nanocomposites as coatings and packaging materials for the postharvest improvement of agricultural product: A review. *Carbohydr. Polym.* **2023**, *295*, 120666. [CrossRef]
30. Youssef, K.; de Oliveira, A.G.; Tischer, C.A.; Hussain, I.; Roberto, S.R. Synergistic effect of a novel chitosan/silica nanocomposites-based formulation against gray mold of table grapes and its possible mode of action. *Int. J. Biol. Macromol.* **2019**, *141*, 247–258. [CrossRef]
31. Mujtaba, M.; Khawar, K.M.; Camara, M.C.; Carvalho, L.B.; Fraceto, L.F.; Morsi, R.E.; Elsabee, M.Z.; Kaya, M.; Labidi, J.; Ullah, H.; et al. Chitosan-based delivery systems for plants: A brief overview of recent advances and future directions. *Int. J. Biol. Macromol.* **2020**, *154*, 683–697. [CrossRef]
32. Niu, X.; Liu, Y.; Song, Y.; Han, J.; Pan, H. Rosin modified cellulose nanofiber as a reinforcing and co-antimicrobial agents in polylactic acid/chitosan composite film for food packaging. *Carbohydr. Polym.* **2018**, *183*, 102–109. [CrossRef]
33. Mansoor, S.; Shahid, S.; Ashiq, K.; Alwadai, N.; Javed, M.; Iqbal, S.; Fátima, U.; Zamand, S.; Sarwar, M.N.; Alshammari, F.H.; et al. Controlled growth of nanocomposite thin layer based on Zn-Doped MgO nanoparticles through Sol-Gel technique for biosensor applications. *Inorg. Chem. Commun.* **2022**, *142*, 109702. [CrossRef]
34. Madni, A.; Kousar, R.; Naeem, N.; Wahid, F. Recent advancements in applications of chitosan-based biomaterials for skin tissue engineering. *J. Bioresour. Bioprod.* **2021**, *6*, 11–25. [CrossRef]
35. *UNE-EN ISO-20200*; Determination of the Degree of Disintegration of Plastic Materials under Simulated Composting Conditions in a Laboratory-Scale Test. ISO: Geneva, Switzerland, 2016.
36. Villegas, C.; Arrieta, M.P.; Rojas, A.; Torres, A.; Faba, S.; Toledo, M.J.; Guitierrez, M.A.; Zavalla, E.; Romero, J.; Galotto, M.J.; et al. PLA/organoclay bionanocomposites impregnated with thymol and cinnamaldehyde by supercritical impregnation for active and sustainable food packaging. *Compos. Pt. B-Eng* **2019**, *176*, 107336. [CrossRef]
37. Faba, S.; Arrieta, M.P.; Agüero, Á.; Torres, A.; Romero, J.; Rojas, A.; Galotto, M.J. Processing Compostable PLA/Organoclay Bionanocomposite Foams by Super-critical CO_2 Foaming for Sustainable Food Packaging. *Polymers* **2022**, *14*, 4394. [CrossRef]
38. Torres, A.; Ilabaca, E.; Rojas, A.; Rodríguez, F.; Galotto, M.J.; Guarda, A.; Villegas, C.; Romero, J. Effect of processing conditions on the physical, chemical and transport properties of polylactic acid films containing thymol incorporated by supercritical impregnation. *Eur. Polym. J* **2017**, *89*, 195–210. [CrossRef]
39. Rebelo, R.C.; Gonçalves, L.P.; Fonseca, A.C.; Fonseca, J.; Rola, M.; Coelho, J.F.; Rola, M.; Serra, A.C. Increased degradation of PLA/PBAT blends with organic acids and derivatives in outdoor weathering and marine environment. *Polymers* **2022**, *256*, 125223. [CrossRef]
40. Borchani, K.E.; Carrot, C.; Jaziri, M. Biocomposites of Alfa fibers dispersed in the Mater-Bi® type bioplastic: Morphology, mechanical and thermal properties. *Compos. Part A* **2015**, *78*, 371–379. [CrossRef]
41. Gubitosa, J.; Rizzi, V.; Fini, P.; Del Sole, R.; Lopedota, A.; Laquintana, V.; Cosma, P. Multifunctional green synthetized gold nanoparticles/chitosan/ellagic acid self-assembly: Antioxidant, sun filter and tyrosinase-inhibitor properties. *Mater. Sci. Eng. C* **2020**, *106*, 110170. [CrossRef]
42. de Souza Tavares, W.; Pena, G.R.; Martin-Pastor, M.; de Sousa, F.F.O. Design and characterization of ellagic acid-loaded zein nanoparticles and their effect on the antioxidant and antibacterial activities. *J. Mol. Liq.* **2021**, *341*, 116915. [CrossRef]
43. Villegas, C.; Torres, A.; Rios, M.; Rojas, A.; Romero, J.; de Dicastillo, C.L.; Valenzuela, X.; Galotto, M.J.; Guarda, A. Supercritical impregnation of cinnamaldehyde into polylactic acid as a route to develop antibacterial food packaging materials. *Food Res. Int.* **2017**, *99*, 650–659. [CrossRef]
44. Antunes, L.R.; Breitenbach, G.L.; Pellá, M.C.G.; Caetano, J.; Dragunski, D.C. Electrospun poly (lactic acid) (PLA)/poly (butylene adipate-co-terephthalate) (PBAT) nanofibers for the controlled release of cilostazol. *Int. J. Biol. Macromol.* **2021**, *182*, 333–342. [CrossRef] [PubMed]
45. Surendren, A.; Mohanty, A.K.; Liu, Q.; Misra, M. A review of biodegradable thermoplastic starches, their blends and composites: Recent developments and opportunities for single-use plastic packaging alternatives. *Green Chem.* **2022**, *24*, 8606–8636. [CrossRef]
46. Lendvai, L.; Apostolov, A.; Karger-Kocsis, J. Characterization of layered silicate-reinforced blends of thermoplastic starch (TPS) and poly (butylene adipate-co-terephthalate). *Carbohydr. Polym.* **2017**, *173*, 566–572. [CrossRef]
47. Bianchi, M.; Dorigato, A.; Morreale, M.; Pegoretti, A. Evaluation of the Physi-cal and Shape Memory Properties of Fully Biodegradable Poly (lactic acid) (PLA)/Poly (butylene adipate terephthalate)(PBAT) Blends. *Polymers* **2023**, *15*, 881. [CrossRef] [PubMed]
48. Jullanun, P.; Yoksan, R. Morphological characteristics and properties of TPS/PLA/cassava pulp biocomposites. *Polym. Test.* **2020**, *88*, 106522. [CrossRef]
49. Pérez-Blanco, C.; Huang-Lin, E.; Abrusci, C. Characterization, biodegradation and cytotoxicity of thermoplastic starch and ethylene-vinyl alcohol copolymer blends. *Carbohydr. Polym.* **2022**, *298*, 120085. [CrossRef]
50. Pulgarin, H.L.C.; Caicedo, C.; López, E.F. Effect of surfactant content on rheological, thermal, morphological and surface properties of thermoplastic starch (TPS) and polylactic acid (PLA) blends. *Heliyon* **2022**, *8*, e10833. [CrossRef] [PubMed]

51. Sepúlveda, J.; Villegas, C.; Torres, A.; Vargas, E.; Rodríguez, F.; Baltazar, S.; Prada, A.; Rojas, A.; Romero, J.; Faba, S.; et al. Effect of functionalized silica nanoparticles on the mass transfer process in active PLA nanocomposite films obtained by supercritical impregnation for sustainable food packaging. *J. Supercrit. Fluids* **2020**, *161*, 104844. [CrossRef]
52. Chang, C.C.; Trinh, B.M.; Mekonnen, T.H. Robust multiphase and multilayer starch/polymer (TPS/PBAT) film with simultaneous oxygen/moisture barrier properties. *J. Colloid Interface Sci.* **2021**, *593*, 290–303. [CrossRef]
53. Li, M.; Jia, Y.; Shen, X.; Shen, T.; Tan, Z.; Zhuang, W.; Zhao, G.; Zhu, C.; Ying, H. Investigation into lignin modified PBAT/thermoplastic starch composites: Thermal, mechanical, rheological and water absorption properties. *Ind. Crops Prod.* **2021**, *171*, 113916. [CrossRef]
54. Zare, E.N.; Makvandi, P.; Tay, F.R. Recent progress in the industrial and biomedical applications of tragacanth gum: A review. *Carbohydr. Polym.* **2019**, *212*, 450–467. [CrossRef] [PubMed]
55. Bonilla, J.; Fortunati, E.L.E.N.A.; Vargas, M.; Chiralt, A.; Kenny, J.M. Effects of chitosan on the physicochemical and antimicrobial properties of PLA films. *J. Food Eng.* **2013**, *119*, 236–243. [CrossRef]
56. Zhang, W.; Cao, J.; Jiang, W. Analysis of film-forming properties of chitosan with different molecular weights and its adhesion properties with different postharvest fruit surfaces. *Food Chem.* **2022**, *395*, 133605. [CrossRef] [PubMed]
57. Fiore, A.; Park, S.; Volpe, S.; Torrieri, E.; Masi, P. Active packaging based on PLA and chitosan-caseinate enriched rosemary essential oil coating for fresh minced chicken breast application. *Food Packag. Shelf Life* **2021**, *29*, 100708. [CrossRef]
58. Tanpichai, S.; Srimarut, Y.; Woraprayote, W.; Malila, Y. Chitosan coating for the preparation of multilayer coated paper for food-contact packaging: Wettability, mechanical properties, and overall migration. *Int. J. Biol. Macromol.* **2022**, *213*, 534–545. [CrossRef]
59. Siripatrawan, U.; Harte, B.R. Physical properties and antioxidant activity of an active film from chitosan incorporated with green tea extract. *Food Hydrocoll.* **2010**, *24*, 770–775. [CrossRef]
60. Sessini, V.; Arrieta, M.P.; Raquez, J.M.; Dubois, P.; Kenny, J.M.; Peponi, L. Thermal and composting degradation of EVA/Thermoplastic starch blends and their nanocomposites. *Polym. Degrad. Stab.* **2019**, *159*, 184–198. [CrossRef]
61. Phothisarattana, D.; Harnkarnsujarit, N. Migration, aggregations and thermal degradation behaviors of TiO_2 and ZnO incorporated PBAT/TPS nanocomposite blown films. *Food Packag. Shelf Life* **2022**, *33*, 100901. [CrossRef]
62. Taiatele, I.; Dal Bosco, T.C.; Faria-Tischer, P.C.; Bilck, A.P.; Yamashita, F.; Bertozzi, J.; Michels, R.N.; Mali, S. Abiotic hydrolysis and compostability of blends based on cassava starch and biodegradable polymers. *J. Polym. Environ.* **2019**, *27*, 2577–2587. [CrossRef]
63. Azevedo, J.V.; Hausnerova, B.; Möginger, B.; Sopik, T. Effect of Chain Extending Cross-Linkers on the Disintegration Behavior of Composted PBAT/PLA Blown Films. *Int. J. Mol. Sci.* **2023**, *24*, 4525. [CrossRef]
64. Costa, A.R.M.; Reul, L.T.; Sousa, F.M.; Ito, E.N.; Carvalho, L.H.; Canedo, E.L. Degradation during processing of vegetable fiber compounds based on PBAT/PHB blends. *Polym. Test* **2018**, *69*, 266–275. [CrossRef]
65. Ruggero, F.; Carretti, E.; Gori, R.; Lotti, T.; Lubello, C. Monitoring of degradation of starch-based biopolymer film under different composting conditions, using TGA, FTIR and SEM analysis. *Chemosphere* **2020**, *246*, 125770. [CrossRef] [PubMed]
66. Sciancalepore, C.; Togliatti, E.; Giubilini, A.; Pugliese, D.; Moroni, F.; Messori, M.; Milanese, D. Preparation and characterization of innovative poly (butylene adipate terephthalate)-based biocomposites for agri-food packaging application. *J. Appl. Polym. Sci.* **2022**, *139*, 52370. [CrossRef]
67. Abraham, A.; Park, H.; Choi, O.; Sang, B.I. Anaerobic co-digestion of bioplastics as a sustainable mode of waste management with improved energy production—A review. *Bioresour. Technol.* **2021**, *322*, 124537. [CrossRef] [PubMed]
68. Xie, L.; Huang, J.; Xu, H.; Feng, C.; Na, H.; Liu, F.; Xue, L.; Zhu, J. Effect of large sized reed fillers on properties and degradability of PBAT composites. *Polymer Compos.* **2023**, *44*, 1752–1761. [CrossRef]

polymers
MDPI

Article

Novel Biocatalysts Based on Bromelain Immobilized on Functionalized Chitosans and Research on Their Structural Features

Marina G. Holyavka [1,2], Svetlana S. Goncharova [1], Andrey V. Sorokin [1,2,3], Maria S. Lavlinskaya [1,2,3], Yulia A. Redko [1], Dzhigangir A. Faizullin [4], Diana R. Baidamshina [5], Yuriy F. Zuev [4,*], Maxim S. Kondratyev [1,6], Airat R. Kayumov [5] and Valeriy G. Artyukhov [1]

1 Biophysics and Biotechnology Department, Voronezh State University, 1 Universitetskaya Square, 394018 Voronezh, Russia
2 Laboratory of Bioresource Potential of Coastal Area, Institute for Advanced Studies, Sevastopol State University, 33 Studencheskaya Street, 299053 Sevastopol, Russia
3 Metagenomics and Food Biotechnologies Laboratory, Voronezh State University of Engineering Technologies, 19 Revolutsii Avenue, 394036 Voronezh, Russia
4 Kazan Institute of Biochemistry and Biophysics, FRC Kazan Scientific Center of the RAS, 2/31 Lobachevsky Street, 420111 Kazan, Russia
5 Institute of Fundamental Medicine and Biology, Kazan Federal University, 18 Kremlevskaya Street, 420008 Kazan, Russia
6 Laboratory of Structure and Dynamics of Biomolecular Systems, Institute of Cell Biophysics of the RAS, 3 Institutskaya Street, 142290 Pushchino, Russia
* Correspondence: yufzuev@mail.ru

Citation: Holyavka, M.G.; Goncharova, S.S.; Sorokin, A.V.; Lavlinskaya, M.S.; Redko, Y.A.; Faizullin, D.A.; Baidamshina, D.R.; Zuev, Y.F.; Kondratyev, M.S.; Kayumov, A.R.; et al. Novel Biocatalysts Based on Bromelain Immobilized on Functionalized Chitosans and Research on Their Structural Features. *Polymers* **2022**, *14*, 5110. https://doi.org/10.3390/polym14235110

Academic Editors: Antonio M. Borrero-López, Concepción Valencia-Barragán, Esperanza Cortés Triviño, Adrián Tenorio-Alfonso and Clara Delgado-Sánchez

Received: 3 November 2022
Accepted: 22 November 2022
Published: 24 November 2022

Abstract: Enzyme immobilization on various carriers represents an effective approach to improve their stability, reusability, and even change their catalytic properties. Here, we show the mechanism of interaction of cysteine protease bromelain with the water-soluble derivatives of chitosan—carboxymethylchitosan, *N*-(2-hydroxypropyl)-3-trimethylammonium chitosan, chitosan sulfate, and chitosan acetate—during immobilization and characterize the structural features and catalytic properties of obtained complexes. Chitosan sulfate and carboxymethylchitosan form the highest number of hydrogen bonds with bromelain in comparison with chitosan acetate and *N*-(2-hydroxypropyl)-3-trimethylammonium chitosan, leading to a higher yield of protein immobilization on chitosan sulfate and carboxymethylchitosan (up to 58 and 65%, respectively). In addition, all derivatives of chitosan studied in this work form hydrogen bonds with His158 located in the active site of bromelain (except *N*-(2-hydroxypropyl)-3-trimethylammonium chitosan), apparently explaining a significant decrease in the activity of biocatalysts. The *N*-(2-hydroxypropyl)-3-trimethylammonium chitosan displays only physical interactions with His158, thus possibly modulating the structure of the bromelain active site and leading to the hyperactivation of the enzyme, up to 208% of the total activity and 158% of the specific activity. The FTIR analysis revealed that interaction between *N*-(2-hydroxypropyl)-3-trimethylammonium chitosan and bromelain did not significantly change the enzyme structure. Perhaps this is due to the slowing down of aggregation and the autolysis processes during the complex formation of bromelain with a carrier, with a minimal modification of enzyme structure and its active site orientation.

Keywords: bromelain; enzyme immobilization; carboxymethylchitosan *N*-(2-hydroxypropyl)-3-trimethylammonium chitosan; chitosan sulfate; chitosan acetate

1. Introduction

Molecules of natural origin are the object of interest for the development of novel drugs and biomaterials due to their physiological activity and the demand for novel therapeutic technologies to be biocompatible and stable in formulations aimed at their preservation and administration [1–5]. Chitosan has been used in various biomedical applications, mainly as a drug carrier for enzymes and peptides, wound-healing accelerator, fat binder, and hemostatic and antimicrobial agent [6]. The molecular structure of this polymer has been offered as an opportunity to add specific mechanical, chemical, or biological characteristics to its conjugations with other molecules [7,8]. The functionalization of chitosan is possible since it possesses functional groups (amino and hydroxy) that are free to bind with other molecules. Chitosan derivatives have demonstrated promising applications in various fields, such as tissue engineering, biomolecule delivery, protection against infections, etc. [9].

Proteases are produced by all organisms—plants, animals, fungi, and bacteria. Bromelain, the protease from the pineapple, is widely used for the treatment of cardiovascular diseases, disorders of blood coagulation and fibrinolysis, infectious diseases, and many types of cancer [10]. Moreover, due to the antiviral, anti-inflammatory, cardioprotective, and anticoagulant activity of bromelain, the enzyme has been suggested as adjunctive therapy for patients with COVID-19 and post-COVID-19. During the spread of new variants of the SARS-CoV-2 virus, such beneficial properties of bromelain could help to prevent the escalation and progression of COVID-19 disease [11]. Bromelain is used in the United States and Europe as an alternative or complementary medication to glucocorticoids, nonsteroidal antirheumatics, and immunomodulatory agents. Very low toxicity ensures its safe use as a remedy for chronic inflammatory diseases, as well as an adjuvant to chemoradiotherapy and perioperative care. Bromelain is capable of enhancing the absorption and tissue permeability of antibiotics after oral, subcutaneous, or intramuscular application. As a result, it can maintain a higher level of a drug in serum and tissue, thus potentiating efficacy and reducing side effects [12].

Stem bromelain (EC.3.4.22.32) is a glycoprotein with one oligosaccharide moiety and one reactive sulfhydryl group per molecule. The highest activity of this enzyme is observed at pH 5.0–8.0 [13]. At a temperature of 21 °C, the aqueous proteolytic activity of bromelain reduces rapidly. Therefore, in a concentrated form (>50 mg/mL), it is stable for a week at room temperature and exhibits minimal inactivation after multiple freeze–thaw cycles [14].

To date, it has been proven that bromelain is well absorbed in the body not only after external but also after oral use and does not have significant side effects even with long-term use. Bromelain has GRAS (Generally Regarded as Safe) status from US federal agencies (CFR 1999, 2009) and can be applied as an enzymatic therapy for humans [15]. Moreover, bromelain has a plant origin, isolated from the extract of *Ananas comosus*, so it is renewable and inexpensive compared with animal-origin enzymes. However, the protease is sensitive to acidity, many chemicals, solvents, and elevated temperatures. Even small conformational changes can reduce enzyme activity, which limits its use in medicine and pharmacology [16,17]. Therefore, one of the most important challenges is the improvement of bromelain stability, mainly by developing new and modifying existing approaches for its stabilization. Among others, the immobilization approach can be used, which was developed to solve the problem of enzyme recovery and reuse [18,19]. Nowadays, a proper immobilization protocol is expected to improve some enzyme features, such as stability, activity, specificity, or selectivity, enlarging the operational conditions [20–22]. Moreover, it is possible to couple enzyme immobilization with enzyme purification, saving costs and time with respect to these processes [23].

Enzymes are immobilized using various carriers, including chitosan and its derivatives [24]. For example, bromelain was immobilized on chitosan films from microbial and animal sources and plasticized with glycerol, wherein the highest enzyme immobilization yield (41%) was observed [25]. Chitosan-based nanoparticles, including lactobionic acid-modified chitosan nanoparticles and linoleic acid-modified carboxymethyl chitosan nanoparticles, have also been used for the immobilization of bromelain [26–29]. Bromelain was covalently immobilized onto the surface of porous chitosan beads without glutaraldehyde [30], with and without alkyl chain spacers of different lengths. The relative activity of immobilized bromelain was found to be high toward a small ester substrate, *N*-benzyl-

L-arginine ethyl ester (BAEE), but rather low toward casein, a high-molecular-weight substrate [31].

It is known that chitosan is characterized by the limited pH range of its solubility (pKa~6.5) [32], which does not fit with its physiological conditions. It is known that aqueous solutions and solid/liquid interfaces constitute quite different microenvironments that profoundly affect the structure and activity of enzymes. The solid surface exerts an electrostatic field extending several nanometers into the bulk solution. Thus, close to the interface, all electrostatic interactions are modified, altering the behavior of solvent molecules, buffer salts, substrates, products, and the enzyme itself. These effects often lead to enhanced enzyme stability, but also, in most cases, result in some loss of enzyme activity, which has been generally accepted as one of the drawbacks of the immobilization process [33]. The modification of chitosan by tuning its properties and obtaining its water-soluble derivatives can solve this problem. In addition, chitosan has polycationic properties and is characterized by certain antibacterial activities. Its modifications, such as carboxymethylchitosan [34], *N*-(2-hydroxypropyl)-3-trimethylammonium chitosan [35], chitosan sulfate [36], and chitosan acetate [37], also possess antimicrobial activity and retain good biocompatibility, mucoadhesiveness, low toxicity, and other important properties of chitosan. Moreover, chitosan sulfate–lysozyme hybrid hydrogels with fine-tuned degradability have demonstrated sustained, inherent antibiotic and antioxidant activities [38]. In creating new antimicrobial formulations, carboxymethylchitosan, *N*-(2-hydroxypropyl)-3-trimethylammonium chitosan, chitosan sulfate, and chitosan acetate look like promising matrices for bromelain immobilization.

In this work, we show the mechanism of bromelain interaction with the water-soluble derivatives of chitosan—carboxymethylchitosan, *N*-(2-hydroxypropyl)-3-trimethylammonium chitosan, chitosan sulfate, and chitosan acetate—and the proteolytic properties of the obtained compositions.

2. Materials and Methods

2.1. Materials

Bromelain (B4882) was purchased from Sigma, Burlington, MA, USA, and was used without any treatments. Azocasein (Sigma-Aldrich, Munich, Germany) was used as a substrate in catalytic activity evaluation experiments. Chitosans with molecular weights of 200, 350, and 600 kDa and degrees of deacetylation ranging from 0.73 to 0.85 were obtained from Bioprogress (Shchelkovo, Russia); glycidyltrimethyl ammonium chloride (>90%) and chloroacetic acid (99%) were purchased from Sigma-Aldrich (Munich, Germany). Sodium hydroxide, isopropyl alcohol, methanol, and acetone (all analytical grade); 2% *w/v* aqueous acetic acid solution prepared from glacial acetic acid (analytical grade) and distilled water; 10% *w/v* aqueous sulfuric acid solution fabricated from sulfuric acid (>98%); and distilled water received by Vekton (Saint-Petersburg, Russia) were used in the synthesis of chitosan derivatives.

2.2. Synthesis of Chitosan Derivatives

Chitosan derivatives were obtained by known methods with some modifications (see below).

2.2.1. Synthesis of Carboxymethyl Chitosan (ChMC)

Synthesis of carboxymethyl chitosan was carried out according to the following protocol: 3.0 g of chitosan was dispersed in 65 mL of isopropyl alcohol; then, a NaOH aqueous solution was introduced drop by drop for 15 min (NaOH:chitosan repeating link = 13:1 mol). After that, a 15% *w/v* solution of chloroacetic acid in *i*ProOH ($CH_2ClCOOH$: chitosan repeating link = 7:1 mol) was added drop by drop to the reaction mixture and stirred for 12 h at room temperature. The resulting precipitate was filtered, suspended, and washed with methanol, and it was dried in a vacuum at 55 ± 2 °C up to constant weight [39]. The yield of products was 79–92%; the degrees of substitution, calculated from the FTIR data, were 0.46,

0.54, and 0.78 for polymers with molecular weights of 600, 350, and 200 kDa, respectively. A significant increase in the degree of substitution for chitosan with a molecular weight of 200 kDa can be explained by the fact that this commercial chemical was in powder form, while the two others were flakes. Carboxymethylation was carried out heterogeneously, so the chitosan form and particle size significantly impacted the resulting products.

2.2.2. Synthesis of *N*-(2-Hydroxy) propyl-3-trimethylammonium Chitosan (HTCCh)

N-(2-hydroxy) propyl-3-trimethylammonium chitosan was obtained with the following process: 3.0 g of chitosan was suspended in 30 mL of distilled water for 30 min at 85 ± 2 °C. Then, the calculated amount of glycidyltrimethyl ammonium chloride (GTMAC) (GTMAC:chitosan repeating link = 3:1 mol) was dropwise added to the reaction mixture for 1 h and kept at 85 ± 2 °C for 10 h. The final product was isolated from the reaction mixtures via precipitation in acetone, washed three times with methanol, and dried in a vacuum oven at 55 ± 2 °C to a constant weight [40]. The product yield was 62–74%; the degrees of substitution calculated from FTIR data were 0.24, 0.19, and 0.57 for chitosans with molecular weights of 600, 350, and 200 kDa, respectively.

2.2.3. Synthesis of Chitosan Sulfate, Chitosan Acetate (ChS)

For the preparation of chitosan sulfate, 5.0 g of chitosan was dissolved in 500 mL of 2% *w/v* aqueous acetic acid solution and then 20 mL of 10% *w/v* aqueous sulfuric acid solution and was stirred constantly for 24 h at 25 ± 2 °C. The formed gel was placed in acetone for 5 days, washed three times with methanol, and dried in a vacuum oven at 55 ± 2 °C to a constant weight. The product yield was 85–96%.

2.2.4. Synthesis of Chitosan Acetate (ChA)

For the chitosan acetate synthesis, 5.0 g of chitosan was dissolved in 500 mL of 2% *w/v* aqueous acetic acid solution and stirred constantly for 24 h at 25 ± 2 °C. The final product was isolated from the reaction mixtures via precipitation in acetone, washed three times with methanol, and dried in a vacuum oven at 55 ± 2 °C to a constant weight. The product yield was 67–81%.

2.3. Immobilization Procedure

The immobilization of bromelain on the synthesized chitosan derivatives was performed using the complexation approach developed previously [41]. Firstly, 20 mL of enzyme solution (2 mg·mL^{-1} in 0.05 M trisodium borate buffer with a pH of 9.0) was added to 1 g of chitosan derivative (ChMC, HTCCh, ChS, or ChA) and incubated for 2 h at 37 °C. After that, the formed precipitate was washed via dialysis using 50 mM of Tris-HCl buffer with a pH of 7.5 through a cellophane membrane with a 25 kDa pore size until there was an absence of protein in the washing water (controlled spectrophotometrically at λ = 280 nm on an SF-2000 spectrophotometer, LOMO-Microsystems, Saint Petersburg, Russia).

2.4. Protein Content Measurement

The protein content in the immobilized enzyme samples was determined using the modified Lowry approach [42]. Before the analysis, the immobilized enzyme was treated with K/Na-tartrate (20 mg·mL^{-1} or 0.7 M) prepared from 1 M NaOH at 50 °C for 10 min to desorb the enzyme from the carrier [43]. The absence of enzyme destruction was controlled by recording its absorption spectrum on a UV-2550PC spectrophotometer (Shimadzu Scientific Instruments Inc., Kyoto, Japan) [44].

2.5. Evaluation of Proteolytic Activity of the Immobilized Enzymes

Azocasein was chosen as the substrate for proteolytic activity measurements [45]. Briefly, the sample was dissolved in 200 µL of buffer (50 mM Tris-HCl, pH 7.5), mixed with 800 µL of azocasein solution (0.5% in the same buffer), and incubated for 30 min at 37 °C [46]. Then, 800 µL of 5% trichloroacetic acid solution was added; after 10 min of incubation at 4 °C, the precipitated unhydrolyzed azocasein was removed via centrifugation (3 min 13,000 rpm). The supernatant (1200 μL) was mixed with 240 μL of 1 M NaOH solution, and its optical density was determined at 410 nm. The unit of catalytic activity was considered as the amount of enzyme hydrolyzing 1 μM of azocasein in 1 min ($\mu M \cdot min^{-1} \cdot mg^{-1}$ of protein).

2.6. Molecular Docking

The structure preparation of bromelain (PDB ID: 1W0Q, accessed on 31 October 2022 https://www.rcsb.org/structure/1W0Q) for docking and the process of interaction modeling using the Autodock Vina program (Accessed on 31 October 2022 https://sourceforge.net/projects/autodock-vina-1-1-2-64-bit/) were carried out as described in [47]. The structural models of ChMC, HTCCh, ChS, or ChA were drawn in the molecular constructor HyperChem (Accessed on 31 October 2022 https://hyperchem.software.informer.com) and, subsequently, were successively optimized in the AMBER forcefield and quantum-chemically in PM3 (Parametric Method 3). The ligand in the docking procedure had maximal conformational freedom: the rotation of functional groups around all single bonds was allowed. The arrangement of the charges of ChMC, HTCCh, ChS, or ChA and the protonation/deprotonation of their molecules was performed automatically in the MGLTools 1.5.6 package (Accessed on 31 October 2022 https://ccsb.scriptps.edu/mgltools/1-5-6).

2.7. Infrared Spectroscopy

Frozen-dried pure protein powders were dissolved in D_2O. Solid preparations of immobilized proteins on chitosan's derivatives, ChMC, HTCCh, ChS, and ChA, were washed with buffer solutions in D_2O. Solutions and solid wet samples were placed on the surface of the ATR working element and equilibrated at 25 °C. The IR spectra of analyzed samples were recorded by an IRAffinity1 spectrometer with an ATR attachment with a single reflection ZnSe working element. The content of protein secondary structures was assessed by fitting absorption spectra in the spectral range 1600–1700 cm^{-1} (amide I absorption band of peptide groups) with the sum of Gaussian components [48]. The position and number of components were determined from the second derivative of the absorption spectra; the fitting was performed using the Fityk 8.0 software.

2.8. Bacterial Strains and Biofilm Assays

Pseudomonas aeruginosa ATCC 27853 and *Staphylococcus aureus* ATCC 29213 were used for the biofilm assays. Bacteria were grown on the LB medium. To obtain rigid biofilms, bacteria were grown for 48 h under static conditions at 37 °C in 24-well, TC-treated, polystyrol plates (1 mL per well in the basal medium (BM) (glucose 5 g, peptone 7 g, $MgSO_4 \times 7H_2O$ 2.0 g and $CaCl_2 \times 2H_2O$ 0.05 g in 1.0 L tap water)) [49]. The mature biofilms were treated for either 3 or 24 h with soluble and immobilized bromelain in PBS, and plates were subjected to crystal violet staining [50].

3. Results and Discussions

3.1. Bromelain's Immobilization

As mentioned above, the proper immobilization protocol will improve enzyme characteristics such as stability and half-life, reducing aggregation, and for proteases, it additionally prevents autolysis [51]. However, the enzyme interactions can significantly affect proteolytic activity [52,53].

We estimated the efficiency of complexation for bromelain as a percentage of the adsorbed enzyme and the total and specific activities of the immobilized enzyme compared with the dissolved enzyme (Figure 1). The maximal amount of bromelain was bound with ChS and ChMC, up to 65% and 58% compared with its amount in solution, respectively (Figure 1A).

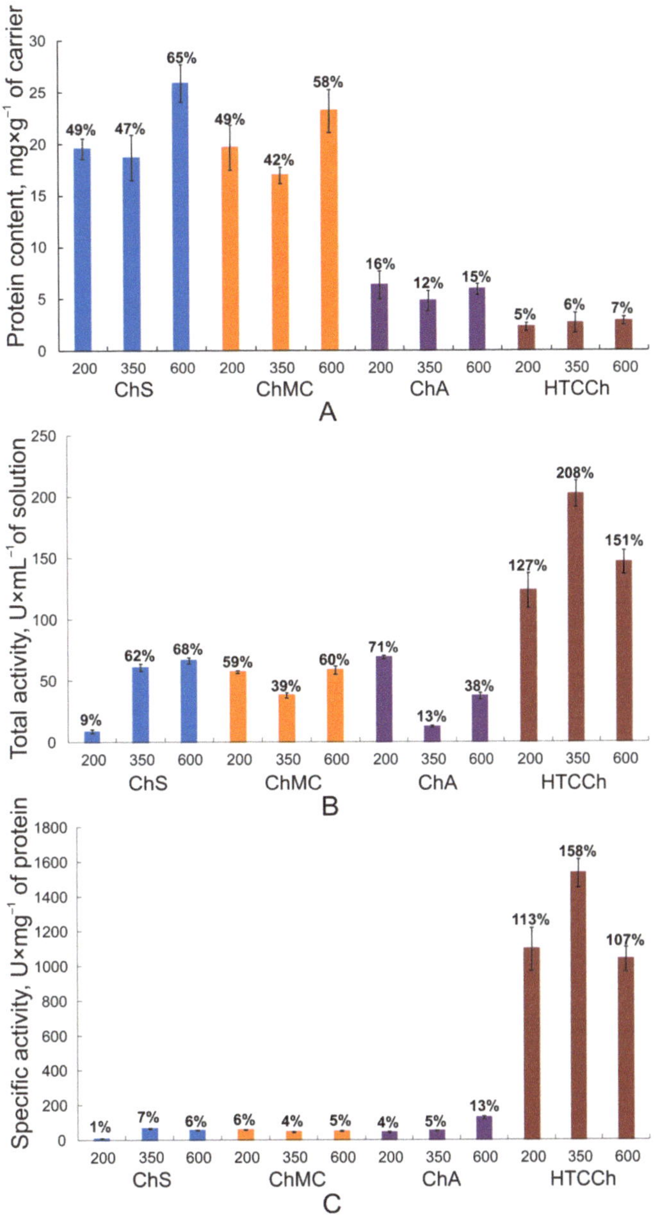

Figure 1. Characteristics of immobilized bromelain: (**A**) Protein content (mg·g^{-1} of carrier) in enzyme complexes with ChS, ChMC, ChA, and HTCCh; (**B**) total activity (U·mL^{-1} of solution) of enzyme complexes with ChS, ChMC, ChA, and HTCCh; (**C**) specific activity (U·mg^{-1} of protein) of enzyme complexes with ChS, ChMC, ChA, and HTCCh. The efficiency of complexation for bromelain expressed as a percentage of the adsorbed enzyme compared with its amount in solution (**A**), the total activity of immobilized enzyme compared with soluble enzyme (**B**), and the specific activity of the immobilized enzyme compared with the dissolved enzyme (**C**) are indicated above the bars. All experiments were performed eight times, and the results represent mean ± confidence interval.

The bromelain activity assay performed on the azocasein as a substrate showed that the total (U·mL^{-1} of solution) and specific (U·mg^{-1} of protein) proteolytic activity of the enzyme immobilized on ChS, ChMC, and ChA was lower compared with the native enzyme. However, binding with HTCCh led to the bromelain hyperactivation phenomenon (Figure 1B,C). Apparently, the interaction with HTCCh promotes the formation of a more catalytically favorable conformation of bromelain globules modulating the active site and increasing the proteolytic activity.

Depending on the desired use, it is known that carefully selecting chitosan and its derivatives is necessary with attention to the degree of substitution, molecular weight, and purity since these characteristics can significantly affect the mechanical and biological properties of the final product [54,55]. However, in this work, we did not find a direct relationship between the molecular weight of chitosan derivatives and the characteristics of immobilized bromelain we obtained.

Below, we tried to explain the results obtained empirically by studying the mechanism of bromelain complexation with ChMC, HTCCh, ChS, and ChA using in silico (molecular docking) and experimental (FTIR spectroscopy) methods.

3.2. Interaction Mechanisms between Bromelain and Chitosan Derivatives

The complexation of enzymes with polysaccharides and their derivatives is a multipronged process that can proceed by using specific bonds and interactions (hydrogen bonds, electrostatic, dipole–dipole, and hydrophobic interactions). The active site orientation regarding the supporting surface plays a critical role, as only properly oriented protease molecules may be accessed by the substrate [56]. Some simple techniques may be used in docking and design studies to account for some of the changes in the conformations of the enzyme during the ligand-binding and complex formation [57].

Like all papain-like proteases, the bromelain molecule is folded into two domains. Domain L is mainly α-helical (α-helices LI, LII, LIII). The key feature of the R domain is its antiparallel β-sheet structure [58]. The R domain also contains two helices: RI and RII, both located on the surface of the protein molecule at opposite ends of the β-sheet structure, which forms the core of this domain [59,60]. The active site of bromelain is located on the border of the L and R domains in a V-shaped cleft formed by Cys26 and His158 [61].

We found that the sorption of bromelain on chitosan derivatives is realized by protein regions located on the border of the L and R domains, including the region of the enzyme's active site (Table 1, Figures 2 and 3), which results in the modification of their catalytic activity in the immobilized state. In addition, the immobilization involves amino acids included to the α-helices and β-sheets of protein molecules (Table 1), which was confirmed by our FTIR spectroscopy experiments (see below). Moreover, the protein content evaluation (Figure 1A) in the immobilized bromelain formulations is in good agreement with our docking calculations. The protein amount and the protein immobilization yield are higher in bromelain complexes with ChS and ChMC (with which, according to in silico calculations, bromelain forms a greater number of hydrogen bonds—13 and 10, respectively) compared with ChA and HTCCh (with which, according to docking results, bromelain forms a smaller number of hydrogen bonds—8 and 7, respectively). The possible number of hydrogen bonds in the complex is one of the main reasons why the protein immobilization yield is higher. ChMC, ChS, and ChA have hydrogen bonds with His158, which is from the active site of bromelain. This can probably explain the significant decrease in enzyme activity after its immobilization. HTCCh forms only physical interactions with His158, which possibly modulate the structure of the active site of bromelain and lead to the hyperactivation of the enzyme (Figure 1B,C).

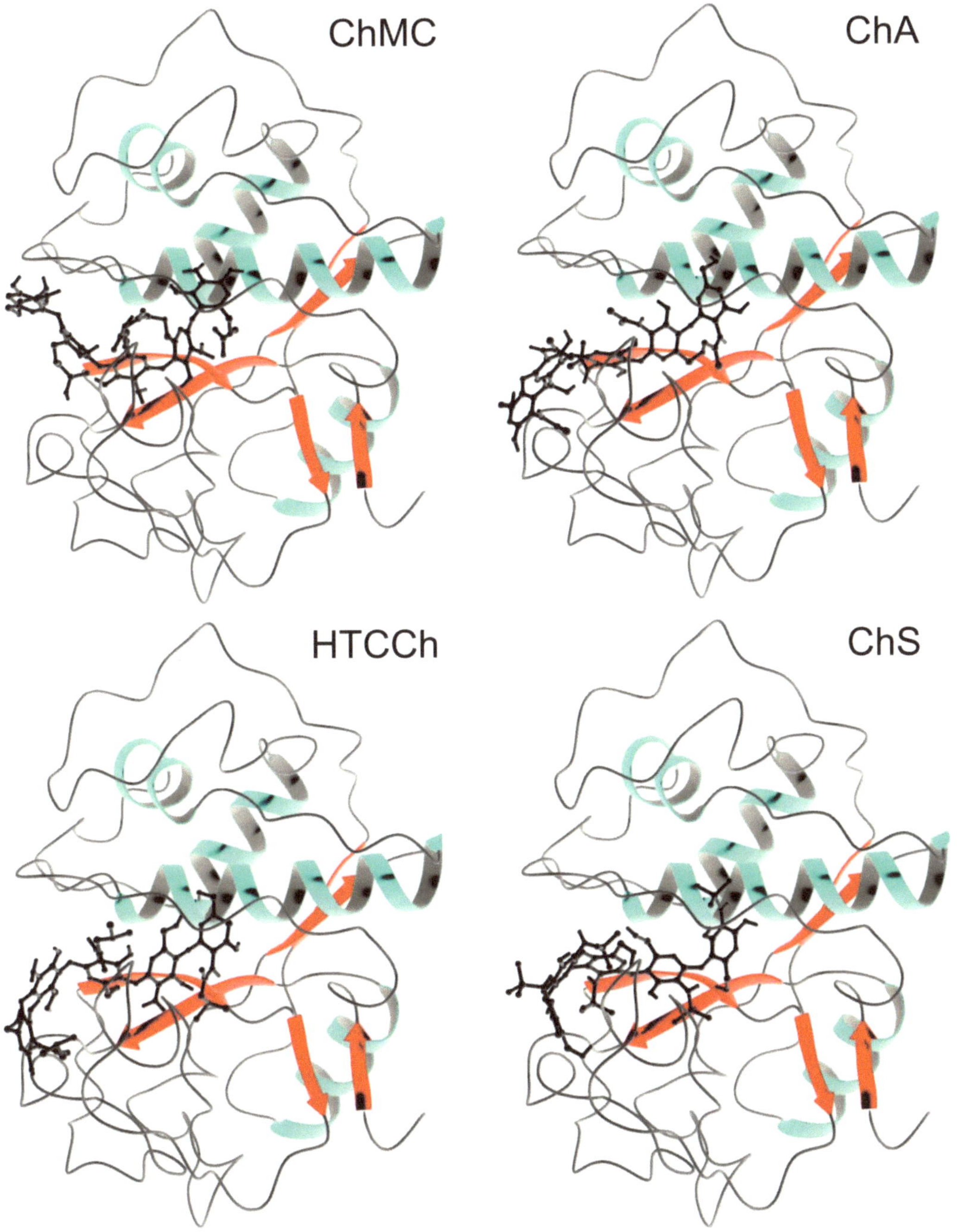

Figure 2. Molecular docking results. Topology of bromelain complexes with ChMC, HTCCh, ChS, and ChA.

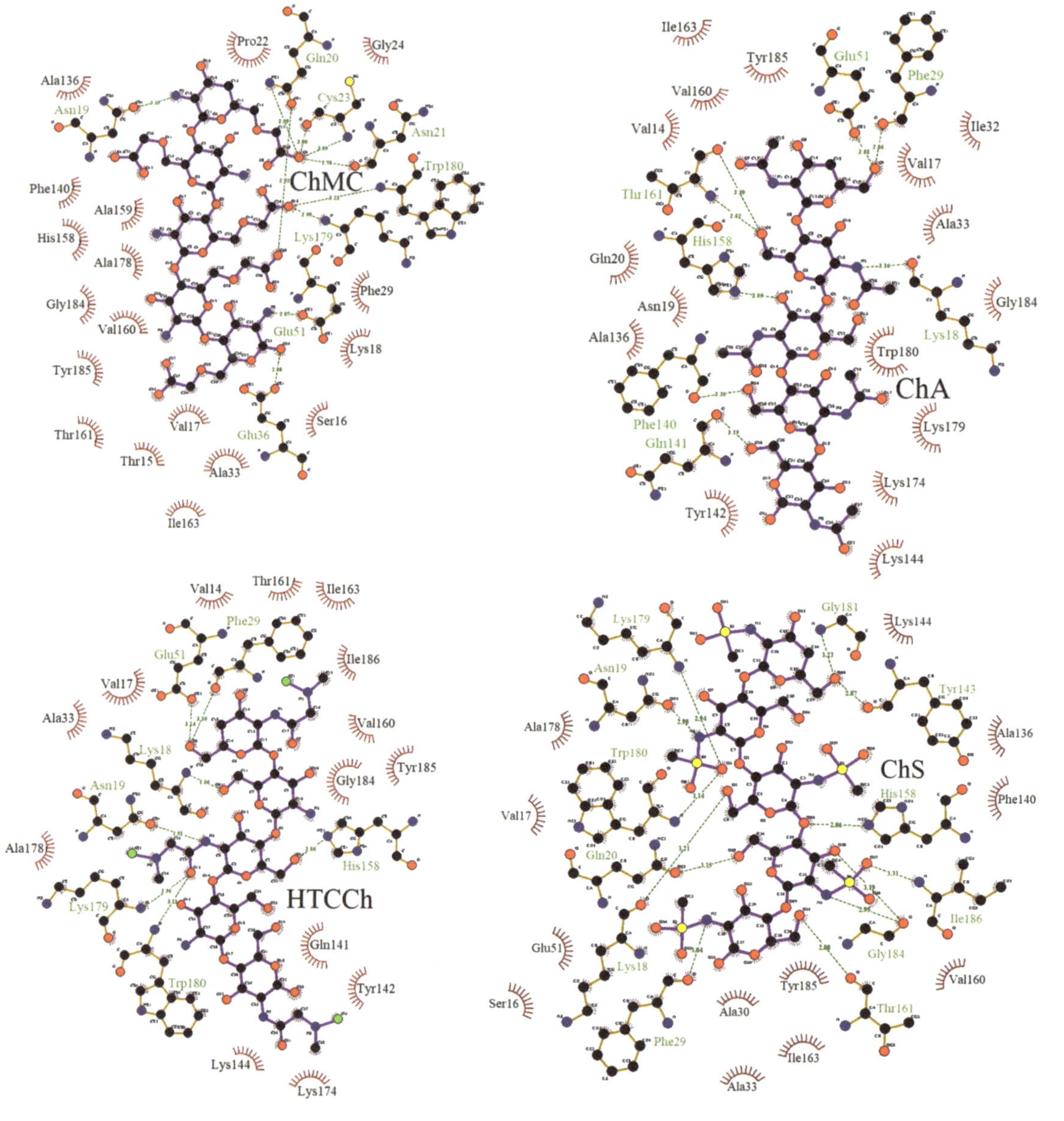

Figure 3. Molecular docking results. Bonds (indicated by dashed lines) and interactions between molecules of bromelain and chitosan derivatives: ChMC, HTCCh, ChS, and ChA.

To study the hydrogen bonding between bromelain and chitosan derivatives in more detail, we analyzed the FTIR spectra of immobilized samples, including the 3000–3600 cm^{-1} region (Figure 4).

Table 1. Bromelain amino acids interacting with ligands *.

Affinity, kcal/mol	Amino Acids Forming	
	Hydrogen Bonds and Their Lengths, Å	Any Other Interactions
	Amino acids of bromelain that form bonds and interactions with ChMC	
−7.8	Asn19, 3.29 Å; Gln20, 2.92 and 2.99 Å; Asn21, 2.70 Å; Cys23, 3.00 and 3.09 Å; Glu36 (αLI), 2.88 Å; Glu51 (αLII), 2.87 Å; Lys179, 2.90 Å; Trp180, 3.23 Å	Thr15, Ser16, Val17, Lys18, Pro22, Gly24, Phe29 (αLI), Ala33 (αLI), Ala136, Phe140, **His158** (βR), Ala159 (βR), Val160 (βR), Thr161 (βR), Ile163 (βR), Ala178, Gly184, Tyr185
	Amino acids of bromelain that form bonds and interactions with HTCCh	
−8	Lys18, 2.86 Å; Asn19, 2.91 Å; Phe29 (αLI), 3.10 Å; Glu51 (αLII), 3.14 Å; **His158** (βR), 2.86 Å; Lys179, 2.96 Å; Trp180, 3.11 Å	Val14, Val17, Ala33 (αLI), Gln141, Tyr142, Lys144, Val160 (βR), Thr161 (βR), Ile163 (βR), Lys174, Ala178, Gly184, Tyr185, Ile186
	Amino acids of bromelain that form bonds and interactions with ChS	
−9.0	Lys18, 3.21 Å; Asn19, 2.90 Å; Gln20, 3.15 Å; Phe29 (αLI), 3.04 Å; Tyr143, 2.87 Å; **His158** (βR), 2.86 Å; Trp161 (βR), 2.80 Å; Lys179, 2.94 Å; Trp180, 3.14 Å; Gly181, 3.23 Å; Gly184, 2.95 and 3.19 Å; Ile186, 3.31 Å	Ser16, Val17, Ala30 (αLI), Ala33 (αLI), Glu51 (αLII), Ala136, Phe140, Lys144, Val160 (βR), Ile163 (βR), Ala178, Tyr185
	Amino acids of bromelain that form bonds and interactions with ChA	
−9.4	Lys18, 3.16 Å; Phe29 (αLI), 2.86 Å; Glu51 (αLII), 2.81 Å; Phe140, 3.26 Å; Gln141, 3.19 Å; **His158** (βR), 3.09 Å; Thr161(βR), 2.62 and 3.30 Å	Val14, Val17, Asn19, Gln20, Ile32 (αLI), Ala33 (αLI), Ala136, Tyr142, Lys144, Val160 (βR), Ile163 (βR), Lys174, Lys179, Trp180, Gly184, Tyr185

* Active site residues are bold; protein secondary structure elements are in brackets.

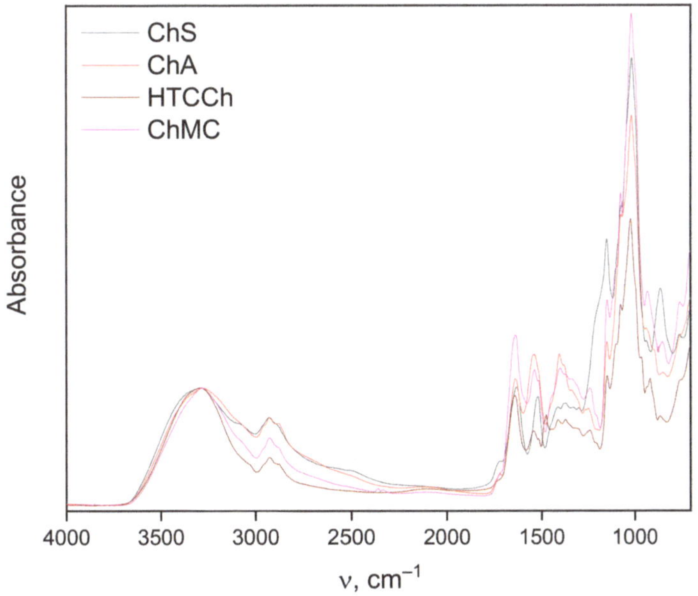

Figure 4. FTIR spectra of dry preparations of bromelain on ChS, ChA, HTCCh, and ChMC matrices.

The intensities of the spectra in the 3000–3600 cm^{-1} region of the stretching vibrations of OH groups are normalized to the maximum of the 3300 cm^{-1} band and allow a rough estimate of the intensity distribution of hydrogen bonds: OH groups with stronger hydrogen bonds absorb at lower frequencies. The number of OH groups absorbing below the center frequency of 3300 cm^{-1} is the smallest in the bromelain/HTCCh system, followed by ChMC, ChA, and ChS in ascending order. The number of OH groups with weak or single hydrogen bonds absorbing above 3300 cm^{-1} is the highest in the bromelain/ChS system, followed in descending order by ChA, HTCCh, and ChMC. It should be noted that it is not possible to exactly estimate the number of hydrogen bonds between the protein and chitosan matrices: the number of these bonds can only be estimated in tens of pieces, while the total number of OH bonds in both the protein and chitosan can reach many hundreds—these differences are completely lost in measurement errors. The above-cited experimental ratios reflect the rearrangement of the entire system of hydrogen bonds as a result of the formation of protein–chitosan complexes, including the secondary structure of the protein and the conformation of the polymer [62], as well as the molecules of hydration water included in their structure.

Figure 5 and Table 2 show the results of an FTIR estimation of the secondary structure of bromelain immobilized on chitosan derivatives: ChMC, ChS, and ChA. It can be seen that after complexation, the number of α-helix contents remains at the same level, the portion of β-sheets significantly increases, and the number of other structures notably decreases.

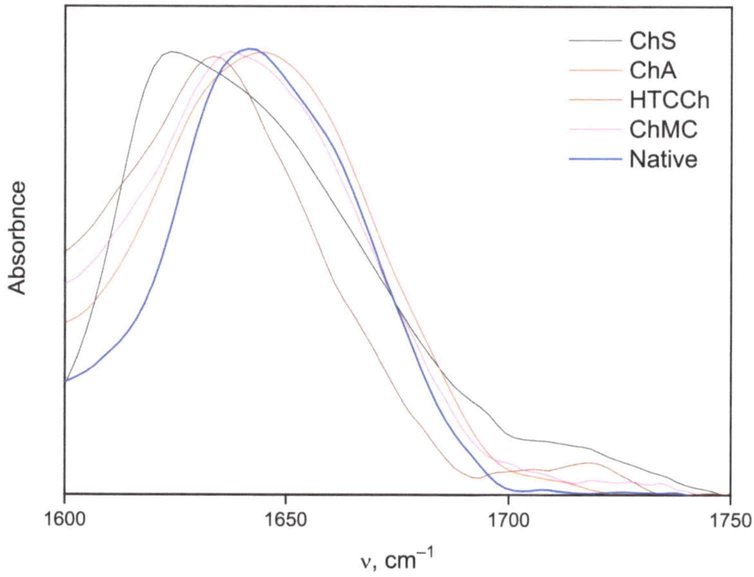

Figure 5. Amide I spectra of bromelain and chitosan derivative complexes are normalized to the maximum.

Table 2. Secondary structure (in%) of bromelain immobilized on chitosan derivatives: ChMC, ChS, and ChA.

Structure	Sample			
	Bromelain in Solution	Bromelain with ChMC	Bromelain with ChS	Bromelain with ChA
α-helix	15	13	13	19
β-sheet	29	43	52	40
other	56	44	35	41

For the complex with HTCCh, such calculations could not be carried out because bromelain was weakly retained on the HTCCh matrix and was washed off with buffer. It was found that complexation with HTCCh did not significantly change the enzyme structure since its structure in the supernatant after washing with buffer was close to the structure of bromelain in solution (without any complexation). Despite the low stability of the bromelain complex with HTCCh, the hyperactivation of the enzyme can be observed. Perhaps this is due to the slowing of aggregation and autolysis processes during the complex formation of bromelain with HTCCh, with minimal modification of enzyme structure and its active site orientation.

Finally, we compare the activity recovery of immobilized bromelain obtained in this work with our previously published results (Figure 6). Earlier, we studied bromelain immobilization on chitosan (Ch) [46], 2-(4-acetamido-2-sulfanilamide) chitosan (SACh) [41], and chitosan copolymer with poly-*N*-vinylimidazole (Ch-g-PVI) [63]. As can be seen, the best activity recovery is achieved for bromelain immobilized on HTCCh and SACh, while for other proposed matrices, its value is dramatically low. Only under the immobilization of chitosan was the phenomenon of bromelain hyperactivation observed. Thus, HTCCh-immobilized bromelain is a promising biocatalayzer for biomedical and biotechnological applications.

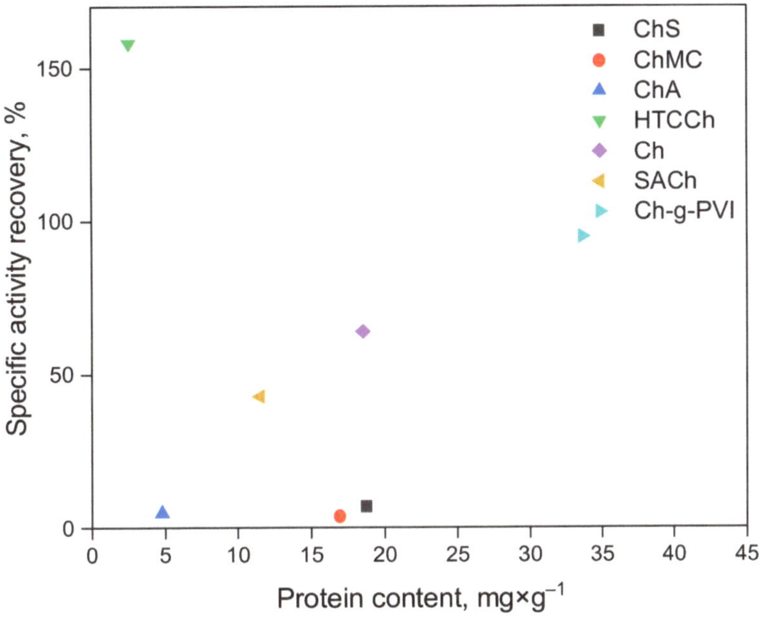

Figure 6. Comparison of specific activity recovery (%) of bromelain immobilized on different chitosan derivatives (values are given for derivatives with a molecular weight of 350 kDa).

3.3. Antibiofilm Properties of Immobilized Bromelain

Various proteases have been suggested as promising antibiofilm tools [64,65], including the cysteine plant proteases ficin and papain in soluble and chitosan-immobilized forms [66–71]. Therefore, the ability of both soluble and immobilized bromelain to disrupt the bacterial biofilms was assessed. For that, 24 h old biofilms were obtained on the surface of 24-well plates washed and treated for either 3 or 24 h with soluble bromelain (0.1 and 0.5 mg·mL^{-1}) or enzyme immobilized on water-soluble derivatives of chitosan—carboxymethylchitosan, *N*-(2-hydroxypropyl)-3-trimethylammonium chitosan, chitosan sulfate, and chitosan acetate. The amount of heterogeneous enzyme corresponded to 0.1 and 0.5 mg·mL^{-1} of soluble enzyme, and as a negative control, pure carriers were taken in similar concentrations. The residual biofilms were subjected to crystal violet staining.

Soluble bromelain destroyed biofilms formed by both *P. aeruginosa* and *S. aureus* (See Figure 7). Among all the variant derivatives of chitosan, only carboxymethylchitosan demonstrated relevant results, and carboxymethylchitosan-immobilized bromelain led to a significant reduction in the biofilms of both pathogens. N-(2-hydroxypropyl)-3-trimethylammonium chitosan, chitosan sulfate, and chitosan acetate by itself and with immobilized protease led to false-negative results, apparently because of the unspecific adherence of plate surfaces (Figures 8 and 9). Taking into account the significant relevance of both *P. aeruginosa* and *S. aureus* on topical surgery wounds and burns, the antibiofilm activity of immobilized bromelain could be a promising approach to coat wound-dressing materials to disrupt and prevent the biofouling of wound-dressings and wounds.

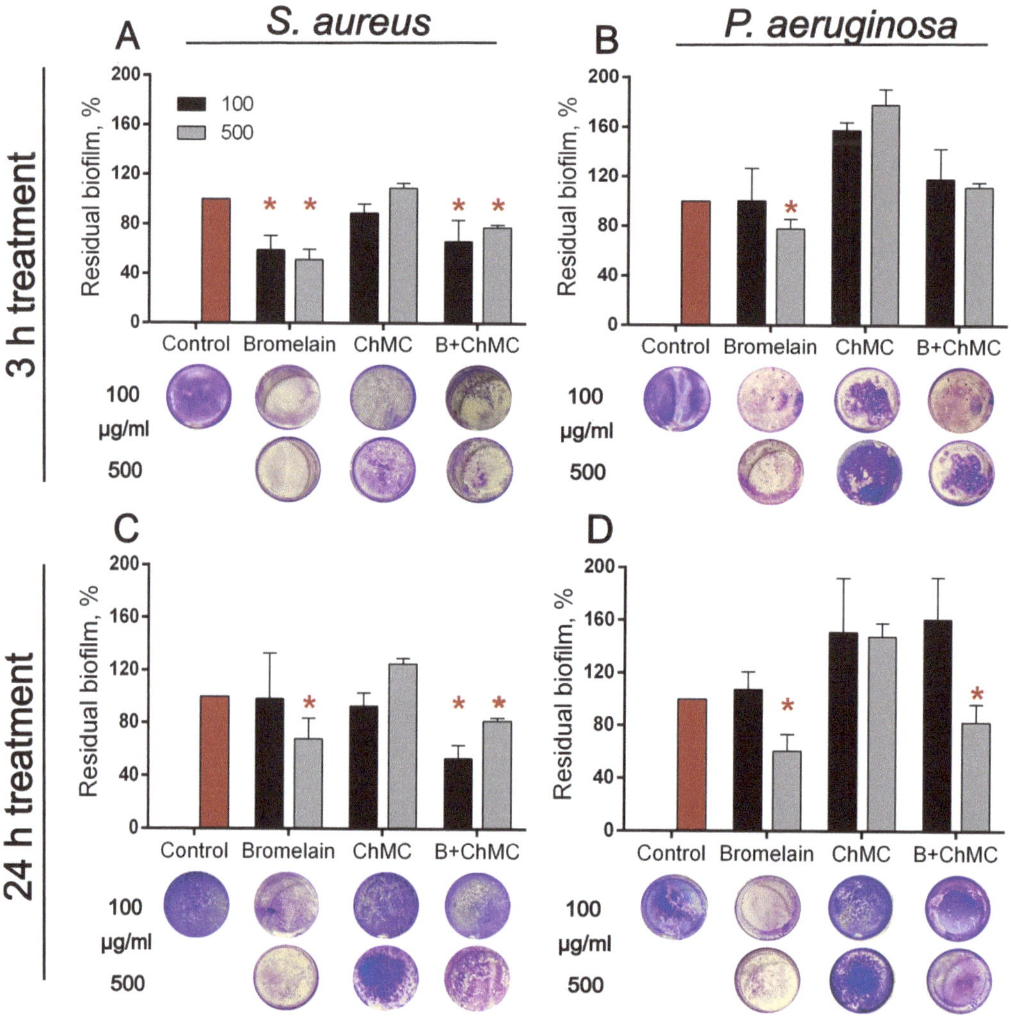

Figure 7. Effect of dissolved bromelain and enzyme immobilized on carboxymethylchitosan (ChMC) on (**A**,**C**) *S. aureus* and (**B**,**D**) *P. aeruginosa* biofilms. The 48 h old biofilms were gently washed and treated for 3 (**A**,**B**) or 24 h (**C**,**D**) with enzymes at concentrations of 0.1 and 0.5 mg·mL^{-1}. The amount of heterogeneous enzyme corresponded to 0.1 and 0.5 mg·mL^{-1} of the soluble enzyme, and as a negative control, pure carriers were taken in similar concentrations. Residual biofilms were quantified with crystal violet staining. Asterisks (*) denote statistically significant differences between untreated and treated samples ($p < 0.05$). Untreated wells were considered to be 100%.

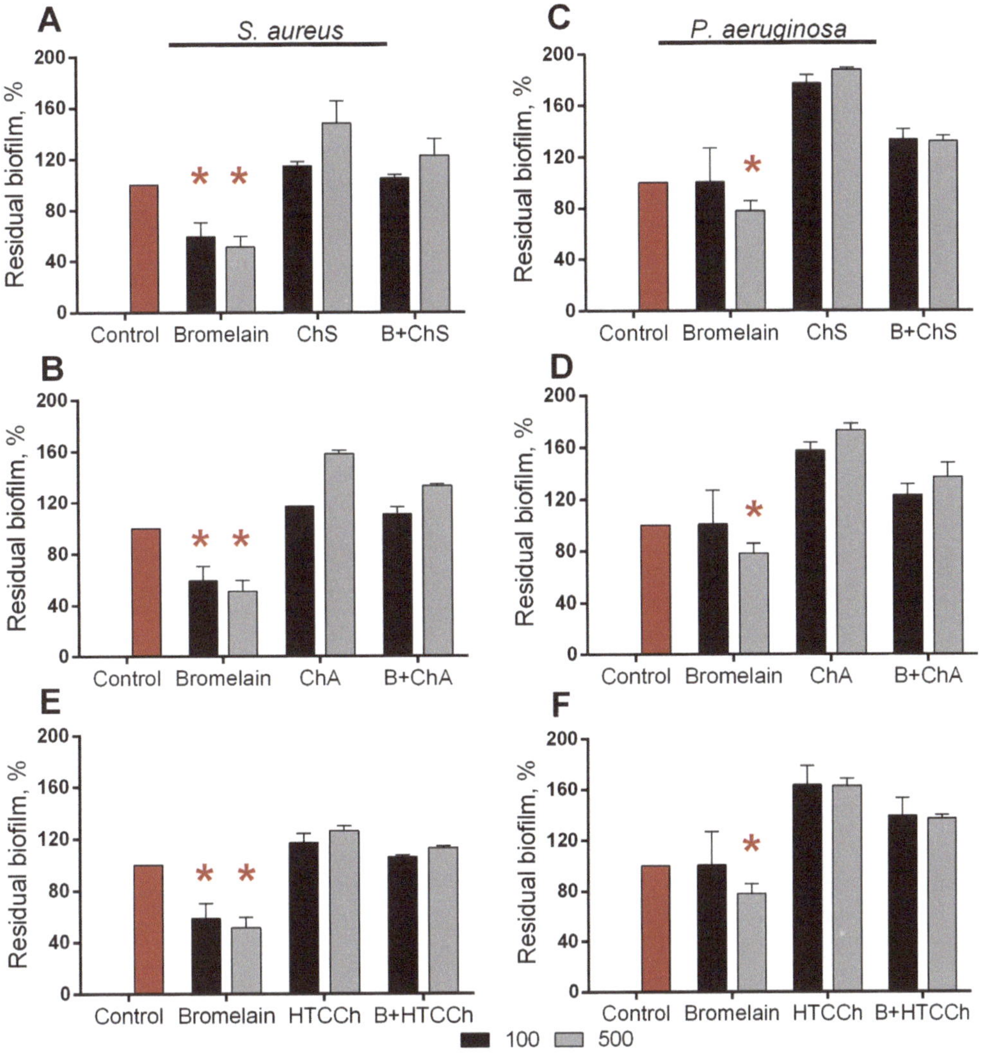

Figure 8. Effect of soluble bromelain and enzyme immobilized on *N*-(2-hydroxypropyl)-3-trimethylammonium chitosan (HTCCh), chitosan sulfate (ChS), and chitosan acetate (ChA) on (**A**,**B**,**E**) *S. aureus* and (**C**,**D**,**F**) *P. aeruginosa* biofilms. The 48 h old biofilms were gently washed and treated for 3 h with enzymes at concentrations of 0.1 and 0.5 mg·mL^{-1}. The amount of heterogeneous enzyme corresponded to 0.1 and 0.5 mg·mL^{-1} of the soluble enzyme; pure carriers in similar concentrations were taken as a negative control. Residual biofilms were quantified with crystal violet staining. Asterisks (*) denote statistically significant differences between untreated and treated samples ($p < 0.05$). Untreated wells were considered to be 100%.

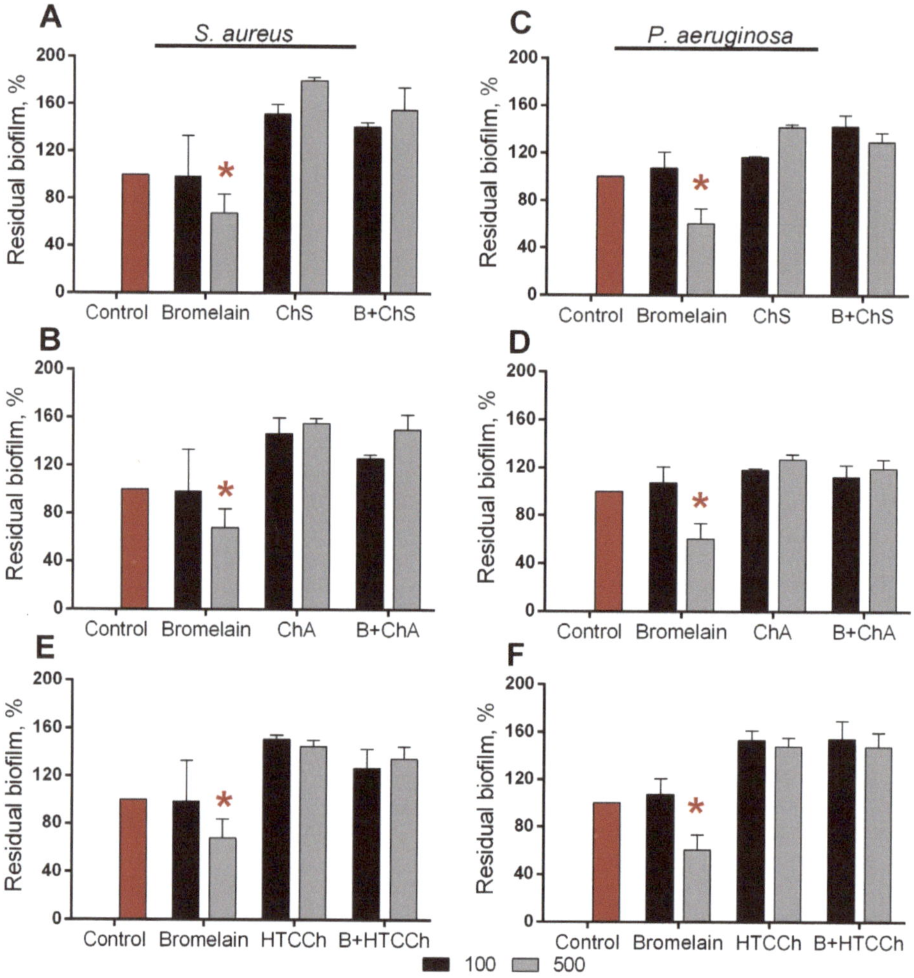

Figure 9. Effect of soluble bromelain and enzyme immobilized on *N*-(2-hydroxypropyl)-3-trimethylammonium chitosan (HTCCh), chitosan sulfate (ChS), and chitosan acetate (ChA) on on (**A**,**B**,**E**) *S. aureus* and (**C**,**D**,**F**) *P. aeruginosa* biofilms. The 48 h old biofilms were gently washed and treated for 24 h with enzymes at concentrations of 0.1 and 0.5 mg·mL^{-1}. The amount of heterogeneous enzyme corresponded to 0.1 and 0.5 mg·mL^{-1} of the soluble enzyme; pure carriers in similar concentrations were taken as a negative control. The residual biofilms were quantified with crystal violet staining. Asterisks (*) denote statistically significant differences between untreated and treated samples ($p < 0.05$). Untreated wells were considered to be 100%.

4. Conclusions

Taken together, our results show that ChS and ChMC form the highest number of hydrogen bonds with bromelain (thirteen and ten, respectively) compared with eight and seven H-bonds for ChA and HTCCh. Apparently, it is one of the main reasons why the protein immobilization yield is higher at complexation with ChS and ChMC (up to 58% and 65%). Moreover, ChMC, ChS, and ChA have H-bonds with His158, which is from the active site of bromelain. This can explain the significant decrease in enzyme activity after its immobilization. HTCCh displays only physical interactions with His158, which possibly modulate the structure of bromelain's active site and lead to enzyme hyperactivation of up to 208% of the total activity (U·mL^{-1} of solution) and up to 158% of the specific activity (U·mg^{-1} of protein). From the FTIR study, it was also found that complexation with HTCCh does not significantly change the bromelain structure. The carboxymethylchitosan-immobilized bromelain provides a significant reduction of the biofilms of both *P. aeruginosa* and *S. aureus*, bacteria often present on topical surgery wounds and burns, suggesting that immobilized bromelain is a promising tool for disrupting and preventing the biofouling of wound-dressing materials and wounds.

Author Contributions: Conceptualization, M.G.H., Y.F.Z. and V.G.A.; methodology, M.G.H. and Y.F.Z.; investigation, A.V.S., S.S.G., Y.A.R., D.R.B. and M.S.K.; resources, M.G.H., Y.F.Z. and D.A.F.; writing—original draft preparation, M.S.L.; writing—review and editing M.G.H., Y.F.Z., D.A.F., V.G.A., A.R.K. and M.S.L.; project administration, M.G.H., Y.F.Z. and V.G.A. All authors have read and agreed to the published version of the manuscript.

Funding: This research was funded by the Russian Science Foundation, Project Number 21-74-20053 (chitosan derivative synthesis and characterization, immobilization of bromelain, M.G.H., A.V.S., M.S.L., D.R.B.); and by the Ministry of Science and Higher Education of Russia under Agreement No. FZGU-2020-0044 (molecular docking, V.G.A.).

Institutional Review Board Statement: This article does not contain any studies with human participants or animals performed by any of the authors.

Informed Consent Statement: Not applicable.

Data Availability Statement: The data presented in this study are available on request from the corresponding author.

Acknowledgments: The authors gratefully acknowledge the Assigned Spectral–Analytical Center of the FRC Kazan Scientific Center of the RAS for enabling them to fulfill their FTIR experiments. Yuriy F. Zuev and Dzhigangir A. Faizullin gratefully acknowledge the government assignment of the FRC Kazan Scientific Center of RAS.

Conflicts of Interest: The authors declare no conflict of interest.

References

1. Atanasov, A.G.; Zotchev, S.B.; Dirsch, V.M.; Supuran, C.T. Natural products in drug discovery: Advances and opportunities. *Nat. Rev. Drug Discov.* **2021**, *20*, 200–216. [CrossRef] [PubMed]
2. Ige, O.O.; Umoru, L.E.; Aribo, S. Natural Products: A Minefield of Biomaterials. *Int. Sch. Res. Not. Mater. Sci.* **2012**, *2012*, 983062. [CrossRef]
3. Joyce, K.; Fabra, G.T.; Bozkurt, Y.; Pandit, A. Bioactive potential of natural biomaterials: Identification, retention and assessment of biological properties. *Signal Transduct. Target. Ther.* **2021**, *6*, 122. [CrossRef] [PubMed]
4. Bogdanova, L.; Zelenikhin, P.; Makarova, A.; Zueva, O.; Salnikov, V.; Zuev, Y.; Ilinskaya, O. Alginate-based hydrogel as delivery system for therapeutic bacterial RNase. *Polymers* **2022**, *14*, 2461. [CrossRef]
5. Makarova, A.O.; Derkach, S.R.; Kadyirov, A.I.; Ziganshina, S.A.; Kazantseva, M.A.; Zueva, O.S.; Gubaidullin, A.T.; Zuev, Y.F. Supramolecular Structure and Mechanical Performance of k-Carrageenan–Gelatin Gel. *Polymers* **2022**, *14*, 4347. [CrossRef] [PubMed]
6. Ramya, R.; Venkatesan, J.; Kim, S.K.; Sudha, P.N. Biomedical Applications of Chitosan: An Overview. *J. Biomater. Tissue Eng.* **2012**, *2*, 100–111. [CrossRef]
7. Wang, W.; Xue, C.; Mao, X. Chitosan: Structural modification, biological activity and application. *Int. J. Biol. Macromol.* **2020**, *164*, 4532–4546. [CrossRef]
8. Makshakova, O.N.; Zuev, Y.F. Interaction-Induced Structural Transformations in Polysaccharide and Protein-Polysaccharide Gels as Functional Basis for Novel Soft-Matter: A Case of Carrageenans. *Gels* **2022**, *8*, 287. [CrossRef]
9. Ojeda-Hernández, D.D.; Canales-Aguirre, A.A.; Matias-Guiu, J.A.; Matias-Guiu, J.; Gómez-Pinedo, U.; Mateos-Díaz, J.C. Chitosan–Hydroxycinnamic Acids Conjugates: Emerging Biomaterials with Rising Applications in Biomedicine. *Int. J. Mol. Sci.* **2022**, *23*, 12473. [CrossRef]
10. Amini, A.; Masoumi-Moghaddam, S.; Morris, D.L. *Utility of Bromelain and N-Acetylcysteine in Treatment of Peritoneal Dissemination of Gastrointestinal Mucin-Producing Malignancies*; Springer: Cham, Switzerland, 2016; pp. 63–80. [CrossRef]
11. Hikisz, P.; Bernasinska-Slomczewska, J. Beneficial Properties of Bromelain. *Nutrients* **2021**, *13*, 4313. [CrossRef] [PubMed]
12. Maurer, H.R. Bromelain: Biochemistry, pharmacology and medical use. *Cell. Mol. Life Sci.* **2001**, *58*, 1234–1245. [CrossRef] [PubMed]

13. Thornhill, S.M.; Kelly, A.M. Natural treatment of perennial allergic rhinitis. *Altern. Med. Rev.* **2000**, *5*, 448–454. [PubMed]
14. Hale, L.P.; Greer, P.K.; Trinh, C.T.; James, C.L. Proteinase activity and stability of natural bromelain preparations. *Int. Immunopharmacol.* **2005**, *5*, 783–793. [CrossRef] [PubMed]
15. Naveena, B.M.; Mendiratta, S.K.; Anjaneyulu, A.S.R. Tenderization of buffalo meat using plant proteases from Cucumis trigonus Roxb (Kachri) and Zingiber officinale roscoe (Ginger rhizome). *Meat Sci.* **2004**, *68*, 363–369. [CrossRef]
16. Varilla, C.; Marcone, M.; Paiva, L.; Baptista, J. Bromelain, a Group of Pineapple Proteolytic Complex Enzymes (Ananas comosus) and Their Possible Therapeutic and Clinical Effects. A Summary. *Foods* **2021**, *10*, 2249. [CrossRef]
17. Hatano, K.-I.; Takahashi, K.; Tanokura, M. Bromein, a Bromelain Inhibitor from Pineapple Stem: Structural and Functional Characteristics. *Protein Pept. Lett.* **2018**, *25*, 838–852. [CrossRef]
18. Di Cosimo, R.; Mc Auliffe, J.; Poulose, A.J.; Bohlmann, G. Industrial Use of Immobilized Enzymes. *Chem. Soc. Rev.* **2013**, *42*, 6437–6474. [CrossRef]
19. Liese, A.; Hilterhaus, L. Evaluation of Immobilized Enzymes for Industrial Applications. *Chem. Soc. Rev.* **2013**, *42*, 6236–6249. [CrossRef]
20. Almeida, F.L.C.; Prata, A.S.; Forte, M.B.S. Enzyme Immobilization: What Have We Learned in the Past Five Years? *Biofuels Bioprod. Biorefining* **2022**, *16*, 587–608. [CrossRef]
21. Rodrigues, R.C.; Berenguer-Murcia, Á.; Carballares, D.; Morellon-Sterling, R.; Fernandez-Lafuente, R. Stabilization of Enzymes via Immobilization: Multipoint Covalent Attachment and Other Stabilization Strategies. *Biotechnol. Adv.* **2021**, *52*, 107821. [CrossRef]
22. Liu, S.; Bilal, M.; Rizwan, K.; Gul, I.; Rasheed, T.; Iqbal, H.M.N. Smart Chemistry of Enzyme Immobilization Using Various Support Matrices—A Review. *Int. J. Biol. Macromol.* **2021**, *190*, 396–408. [CrossRef] [PubMed]
23. Barbosa, O.; Ortiz, C.; Berenguer-Murcia, Á.; Torres, R.; Rodrigues, R.C.; Fernandez-Lafuente, R. Strategies for the One-Step Immobilization-Purification of Enzymes as Industrial Biocatalysts. *Biotechnol. Adv.* **2015**, *33*, 435–456. [CrossRef]
24. Nwagu, T.N.; Ugwuodo, C.J. Stabilizing bromelain for therapeutic applications by adsorption immobilization on spores of probiotic Bacillus. *Int. J. Biol. Macromol.* **2019**, *127*, 406–414. [CrossRef] [PubMed]
25. Zappino, M.; Cacciotti, I.; Benucci, I.; Nanni, F.; Liburdi, K.; Valentini, F.; Esti, M. Bromelain immobilization on microbial and animal source chitosan films, plasticized with glycerol, for application in wine-like medium: Microstructural, mechanical and catalytic characterisations. *Food Hydrocoll.* **2015**, *45*, 41–47. [CrossRef]
26. Wei, B.; He, L.; Wang, X.; Yan, G.Q.; Wang, J.; Tang, R. Bromelain-decorated hybrid nano-particles based on lactobionic acid-conjugated chitosan for in vitro anti-tumor study. *J. Biomater. Appl.* **2017**, *32*, 206–218. [CrossRef] [PubMed]
27. Ataide, J.A.; Geraldes, D.C.; Gérios, E.F.; Bissaco, F.M.; Cefali, L.C.; Oliveira-Nascimento, L.; Mazzola, P.G. Freeze-dried chitosan nanoparticles to stabilize and deliver bromelain. *J. Drug Deliv. Sci. Technol.* **2021**, *61*, 102225. [CrossRef]
28. Wang, X.; He, L.; Wei, B.; Yan, G.; Wang, J.; Tang, R. Bromelain-immobilized and lactobionic acid-modified chitosan nanoparticles for enhanced drug penetration in tumor tissues. *Int. J. Biol. Macromol.* **2018**, *115*, 129–142. [CrossRef]
29. Tan, Y.-L.; Liu, C.-G.; Yu, L.-J.; Chen, X.-G. Effect of linoleic-acid modified carboxymethyl chitosan on bromelain immobilization onto self-assembled nanoparticles. *Front. Mater. Sci. China* **2008**, *2*, 209–213. [CrossRef]
30. Ilaria, B.; Marco, E.; Katia, L.; Vittoria, G.A.M. Pineapple stem bromelain immobilized on different supports: Catalytic properties in model wine. *Biotechnol. Prog.* **2012**, *28*, 1472–1477. [CrossRef] [PubMed]
31. Seo, H.; Itoyama, K.; Morimoto, K.; Takagishi, T.; Oka, M.; Hayashi, T. Spacer effects on enzymatic activity of bromelain immobilized onto porous chitosan beads. *Eur. Polym. J.* **1998**, *34*, 917–922. [CrossRef]
32. Sorokin, A.; Lavlinskaya, M. Synthesis of the superabsorbents enriched in chitosan derivatives with excellent water absorption properties. *Polym. Bull.* **2022**, *79*, 407–427. [CrossRef]
33. Wade, R.C.; Sobolev, V.; Ortiz, A.R.; Peters, G. Computational approaches to modeling receptor flexibility upon ligand binding: Application to interfacially activated enzymes. *Struct.-Based Drug Des.* **1998**, *352*, 223–232. [CrossRef]
34. Rathinam, S.; Solodova, S.; Kristjánsdóttir, I.; Hjálmarsdóttir, M.A.; Másson, M. The antibacterial structure-activity relationship for common chitosan derivatives. *Int. J. Biol. Macromol.* **2020**, *165*, 1686–1693. [CrossRef] [PubMed]
35. Chi, W.; Qin, C.; Zeng, L.; Li, W.; Wang, W. Microbiocidal activity of chitosan-N-2-hydroxypropyl trimethyl ammonium chloride. *J. Appl. Polym. Sci.* **2006**, *103*, 3851–3856. [CrossRef]
36. Pires, N.R.; Cunha, P.L.R.; Maciel, J.S.; Angelim, A.L.; Melo, V.M.M.; Paula, R.C.M.; Feitosa, J.P.A. Sulfated chitosan as tear substitute with no antimicrobial activity. *Carbohydr. Polym.* **2013**, *91*, 92–99. [CrossRef]
37. Dai, T.; Tegos, G.P.; Burkatovskaya, M.; Castano, A.P.; Hamblin, M.R. Chitosan Acetate Bandage as a Topical Antimicrobial Dressing for Infected Burns. *Antimicrobal Agents Chemother.* **2009**, *3*, 393–400. [CrossRef]
38. Aguanell, A.; Pozo, M.L.; Pérez-Martín, C.; Pontes, G.; Bastida, A.; Fernández-Mayoralas, A.; García-Junceda, E.; Revuelta, J. Chitosan sulfate-lysozyme hybrid hydrogels as platforms with fine-tuned degradability and sustained inherent antibiotic and antioxidant activities. *Carbohydr. Polym.* **2022**, *291*, 119611. [CrossRef]
39. Chen, S.C.; Wu, Y.C.; Mi, F.L.; Lin, Y.H.; Yu, L.C.; Sung, H.W. A novel pH-sensitive hydrogel composed of N,O-carboxymethyl chitosan and alginate cross-linked by genipin for protein drug delivery. *J. Control. Release* **2004**, *96*, 285–300. [CrossRef] [PubMed]
40. Gorshkova, M.Y.; Volkova, I.F.; Alekseeva, S.G.; Molotkova, N.N.; Skorikova, E.E.; Izumrudov, V.A. Water-soluble modified chitosan and its interaction with a polystyrenesulfonate anion. *Polym. Sci. Ser. A* **2011**, *53*, 57–66. [CrossRef]

41. Olshannikova, S.S.; Malykhina, N.V.; Lavlinskaya, M.S.; Sorokin, A.V.; Yudin, N.E.; Vyshkvorkina, Y.M.; Lukin, A.N.; Holyavka, M.G.; Artyukhov, V.G. Novel Immobilized Biocatalysts Based on Cysteine Proteases Bound to 2-(4-Acetamido-2-sulfanilamide) Chitosan and Research on Their Structural Features. *Polymers* **2022**, *14*, 3223. [CrossRef]
42. Lowry, O.H.; Rosebrough, N.J.; Faar, A.L.; Randall, R.J. Protein measurement with Folin-phenol reagent. *J. Biol. Chem.* **1951**, *193*, 265–275. [CrossRef] [PubMed]
43. Artyukhov, V.G.; Kovaleva, T.A.; Kholyavka, M.G.; Bityutskaya, L.A.; Grechkina, M.V. Thermal inactivation of free and immobilized inulinase. *Appl. Biochem. Microbiol.* **2010**, *46*, 385–389. [CrossRef]
44. Pankova, S.M.; Sakibaev, F.A.; Holyavka, M.G.; Vyshkvorkina, Y.M.; Lukin, A.N.; Artyukhov, V.G. Studies of the Processes of the Trypsin Interactions with Ion Exchange Fibers and Chitosan. *Russ. J. Bioorganic Chem.* **2021**, *47*, 765–776. [CrossRef]
45. Sabirova, A.R.; Rudakova, N.L.; Balaban, N.P.; Ilyinskaya, O.N.; Demidyuk, I.V.; Kostrov, S.V.; Rudenskaya, G.N.; Sharipova, M.R. A novel secreted metzincin metalloproteinase from Bacillus intermedius. *FEBS Lett.* **2010**, *584*, 4419–4425. [CrossRef] [PubMed]
46. Holyavka, M.; Faizullin, D.; Koroleva, V.; Olshannikova, S.; Zakhartchenko, N.; Zuev, Y.; Kondtatyev, M.; Zakharova, E.; Artyukhov, V. Novel biotechnological formulations of cysteine proteases, immobilized on chitosan. Structure, stability and activity. *Int. J. Biol. Macromol.* **2021**, *180*, 161–176. [CrossRef] [PubMed]
47. Holyavka, M.G.; Kondratyev, M.S.; Samchenko, A.A.; Kabanov, A.V.; Komarov, V.M.; Artyukhov, V.G. In silico design of high-affinity ligands for the immobilization of inulinase. *Comput. Biol. Med.* **2016**, *71*, 198–204. [CrossRef]
48. Byler, D.M.; Susi, H. Examination of the secondary structure of proteins by deconvolved FTIR spectra. *Biopolymers* **1986**, *25*, 469–487. [CrossRef] [PubMed]
49. Kayumov, A.R.; Khakimullina, E.N.; Sharafutdinov, I.S.; Trizna, E.Y.; Latypova, L.Z.; Lien, H.T.; Margulis, A.B.; Bogachev, M.I.; Kurbangalieva, A.R. Inhibition of biofilm formation in Bacillus subtilis by new halogenated furanones. *J. Antibiot.* **2015**, *68*, 297–301. [CrossRef] [PubMed]
50. O'Toole, G.A.; Kolter, R. Initiation of biofilm formation in Pseudomonas fluorescens WCS365 proceeds via multiple, convergent signalling pathways: A genetic analysis. *Mol. Microbiol.* **1998**, *28*, 449–461. [CrossRef]
51. Cui, J.; Ren, S.; Lin, T.; Feng, Y.; Jia, S. Shielding effects of Fe3+-tannic acid nanocoatings for immobilized enzyme on magnetic 611 Fe3O4@silica core shell nanosphere. *Chem. Eng. J.* **2018**, *343*, 629–637. [CrossRef]
52. Ramezani, R.; Aminlari, M.; Fallahi, H. Effect of chemically modified soy proteins and ficin-tenderized meat on the quality attributes of sausage. *J. Food Sci.* **2006**, *68*, 85–88. [CrossRef]
53. Marques, A.C.; Marostica, M.R.; Pastore, G.M. Some nutritional, technological and environmental advances in the use of enzymes in meat products. *Enzym. Res.* **2010**, *2010*, 480923. [CrossRef] [PubMed]
54. Farhadihosseinabadi, B.; Zarebkohan, A.; Eftekhary, M.; Heiat, M.; Moghaddam, M.M.; Gholipourmalekabadi, M. Crosstalk between chitosan and cell signaling pathways. *Cell. Mol. Life Sci.* **2019**, *76*, 2697–2718. [CrossRef] [PubMed]
55. Nunes, Y.L.; de Menezes, F.L.; de Sousa, I.G.; Cavalcante, A.L.G.; Cavalcante, F.T.T.; da Silva Moreira, K.; de Oliveira, A.L.B.; Mota, G.F.; da Silva Souza, J.E.; de Aguiar Falcão, I.R.; et al. Chemical and physical Chitosan modification for designing enzymatic industrial biocatalysts: How to choose the best strategy? *Int. J. Biol. Macromol.* **2021**, *181*, 1124–1170. [CrossRef] [PubMed]
56. Tavano, O.L.; Berenguer-Murcia, A.; Secundo, F.; Fernandez-Lafuente, R. Biotechnological applications of proteases in food technology. *Compr. Rev. Food Sci. Food Saf.* **2018**, *17*, 412–436. [CrossRef] [PubMed]
57. Hoarau, M.; Badieyan, S.; Marsh, E.N.G. Immobilized enzymes: Understanding enzyme–surface interactions at the molecular level. *Org. Biomol. Chem.* **2017**, *15*, 9539–9551. [CrossRef]
58. Drenth, J.; Jansonius, J.N.; Koekoek, R.; Wolthers, B.G. The structure of papain. *Adv. Protein Chem.* **1971**, *25*, 79–115. [CrossRef]
59. Kamphuis, I.G.; Kalk, K.H.; Swarte, M.B.A.; Drenth, J. Structure of papain refined at 1.65 Å resolution. *J. Mol. Biol.* **1984**, *179*, 233–256. [CrossRef]
60. Novinec, M.; Lenarcic, B. Papain-like peptidases: Structure, function, and evolution. *Biomol. Concepts* **2013**, *4*, 287–308. [CrossRef]
61. Holyavka, M.; Pankova, S.; Koroleva, V.; Vyshkvorkina, Y.; Lukin, A.; Kondratyev, M.; Artyukhov, V. Influence of UV radiation on molecular structure and catalytic activity of free and immobilized bromelain, ficin and papain. *J. Photochem. Photobiol. B Biol.* **2019**, *201*, 111681. [CrossRef]
62. Makshakova, O.N.; Faizullin, D.A.; Zuev, Y.F. Interplay between secondary structure and ion binding upon thermoreversible gelation of κ-carrageenan. *Carbohydr. Polym.* **2020**, *227*, 115342. [CrossRef]
63. Sorokin, A.V.; Olshannikova, S.S.; Lavlinskaya, M.S.; Holyavka, M.G.; Faizullin, D.A.; Zuev, Y.F.; Artukhov, V.G. Chitosan Graft Copolymers with N-Vinylimidazole as Promising Matrices for Immobilization of Bromelain, Ficin, and Papain. *Polymers* **2022**, *14*, 2279. [CrossRef] [PubMed]
64. Nahar, S.; Mizan, M.F.R.; Ha, A.J.W.; Ha, S.D. Advances and Future Prospects of Enzyme-Based Biofilm Prevention Approaches in the Food Industry. *Compr. Rev. Food Sci. Food Saf.* **2018**, *17*, 1484–1502. [CrossRef] [PubMed]
65. Thallinger, B.; Prasetyo, E.N.; Nyanhongo, G.S.; Guebitz, G.M. Antimicrobial enzymes: An emerging strategy to fight microbes and microbial biofilms. *Biotechnol. J.* **2013**, *8*, 97. [CrossRef]
66. Mohamed, S.H.; Mohamed, M.S.; Khalil, M.S.; Mohamed, W.S.; Mabrouk, M.I. Antibiofilm activity of papain enzyme against pathogenic Klebsiella pneumoniae. *J. Appl. Pharm. Sci.* **2018**, *6*, 163–168. [CrossRef]
67. Atacan, K.; Ozacar, M. Investigation of antibacterial properties of novel papain immobilized on tannic acid modified Ag/CuFe2O4 magnetic nanoparticles. *Int. J. Biol. Macromol.* **2018**, *109*, 720–731. [CrossRef] [PubMed]

68. Trizna, E.; Baydamshina, D.; Kholyavka, M.; Sharafutdinov, I.; Hairutdinova, A.; Khafizova, F.; Zakirova, E.; Hafizov, R.; Kayumov, A. Soluble and immobilized papain and trypsin as destroyers of bacterial biofilms. *Genes Cells* **2016**, *10*, 106–112.
69. Baidamshina, D.R.; Koroleva, V.A.; Trizna, E.Y.; Pankova, S.M.; Agafonova, M.N.; Chirkova, M.N.; Vasileva, O.S.; Akhmetov, N.; Shubina, V.V.; Porfiryev, A.G.; et al. Anti-biofilm and wound-healing activity of chitosan-immobilized Ficin. *Int. J. Biol. Macromol.* **2020**, *164*, 4205–4217. [CrossRef]
70. Baidamshina, D.R.; Koroleva, V.A.; Olshannikova, S.S.; Trizna, E.Y.; Bogachev, M.I.; Artyukhov, V.G.; Holyavka, M.G.; Kayumov, A.R. Biochemical Properties and Anti-Biofilm Activity of Chitosan-Immobilized Papain. *Mar. Drugs* **2021**, *19*, 197. [CrossRef]
71. Baidamshina, D.R.; Trizna, E.Y.; Holyavka, M.G.; Bogachev, M.I.; Artyukhov, V.G.; Akhatova, F.S.; Rozhina, E.V.; Fakhrullin, R.F.; Kayumov, A.R. Targeting microbial biofilms using Ficin, a nonspecific plant protease. *Sci. Rep.* **2017**, *7*, srep46068. [CrossRef]

polymers

MDPI

Review

The Role of Mucoadhesion and Mucopenetration in the Immune Response Induced by Polymer-Based Mucosal Adjuvants

Nathaly Vasquez-Martínez [1,2], Daniel Guillen [1], Silvia Andrea Moreno-Mendieta [1,2,3], Sergio Sanchez [1] and Romina Rodríguez-Sanoja [1,*]

[1] Instituto de Investigaciones Biomédicas, Universidad Nacional Autónoma de México, Circuito, Mario de La Cueva s/n, C.U., Coyoacán, Mexico City 04510, Mexico; vasquezmarnathaly@gmail.com (N.V.-M.)
[2] Programa de Doctorado en Ciencia Bioquímicas, Universidad Nacional Autónoma de México, Circuito de Posgrado, C.U., Coyoacán, Mexico City 04510, Mexico
[3] Consejo Nacional de Ciencia y Tecnología, Benito Juárez, Mexico City 03940, Mexico
* Correspondence: romina@iibiomedicas.unam.mx

Abstract: Mucus is a viscoelastic gel that acts as a protective barrier for epithelial surfaces. The mucosal vehicles and adjuvants need to pass through the mucus layer to make drugs and vaccine delivery by mucosal routes possible. The mucoadhesion of polymer particle adjuvants significantly increases the contact time between vaccine formulations and the mucosa; then, the particles can penetrate the mucus layer and epithelium to reach mucosa-associated lymphoid tissues. This review presents the key findings that have aided in understanding mucoadhesion and mucopenetration while exploring the influence of physicochemical characteristics on mucus–polymer interactions. We describe polymer-based particles designed with mucoadhesive or mucopenetrating properties and discuss the impact of mucoadhesive polymers on local and systemic immune responses after mucosal immunization. In future research, more attention paid to the design and development of mucosal adjuvants could lead to more effective vaccines.

Keywords: mucoadhesion; polymeric particles; immune response; mucosal vaccines; mucosal adjuvants

check for updates

Citation: Vasquez-Martínez, N.; Guillen, D.; Moreno-Mendieta, S.A.; Sanchez, S.; Rodríguez-Sanoja, R. The Role of Mucoadhesion and Mucopenetration in the Immune Response Induced by Polymer-Based Mucosal Adjuvants. *Polymers* **2023**, *15*, 1615. https://doi.org/10.3390/polym15071615

Academic Editors: Antonio M. Borrero-López, Concepción Valencia-Barragán, Esperanza Cortés Triviño, Adrián Tenorio-Alfonso and Clara Delgado-Sánchez

Received: 17 February 2023
Revised: 9 March 2023
Accepted: 10 March 2023
Published: 24 March 2023

1. Introduction

The development of mucosal vaccines continues to be a priority in the fight against microorganisms whose entry is the mucosa. However, the mucus layer limits the passage of antigens across the epithelium to reach mucosa-inductive sites. This protective barrier facilitates the clearance of foreign pathogens and particles. The mucosal compartments have the epithelium and the mucosal immune system as barriers to defense. The epithelial barrier interconnected by tight junctions, the mucus layer, antimicrobial peptides, and immunoglobulin A (IgA) production prevent, as a whole, access to pathogenic microorganisms, foreign particles, and toxins [1–3].

Furthermore, the mucosal immune system, comprising a network of mucosa-associated lymphoid tissues (MALT), is responsible for initiating and establishing the antigen-specific innate and adaptive immune response following infection or vaccination [4]. Both inducing and effector sites are found in MALT. For example, in the small intestine, immune-inductive gut-associated lymphoid tissue (GALT) comprises the Peyer patches, mesenteric lymph nodes, and isolated lymphoid follicles (Figure 1). The GALT is covered by a follicle-associated epithelium, predominantly composed of enterocytes and membranous cells (M cells) [5]. M cells, surrounded by a thin layer of mucus, are responsible for transferring, via phagocytosis or transcytosis, bacteria and particulate antigens from the luminal side to the basal side of the epithelium and to the subepithelial dome (SED) [6,7], along with the other intestinal epithelial cells [8,9]. Regardless of the sampling mechanism, antigen-specific primed cells in the mucosa leave the encounter site to enter the lymph, then the bloodstream, and re-enter the mucosal tissues of origin, where they differentiate into effector or memory B and T cells, a process mediated by integrins [10]. In the small intestine, this effector site is the lamina propria. For a deeper understanding of the inductive and effector sites in the mucosal immune responses other than GALT and NALT, the following reviews are suggested [11–13].

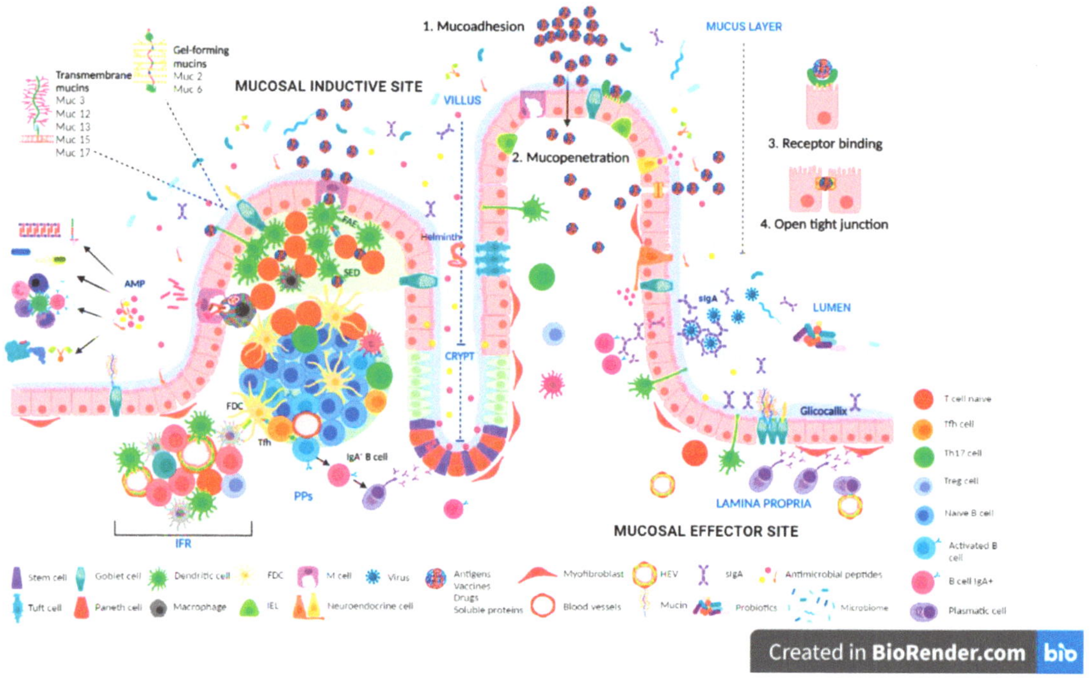

Figure 1. Components of gut-associated lymphoid tissues. The intestinal epithelium comprises multiple cell types derived from intestinal stem cells (IECs), including absorptive enterocytes, Paneth cells, goblet cells, tuft cells, and enteroendocrine cells. The IECs maintain gut homeostasis by synthesizing and secreting mucins, antimicrobial peptides, hormones, and soluble proteins. Furthermore, the IECs participate in the recognition of microorganisms via PPRs. The inner mucus layer or glycocalyx that lines the intestinal epithelium contains many antimicrobial peptides (AMPs) and IgA, both with effector functions. The outer mucus layer is also colonized by commensal microbiota. AMPs bind to glycosylated proteins, neutralize bacterial toxins, participate in the recruitment of effectors cells, and directly kill bacteria. Antigen uptake can occur by M-cell-mediated transcytosis, via macrophage/dendritic cells extending transepithelial dendrites into the gut lumen, paracellular pathway, or through goblet-cells-associated antigen passages. The passage of the particles is further facilitated by mucoadhesion and mucopenetration phenomena. In gut-associated lymphoid tissues, antigens are presented to naive T and B cells with subsequent antigen-specific immune responses. PPs are mucosa-inductive sites for immune responses in the gastrointestinal tract, while the lamina propria and epithelial compartment constitute effector sites. FAE, follicle associated epithelial; SED, subepithelial dome; PPs, Peyer's patches; IFR, interfollicular regions; FDC, follicular dendritic cells; Tfh, T follicular helper cells.

Considering the barrier properties of mucosal surfaces, extensive studies have been performed to develop strategies for prolonging the residence time of vaccines in epithelial tissues; one of the most relevant is the use of mucoadhesive formulations. This review provides an overview of mucus, mucin, and how the interactions between mucus and particles occur. We discuss the physicochemical characteristics of particles that lead to improved mucoadhesion and/or mucopenetration.

The paper aims to show solid scientific evidence to re-evaluate the correlation between improving the mucoadhesive or mucopenetrating characteristics of polymer-based particles used as mucosal vaccine adjuvants and the increase in specific systemic and mucosal immune responses. Likewise, we address potential strategies for mucus penetration, highlighting the importance of incorporating them to design more effective mucosal vaccines. For these purposes, most of the included research papers are those in which the mucoadhesion test of polymer particles was published together with the assessment of adjuvanted capacity following mucosal vaccination.

2. Mucosal Vaccination

The mucosal surfaces cover a vast extension of the body surface area. Due to continuous environmental exposure, many pathogens, or antigens, as well as particles found in the air and toxins, have the mucosal surfaces as portals of entry to the body. Therefore, the mucosal tissues play a fundamental role in protecting from invasion by harmful microorganisms through physical and biological barriers. The characteristic induction of antigen-specific sIgA antibodies, both local and distant, as well as protective immunity in systemic and mucosal compartments, makes mucosal vaccination the best tool for reducing mortality and morbidity caused by infectious pathogens that enter the body through the mucosal surfaces [14,15].

However, most vaccines licensed for use in humans are currently administered parenterally. Although parenteral immunization successfully induces a protective systemic immune response, it hardly induces an effective mucosal immune response, and the cellular mechanisms underlying this response remain largely unknown [16–18].

The SARS-CoV-2 pandemic highlighted mucosal vaccination's importance in triggering an immune response at the predominant sites of pathogen infections and protecting against mucosal invasion. Several nasal and oral vaccines are currently in the clinical phase (Table 1), thanks to numerous researchers who have focused on developing mucosal vaccination platforms for other diseases in recent decades. It is essential to clarify that there are already authorized vaccines for application through these routes for other diseases. For example, oral vaccines are currently on the market for *Vibrio cholerae* (Dukoral®, ShanChol™ OCV, Euvichol-Plus®/Euvichol®, and Vaxchora®) [19], poliovirus (BIOPOLIO™ B1/3), rotavirus (RotaRix®, RotaTeq®, Rotavac®, and RotaSiil®) [20], *Salmonella typhi* (Vivotif® Ty21A) [21], and the adenovirus vaccine (Adenovirus types 4 and 7) approved for military use only [22]. On the other hand, presented in the form of a spray for nasal administration, there are the influenza vaccines against type A and B viruses (FluMist® Quadrivalent and Nasovac-S™) [23,24]. Mucosal vaccines induce diverse immune responses in strength, efficiency, and long-term protection [25–27]. Most of these formulations contain live attenuated or inactivated whole-cell organisms or viruses [28,29]; consequently, the limitations and challenges are many, especially those related to safety. In this regard, developing subunit vaccines becomes a promissory strategy because they are safer but, unfortunately, less immunogenic. Consequently, most of the time, subunit vaccines demand the use of adjuvants that require specific characteristics for mucosal routes; regrettably, there are no approved vaccines for these routes.

Table 1. Current mucosal COVID-19 vaccine candidates in clinical trials. Adapted from WHO vaccine tracker and landscape, February 2023 (Available online: https://www.who.int/publications/m/item/draft-landscape-of-COVID-19-candidate-vaccines, accessed on: 10 February 2023). RBD, receptor-binding domain; Ad, adenovirus; MVA, modified vaccinia Ankara; LAIV, live attenuated influenza virus; NDV, Newcastle disease virus. **R**, route of administration: in, intranasal; or, oral; ae, aerosol; ih, inhale.

Protein Subunit				
Antigen	**Name**	**Developer**	**Clinical Trial**	**R**
Spike β-variante	ACM-001	ACM Biolabs	Phase 1 NTC05385991	in
RBD of S protein+ AgnHB	CIBG-669	CIGB, Cuba	Phase 1/2 RPCE0000345	in
RBD of S protein	Cov2-OGEN1	USSF/VaxForm	Phase 1 NCT04893512	or
Spike	OMV-linked Hexapro	Intravacc B. V	Phase 1 NCT05604690	In
Live attenuated virus				
S protein	Mv-014-212	Meissa Vaccines, Inc.	Phase 1 NCT04798001	in
S protein	COVI -VAC	CODAGENIX Inc./Serum Institute of India	Phase 3 ISRCTN15779782	in
RBD of S protein	Razi Cov Pars	Razi Vaccine and Serum Research Institute	Phase2IRCT20201214049709N2	In
Viral vector				
NON-REPLICANT				
▪ *Adenoviral vector*				
S protein	ChAdOx1 nCoV-19	University of Oxford	Phase 1 NCT04816019	in
S + nucleocapside	VXA-CoV-2-1-Ad5	Vaxart	Phase 2 NCT04563702	or
S protein	BBV154	Bharat Biotech International Limited	Phase 3 CTRI/2022/02/040065	in
RBD	Ad5-nCoV	CanSino Biological Inc./Beijing Institute of Biotechnology	Phase 4 NTC05303584	ih
Spike + nucleocapsid	Ad5-triCoVMac	McMaster University	Phase 1 NTC05094609	ae
		PIV5 vector		
Spike	CVXGA1	CyanVac LLC	Phase 1 NCT04954287	in
▪ *Ankara vector*				
Spike	MVA-SARS-2-ST	Hannover Medical School	Phase 1 NCT05226390	ih
▪ *Influenza vector*				
N protein fragment	Corfluvec	Research Institute of Influenza	Phase 1/2 NCT05696067	In
REPLICANT				
▪ *Intranasal flu-based RBD*				
RBD of S protein	DelNS1-2019-nCoV-RBD-OPT1	University of Hong Kong, Xiamen University, and Bejing Wantai Biological Pharmacy	Phase 3 ChiCTR2000037782	in
▪ *Live recombinant Newcastle-disease-virus-vectored*				
Spike	NDV-HXP-S	Sean Liu, Icahn School of Medicine at Mount Sinai	Phase 2/3 NTC05354024	In
Others				
▪ *DNA based vaccine*				
S protein	BacTRL-Spike DNA vaccine	Symvivo Corporation	Phase 1/2 NTC04845191	or
Bacterial antigen-spore expression vector				
Spike	*B. subtilis spores*	DreamTec Research Limited	Phase 1 NA	or

2.1. Mucosal Vaccine Adjuvants

The induction of robust immune responses after immunization by mucosal routes requires the antigen on the mucosal surface for its transport across the epithelial barrier. For this reason, the rational design of mucosal vaccines and adjuvants must consider the specific problems related to the route of administration, such as changes in pH, enzymatic degradation, or entrapment in the mucus layer that limit the absorption of antigens. Consequently, it is a challenge to find effective mucosal adjuvants that allow overcoming the limitations associated with the mucosal barrier, enhancing local and systemic immune responses.

Several parenteral adjuvants have been approved for use in humans, including adjuvants based on aluminum hydroxide salts and gels, virosomes, oil-in-water emulsions (MF59®, AS03, MontanideTM ISA 51 VG, and ISA 720 VG) [30–32], monophosphoryl Lipid A (MPL) [33], Adjuvant Systems® (an immunostimulant combination of AS01 (MPL and QS-21) [34] and AS04 (MPL and aluminum hydroxide)) [35], CpG1018 [36] (a synthetic TLR-9 agonist adjuvant (Dynavax®)), and recently, Matrix-MTM (saponins from *Quillaja saponaria*), which were authorized in a recombinant vaccine for SARS-CoV-2 [37]. However, none of these adjuvants have been licensed for human mucosal use, although several mucosal vaccine adjuvants are currently under pre-clinical evaluation [38–40]. The rapid progress in SARS-CoV-2 mucosal vaccine development has allowed more adjuvants to be taken to the clinical phase, such as adenoviral vector as self-adjuvant [41], artificial-cell-membrane polymersome-encapsulated CpG (NCT05385991) or membrane vesicles from *Neisseria meningitis* [42] (Table 1). In the pre-clinical phase, N-N-N-trimethyl-chitosan particles designed for mucosal administration are under investigation [43].

Use of Particulate Systems as Mucosal Vaccine Adjuvants

Encapsulation, entrapment, or conjugation of antigens within particulate systems is one of the most promising approaches for the mucosal administration of vaccines. Particles protect the antigens from in vivo enzymatic degradation, prolong the residence time in mucosa favoring delivery and absorption [44], promote the transport of antigens and cells to lymph nodes [45], improve antigen presentation, and enhance immunogenicity [46]. All these benefits are reflected in greater vaccine efficacy [47–51]. Scheme 1 summarizes the general classification of particle-based systems used commonly as vaccine adjuvants.

The performance of these particles as adjuvants is usually determined by their intrinsic characteristics, such as size [52–54], shape [55,56], surface charge [57], and hydrophobicity [58,59], and also are influenced by the methods used for antigen loading [60], the density of antigen on the surface [61], the ability of controlled-release kinetics [62], functionalization [63–65], and importantly mucoadhesion [66,67]. In many cases, particulate vaccine adjuvants mimic the size, shape, and surface molecule organization of pathogenic microorganisms and can contain molecules such as pathogen-associated molecular patterns, which directly impact the recognition, interaction, phagocytosis, and processing of antigens by antigen-presenting cells (APCs), affecting vaccine efficacy [68–70].

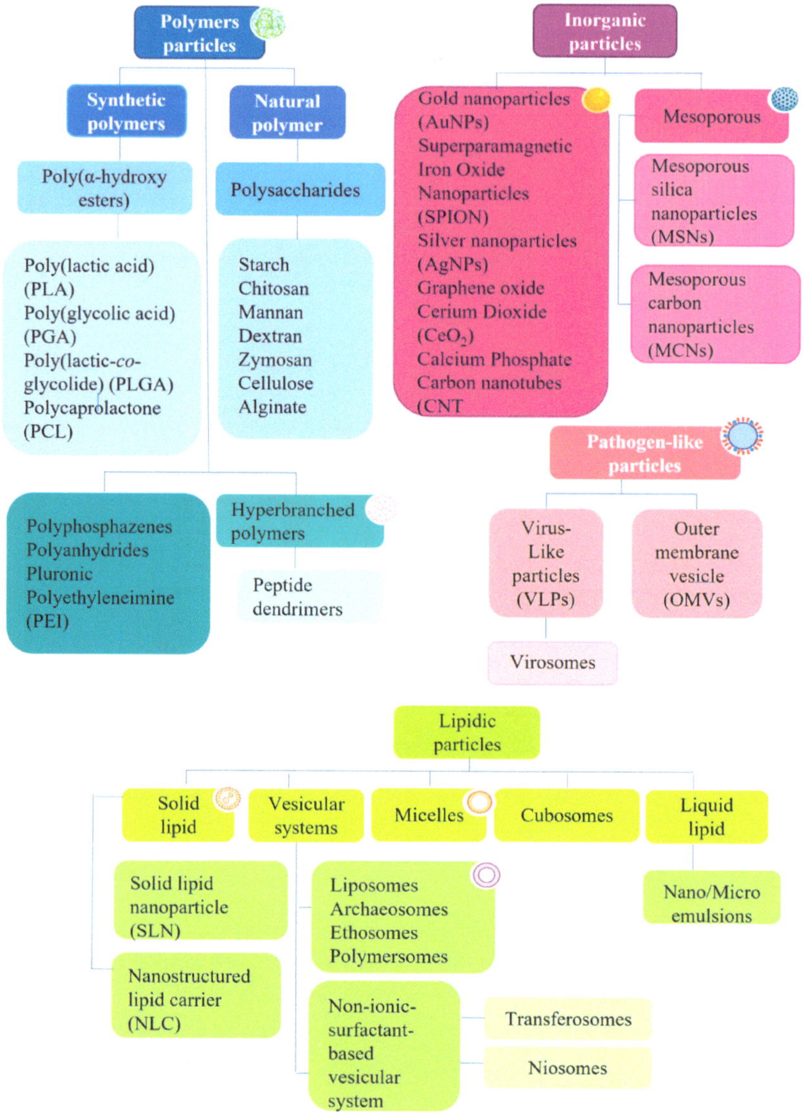

Scheme 1. Material types of particulate systems used as mucosal adjuvants. PCL, polycaprolactone; PLGA, poly (lactic-co-glycolide); PGA, poly (α-L-glycolic acid); PLA, poly (lactic acid); γ- PGA, poly (glutamic acid); PEI, polyethyleneimine; VLPs, virus-like particles; MSNs, mesoporous silica nanoparticles; MCNs, mesoporous carbon nanoparticles.

3. Mucoadhesion

The mucus layer covering the mucosal epithelium is mainly synthesized and secreted by goblet cells [71,72]. Mucus is composed of water (>95%), electrolytes, enzymes, salts, DNA, lipids, growth factors, antimicrobial peptides, immunoglobulins, and mucins, the most abundant high-molecular-weight glycoproteins of the extracellular mucus [73,74]. The mucus layer's composition, thickness, viscosity, and rheological properties vary widely among mucosal tissues. For instance, the nasal mucus is thinner (10 μm thick), making it highly permeable compared to the mucus layer along the gastrointestinal tract, whose thicknesses range from 180 μm to 800 μm from the antrum to the colon, respectively [75,76]. The mucus's rheological properties also vary according to the anatomical site and the type, composition, and properties of the mucins [77]; therefore, the transport of microorganisms, molecules, particulate matter, drugs, exogenous, and endogenous agents through the mucus is also different. For a better understanding of essential functions, general features, and distribution according to the anatomical location of mucins, see Table 2.

Table 2. Overview of mucins. The structural and functional details of mucins are described. Mucins (MUC) are differentially expressed in mucosal tissues. PTS domains: proline/threonine/serine-rich domains; VNTRs: variable number of tandem repeats; GalNAc: N-acetylgalactosamine; OH-: hydroxyl groups; S: serine; T: threonine.

MUCINS	
General characteristics	▪ High-molecular-weight glycoproteins (10–40 MDa) ▪ Molecular organization: a central region with PTS domains with a variable number of VNTRs ▪ Domains: Highly glycosylated via O-glycosidic linkages between GalNAc and the OH- of S and T residues from the protein core ▪ In the gastric mucus, some of these sugars bear sulfate ester groups ▪ A high proportion of cysteine domains participate in dimer formation and form multimers
Classification	▪ Membrane-bound/transmembrane ▪ Secreted/gel-forming ▪ Soluble
Biological functions	▪ Prevents the adhesion of pathogens, foreign debris, or cells to the mucosal epithelium by steric hindrance ▪ Establishes a selective filter for the diffusion of particles and small molecules ▪ Bound by and enriched with nutrients, allowing them to harbor beneficial microbial communities ▪ Attenuates the virulence of opportunistic microorganisms ▪ Regulates signal transduction
Anatomical location	**Type**
Oral cavity	MUC1, MUC4, MUC5B, MUC7, MUC19
Eye	MUC1, MUC4, MUC16, MUC20, MUC21, MUC22. MUC2, MUC5B, MUC5A, MUC5B, MUC5AC, MUC7, MUC19
Respiratory tract	MUC1, MUC4, MUC16, MUC20, MUC5AC, MUC 5B, MUC19
Stomach	MUC1, MUC5C, MUC6
Small intestine	MUC13, MUC17, MUC2
Colon	MUC2
Female reproductive tract	MUC1, MUC4, MUC5
Male urogenital tract	MUC1, MUC19, MUC20, MUC5B, MUC6

Some synthetic and natural materials/macromolecules and hydrocolloids adhere to biological surfaces [78] and remain attached for an extended period of time by interfacial forces. When the adhesive attachment occurs with mucus or epithelial tissues, the phenomenon is considered mucoadhesion (Figure 2) [79] and involves an interaction with mucin.

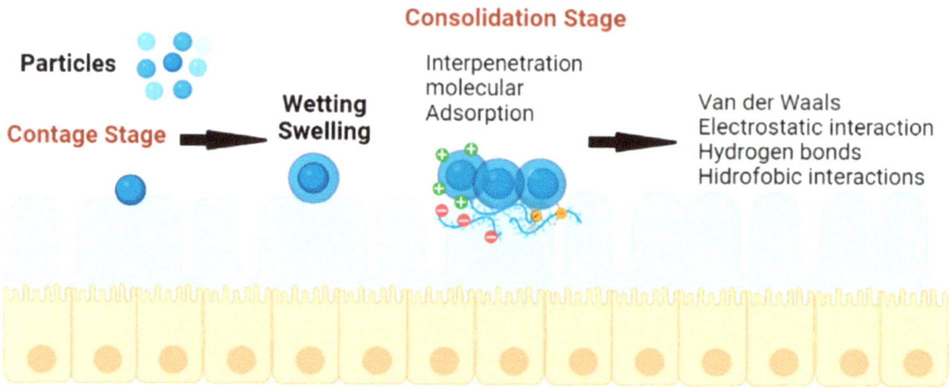

Figure 2. Mucoadhesion of polymeric particles to mucous membranes. First, an intimate interaction between polymeric particles and the mucous membrane occurs in the contact stage, with subsequent wetting and swelling. Later, in the state of consolidation, physicochemical interactions enable the adsorption and molecular interpenetration of polymer chains into the mucus layer that covers the epithelial surface, eventually leading to prolonged adhesion.

Different theories try to explain the interactions between bioadhesive polymers and mucosal surfaces from both physical and chemical perspectives: the electronic theory [80], the adsorption theory [78], wetting theory [81,82], diffusion theory [83], and fracture and mechanic theory [84] (details in Figure 3). In all of them, the molecules must bind through the interface, an interfacial layer formed between the adhesive and the mucosal tissue. However, the links between the polymers and the mucus differ in each theory.

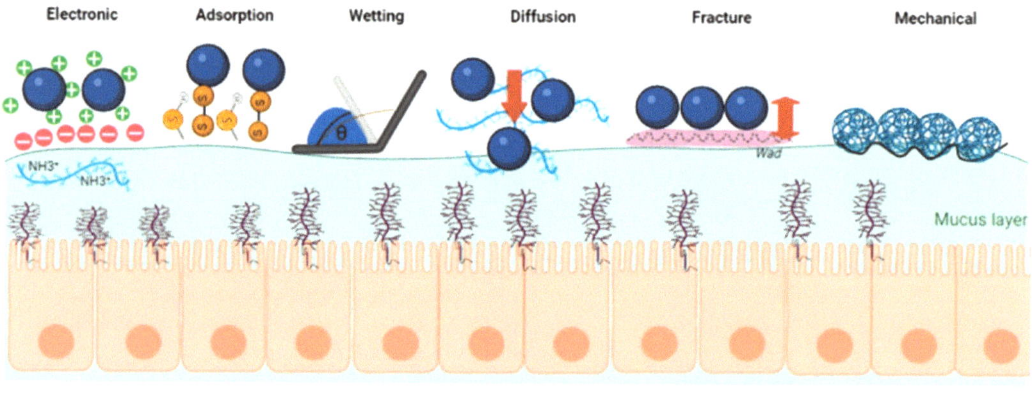

Figure 3. Theories of mucoadhesion. The electronic theory describes mucoadhesion mediated by electrostatic forces. Adsorption assumes intermolecular contact at the interface and adhesion of materials by intermolecular forces. Wetting is described in terms of the spreading coefficient; therefore, the requirement for mucoadhesive materials to adhere is that these can spread spontaneously onto a surface. The diffusion theory: interdiffusion and interpenetration through the polymer-mucin interface according to their concentration gradient. Fracture describes the forces required to separate two surfaces after bonding and assumes that adhesion bond failure occurs at the interface (*Wad*: work adhesion). Mechanical assumes that the irregular rough or abrasive substrate surface provides mechanical keying.

In addition to pharmacokinetic studies, in vivo and ex vivo methods for assessing mucoadhesion allow for the direct investigation of the particulate systems' adhesion to mucosal tissues [85,86] or mucosa-mimetic materials [87]. Likewise, it is possible to perform in vitro determinations that depend mainly on the physical properties of the polymers [88]. Several detailed reviews can be consulted to learn more about the methods used in mucoadhesion [89–91].

Determining Factors in the Mucoadhesion and Mucopenetration of Polymeric Particulate Systems

The mucoadhesive properties of particulate adjuvants can be affected by the physicochemical characteristics of polymers, such as size, ζ-potential, elasticity, molecular weight, or spatial conformation, as well as by environmental factors, such as pH and presence of metal ions, and physiological factors, including mucin turnover. The particles are trapped in mucus networks through polyvalent adhesive interactions [92]. Sulfate groups on N-acetylglucosamine and galactose and carboxylic groups on sialic acid sugars confer negative charges to the mucin under most pH conditions [93]. Hydrophilic particles adhere to the negatively charged moieties, while hydrophobic particles are captured by low-affinity bonds between hydrophobic protein residues and particle surfaces. Although cationic polymers have shown better mucoadhesive properties [94], anionic polymers also attach to mucin just as much as cationic or nonionic polymers. This occurs thanks to surface carboxyl groups in mucin that permit interaction via hydrogen bonds with the oligosaccharide chains [95]. Mucus limits the diffusion of particles of any size, so it seems that size has a more significant contribution to mucopenetration than to mucoadhesion. Thus, the transport rate of particles in mucus decreases with increasing particle size [96,97].

The surface functionalization of particles affects adhesion and permeation across the mucus of particles of equal size, i.e., as expected, polystyrene NPs of a small size (100 nm) penetrate the mucus better than large particles (500 nm); however, among 500 nm particles, sulfate-functionalized particles were 1.7 times more permeable than amino-functionalized particles [98].

On the other hand, cylindrical-shaped NPs and rod-shaped nanocrystals have been shown to penetrate the mucus more efficiently compared to their spherical counterparts with similar particle sizes and surface charges [99]. Similarly, compared to their soft and hard counterparts, particles with moderate stiffness have a higher diffusivity through the mucus [100].

4. Mucoadhesive and Mucopenetrating Polymer-Based Adjuvants

For at least the past four decades, multiple research groups have searched for materials for the development of mucoadhesive and/or mucopenetrating pharmaceutical formulations to improve the bioavailability of active constituents [101–104]. Both the adhesive and mucopenetrating properties of particulate adjuvants allow them to reach the epithelial barrier. Once there, the particulate can be transported to the basolateral side to initiate mucosal immune responses.

Several materials commonly used in the pharmaceutical industry are also used as mucosal vaccine adjuvants; however, few studies have been devoted to evaluating the direct influence of mucoadhesion and mucopenetration on the strength and quality of the antigen-specific immune responses stimulated after mucosal vaccination. Consequently, we provide an overview of polymer-based particles in the following sections. Based on the available experimental findings, we analyzed the association between increased mucoadhesive strength, mucosal penetrability, and enhanced immune response quality after mucosal vaccination.

4.1. Chitosan and Chitosan Derivatives

Chitosan (CS) is a natural cationic polysaccharide obtained by the deacetylation of chitin. CS has been widely used in various biomedical applications due to its biodegradability, biocompatibility, low toxicity, immunogenicity, and mucoadhesive properties [105–107].

The mucoadhesive properties of CS are attributed to the protonation of the amino groups in weakly acidic media, which interact with the negatively charged sialic acid moieties of mucin. However, different chemical processes have been applied to CS to improve its application limitations, such as high hydrophilicity, low solubility from pH 7.4, high degree of swelling, and thermal stability [108,109]. These modifications, in turn, are favorable to promoting adhesion to mucosal surfaces and, as a result, enhance mucosal contact time.

For example, the chemical conjugation of CS with hydrophilic ethylene glycol branches improves solubility in water at neutral and acid pH values and its steric stability [110,111]. Pawar and Jaganathan (2016) compared the immunogenicity of CS NPs and glycol chitosan (GC) NPs loaded with a hepatitis B surface antigen (HBsAg) after nasal administration in Balb/c mice. While the anti-HBsAg antibody titer induced by HBsAg alone was minimal, HBsAg conjugated with GC NPs significantly increased serum IgG and IgA antibody titers in nasal, saliva, and vaginal secretions, compared to the CS-conjugated group. Splenocytes isolated from mice immunized with GC NPs and CS NPs secreted significantly higher amounts of IL-2 and IFN-γ than the control mice immunized with Alum-adsorbed HBsAg. Nasal clearance studies of radiolabeled particles in rabbits showed a nasal cavity retention time of up to 240 min for GC NPs (20% radioactivity) compared to 180 min for CS NPs (20% radioactivity) and 90 min for HBsAg alone (<20% radioactivity). In confirming nasal deposition after nasal administration in mice, only formulations with GC and CS NPs were retained in the NALT at 30 min, with higher fluorescence intensity for GC NPs than FITC-BSA [112].

Similarly, adding cross-linking agents, such as tripolyphosphate (TPP), improves the encapsulation efficiency during the elaboration of CS NPs [113]. Co-crosslinked vanillin/TPP was used for developing a trivalent oral vaccine (DwPT). Studies of the adhesion behavior of the microspheres were related to the ζ-potential of the groups, the electrostatic interaction between the positively charged CS and the negatively charged sialic acid of mucin, and the degree of cross-linking. Thus, the highest swelling index was for the group with the lowest degree of cross-linking. Batches with electropositive charge (placebo CS microspheres, diphtheria toxoid (DT) CS microspheres, and tetanus toxoid (TT) CS microspheres: $\sim$+30 mV) showed a higher adhesion to mucin than those with ζ-potential around +10 mV (whole-cell pertussis (wP) CS microsphere and trivalent (DwPT)). Antibody response in serum corresponded to the mucoadhesion of the microspheres, developing a higher IgG antibody titer in TT and DT batches on days 28 and 35 after immunization, followed by batches with a lower adherence (PT: pertussis toxin). This response was consistent with that observed in saliva and intestinal secretions [114].

Other derivatives of CS have been developed to improve, specifically, absorption and bioadhesion properties. Currently, the most used are obtained by quaternization, acylation, thiolation, and carboxymethylation [109]. Trimethyl chitosan (TMC), a quaternized derivative of CS with polyampholytic properties, improves CS solubility without affecting its mucoadhesive cationic nature, reduces cytotoxicity, and enhances absorption on mucosal surfaces in a wide range of pH values, increasing the carrying capacity [115–117].

In 2010, Vyas laboratory used PLGA microparticles (MPs) coated with CS and TMC for the intranasal administration of HBsAg to mice. While unmodified PLGA MPs had a negative ζ-potential (-14.4 ± 1.2), the coating with CS and TMC increased the ζ-potential to values between +5 mV and +10 mV for PLGA/CS MP and +10 mV and +20 mV for PLGA/TMC MP. The authors also indicated that the ζ-potential directly influenced the adsorption capacity of MPs to mucin, i.e., PLGA MPs showed insignificant mucin retention, while CS-PLGA and TMC-coated MPs had significantly higher mucoadhesive properties. Remarkably, this increase in mucoadhesion improved the immunogenicity of the formulation. However, PLGA/TMC MPs induced substantially higher antibody IgG titers throughout the study than PLGA/CS MPs, both in serum and distal mucosal sites [118]. A second study found the same results with PLGA/TMC NPs and demonstrated the adjuvanticity effect of TMC through the stimulation of dendritic cell maturation. Furthermore, TMC-coated MPs were selectively taken up by M cells in the NALT following nasal administration compared to the FITC-BSA solution, which would substantially explain the enhancement of vaccine formulations' immunogenicity [119].

Another quaternized CS derivative is N-[(2-hydroxyl-3-trimethyl ammonium) propyl] CS or HTCC. HTCC polymers have different degrees of quaternization or extent of positive charge [116]. Zhang et al. prepared OVA-loaded curdlan sulfate-O-HTCC NPs as an intranasal vaccine system. Although the inclusion of curdlan, a β-glucan capable of activating innate immune cells via Dectin-1 receptors and TLR-4 [120], could promote the antigen-specific immune response, its negative surface charge was considered a limitation for mucosal application. For this reason, O-HTCC was added, which, in addition to conferring a positive ζ-potential on the particle, improved its adhesion and subsequent internalization by epithelial cells due to its high viscosity. The OVA–curdlan–O-HTCC complex led to higher OVA-specific CD4+ T-cell, CD8+ T-cell, and B-cell proliferation when nasally administered to mice, compared with the proliferation induced by OVA, OVA–curdlan, OVA–CS, or CS–curdlan [121].

Carboxymethyl chitosan (CMCS) is another water-soluble CS derivative with an improved degradation rate, a desired characteristic for its use in vivo [122]. Recently, CMCS was also used to coat the surface of calcium phosphate (CaP) NPs. The electrostatic interactions and hydrogen bonds between mucin and CaP–CMCS–BSA allowed in vitro adhesion close to 90% compared to CaP–BSA adhesion (60%). Additionally, the diffusion efficiency was higher for CaP–CMCS–BSA than for CaP–CS–BSA, CaP–BSA, and BSA alone. The coating with CMCS and CS improved the apparent permeability coefficient in the mucus layer at 2 h, an index of apical to basolateral transport. Ex vivo biodistribution in a rat study showed that CaP–CMCS–BSA/FITC absorption was improved in the small intestine at 2 h compared to CaP–CS–BSA/FITC, attributed to the change in surface charge caused by coating with CS and its derivative (CaP–CMCS–BSA, ζ-potential: −4.7 mV vs. CaP–CS–BSA, ζ-potential: 8.5 mV). These findings are correlated with the efficacy of oral vaccination since the levels of IgG and sIgA antibodies in sera and feces, respectively, increased after each boost in the animals that received CaP–CMCS–OVA compared to OVA alone [123].

For their part, methyl CS has been studied for diverse biological activities, including as tissue regeneration activator, absorption enhancer, and mucoadhesive [124]. Suksamran et al. evaluated methylated CS MPs for entrapping OVA. Calcium alginate MPs–OVA, calcium alginate–yam starch microparticles (YMP)–OVA, and (YMP)–OVA coated with methylated N-(4-N, N-dimethylaminocinnamyl) CS ($TM_{65}CM_{50}CS$) were used in this work. The evaluation of swelling showed that the degree and rate of swelling of the $TM_{65}CM_{50}CS$-coated MPs were higher than those uncoated, both in HCL pH 1.2 and in PBS pH 7.4. Similarly, the in vitro mucoadhesion study using the everted gut sac with porcine jejunum showed that, while the adherence percentages of calcium alginate MPs and YMP MPs were low (29.62% and 11.29%, respectively), the coating with $TM_{65}CM_{50}CS$ of both particles increased mucosal adhesion during the first hour (45.64% and 43.38%, respectively). Oral immunization resulted in significantly higher IgG and IgA levels in mice receiving OVA-loaded $TM_{65}CM_{50}CS$-coated MPs, which again confirms the role of mucoadhesive polymers in immunogenicity [125].

The ζ-potential of the CS-based vaccines significantly influences the induction of an immune response affecting more than one mechanism. Jesus et al. demonstrated that. after the intranasal administration of polycaprolactone/CS (PCL/CS) NPs in C57BL/6 mice, the lowest dose of adsorbed antigen (1.5 μg HBsAg) induced antibody titers comparable to the dose containing six times more adsorbed antigen (10 μg HBsAg). Furthermore, this group had the highest number of responding animals. However, serum IgG titers were significantly low compared to previous studies with the same dose of antigen (1.5 μg HBsAg), so the authors suggested that the decrease in ζ-potential (CS: +30 mV) to values close to neutrality generated by antigen interaction (PCL/CS: +26 mV; PCL/CS: 1.5 μg HBsAg: +22 mV; PCL/CS:10 μg HBsAg: +5.7 mV) leads to a reduced uptake in the epithelial barrier. These observations were independent of the mucoadhesive behavior of the particles without antigen evaluated in vitro. Therefore, the authors suggested that the antigen on the particle's surface reduces the ζ-potential and hinders the interaction with mucin in vivo, avoiding particle–cell interactions and ultimately impacting the immune response [126]. Although this finding contradicts what was observed for other CS-based particles reviewed, it highlights the importance of assessing the mucoadhesion of the polymeric system alone, as well as the particle-entrapped antigens of interest.

4.2. Cellulose Derivatives

Carboxymethylcellulose (CMC), an anionic and water-soluble cellulose derivative [127], has been successfully used as a mucoadhesive polymer to enhance immune responses. Hanson et al. developed CMC and alginate (ALG) wafers loaded with the HIV gp140 protein and with α-GalSer as an adjuvant. In ex vivo tests with porcine sublingual mucosa, wafers with a higher CMC content withstood intense mucosal washings and had a higher tissue penetration of the coupled protein (fluorescently labeled bovine serum albumin (BSA)) compared to wafers with a higher ALG content and the free protein. However, the presence of ALG in the formulation was necessary to maintain protein stability on the wafer. Following sublingual administration in mice, most mucoadhesive wafers generated a greater T-cell response in the lungs and cervical lymph nodes [128]. In other studies, it has been suggested that CMC's viscosity and anionic structure allows the formation of ionic bonding and hydrogen bonds with mucin layers [129–131].

4.3. Mannan-Decorated Polymeric Particles

Similar results have also been achieved using the dual immunostimulant and mucoadhesive capacity of mannan isolated from the cell wall of *Saccharomyces cerevisiae* [132]. Mannans present immunostimulatory activity via pathogen recognition receptors (PRRs) in APCs. An in vivo optical imaging system, following the intranasal administration of thiolated hydroxypropylmethylcellulose phthalate microspheres (Man-THM), showed that mannan decoration increased the residence time of Cy5.5-conjugated OVA-loaded Man-THM in the respiratory mucosa compared to OVA alone or OVA-loaded THM. Subsequently, the mucosal immune response was evaluated following the nasal immunization of the ApxIIA toxin from *Actinobacillus pleuropneumoniae* loaded in the MPs groups. The findings also demonstrated that the microspheres reached the lungs and secondary lymphoid tissues and induced systemic IgG and secretory IgA responses to the ApxIIA in bronchoalveolar lavage (BAL) and nasal and vaginal washes. Although the immunostimulatory role of mannosylation in enhancing immunogenicity has been reported [133,134], in this work, the authors highlighted the mucoadhesion of the mannosylated microspheres to explain the improved immunogenicity in vivo.

4.4. Alginate Coating

Vyas and his team (2014) assessed the coating of CS MPs with alginate (A-CSMp). In contrast to most of the works reviewed up to this point, where the positive surface charge plays a fundamental role in adhesion to mucin, alginate as an anionic polyelectrolyte changes the ζ-potential of the particle to an electronegative value (−29.7 mV). FITC-BSA was rapidly washed from rat jejunal tissues; however, the in vitro retention time in the mucosa was prolonged when FITC-BSA was associated with A-CSMp. In the same way, in the in vivo assays, only A-CSM loaded with FITC-BSA successfully generated uptake by M cells in Peyer's patches. When evaluating the efficacy of the particulate system in an oral anthrax vaccination model, high-titer anti-PA serum IgA and IgG antibodies were observed in animals receiving particles loaded with antigens compared to the free *Bacillus anthracis* protective antigen [135].

Similarly, Saraf et al. loaded alginate-coated CS NPs (ACNPs) with HBsAg anchored to E. coli EH-100 lipopolysaccharide (LPS) (LPS-HBs-ACNPs) as an adjuvant for oral administration. As expected, the alginate coating changed the ζ-potential of the NPs from +45.2 mV (0.5% CS-0.1% TPP) to −26.2 mV (0.5% CS-0.1% TPP-2% alginate-2%

LPS) due to the negatively charged -COO- electrostatic interaction of the alginate on the positively charged -NH3 of the CS. Despite the ζ-potential's more negative values, in vitro mucoadhesion studies showed that alginate-coated NPs were more mucoadhesive than CS NPs alone. Although anti-HBsAg serum IgG titers were higher for HB-ACNPs after oral administration, sIgA antibody titers in mucosal secretions were higher for LPS-HBs-ACNPs. The anchoring of LPS targeted the NPs to M cells, conferring immunogenicity to the system [136] independently of the mucoadhesive properties of ACNPs. As in the case of LPS, any ligand can be anchored to the particulate system to target it and to allow specific binding to M cells or mucosal epithelial cells. Excellent reviews have been conducted on this topic [137–139].

On the other hand, sodium alginate protects the NPs from the hostile environment of the gastrointestinal tract, the same as the introduction of hydrophilic groups, such as hydroxyalkyls, carboxyalkyls, succinyls, and thiols, or polymer grafts, such as PEG. In this way, Amin and Boateng (2022) proposed a system based on OVA-loaded CS NPs coated with sodium alginate or PEG for oral vaccine administration. Both sodium alginate and PEG coatings increased the stability of NPs upon exposure to gastric fluids with the protection of the encapsulated protein (4 h and 1 h, respectively), compared to uncoated NPs (<30 min). After transfer into simulated intestinal fluid, both coatings showed stability for 120 h, although with different release profiles of OVA. Increased alginate concentrations were related to a higher level of mucin binding. According to the authors, the alginate coating ensures stability, allows a higher antigen load to reach the site for mucosal immune response, improves mucoadhesive properties, and enhances the sustained release of antigen-loaded NPs [140].

4.5. Xyloglucan

Xyloglucan (XG), a non-anionic polysaccharide and the main hemicellulose component, has been applied with *Quillaja* saponins to vaccine formulations against brucellosis. While *Brucella* LPS was weakly immunogenic, when *B. abortus* LPS-loaded XG NPs were administered nasally to Balb/c mice, higher systemic and mucosal IgG antibody levels and mucosal IgA were induced. Increased immunogenicity was associated with a greater mucoadhesion force of the XG and the LPS-XG NPs compared to the LPS alone, as well as the ex vivo retention of LPS-XG NPs over 24 h in goat mucosa [141].

As in the case of CS, XG has been previously used in pharmaceutical applications in different formulations and by different routes, including mucosal, transdermal, and intraperitoneal, due to its biodegradability, cost-effectiveness, and non-toxicity. Some authors have suggested that the XG molecular structure, "mucin-like," is responsible for mucoadhesive properties, including swelling capacity and increasing concentration-dependent viscosity [142,143]. All these characteristics, taken together, expand the possibility of the future use of XG in mucosal vaccinations [144].

4.6. Poly (Acrylic Acid) and its Derivatives

Poly (acrylic acid) and its derivatives have excellent mucoadhesive capacity compared to cellulose, polycarbophil, chitosan, and pectin [145–147]. An example is Carbopol®, a highly cross-linked hydrophilic polymer, which provides it with mucoadhesive and viscoelastic properties. Coucke et al. used spray-dried powders of amylopectin (Amioca®) with polyacrylic acid (Carbopol® 974P) in different proportions (SD 0/100, 25/75, 50/50, 85/15, and 100/0) for the intranasal administration of H3N2-inactivated influenza virus and in combination with the LTR192G adjuvant in rabbits. The formulation SD25/75 induced the highest serum response of IgG anti-haemagglutinin compared to the formulation SD100/0, thus highlighting the importance of polyacrylic acid. Despite this, neither SD25/75 nor SD0/100 induce a local mucosal response. The immune response was directly related to the negative ζ-potential of Carbopol® 974P and the mucoadhesive properties of the formulations. The reticulated, predominantly elastic, or highly structured characteristics of SD25/75 (G′ >> G″) increased the residence time in the nasal cavity. In contrast, the lowest viscosity and cross-linking of SD100/0 were associated with a low mucosal retention [148].

4.7. γ-PGA

The poly-γ-glutamic acid (γ-PGA)-based vaccine adjuvant, an anionic biopolymer, was used for the intranasal delivery of the influenza fusion protein sM2HA2 and OVA, co-administered with 3-O-desacyl-4′-monophosphoryl lipid A (MPL) and QS21 in a system denominated γ-PGA/MPL/QS21(CA-PMQ). Using in vivo single-positron-emission computed tomography imaging, it was possible to determine that γ-PGA increased the OVA residence time by up to 12 h in the nasal cavity. This signal decreased at 6 h when OVA was administered alone. This result is correlated with the higher serum IgG, IgG1, and IgG2a antibody responses in the groups vaccinated with OVA/CA-PMQ and sM2HA2/CA-PMQ compared to the groups that received OVA and sM2HA2 alone, as well as being superior to that induced by the cholera toxin used as a mucosal adjuvant. Likewise, animals vaccinated with the antigen/CA-PMQ induced more IL-4 and IFN-γ–secreting cell populations in the spleens stimulated with OVA, sM2, and HA2 protein than mice immunized with proteins alone or the control group. Additionally, the CA-PMQ induced high titers of sM2HA2-specific IgA antibodies at the administration and distal sites, along with an increased survival time (80–100%) following the challenge with influenza A subtypes and cleared pulmonary viral titers [149]. The presence of carboxyl groups within γ-PGA can provide a strong interaction with the mucus layer.

The anionic model (−35.5 mV) of Kurosaki et al. with benzalkonium chloride (BK) and γ-PGA NPs in a complex with OVA (OVA/BK/γ-PGA) was used for pulmonary administration. They observed an increased fluorescence intensity in the lung (Alexa647-OVA/BK/γ-PGA) indicative of lung deposition compared to Alexa647-OVA. OVA/BK/γ-PGA increased the levels of specific IgG antibodies, while in the animals that received OVA or the vehicle (BK/γ-PGA), anti-OVA IgG was not detected. The induction of immune responses at the mucosal site was also significantly higher in the OVA/BK/γ-PGA group [150]. Their study did not discuss the role of γ-PGA mucoadhesion in the results obtained. However, the authors suggest the uptake efficiency of BK/γ-PGA NPs by the antigen-presenting cells in the alveolar region. Due to the high capture efficiency of particles <2 μm in the lung [151], the adhesion phenomenon could favor the increased particle residence time in the lung mucosa. Evaluating bioadhesive properties in these systems could help to improve rational vaccine design using polymeric particles.

4.8. Thiolated Polymers

The previously reviewed polymer-based adjuvants could be thiolated to improve mucoadhesion. In the past two decades, important research has been conducted using thiolated polymers or so-call "thiomers", mainly in excipients for drug delivery. Thiomers can interact with mucin through disulfide bonds with the cysteine-rich subdomains of mucus glycoproteins [152]. These covalent bonds are supposed to have stronger binding than the non-covalent interactions that are formed between the polymers and the sialic acid of the mucus layer [153], improving the mucoadhesive properties of the polymers.

Using a tensile test and rotating cylinder method to obtain compressed tablets, Roldo et al. demonstrated that increasing the number of thiol groups covalently attached to chitosan-4-thio-butyl-amidine conjugated significantly improves mucoadhesion compared to unmodified CS. Thiolation increased the total adhesion work (TWA, μJ) up to 100 times [154]. Similarly, thiol reactivity impacts mucopenetration. When the thiol reactivity is medium to low, extensive interpenetration occurs in the mucus layer, with a larger interface for disulfide bond formation. Conversely, highly reactive thiols have difficulty penetrating through the mucus because they form disulfide bonds with the mucins on the surface of the mucus layer, facilitating their rapid removal through mucus turnover [155].

In a recent study, Sinani et al. immunized Balb/c mice with BSA-loaded NPs prepared using aminated CS (aCS) and aminated and thiolated CS (atCS) polymers; mice were nasally immunized at 14-day intervals. At the end of the experiment (day 253), the nanoparticles (aCS and atCS) induced a more robust systemic response, resulting in an almost two orders of magnitude higher systemic IgG titer than the BSA/CpG ODN control, with atCS being the best. These results are correlated with the increased mucoadhesion observed in the aCS and the atCS. Both aCS and atCS modulated the Th2 immune response and enabled immune response at distal mucosal sites [156].

Cellulose acetate phthalate (CAP) is widely used as an enteric coating for pharmaceutical dosage forms due to its solubility at pH values above 6 (such as in the intestines) but poor water solubility at a low pH (such as in the stomach). After exposure to intestinal fluids, the polymer swells, with the subsequent softening or complete dissolution of the phthalate, allowing the release of the biologically active compounds [157]. Lee et al. orally immunized mice with M5BT, a chimerical multi-epitope recombinant protein of foot-and-mouth disease virus (FMDV), alone, loaded in thiolated CAP MPs (T-CAP), or loaded in non-thiolated MPs (CAP). In ex vivo studies in the porcine intestinal mucosa, T-CAP mucoadhesion was 1.48-fold higher than CAP. The improvement in the mucoadhesion properties was reflected in the highest production of antigen-specific IgG antibodies in animals that received M5BT/T CAP. Similarly, this group of animals had significantly higher levels of anti-M5BT IgA in fecal samples at 2 and 4 weeks due to the longer transit time of antigens in the mucosa and increased MHC class II- expression on APC in PPs, related to IgA production [158].

For cationic thiomers such as atCS, the interactions are predominantly driven by electrostatic forces. In contrast, for anionic thiomers, such as T-CAP, interaction with the mucus occurs through hydrogen bonds, van der Waals interactions, and chain entanglement. In both cases, the bioavailability is improved by the extension of the residence time [159]. Notably, regardless of the surface charge of the polymer particles and resulting surface forces, the thiolation of both polymers improved in vivo immune response.

Further evidence has shown that thiomers are susceptible to thiol oxidation at pH ≥ 5, with their effectiveness being reduced following oral administration. Typically, thiol groups (R-SH) can form disulfide bonds with mercaptopyridine substructures, whereby thiol groups are stabilized against oxidation and increase their reactivity. S-protected thiomers, so-called "pre-activated", have shown greater mucoadhesion than unprotected thiomers, according to Iqbal et al. (2012) [153]. In this work, Iqbal et al. synthesized a polymer with improved mucoadhesive, cohesive, and in situ gelling properties. For this purpose, poly (acrylic acid) (PAA), PAA-cysteine (PAA-cys), and 2-mercaptonicotinic acid (2MNA) coupled with PAA-cys (PAA-cys-2MNA) were compressed into tablets, and the mucoadhesion strength was determined by the rotating cylinder method. Adding thiol groups improved the mucoadhesive properties 456-fold, while the S-protected thiomer increased the contact time to 960-fold compared to unmodified PAA. These thiolated nanosized carriers and others, such as thiolated cyclodextrins [155,160], are research fields that may be explored further for mucosal vaccine development.

5. Enhancement of Epithelial Permeability by Polymer-Based Adjuvants

Although mucoadhesive molecules improve the bioavailability of drugs and antigens administered via the mucosa, the mucus layer still limits passage into the epithelium. The transit time of particles in the mucosa is determined by the physiological renewal time of the secreted mucus layer [161]. Mucus turnover reduces the mucosal residence time of particulate delivery systems because they can be trapped by the mucus and rapidly eliminated [162], which could compromise their effectiveness as mucosal adjuvants.

Therefore, polymer-based adjuvants are expected to adhere to the mucous layer, penetrate the epithelium, and reach the inductive sites for mucosal immune responses before being removed. Hence, this section briefly describes the strategies to facilitate mucus barrier penetration and improve the permeability of polymer-based adjuvants once they are in the mucosa.

5.1. Mucus-Penetrating Particles

Particles with a low adhesion and small size, thus with few steric hindrances to the mucin network, are often referred to as mucus-penetrating particles. Unlike mucoadhesive particles, mucus-penetrating particles seek to minimize the strength of electrostatic and hydrophobic interactions with the mucin. Polymers with neutral or low positive charges are generally included in the design of mucus-penetrating formulations. Several studies have reported the surface coating of particles with PEG. PEG is used as an adhesion promoter acting at the interface to improve adhesion. Hence, PEG chains tethered or grafted are covalently attached at one end on the polymer surface while the other is free, allowing PEG to diffuse from the polymer network to the mucus and enhancing interpenetration [163]. Wang et al. further demonstrated the formation of hydrogen bonds between the ether oxygen atoms of the PEG chain and glycosylated proteins of mucins. Additionally, they reported PEG with a low molecular weight (2 and 10 kDa), near-neutral surface charge (ζ-potential of 2 ± 4 and 1 ± 3 mV, respectively), minimized mucoadhesion by reducing hydrophobic hydrogen bonding, and electrostatic interactions to have better mucus-penetrating properties. The authors even proposed that PEG-covered particles between -10 and -7 mV are within the interval that defines mucoadhesive vs. mucoinert characteristics [164].

Despite its widespread use in over-the-counter drugs and vaccines, recent approaches suggest that PEG is not immunologically inert [165–167]. Several authors demonstrated that introducing PEG to mucosal vaccine formulations increases their protective efficacy [168,169]. Similarly, an extensive recent review explained the impact of PEGylation in terms of biodistribution for anticipating safety and efficacy [170]. Therefore, it is essential to study the tolerability and safety profile of PEG, despite being an alternative to increased mucopenetration.

Some works have also raised doubts about coating particles with PEG due to surface modifications that can alter the linked polymers' physical and biological properties. Bamberger et al. evaluated the effects on APC response after functionalizing spermine NPs with acetylated dextran (Sp-Ac-DEX) through a process called DEXylation and PEGylation. The average particle size was considerably increased by DEXylation, with subsequent aggregation. PEGylation and DEXylation decreased the primary amines and, therefore, the ζ-potential. This was reflected in the 20% reduction in the cell viability of bone-marrow-derived dendritic cells and macrophages treated with DEXylated NPs, whereas PEGylation treatment increased viability by 10–20% compared to unmodified NPs. However, the binding and cellular uptake of surface-modified NPs was lower in PEGylated particles [171].

Other polymers with mucopenetration ability are poloxamers, also known as Pluronic®. These block copolymers consist of hydrophilic poly (ethylene oxide) (PEO), and hydrophobic block-poly (propylene oxide) (PPO) ordered in an A-B-A triblock structure: PEO-PPO-PEO [172–174]. Díaz et al. demonstrated that the addition of mucoadhesive and thermosensitive poloxamer 407(F127)-based hydrogels to CS microspheres in a formulation for nasal and conjunctival ram immunization improved both local and systemic humoral immune responses against the BLSOmp31 antigen, an outer membrane protein of *Brucella* spp., along with the reduced excretion of *Brucella ovis* [175]. Pastor et al. proposed a Pluronic® (PF127) and Gantrez® AN119 thermally sensitive hydrogel for intranasal vaccine delivery since the hydrogel increases the residence time of the antigens in the nasal epithelium, allowing their penetration into the deep skin layers of the nose thus reaching the submucosa, where they can trigger an immune response [176].

Another type of mucopenetrant includes nanoemulsions (NEs). Di Cola et al. evaluated PEG-coated O/W NEs with emulsified, added CS as a proposal for the nasal administration of drugs or vaccines. They observed that CS-added NE led to a local shrinking of the mucin gel network, forming larger pores between the mucin bundles. This phenomenon does not occur in the absence of CS. The SAXS (small-angle X-ray) monitoring of the penetration of solute CS-added NE into the PGM showed a higher diffusion over time (20 min) through the mucus mesh. SANS (neutron scattering) confirmed that, unlike the steric hindrance caused by the pore-like size of mucus caused by mucoinert NPs, the CS-added NE based on Solutol® mucopenetrates by the collapse of the mucus mesh [177].

Coating dextran particles with mucopenetration properties have also been explored to improve drug administration performance [178–181] and enhance immunoadjuvant activity in vivo [182]. Other strategies, such as coating polymeric particles with polydopamine (PDA) [183] or cell-penetrating peptides [184] used in drug delivery, might be explored and characterized in mucosal vaccines, as well as continuing the search for new adjuvants with mucopenetrating properties.

5.2. Permeation of Polymeric Particles via the Mucus Layer

An additional consideration for the design of polymer-based particles is passing through the second barrier, the epithelial cell membrane. The permeability of peptides, proteins, and drugs is often deficient. In this sense, absorption enhancers have been developed, which, in addition to preventing enzymatic degradation, facilitate the opening of the epithelial barrier and improve absorption through intracellular or paracellular mechanisms [185]. Absorption and permeation enhancers include surfactants, such as bile salts, fatty acids, phospholipids, tight junction modulators, cyclodextrins, and detergents [186–189]. This group also includes mucolytic agents, such as acetylcysteine or enzymes, which can decrease the elastic properties and dynamic viscosity of the mucus, influencing the integrity of the mucus layer [190]. For example, Zhang et al. reported the oral administration, in mice, of self-assembled nanoparticles with recombinant urease subunit B from *Helicobacter pylori*, coated with a cell-penetrating peptide, and coated with PEG derivative. NPs were transported transepithelially, improving the systemic and mucosal antibody response and the protection against *H. pylori* after the challenge [191]. It will be essential to continue studying absorption enhancers in mucosal vaccine formulations to improve the immune response.

6. Challenges and Opportunities

For several decades, many polymer-based particles have shown promise as potential human mucosal vaccine adjuvants due to their biodegradability, biocompatibility, and nontoxicity characteristics. Added to this is the extensive study of the adjuvant mechanisms of particulate systems. However, in the mucosa, the mucin networks that cover the compartments are often considered a barrier for the particles, so the mucoadhesive and mucopenetrating capacity of the polymer-based particles often defines their adjuvant mechanism of action.

The search for polymers with better mucoadhesive properties, regardless of the polymer's source, but focusing on the physicochemical characterization of polymeric particles and the contribution of these properties to mucoadhesion, will allow the rational design of mucosal vaccines. However, it is not an easy task because, on the one hand, the smallest nanoparticles are the most mucopenetrating. Still, on the other, there is a lack of studies that suggest an ideal surface charge or a hydrophobicity that favors adhesion. At the same time, it cannot be ignored that there are multiple other cellular mechanisms to elicit the immune response triggered by the polymeric particles, i.e., enhanced antigen uptake, immune cell presentation and recruitment, and traffic to lymph nodes.

Studies demonstrating the correlation between the observed immune response, the physicochemical characteristics, the mucoadhesion, and the mucopenetration ability are scarce. More studies that examine all these factors simultaneously are required to position mucoadhesion as another immune response mechanism necessary for designing more efficient polymer-based particulate adjuvants.

7. Conclusions

The COVID-19 pandemic highlighted the need for mucosal vaccination as an effective strategy to eradicate infectious diseases that have the mucosa as a natural route of infection. Mucoadhesion is probably the most important feature to improve local and systemic immune responses since, by prolonging the residence time of particulate polymers in mucosal tissues, the absorption and sometimes penetration through the mucosal epithelia are allowed and improved. In this sense, studying the physicochemical characteristics of the polymeric particles used as mucosal vaccine adjuvants and how they affect mucoadhesion is crucial to developing new mucosal vaccines.

Author Contributions: N.V.-M. conceptualization, writing—original draft. D.G., S.A.M.-M. and S.S. reviewed the document, and R.R.-S. reviewed, discussed, and edited the article. All authors have read and agreed to the published version of the manuscript.

Funding: This work was funded by the Consejo Nacional de Ciencia y Tecnología (CONACYT), México (grant number A1-S-9849) and PAPIIIT/DGAPA/UNAM (grant numbers IN 216419 and IN 216722). Nathaly Vasquez Martínez is a doctoral student from Doctorado en Ciencias Bioquímicas, UNAM. She received a fellowship from CONACYT.

Institutional Review Board Statement: Not applicable.

Informed Consent Statement: Not applicable.

Data Availability Statement: Not applicable.

Acknowledgments: The authors are grateful for the financial support of the CONACYT and PAPIIT/DGAPA/UNAM.

Conflicts of Interest: The authors declare no conflict of interest.

References

1. Lycke, N.Y.; Bemark, M. The regulation of gut mucosal IgA B-cell responses: Recent developments. *Mucosal Immunol.* **2017**, *10*, 1361–1374. [CrossRef] [PubMed]
2. Tezuka, H.; Ohteki, T. Regulation of IgA production by intestinal dendritic cells and related. *Front. Immunol.* **2019**, *10*, 1891. [CrossRef] [PubMed]
3. Ornelas, A.; Dowdell, A.S.; Lee, J.S.; Colgan, S.P. Microbial metabolite regulation of epithelial cell-cell interactions and barrier function. *Cells* **2022**, *11*, 944. [CrossRef] [PubMed]
4. Kiyono, H.; Fukuyama, S. Nalt-versus Peyer's-patch-mediated mucosal immunity. *Nat. Rev. Immunol.* **2004**, *4*, 699–710. [CrossRef] [PubMed]
5. Nakamura, Y.; Kimura, S.; Hase, K. M Cell-dependent antigen uptake on follicle-associated epithelium for mucosal immune surveillance. *Inflamm. Regen.* **2018**, *38*, 15. [CrossRef]
6. Ohno, H.J.B. Special review—Crosstalk between the intestinal immune system and gut commensal microbiota intestinal M. *Cells* **2016**, *159*, 151–160. [CrossRef]
7. Kimura, S. Molecular Insights into the Mechanisms of M-Cell Differentiation and transcytosis in the mucosa-associated lymphoid tissues. *Anat. Sci. Int.* **2018**, *93*, 23–34. [CrossRef]
8. Howe, S.E.; Lickteig, D.J.; Plunkett, K.N.; Ryerse, J.S.; Konjufca, V. The uptake of soluble and particulate antigens by epithelial cells in the mouse small intestine. *PLoS ONE* **2014**, *9*, e86656. [CrossRef]
9. Kulkarni, P.; Rawtani, D.; Barot, T. Design, development and in-vitro/in-vivo evaluation of intranasally delivered rivastigmine and N-acetyl cysteine loaded bifunctional niosomes for applications in combinative treatment of alzheimer's disease. *Eur. J. Pharm. Biopharm.* **2021**, *163*, 1–15. [CrossRef]
10. Agace, W.W. T-Cell Recruitment to the Intestinal Mucosa. *Trends Immunol.* **2008**, *29*, 514–522. [CrossRef]

11. Oya, Y.; Kimura, S.; Nakamura, Y.; Ishihara, N.; Takano, S.; Morita, R.; Endo, M.; Hase, K. Characterization of M cells in tear duct-associated lymphoid tissue of mice: A potential role in immunosurveillance on the ocular surface. *Front. Immunol.* **2021**, *12*, 779709. [CrossRef]
12. Rivellese, F.; Pontarini, E.; Pitzalis, C. *Inducible Lymphoid Organs*; Springer: Berlin/Heidelberg, Germany, 2020; Volume 426, ISBN 9783030517465.
13. Zhou, J.Z.; Way, S.S.; Chen, K. Immunology of the uterine and vaginal mucosae. *Trends Immunol.* **2018**, *39*, 302–314. [CrossRef]
14. Brandtzaeg, P. Secretory IgA: Designed for anti-microbial defense. *Front. Immunol.* **2013**, *4*, 222. [CrossRef]
15. Lavelle, E.C.; Ward, R.W. Mucosal vaccines—Fortifying the frontiers. *Nat. Rev. Immunol.* **2022**, *22*, 236–250. [CrossRef]
16. Su, F.; Patel, G.B.; Hu, S.; Chen, W. Induction of mucosal immunity through systemic immunization: Phantom or reality? *Hum. Vaccines Immunother.* **2016**, *12*, 1070–1079. [CrossRef]
17. Clements, J.D.; Freytag, L.C. Parenteral vaccination can be an effective means of inducing protective mucosal responses. *Clin. Vaccine Immunol.* **2016**, *23*, 438–441. [CrossRef]
18. Garziano, M.; Utyro, O.; Strizzi, S.; Vanetti, C.; Saulle, I.; Conforti, C.; Cicilano, F.; Ardizzone, F.; Cappelletti, G.; Clerici, M.; et al. Saliva and plasma neutralizing activity induced by the administration of a third Bnt162b2 vaccine dose. *Int. J. Mol. Sci.* **2022**, *23*, 14341. [CrossRef]
19. Shaikh, H.; Lynch, J.; Kim, J.; Excler, J.L. Current and future cholera vaccines. *Vaccine* **2020**, *38*, A118–A126. [CrossRef]
20. Varghese, T.; Kang, G.; Steele, A.D. Understanding rotavirus vaccine efficacy and effectiveness in countries with high child mortality. *Vaccines* **2022**, *10*, 346. [CrossRef]
21. Booth, J.S.; Goldberg, E.; Barnes, R.S.; Greenwald, B.D.; Sztein, M.B. Oral typhoid vaccine Ty21a elicits antigen-specific resident memory CD_4^+ T cells in the human terminal ileum lamina propria and epithelial compartments. *J. Transl. Med.* **2020**, *18*, 102. [CrossRef]
22. Collins, N.D.; Adhikari, A.; Yang, Y.; Kuschner, R.A.; Karasavvas, N.; Binn, L.N.; Walls, S.D.; Graf, P.C.F.; Myers, C.A.; Jarman, R.G.; et al. Live oral adenovirus type 4 and type 7 vaccine induces durable antibody response. *Vaccines* **2020**, *8*, 411. [CrossRef]
23. Carter, N.J.; Curran, M.P. Live attenuated influenza vaccine (FluMist®; Fluenz™). *Drugs* **2011**, *71*, 1591–1622. [CrossRef]
24. Kulkarni, P.S.; Raut, S.K.; Dhere, R.M. A post-marketing surveillance study of a human live-virus pandemic influenza a (H1N1) vaccine (Nasovac®) in India. *Hum. Vaccines Immunother.* **2013**, *9*, 122–124. [CrossRef]
25. Czerkinsky, C.; Holmgren, J. Mucosal delivery routes for optimal immunization: Targeting immunity to the right tissues. In *Current Topics in Microbiology and Immunology*; Springer: Berlin/Heidelberg, Germany, 2012; Volume 354, pp. 1–18. ISBN 9783540767756. [CrossRef]
26. Delph, K.M.; Davis, E.G.; Bello, N.M.; Hankins, K.; Wilkerson, M.J.; Ewen, C.L. Journal of equine veterinary science comparison of immunologic responses following intranasal and oral administration of a USDA-approved, live-attenuated *Streptococcus Equi* Vaccine. *J. Equine Vet. Sci.* **2018**, *60*, 29–34.e1. [CrossRef]
27. Shillova, N.; Howe, S.E.; Hyseni, B.; Ridgell, D. Crossm chlamydia-specific IgA secretion in the female reproductive tract induced via per-oral immunization confers protection against primary *Chlamydia* challenge. *Infect. Immun.* **2021**, *89*, 1–16. [CrossRef]
28. Jahnmatz, M.; Richert, L.; Storsaeter, J.; Colin, C.; Bauduin, C.; Thalen, M.; Solovay, K.; Rubin, K. Safety and immunogenicity of the live attenuated intranasal pertussis vaccine BPZE1: A phase 1b, double-blind, randomised, placebo-controlled dose-escalation study. *Lancet Infect. Dis.* **2020**, *20*, 1290–1301. [CrossRef]
29. Li, L.; Shi, N.; Xu, N.; Wang, H.; Zhao, H.; Xu, H.; Liu, D.; Zhang, Z. Safety and viral shedding of live attenuated influenza vaccine (LAIV) in Chinese healthy juveniles and adults: A phase I. *Vaccines* **2022**, *10*, 1796. [CrossRef]
30. Ascarateil, S.; Puget, A.; Koziol, M. Safety data of montanide ISA 51 VG and montanide ISA 720 VG, two adjuvants dedicated to human therapeutic vaccines. *J. Immunother. Cancer* **2015**, *3*, P428. [CrossRef]
31. Cohet, C.; Van Der Most, R.; Bauchau, V.; Bekkat-berkani, R.; Doherty, T.M.; Schuind, A.; Tavares, F.; Silva, D.; Rappuoli, R.; Garçon, N.; et al. Safety of AS03-adjuvanted influenza vaccines: A review of the evidence. *Vaccine* **2019**, *37*, 3006–3021. [CrossRef]
32. Li, A.P.Y.; Cohen, C.A.; Leung, N.H.L.; Fang, V.J.; Gangappa, S.; Sambhara, S.; Levine, M.Z.; Iuliano, A.D.; Perera, R.A.P.M.; Ip, D.K.M.; et al. Immunogenicity of standard, high-dose, MF59-adjuvanted, and Recombinant-HA Seasonal Influenza Vaccination in Older Adults. *Npj Vaccines* **2021**, *6*, 25. [CrossRef]
33. Wang, Y.Q.; Bazin-Lee, H.; Evans, J.T.; Casella, C.R.; Mitchell, T.C. MPL adjuvant contains competitive antagonists of human TLR4. *Front. Immunol.* **2020**, *11*, 577823. [CrossRef] [PubMed]
34. Didierlaurent, A.M.; Laupèze, B.; Di Pasquale, A.; Hergli, N.; Collignon, C.; Garçon, N. Adjuvant system AS01: Helping to overcome the challenges of modern vaccines. *Expert Rev. Vaccines* **2017**, *16*, 55–63. [CrossRef] [PubMed]
35. Schwarz, T.F.; Huang, L.M.; Lin, T.Y.; Wittermann, C.; Panzer, F.; Valencia, A.; Suryakiran, P.V.; Lin, L.; Descamps, D. Long-term immunogenicity and safety of the HPV-16/18 AS04-adjuvanted vaccine in 10- to 14-year-old girls open 6-year follow-up of an initial observer-blinded, randomized trial. *Pediatr. Infect. Dis. J.* **2014**, *33*, 1255–1261. [CrossRef]
36. Hsieh, S.; Liu, M.; Chen, Y.; Lee, W.; Hwang, S.; Cheng, S.; Ko, W.; Hwang, K. Safety and immunogenicity of CpG 1018 and aluminium hydroxide-adjuvanted SARS-CoV-2 S-2P protein vaccine MVC-COV1901: Interim results of a large-scale, double-blind, randomised, placebo-controlled phase 2 trial in Taiwan. *Lancet Respir.* **2021**, *9*, 1396–1406. [CrossRef]

37. Parums, D.V. Editorial: First approval of the protein-based adjuvanted nuvaxovid (NVX-CoV2373) novavax vaccine for SARS-CoV-2 could increase vaccine uptake and provide immune protection from viral variants. *Med. Sci. Monit.* **2022**, *28*, e936523-1–e936523-3. [CrossRef]
38. Pan, S.; Hsieh, S.; Lin, C.; Hsu, Y.; Chang, M.; Chang, S. A randomized, double-blind, controlled clinical trial to evaluate the safety and immunogenicity of an intranasally administered trivalent inactivated influenza vaccine with adjuvant LTh(aK): A phase I study. *Vaccine* **2019**, *37*, 1994–2003. [CrossRef]
39. Schussek, S.; Bernasconi, V.; Mattsson, J.; Wenzel, U.A.; Strömberg, A.; Gribonika, I.; Schön, K.; Lycke, N.Y. The CTA1-DD adjuvant strongly potentiates follicular dendritic cell function and germinal center formation, which results in improved neonatal immunization. *Mucosal Immunol.* **2020**, *13*, 545–557. [CrossRef]
40. Van Herck, S.; Feng, B.; Tang, L. Delivery of STING agonists for adjuvanting subunit vaccines. *Adv. Drug Deliv. Rev.* **2021**, *179*, 114020. [CrossRef]
41. Kantarcioglu, B.; Iqbal, O.; Lewis, J.; Carter, A.; Singh, M.; Lievano, F.; Ligocki, M.; Jeske, W.; Adiguzel, C.; Gerotziafas, T.; et al. An update on the status of vaccine development for SARS-CoV-2 including variants. practical considerations for COVID-19 special populations. *Clin. Appl. Thromb. Hemost.* **2022**, *28*, 10760296211056648. [CrossRef]
42. Van der Ley, P.A.; Zariri, A.; Van Riet, E.; Oosterhoff, D.; Kruiswijk, C.P. An intranasal OMV-based vaccine induces high mucosal and systemic protecting immunity against a SARS-CoV-2 infection. *Front. Immunol.* **2021**, *12*, 781280. [CrossRef]
43. Jearanaiwitayakul, T.; Seesen, M.; Chawengkirttikul, R.; Limthongkul, J.; Apichirapokey, S.; Sapsutthipas, S.; Phumiamorn, S.; Sunintaboon, P.; Ubol, S. Intranasal administration of RBD nanoparticles confers induction of mucosal and systemic immunity against SARS-CoV-2. *Vaccines* **2021**, *9*, 768. [CrossRef] [PubMed]
44. Li, J.; Qiang, H.; Yang, W.; Xu, Y.; Feng, T.; Cai, H.; Wang, S.; Liu, Z.; Zhang, Z.; Zhang, J. Oral insulin delivery by epithelium microenvironment-adaptive nanoparticles. *J. Control. Release* **2022**, *341*, 31–43. [CrossRef] [PubMed]
45. McCright, J.; Skeen, C.; Yarmovsky, J.; Maisel, K. Acta biomaterialia nanoparticles with dense poly (ethylene glycol) coatings with near neutral charge are maximally transported across lymphatics and to the lymph nodes. *Acta Biomater.* **2022**, *145*, 146–158. [CrossRef]
46. Nguyen, B.; Tolia, N.H. Protein-based antigen presentation platforms for nanoparticle vaccines. *Npj Vaccines* **2021**, *6*, 70. [CrossRef]
47. Guillén, D.; Moreno-Mendieta, S.; Pérez, R.; Espitia, C.; Sánchez, S.; Rodríguez-Sanoja, R. Starch granules as a vehicle for the oral administration of immobilized antigens. *Carbohydr. Polym.* **2014**, *112*, 210–215. [CrossRef] [PubMed]
48. Moreno-Mendieta, S.A.; Guillén, D.; Espitia, C.; Hernández-Pando, R.; Sanchez, S.; Rodríguez-Sanoja, R. A novel antigen-carrier system: The *Mycobacterium tuberculosis* acr protein carried by raw starch microparticles. *Int. J. Pharm.* **2014**, *474*, 241–248. [CrossRef]
49. Barnowski, C.; Kadzioch, N.; Damm, D.; Yan, H.; Temchura, V. Advantages and limitations of integrated flagellin adjuvants for HIV-based nanoparticle B-Cell vaccines. *Pharmaceutics* **2019**, *11*, 204. [CrossRef]
50. Vijayan, V.; Mohapatra, A.; Uthaman, S.; Park, I.K. Recent advances in nanovaccines using biomimetic immunomodulatory materials. *Pharmaceutics* **2019**, *11*, 534. [CrossRef]
51. Vu, M.N.; Kelly, H.G.; Kent, S.J.; Wheatley, A.K. Current and future nanoparticle vaccines for COVID-19. *EBioMedicine* **2021**, *74*, 103699. [CrossRef]
52. Oyewumi, M.O.; Kumar, A.; Cui, Z.A. Nano-microparticles as immune adjuvants: Correlating particle sizes and the resultant immune responses. *Expert Rev. Vaccines* **2011**, *9*, 1095–1107. [CrossRef]
53. Shah, R.R.; Amiji, M.M.; Brito, L.A. The impact of size on particulate vaccine adjuvants. *Nanomedicine* **2014**, *9*, 2671–2681. [CrossRef] [PubMed]
54. Nakamura, T.; Kawai, M.; Sato, Y.; Maeki, M.; Tokeshi, M.; Harashima, H. The Effect of size and charge of lipid nanoparticles prepared by microfluidic mixing on their lymph node transitivity and distribution. *Mol. Pharm.* **2020**, *17*, 944–953. [CrossRef] [PubMed]
55. Tazaki, T.; Tabata, K.; Ainai, A.; Ohara, Y.; Kobayashi, S. Conjugated RNA Adjuvants for Intranasal Inactivated Influenza Vaccines. *RSC Adv.* **2018**, *8*, 16527–16536. [CrossRef] [PubMed]
56. Gogoi, H.; Mani, R.; Aggarwal, S.; Malik, A.; Munde, M.; Bhatnagar, R. Crystalline and amorphous preparation of aluminum hydroxide nanoparticles enhances protective antigen domain 4 specific immunogenicity and provides protection against anthrax. *Int. J. Nanomed.* **2020**, *15*, 239–252. [CrossRef]
57. O'Hagan, D.T. New generation vaccine adjuvants. In *Encyclopedia of Life Sciences*; John Wiley & Sons, Ltd.: Hoboken, NJ, USA, 2007; pp. 1–7. ISBN 9780470015902. [CrossRef]
58. Liu, Y.; Wang, L.; Weifeng, Z.; Chen, X. Surface hydrophobicity of microparticles modulates adjuvanticity. *J. Mater. Chem. B* **2013**, *1*, 3888–3896. [CrossRef]
59. Shima, F.; Akagi, T.; Akashi, M. Effect of Hydrophobic side chains in the induction of immune responses by nanoparticle adjuvants consisting of amphiphilic poly (γ-glutamic acid). *Bioconjug. Chem.* **2015**, *26*, 890–898. [CrossRef]
60. Katare, Y.K.; Muthukumaran, T.; Panda, A.K. Influence of particle size, antigen load, dose and additional adjuvant on the immune response from antigen loaded PLA microparticles. *Int. J. Pharm.* **2005**, *301*, 149–160. [CrossRef]

61. Kapadia, C.H.; Tian, S.; Perry, J.L.; Luft, J.C.; Desimone, J.M. Role of linker length and antigen density in nanoparticle peptide vaccine. *ACS Omega* **2019**, *4*, 5547–5555. [CrossRef]
62. Zhang, W.; Zhu, C.; Xiao, F.; Liu, X.; Xie, A.; Chen, F.; Dong, P.; Lin, P.; Zheng, C.; Zhang, H.; et al. PH-controlled release of antigens using mesoporous silica nanoparticles delivery system for developing a fish oral vaccine. *Front. Immunol.* **2021**, *12*, 644396. [CrossRef]
63. Jahan, S.T.; Sadat, S.M.; Haddadi, A. Design and immunological evaluation of anti-CD205-tailored PLGA-based nanoparticulate cancer vaccine. *Int. J. Nanomed.* **2018**, *ume 13*, 367–386. [CrossRef]
64. Schmidt, S.T.; Olsen, C.L.; Franzyk, H.; Wørzner, K.; Korsholm, K.S.; Rades, T.; Andersen, P.; Foged, C.; Christensen, D. Comparison of two different pegylation strategies for the liposomal adjuvant CAF09: Towards induction of CTL responses upon subcutaneous vaccine administration. *Eur. J. Pharm. Biopharm.* **2019**, *140*, 29–39. [CrossRef]
65. Genito, C.J.; Batty, C.J.; Bachelder, E.M.; Ainslie, K.M. Considerations for size, surface charge, polymer degradation, co-delivery, and manufacturability in the development of polymeric particle vaccines for infectious diseases. *Adv. NanoBiomed. Res.* **2021**, *1*, 2000041. [CrossRef] [PubMed]
66. Yan, S.; Gu, W.; Xu, Z.P. Re-Considering How Particle Size and Other Properties of Antigen-Adjuvant Complexes Impact on the Immune Responses. *J. Colloid. Interface Sci.* **2013**, *395*, 1–10. [CrossRef] [PubMed]
67. Bastola, R.; Lee, S. Physicochemical properties of particulate vaccine adjuvants: Their pivotal role in modulating immune responses. *J. Pharm. Investig.* **2018**, *49*, 279–285. [CrossRef]
68. Ulery, B.D.; Petersen, L.K.; Phanse, Y.; Kong, C.S.; Broderick, S.R.; Kumar, D.; Ramer-Tait, A.E.; Carrillo-Conde, B.; Rajan, K.; Wannemuehler, M.J.; et al. Rational design of pathogen-mimicking amphiphilic materials as nanoadjuvants. *Sci. Rep.* **2011**, *1*, 198. [CrossRef] [PubMed]
69. Wu, J.; Ma, G. Biomimic strategies for modulating the interaction between particle adjuvants and antigen-presenting. *Cells Biomater. Sci.* **2020**, *8*, 2366–2375. [CrossRef] [PubMed]
70. Moreno-Mendieta, S.; Guillén, D.; Vasquez-Martínez, N.; Hernández-Pando, R.; Sánchez, S.; Rodríguez-Sanoja, R. Understanding the phagocytosis of particles: The key for rational design of vaccines and therapeutics. *Pharm. Res.* **2022**, *39*, 1823–1849. [CrossRef] [PubMed]
71. Jackson, A.D.; Jackson, A.D. Airway Goblet-Cell Mucus Secretion. *Trends. Pharmacol. Sci.* **2001**, *22*, 39–45. [CrossRef]
72. Birchenough, G.; Johansson, M.; Gustafsson, J.; Bergström, J.H.; Hansson, G.C. New developments in goblet cell mucus secretion and function. *Mucosal Immunol.* **2015**, *8*, 712–719. [CrossRef]
73. Creeth, J.M. Constituents of mucus and their separation. *Br. Med. Bull.* **1978**, *34*, 17–24. [CrossRef]
74. Dupont, A.; Heinbockel, L.; Brandenburg, K.; Hornef, M.W. To protect the intestinal mucosa. *Gut Microbes* **2014**, *5*, 761–765. [CrossRef] [PubMed]
75. Atuma, C.; Strugala, V.; Allen, A.; Holm, L. The adherent gastrointestinal mucus gel layer: Thickness and physical state in vivo. *Am. J. Physiol. Gastrointest. Liver. Physiol.* **2014**, *280*, G922–G929. [CrossRef] [PubMed]
76. Gustafsson, J.K.; Ermund, A.; Johansson, M.E.; Schütte, A.; Hansson, G.C.; Sjövall, H. An ex vivo method for studying mucus formation, properties, and thickness in human colonic biopsies and mouse small and large intestinal explants. *Am. J. Physiol. Gastrointest. Liver. Physiol.* **2012**, *302*, G430–G438. [CrossRef]
77. McGuckin, M.A.; Thornton, D.J.; Whitsett, J.A. *Mucins and Mucus*, 4th ed.; Elsevier: Amsterdam, The Netherlands, 2015; pp. 231–250. ISBN 9780124158474.
78. Peppas, N.A.; Buri, P.A. Surface, interfacial and molecular aspects of polymer bioadhesion on soft tissues. *J. Control. Release* **1985**, *2*, 257–275. [CrossRef]
79. Leung, S.S.; Robinson, J.R. Polymer structure features contributing to mucoadhesion. II. *J. Control. Release* **1990**, *12*, 187–194. [CrossRef]
80. Nyström, B.; Kjøniksen, A.; Beheshti, N.; Maleki, A.; Zhu, K.; Knudsen, K.D.; Pamies, R.; Hernández, J.G.; García, J.; Torre, D. Characterization of polyelectrolyte features in polysaccharide systems and mucin. *Adv. Colloid Interface Sci.* **2010**, *158*, 108–118. [CrossRef] [PubMed]
81. Jabbari, E.; Peppas, N. Polymer-polymer interdiffusion and adhesion. *Polym. Rev. Part C* **1994**, *34*, 205–241. [CrossRef]
82. Sriamornsak, P.; Wattanakorn, N.; Nunthanid, J.; Puttipipatkhachorn, S. Mucoadhesion of Pectin as evidence by wettability and chain interpenetration. *Carbohydr. Polym.* **2008**, *74*, 458–467. [CrossRef]
83. Shaikh, R.; Raghu, T.; Singh, R.; Garland, M.J.; David, A.; Donnelly, R.F. Mucoadhesive drug delivery systems. *J. Pharm. Bioallied. Sci.* **2011**, *3*, 89–100. [CrossRef]
84. Saini, H.K.; Nautiyal, U. Pioneering and encouraging approach–mucoadhesive drug delivery system. *Int. J. Pharm. Med. Res* **2017**, *5*, 455–463.
85. Pereira, M.N.; Reis, T.A.; Matos, B.N.; Cunha-filho, M.; Gratieri, T.; Gelfuso, G.M. Colloids and surfaces B: Biointerfaces novel ex vivo protocol using porcine vagina to assess drug permeation from mucoadhesive and colloidal pharmaceutical systems. *Colloids Surf. B Biointerfaces* **2017**, *158*, 222–228. [CrossRef] [PubMed]
86. Gidvall, S.; Björklund, S.; Feiler, A.; Dahlström, B.; Rönn, R.; Engblom, J.; Valetti, S. A novel versatile flow-donor chamber as biorelevant ex-vivo test assessing oral mucoadhesive formulations. *Eur. J. Pharm. Sci.* **2021**, *166*, 105983. [CrossRef] [PubMed]

87. Falavigna, M.; Pattacini, M.; Wibel, R.; Sonvico, F.; Škalko-Basnet, N.; Flaten, G.E. The vaginal-PVPA: A vaginal mucosa-mimicking in vitro permeation tool for evaluation of mucoadhesive formulations. *Pharmaceutics* **2020**, *12*, 568. [CrossRef]
88. Rossi, S.; Vigani, B.; Bonferoni, M.C.; Sandri, G.; Caramella, C.; Ferrari, F. Journal of pharmaceutical and biomedical analysis rheological analysis and mucoadhesion: A 30 year-old and still active combination. *J. Pharm. Biomed. Anal.* **2018**, *156*, 232–238. [CrossRef] [PubMed]
89. Bassi, J.; Barbosa, S.; Ferreira, D.S.; De Freitas, O.; Bruschi, M.L. A critical review about methodologies for the analysis of mucoadhesive properties of drug delivery systems. *Drug Dev. Ind. Pharm.* **2017**, *43*, 1053–1070. [CrossRef] [PubMed]
90. Drumond, N.; Stegemann, S. Colloids and surfaces B: Biointerfaces polymer adhesion predictions for oral dosage forms to enhance drug administration safety. Part 3: Review of in vitro and in vivo methods used to predict esophageal adhesion and transit time. *Colloids Surf. B Biointerfaces* **2018**, *165*, 303–314. [CrossRef]
91. Bayer, I.S. Recent advances in mucoadhesive interface materials, mucoadhesion characterization, and technologies. *Adv. Mater. Interfaces* **2022**, *9*, 2200211. [CrossRef]
92. Lai, S.K.; Wang, Y.; Hanes, J. Mucus-penetrating nanoparticles for drug and gene delivery to mucosal tissues. *Adv. Drug Deliv. Rev.* **2009**, *61*, 158–171. [CrossRef]
93. Petrou, G.; Crouzier, T. Mucins as multifunctional building blocks of biomaterials. *Biomater. Sci.* **2018**, *6*, 2282–2297. [CrossRef]
94. Jelkmann, M.; Leichner, C.; Menzel, C.; Kreb, V.; Bernkop-Schnürch, A. Cationic starch derivatives as mucoadhesive and soluble excipients in drug delivery. *Int. J. Pharm.* **2019**, *570*, 118664. [CrossRef]
95. Martău, G.A.; Mihai, M.; Vodnar, D.C. The use of chitosan, alginate, and pectin in the biomedical and food sector-biocompatibility, bioadhesiveness, and biodegradability. *Polymers* **2019**, *11*, 1837. [CrossRef] [PubMed]
96. Yildiz, H.M.; Mckelvey, C.A.; Marsac, P.J.; Carrier, R.L.; Point, W. Size selectivity of intestinal mucus to diffusing particulates is dependent on surface chemistry and exposure to lipids. *J. Drug Target.* **2016**, *23*, 768–774. [CrossRef] [PubMed]
97. Lamson, N.G.; Berger, A.; Fein, K.C.; Whitehead, K.A. Proteins by enhancing intestinal permeability. *Nat. Biomed. Eng.* **2020**, *4*, 84–96. [CrossRef] [PubMed]
98. Bandi, S.P.; Kumbhar, Y.S. Effect of particle size and surface charge of nanoparticles in penetration through intestinal mucus barrier. *J. Nanopart. Res.* **2020**, *22*, 62. [CrossRef]
99. Guo, S.; Sun, J.; Gan, Y.; Shi, X.; Gao, H. Rotation-facilitated rapid transport of nanorods in mucosal tissues. *J. Control. Release* **2016**, *307*, 64–75. [CrossRef]
100. Yu, M.; Xu, L.; Tian, F.; Su, Q.; Zheng, N.; Yang, Y.; Wang, J.; Wang, A.; Zhu, C.; Guo, S.; et al. Rapid transport of deformation-tuned nanoparticles across biological hydrogels and cellular barriers. *Nat. Commun.* **2018**, *9*, 2607. [CrossRef]
101. Gómez-Guillén, M.C.; Montero, M.P. Food hydrocolloids enhancement of oral bioavailability of natural compounds and probiotics by mucoadhesive tailored biopolymer-based nanoparticles: A review. *Food Hydrocoll.* **2021**, *118*, 106772. [CrossRef]
102. Das, S.; Kaur, S.; Rai, V.K. Gastro-retentive drug delivery systems: A recent update on clinical pertinence and drug delivery. *Drug Deliv. Transl. Res.* **2021**, *11*, 1849–1877. [CrossRef]
103. Tan, S.J.L.; Billa, N. Improved bioavailability of poorly soluble drugs through gastrointestinal muco-adhesion of lipid nanoparticles. *Pharmaceutics* **2021**, *13*, 1817. [CrossRef]
104. De Lima, C.S.A.; Varca, J.P.R.O.; Alves, M.; Nogueira, K.M.; Cruz, C.P.C.; Rial-Hermida, M.I.; Kadłubowski, S.S.; Varca, G.H.C.; Lug, A.B. Mucoadhesive polymers and their applications in drug delivery systems for the treatment of bladder cancer. *Gels* **2022**, *8*, 587. [CrossRef]
105. Liu, Q.; Zheng, X.; Zhang, C.; Shao, X.; Zhang, X.; Zhang, Q.; Jiang, X. Antigen-conjugated N-trimethylaminoethylmethacrylate chitosan nanoparticles induce strong immune responses after nasal administration. *Pharm. Res.* **2015**, *32*, 22–36. [CrossRef]
106. Collado-González, M.; Espinosa, Y.G.; Goycoolea, F.M. Interaction between chitosan and mucin: Fundamentals and applications. *Biomimetics* **2019**, *4*, 32. [CrossRef]
107. Gong, X.; Gao, Y.; Shu, J.; Zhang, C.; Zhao, K. Chitosan-based nanomaterial as immune adjuvant and delivery carrier for vaccines. *Vaccines* **2022**, *10*, 1906. [CrossRef]
108. Szymańska, E.; Winnicka, K. Stability of chitosan—A challenge for pharmaceutical and biomedical applications. *Mar. Drugs* **2015**, *13*, 1819–1846. [CrossRef]
109. Safdar, R.; Aziz, A.; Arunagiri, A.; Regupathi, I.; Thanabalan, M.; Engineering, C.; Petronas, T.; Iskandar, B.S.; Ridzuan, P.D. Potential of chitosan and its derivatives for controlled drug release applications—A review. *J. Drug Deliv. Sci. Technol.* **2019**, *49*, 642–659. [CrossRef]
110. Trapani, A.; Sitterberg, J.; Bakowsky, U.; Kissel, T. The potential of glycol chitosan nanoparticles as carrier for low water soluble drugs. *Int. J. Pharm.* **2009**, *375*, 97–106. [CrossRef]
111. Lin, F.; Jia, H.; Wu, F. Glycol chitosan: A water-soluble polymer for cell imaging and drug delivery. *Molecules* **2019**, *24*, 4371. [CrossRef]
112. Pawar, D.; Jaganathan, K.S. Mucoadhesive glycol chitosan nanoparticles for intranasal delivery of hepatitis B vaccine: Enhancement of mucosal and systemic immune response. *Drug Deliv.* **2016**, *23*, 185–194. [CrossRef]
113. Kim, E.S.; Baek, Y.; Yoo, H.; Lee, J.; Lee, H.G. Chitosan-tripolyphosphate nanoparticles prepared by ionic gelation improve the antioxidant activities of astaxanthin in the in vitro and in vivo model. *Antioxidants* **2022**, *11*, 479. [CrossRef]

114. Walke, S.; Srivastava, G.; Routaray, C.B.; Dhavale, D.; Pai, K.; Doshi, J.; Kumar, R.; Doshi, P. Preparation and characterization of microencapsulated DwPT trivalent vaccine using water soluble chitosan and its in-vitro and in-vivo immunological properties. *Int. J. Biol. Macromol.* **2018**, *107*, 2044–2056. [CrossRef]
115. Snyman, D.; Hamman, J.H.; Kotze, A.F. Evaluation of the mucoadhesive properties of N-trimethyl chitosan chloride. *Drug Dev. Ind. Pharm.* **2003**, *29*, 61–69. [CrossRef]
116. Pathak, K.; Misra, S.K.; Sehgal, A.; Singh, S.; Bungau, S.; Najda, A.; Gruszecki, R.; Behl, T. Biomedical applications of quaternized chitosan. *Polymers* **2021**, *13*, 2514. [CrossRef]
117. Kim, Y.H.; Yoon, K.S.; Lee, S.; Park, E.; Rhim, J. Synthesis of fully deacetylated quaternized chitosan with enhanced antimicrobial activity and low cytotoxicity. *Antioxidants* **2022**, *11*, 1644. [CrossRef]
118. Pawar, D.; Goyal, A.K.; Mangal, S.; Mishra, N.; Vaidya, B.; Tiwari, S.; Jain, A.K.; Vyas, S.P. Evaluation of mucoadhesive PLGA microparticles for nasal immunization. *AAPS J.* **2010**, *12*, 130–137. [CrossRef]
119. Krishnakumar, D.; Kalaiyarasi, D.; Bose, J.C.; Jaganathan, K.S. Evaluation of Mucoadhesive nanoparticle based nasal vaccine. *J. Pharm. Investig.* **2012**, *42*, 315–326. [CrossRef]
120. Kim, H.S.; Park, K.H.; Lee, H.K.; Kim, J.S.; Kim, Y.G.; Lee, J.H.; Kim, K.H.; Yun, J.; Hwang, B.Y.; Hong, J.T.; et al. Curdlan activates dendritic cells through dectin-1 and toll-like receptor 4 signaling. *Int. Immunopharmacol.* **2016**, *39*, 71–78. [CrossRef]
121. Zhang, S.; Huang, S.; Lu, L.; Song, X.; Li, P.; Wang, F. Curdlan sulfate-O-linked quaternized chitosan nanoparticles: Potential adjuvants to improve the immunogenicity of exogenous antigens via intranasal vaccination. *Int. J. Nanomed.* **2018**, *13*, 2377–2394. [CrossRef]
122. Lu, G.; Sheng, B.; Wang, G.; Wei, Y.; Gong, Y.; Zhang, X.; Zhang, L. Controlling the degradation of covalently cross-linked carboxymethyl chitosan utilizing bimodal molecular weight distribution. *J. Biomater. Appl.* **2009**, *23*, 435–451. [CrossRef]
123. Cao, P.; Wang, J.; Sun, B.; Rewatkar, P.; Popat, A.; Fu, C.; Peng, H.; Xu, Z.P.; Li, L. Enhanced mucosal transport of polysaccharide-calcium phosphate nanocomposites for oral vaccination. *ACS Appl. Bio. Mater.* **2021**, *4*, 7865–7878. [CrossRef]
124. Rúnarsson, Ö.V.; Holappa, J.; Nevalainen, T.; Hjálmarsdóttir, M.; Järvinen, T.; Loftsson, T.; Einarsson, J.M.; Jónsdóttir, S.; Valdimarsdóttir, M.; Másson, M. Antibacterial activity of methylated chitosan and chitooligomer derivatives: Synthesis and structure activity relationships. *Eur. Polym. J.* **2007**, *43*, 2660–2671. [CrossRef]
125. Suksamran, T.; Ngawhirunpat, T.; Rojanarata, T.; Sajomsang, W.; Pitaksuteepong, T.; Opanasopit, P. Methylated N-(4-N,N-Dimethylaminocinnamyl) chitosan-coated electrospray OVA-loaded microparticles for oral vaccination. *Int. J. Pharm.* **2013**, *448*, 19–27. [CrossRef] [PubMed]
126. Jesus, S.; Soares, E.; Costa, J.; Borchard, G.; Borges, O. Immune response elicited by an intranasally delivered HBsAg low-dose adsorbed to poly-ε-caprolactone based nanoparticles. *Int. J. Pharm.* **2016**, *504*, 59–69. [CrossRef] [PubMed]
127. Rahman, M.S.; Hasan, M.S.; Nitai, A.S.; Nam, S.; Karmakar, A.K.; Ahsan, M.S.; Shiddiky, M.J.A.; Ahmed, M.B. Recent developments of carboxymethyl cellulose. *Polymers* **2021**, *13*, 1345. [CrossRef] [PubMed]
128. Hanson, S.M.; Singh, S.; Tabet, A.; Sastry, K.J.; Barry, M.; Wang, C. Mucoadhesive wafers composed of binary polymer blends for sublingual delivery and preservation of protein vaccines. *J. Control. Release* **2021**, *330*, 427–437. [CrossRef] [PubMed]
129. Mishra, M.; Mishra, B. Mucoadhesive microparticles as potential carriers in inhalation delivery of doxycycline hyclate: A comparative study. *Acta Pharm. Sin. B.* **2012**, *2*, 518–526. [CrossRef]
130. Cook, S.L.; Woods, S.; Methven, L.; Parker, J.K.; Khutoryanskiy, V.V. Mucoadhesive polysaccharides modulate sodium retention, release and taste perception. *Food Chem.* **2018**, *240*, 482–489. [CrossRef]
131. Baus, R.A.; Zahir-Jouzdani, F.; Dünnhaupt, S.; Atyabi, F.; Bernkop-Schnürch, A. Mucoadhesive hydrogels for buccal drug delivery: In vitro-in vivo correlation study. *Eur. J. Pharm. Biopharm.* **2019**, *142*, 498–505. [CrossRef]
132. Li, H.S.; Shin, M.K.; Singh, B.; Maharjan, S.; Park, T.E.; Kang, S.K.; Yoo, H.S.; Hong, Z.S.; Cho, C.S.; Choi, Y.J. Nasal immunization with mannan-decorated mucoadhesive HPMCP microspheres containing ApxIIA toxin induces protective immunity against challenge infection with *Actinobacillus Pleuropneumoiae* in mice. *J. Control. Release* **2016**, *233*, 114–125. [CrossRef]
133. Luong, M.; Lam, J.S.; Chen, J.; Levitz, S.M. Effects of fungal N- and O-linked mannosylation on the immunogenicity of model vaccines. *Vaccine* **2007**, *25*, 4340–4344. [CrossRef]
134. Kreer, C.; Kuepper, J.M.; Zehner, M.; Quast, T.; Kolanus, W.; Schumak, B.; Burgdorf, S. N-glycosylation converts non-glycoproteins into mannose receptor ligands and reveals antigen-specific T cell responses in vivo. *Oncotarget* **2017**, *8*, 6857–6872. [CrossRef]
135. Mangal, S.; Pawar, D.; Agrawal, U.; Jain, A.K.; Vyas, S.P. Evaluation of Mucoadhesive carrier adjuvant: Toward an oral anthrax vaccine. *Artif. Cells Nanomed. Biotechnol.* **2014**, *42*, 47–57. [CrossRef]
136. Saraf, S.; Jain, S.; Sahoo, R.N.; Mallick, S. Lipopolysaccharide derived alginate coated hepatitis B antigen loaded chitosan nanoparticles for oral mucosal immunization. *Int. J. Biol. Macromol.* **2020**, *154*, 466–476. [CrossRef]
137. Des Rieux, A.; Pourcelle, V.; Cani, P.D.; Marchand-Brynaert, J.; Préat, V. Targeted nanoparticles with novel non-peptidic ligands for oral delivery. *Adv. Drug Deliv. Rev.* **2013**, *65*, 833–844. [CrossRef]
138. Longet, S.; Lundahl, M.L.E.; Lavelle, E.C. Targeted strategies for mucosal vaccination. *Bioconjug. Chem.* **2018**, *29*, 613–623. [CrossRef]
139. Lee, N.K.; Kim, S.N.; Park, C.G. Immune cell targeting nanoparticles: A review. *Biomater. Res.* **2021**, *25*, 44. [CrossRef]
140. Amin, M.K.; Boateng, J.S. Enhancing stability and mucoadhesive properties of chitosan nanoparticles by surface modification with sodium alginate and polyethylene glycol for potential oral mucosa vaccine delivery. *Mar. Drugs* **2022**, *20*, 156. [CrossRef]

141. Vyas, S.; Dhoble, S.; Ghodake, V.; Patravale, V. Xyloglucan based mucosal nanovaccine for immunological protection against brucellosis developed by supercritical fluid technology. *Int. J. Pharm. X* **2020**, *2*, 100053. [CrossRef]
142. Piqué, N.; Gómez-Guillén, M.d.C.; Montero, M.P. Xyloglucan, a plant polymer with barrier protective properties over the mucous membranes: An overview. *Int. J. Mol. Sci.* **2018**, *19*, 673. [CrossRef]
143. Campolo, M.; Lanza, M.; Filippone, A.; Paterniti, I.; Casili, G.; Scuderi, S.A.; Ardizzone, A.; Cuzzocrea, S.; Esposito, E. Evaluation of a product containing xyloglucan and pea protein on skin barrier permeability. *Skin Pharmacol. Physiol.* **2020**, *33*, 231–236. [CrossRef]
144. Dutta, P.; Giri, S.; Giri, T.K. Xyloglucan as green renewable biopolymer used in drug delivery and tissue engineering. *Int. J. Biol. Macromol.* **2020**, *160*, 55–68. [CrossRef]
145. Grabovac, V.; Guggi, D.; Bernkop-Schnürch, A. Comparison of the mucoadhesive properties of various polymers. *Adv. Drug Deliv. Rev.* **2005**, *57*, 1713–1723. [CrossRef] [PubMed]
146. Pérez-González, G.L.; Villarreal-Gómez, L.J.; Serrano-Medina, A.; Torres-Martínez, E.J.; Cornejo-Bravo, J.M. Mucoadhesive electrospun nanofibers for drug delivery systems: Applications of polymers and the parameters' roles. *Int. J. Nanomed.* **2019**, *14*, 5271–5285. [CrossRef] [PubMed]
147. Lam, H.T.; Zupančič, O.; Laffleur, F.; Bernkop-Schnürch, A. Mucoadhesive properties of polyacrylates: Structure–function. *Int. J. Adhes. Adhes.* **2021**, *107*, 102857. [CrossRef]
148. Coucke, D.; Schotsaert, M.; Libert, C.; Pringels, E.; Vervaet, C.; Foreman, P.; Saelens, X.; Remon, J.P. Spray-dried powders of starch and crosslinked poly(acrylic acid) as carriers for nasal delivery of inactivated influenza vaccine. *Vaccine* **2009**, *27*, 1279–1286. [CrossRef]
149. Noh, H.J.; Chowdhury, M.Y.E.; Cho, S.; Kim, J.-H.; Park, H.S.; Kim, C.-J.; Poo, H.; Sung, M.-H.; Lee, J.-S.; Lim, Y.T. Programming of influenza vaccine broadness and persistence by mucoadhesive polymer-based adjuvant systems. *J. Immunol.* **2015**, *195*, 2472–2482. [CrossRef]
150. Kurosaki, T.; Katafuchi, Y.; Hashizume, J.; Harasawa, H.; Nakagawa, H. Induction of mucosal immunity by pulmonary administration of a cell-targeting nanoparticle. *Drug Deliv.* **2021**, *28*, 1585–1593. [CrossRef]
151. Fedorovitch, G. *Aerosol Particles in Lungs: Theoretical Modeling of Deposition and Mucociliary Clearance*; IntechOpen: London, UK, 2019; Open Access Books Built by Scientists; pp. 1–16. [CrossRef]
152. Leitner, V.M.; Walker, G.F.; Bernkop-Schnürch, A. Thiolated polymers: Evidence for the formation of disulphide bonds with mucus glycoproteins. *Eur. J. Pharm. Biopharm.* **2003**, *56*, 207–2114. [CrossRef]
153. Iqbal, J.; Shahnaz, G.; Dünnhaupt, S.; Müller, C.; Hintzen, F.; Bernkop-schnürch, A. Biomaterials preactivated thiomers as mucoadhesive polymers for drug delivery. *Biomaterials* **2012**, *33*, 1528–1535. [CrossRef]
154. Roldo, M.; Hornof, M.; Caliceti, P.; Bernkop-schnu, A. Mucoadhesive thiolated chitosans as platforms for oral controlled drug delivery: Synthesis and in vitro evaluation. *Eur. J. Pharm. Biopharm.* **2004**, *57*, 115–121. [CrossRef]
155. Leichner, C.; Jelkmann, M.; Bernkop-schnürch, A. Thiolated polymers: Bioinspired polymers utilizing one of the most important bridging structures in nature. *Adv. Drug Deliv. Rev.* **2019**, *151–152*, 191–221. [CrossRef]
156. Sinani, G.; Sessevmez, M.; Gök, M.K.; Özgümüş, S.; Alpar, H.O.; Cevher, E. Modified chitosan-based nanoadjuvants enhance immunogenicity of protein antigens after mucosal vaccination. *Int. J. Pharm.* **2019**, *569*, 118592. [CrossRef]
157. Malm, C.J.; Emerson, J.; Hiatt, G.D. Cellulose acetate phthalate as an enteric coating material. *J. Am. Pharm. Assoc.* **1951**, *40*, 520–525. [CrossRef]
158. Lee, H.B.; Yoon, S.Y.; Singh, B.; Oh, S.H.; Cui, L.; Yan, C.; Kang, S.K.; Choi, Y.J.; Cho, C.S. Oral immunization of FMDV vaccine using PH-sensitive and mucoadhesive thiolated cellulose acetate phthalate microparticles. *Tissue Eng. Regen. Med.* **2018**, *15*, 1–11. [CrossRef] [PubMed]
159. Singh, I.; Rana, V. Enhancement of mucoadhesive property of polymers for drug delivery applications: A critical review. *Rev. Adhe. Adhes.* **2013**, *1*, 271. [CrossRef]
160. Schneider, H.; Pelaseyed, T.; Svensson, F.; Johansson, M.E.V. Study of mucin turnover in the small intestine by in vivo labeling. *Sci. Rep.* **2018**, *8*, 5760. [CrossRef] [PubMed]
161. Asim, M.H.; Nazir, I.; Jalil, A.; Matuszczak, B.; Bernkop-schnürch, A. Tetradeca-thiolated cyclodextrins: Highly mucoadhesive and in-situ gelling oligomers with prolonged mucosal adhesion. *Int. J. Pharm.* **2020**, *577*, 119040. [CrossRef]
162. Bansil, R.; Turner, B.S. The biology of mucus: Composition, synthesis and organization. *Adv. Drug Deliv. Rev.* **2018**, *124*, 3–15. [CrossRef]
163. Serra, L.; Doménech, J.; Peppas, N.A. Design of poly (ethylene glycol)-tethered copolymers as novel mucoadhesive drug delivery systems. *Eur. J. Pharm. Biopharm.* **2006**, *63*, 11–18. [CrossRef]
164. Wang, Y.; Lai, S.K.; Pace, A.; Cone, R.; Hanes, J. Addressing the PEG mucoadhesivity paradox to engineer nanoparticles that "slip" through the human mucus barrier. *Angew. Chem. Int. Ed. Engl.* **2009**, *47*, 9726–9729. [CrossRef]
165. Karabasz, A.; Szczepanowicz, K.; Cierniak, A.; Mezyk-Kopec, R.; Dyduch, G.; Szczęch, M.; Bereta, J.; Bzowska, M. In vivo studies on pharmacokinetics, toxicity and immunogenicity of polyelectrolyte nanocapsules functionalized with two different polymers: Poly-L-glutamic acid or PEG. *Int. J. Nanomed.* **2019**, *14*, 9587–9602. [CrossRef]
166. Chen, B.M.; Cheng, T.L.; Roffler, S.R. Polyethylene glycol immunogenicity: Theoretical, clinical, and practical aspects of anti-polyethylene glycol antibodies. *ACS Nano* **2021**, *15*, 14022–14048. [CrossRef] [PubMed]

167. Estapé Senti, M.; de Jongh, C.A.; Dijkxhoorn, K.; Verhoef, J.J.F.; Szebeni, J.; Storm, G.; Hack, C.E.; Schiffelers, R.M.; Fens, M.H.; Boross, P. Anti-PEG antibodies compromise the integrity of pegylated lipid-based nanoparticles via complement. *J. Control. Release* **2022**, *341*, 475–486. [CrossRef]
168. Chang, X.; Yu, W.; Ji, S.; Shen, L.; Tan, A.; Hu, T. Conjugation of PEG-hexadecane markedly increases the immunogenicity of pneumococcal polysaccharide conjugate vaccine. *Vaccine* **2017**, *24*, 1698–1704. [CrossRef] [PubMed]
169. Abhyankar, M.M.; Orr, M.T.; Lin, S.; Suraju, M.O.; Simpson, A.; Blust, M.; Pham, T.; Guderian, J.A.; Tomai, M.A.; Elvecrog, J.; et al. Adjuvant composition and delivery route shape immune response quality and protective efficacy of a recombinant vaccine for *Entamoeba histolytica*. *npj Vaccines*. **2018**, *3*, 22. [CrossRef] [PubMed]
170. Shi, D.; Beasock, D.; Fessler, A.; Szebeni, J.; Ljubimova, J.Y.; Afonin, K.A.; Dobrovolskaia, M.A. To PEGylate or not to PEGylate: Immunological properties of nanomedicine's most popular component, polyethylene glycol and its alternatives. *Adv. Drug Deliv. Rev.* **2023**, *180*, 114079. [CrossRef]
171. Bamberger, D.; Hobernik, D.; Konhäuser, M.; Bros, M.; Wich, P.R. Surface modification of polysaccharide-based nanoparticles with PEG and dextran and the effects on immune cell binding and stimulatory characteristics. *Mol. Pharm.* **2017**, *14*, 4403–4416. [CrossRef]
172. Batrakova, E.V.; Kabanov, A.V. Pluronic block copolymers: Evolution of drug delivery concept from inert nanocarriers to biological response modifiers. *J. Control. Release* **2008**, *130*, 98–106. [CrossRef]
173. Liu, D.; Yang, M.; Wang, D.; Jing, X.; Lin, Y.; Feng, L.; Duan, X. Dpd Study on the interfacial properties of PEO/PEO-PPO-PEO/PPO ternary blends: Effects of pluronic structure and concentration. *Polymers* **2021**, *13*, 2866. [CrossRef]
174. Petit, B.; Bouchemal, K.; Vauthier, C.; Djabourov, M.; Ponchel, G. The counterbalanced effect of size and surface properties of chitosan-coated poly (isobutylcyanoacrylate) nanoparticles on mucoadhesion due to pluronic F68 addition. *Pharm. Res.* **2012**, *29*, 943–952. [CrossRef]
175. Díaz, A.G.; Quinteros, D.A.; Paolicchi, F.A.; Rivero, M.A.; Palma, S.D.; Pardo, R.P.; Clausse, M.; Zylberman, V.; Goldbaum, F.A.; Estein, S.M. Mucosal immunization with polymeric antigen BLSOmp31 using alternative delivery systems against *Brucella Ovis* in rams. *Vet. Immunol. Immunopathol.* **2019**, *209*, 70–77. [CrossRef]
176. Pastor, Y.; Ting, I.; Luisa, A.; Manuel, J.; Gamazo, C. Intranasal delivery system of bacterial antigen using thermosensitive hydrogels based on a pluronic-gantrez conjugate. *Int. J. Pharm.* **2020**, *579*, 119154. [CrossRef] [PubMed]
177. Di Cola, E.; Cantu, L.; Brocca, P.; Rondelli, V.; Fadda, G.C.; Canelli, E.; Martelli, P.; Clementino, A.; Sonvico, F.; Bettini, R.; et al. Novel O/W nanoemulsions for nasal administration: Structural hints in the selection of performing vehicles with enhanced mucopenetration. *Colloids Surf. B Biointerfaces* **2019**, *183*, 110439. [CrossRef]
178. Lopes, M.; Shrestha, N.; Correia, A.; Shahbazi, M.; Sarmento, B.; Hirvonen, J.; Veiga, F.; Seic, R. Dual chitosan/albumin-coated alginate/dextran sulfate nanoparticles for enhanced oral delivery of insulin. *J. Control. Release* **2016**, *232*, 29–41. [CrossRef]
179. Manchanda, S.; Sahoo, P.K.; Majumdar, D.K. Mucoadhesive chitosan-dextran sulfate nanoparticles of acetazolamide for ocular hypertension. *Nanotechnol. Rev.* **2016**, *5*, 445–453. [CrossRef]
180. Ferreira, L.M.B.; Alonso, J.D.; Kiill, C.P.; Ferreira, N.N.; Buzzá, H.H.; Martins de Godoi, D.R.; de Britto, D.; Assis, O.B.G.; Seraphim, T.V.; Borges, J.C.; et al. Exploiting supramolecular interactions to produce bevacizumab-loaded nanoparticles for potential mucosal delivery. *Eur. Polym. J.* **2018**, *103*, 238–250. [CrossRef]
181. Elmowafy, E.; Soliman, M.E. International journal of biological macromolecules losartan-chitosan/dextran sulfate microplex as a carrier to lung therapeutics: Dry powder inhalation, aerodynamic profile and pulmonary tolerability. *Int. J. Biol. Macromol.* **2019**, *136*, 220–229. [CrossRef] [PubMed]
182. Pirouzmand, H.; Khameneh, B.; Tafaghodi, M. Immunoadjuvant potential of cross-linked dextran microspheres mixed with chitosan nanospheres encapsulated with tetanus toxoid. *Pharm. Biol.* **2017**, *55*, 212–217. [CrossRef] [PubMed]
183. Poinard, B.; Lam, S.A.E.; Neoh, K.G.; Kah, J.C.Y. Mucopenetration and biocompatibility of polydopamine surfaces for delivery in an ex vivo porcine bladder. *J. Control. Release* **2019**, *300*, 161–173. [CrossRef]
184. Uhl, P.; Grundmann, C.; Sauter, M.; Storck, P.; Tursch, A.; Özbek, S.; Leotta, K.; Roth, R.; Witzigmann, D.; Kulkarni, J.A.; et al. Coating of PLA-nanoparticles with cyclic, arginine-rich cell penetrating peptides enables oral delivery of liraglutide. *Nanomedicine* **2020**, *24*, 102132. [CrossRef]
185. Ghadiri, M.; Young, P.M.; Traini, D. Strategies to enhance drug absorption via nasal and pulmonary routes. *Pharmaceutics* **2019**, *11*, 113. [CrossRef]
186. Suzuki, H.; Kondoh, M.; Li, X.; Takahashi, A.; Matsuhisa, K.; Matsushita, K.; Kakamu, Y.; Yamane, S.; Kodaka, M.; Isoda, K.; et al. A toxicological evaluation of a claudin modulator, the c-terminal fragment of *Clostridium Perfringens* enterotoxin, in mice. *Pharmazie* **2011**, *66*, 543–546. [CrossRef]
187. Moghimipour, E.; Ameri, A.; Handali, S. Absorption-enhancing effects of bile salts. *Molecules* **2015**, *20*, 14451–14473. [CrossRef]
188. Zhang, H.; Huang, X.; Sun, Y.; Lu, G.; Wang, K.; Wang, Z.; Xing, J.; Gao, Y. Improvement of pulmonary absorption of poorly absorbable macromolecules by hydroxypropyl-β-cyclodextrin grafted polyethylenimine (HP-β-CD-PEI) in rats. *Int. J. Pharm.* **2015**, *489*, 294–303. [CrossRef]
189. Zhang, T.; Li, M.; Han, X.; Nie, G.; Zheng, A. Effect of different absorption enhancers on the nasal absorption of nalmefene hydrochloride. *AAPS PharmSciTech.* **2022**, *23*, 143. [CrossRef] [PubMed]

190. Oh, D.W.; Kang, J.H.; Kim, Y.J.; Na, S.B.; Kwan Kwon, T.; Kim, S.; Hwan Shin, D.; Jie, G.; Shin, M.S.; Sung Kang, K.; et al. Preparation of inhalable N-acetylcysteine-loaded magnetite chitosan microparticles for nitrate adsorption in particulate matter. *Int. J. Pharm.* **2023**, *630*, 122454. [CrossRef] [PubMed]
191. Zhang, Y.; Li, H.; Wang, Q.; Hao, X.; Li, H.; Sun, H.; Han, L.; Zhang, Z.; Zou, Q.; Sun, X. Rationally designed self-assembling nanoparticles to overcome mucus and epithelium transport barriers for oral vaccines against *Helicobacter Pylori. Adv. Funct. Mater.* **2018**, *28*, 1802675. [CrossRef]

polymers

MDPI

Article

Melanin Nanoparticles Obtained from Preformed Recombinant Melanin by *Bottom-Up* and *Top-Down* Approaches

Sergio Alcalá-Alcalá [1], José Eduardo Casarrubias-Anacleto [1], Maximiliano Mondragón-Guillén [2], Carlos Alberto Tavira-Montalvan [2], Marcos Bonilla-Hernández [1], Diana Lizbeth Gómez-Galicia [3], Guillermo Gosset [4] and Angélica Meneses-Acosta [2,*]

1 Laboratorio de Investigación en Tecnología Farmacéutica, Facultad de Farmacia, Universidad Autónoma del Estado de Morelos, Cuernavaca 62209, Morelos, Mexico; sergio.alcala@uaem.mx (S.A.-A.)
2 Laboratorio de Biotecnología Farmacéutica, Facultad de Farmacia, Universidad Autónoma del Estado de Morelos, Cuernavaca 62209, Morelos, Mexico
3 Farmacia Hospitalaria, Facultad de Farmacia, Universidad Autónoma del Estado de Morelos, Cuernavaca 62209, Morelos, Mexico
4 Departamento de Ingeniería Celular y Biocatálisis, Instituto de Biotecnología, Universidad Nacional Autónoma de México, Cuernavaca 62209, Morelos, Mexico
* Correspondence: angelica_meneses@uaem.mx; Tel.: +52-7773297000 (ext. 3366)

Abstract: Melanin is an insoluble, amorphous polymer that forms planar sheets that aggregate naturally to create colloidal particles with several biological functions. Based on this, here, a preformed recombinant melanin (PRM) was utilized as the polymeric raw material to generate recombinant melanin nanoparticles (RMNPs). These nanoparticles were prepared using *bottom-up* (nanocrystallization—NC, and double emulsion–solvent evaporation—DE) and *top-down* (high-pressure homogenization—HP) manufacturing approaches. The particle size, Z-potential, identity, stability, morphology, and solid-state properties were evaluated. RMNP biocompatibility was determined in human embryogenic kidney (HEK293) and human epidermal keratinocyte (HEKn) cell lines. RMNPs prepared by NC reached a particle size of 245.9 ± 31.5 nm and a Z-potential of −20.2 ± 1.56 mV; 253.1 ± 30.6 nm and −39.2 ± 0.56 mV compared to that obtained by DE, as well as RMNPs of 302.2 ± 69.9 nm and −38.6 ± 2.25 mV using HP. Spherical and solid nanostructures in the *bottom-up* approaches were observed; however, they were an irregular shape with a wide size distribution when the HP method was applied. Infrared (IR) spectra showed no changes in the chemical structure of the melanin after the manufacturing process but did exhibit an amorphous crystal rearrangement according to calorimetric and PXRD analysis. All RMNPs presented long stability in an aqueous suspension and resistance to being sterilized by wet steam and ultraviolet (UV) radiation. Finally, cytotoxicity assays showed that RMNPs are safe up to 100 μg/mL. These findings open new possibilities for obtaining melanin nanoparticles with potential applications in drug delivery, tissue engineering, diagnosis, and sun protection, among others.

Keywords: melanin nanoparticles; preformed recombinant melanin; nanocrystallization; high-pressure homogenization; double emulsion–solvent evaporation

check for updates

Citation: Alcalá-Alcalá, S.; Casarrubias-Anacleto, J.E.; Mondragón-Guillén, M.; Tavira-Montalvan, C.A.; Bonilla-Hernández, M.; Gómez-Galicia, D.L.; Gosset, G.; Meneses-Acosta, A. Melanin Nanoparticles Obtained from Preformed Recombinant Melanin by *Bottom-Up* and *Top-Down* Approaches. *Polymers* **2023**, *15*, 2381. https://doi.org/10.3390/polym15102381

Academic Editors: Antonio M. Borrero-López, Concepción Valencia-Barragán, Esperanza Cortés Triviño, Adrián Tenorio-Alfonso, Clara Delgado-Sánchez and Helena Felgueiras

Received: 17 March 2023
Revised: 15 May 2023
Accepted: 16 May 2023
Published: 19 May 2023

1. Introduction

Melanin is an irregular, amorphous polymer of high molecular weight, generated by the oxidative polymerization of various indole and phenolic compounds such as L-tyrosine or levodopa (L-DOPA), which forms graphite-like planar sheets that aggregate naturally in a hierarchal fashion to form colloidal structures that can reach diameters of hundreds of nanometers [1]. The physical properties that make melanin unique are its light-absorbing ability, chemical resistance, electrical conductivity, and its low or null solubility in organic solvents or acidic aqueous media [1,2]. This natural polymeric pigment is distributed in all taxa of nature in compact granules integrated into the cytoplasm of different cells such as plant cell walls, some bacteria, fungi, the feathers of birds, and in the skin and hair of mammals [3]. Melanin has several biological functions, including protection against oxidizing agents [4,5], thermoregulation, the promotion of cell proliferation and differentiation [6]; it aids against osmotic and temperature changes [7], the capture of ultraviolet radiation [8], and it possesses chelating properties and immunomodulatory activities [9]. Several melanin types are derived from the evolutionary process: eumelanin is present in animals and humans as part of dark colors; pheomelanin is regularly associated with proteins to form chromoproteins that give a reddish-yellow coloration, and pyomelanin that is found in some fungi [10]. In humans, pheomelanin and pyomelanin are also found in other tissues and organs such as the retina, hair, brain, and liver [11]. Eumelanin is a brown-to-black pigment produced in humans by melanocytes (cells found in the basal layer of the epidermis in the skin) and it is distributed in granules ranging in size from 200 to 300 nm in the keratinocytes of the skin [12]. Knowledge of the melanin-formation route and of its properties as a polymer has enabled synthetic or biotechnological melanin to be obtained, which can be controlled in the physical characteristics or handled to generate new materials, such as copolymers, composites, or nanostructures [13,14].

Several studies have highlighted the potential use of melanin nanoparticles (MNPs) in different areas; their semiconductor properties have led to the generation of electronic films [15,16] and their high radiation-absorption capacity and antioxidant properties have been exploited to serve as adjuvants in cancer-radiation therapy [17,18] and sun protection against UV radiation [19,20]. In addition, they have also been suggested as drug carriers [21,22]. Other findings have shown an MNP affinity for metal ions such as Mg^{2+}, Ca^{2+}, and Na^{1+}, acting as chelating agents, or with the ability to trap metal ions with a potential toxic effect, such as Cd^{2+} and Pb^{2+} [23]. Furthermore, MNPs present good photoacoustic properties that have been used for the diagnosis of breast tumors [24,25] and as a contrast agent in optoacoustic tomography [26]. Other studies have demonstrated that MNPs could protect hematopoietic cells from free radicals induced by gamma radiation in mice [27]. Additionally, along with developing new biomaterials, such as biodegradable polymers, MNPs have become useful in tissue engineering, serving as a support and guide for proper cell growth by forming biofunctional structures [28].

The most common manufacturing process for obtaining synthetic MNPs reported in the literature has mainly been the *in-situ* polymerization method by the oxidation of L-tyrosine, its natural precursor. However, biotechnology and genetic engineering have made it possible to obtain melanin in industrial quantities from recombinant sources, with the advantage of knowing its chemical, physical, and physicochemical properties prior to material manipulation in a nanoparticle manufacturing process [29,30], favoring homogeneous production and improved quality control. This preformed raw material can be controlled in molecular weight and chemical structure, and due to its mechanical and thermal resistance, may be nanosized using different techniques without the necessity of polymerization reactions, allowing its use in large quantities.

On the other hand, there are two different approaches to address the nanotechnology, that is, *bottom-up* and *top-down*, which include all the existing manufacturing methods to generate nanoparticles. For the selection of the best approach, it is necessary to know the raw material in terms of the solid state, chemical behavior, solubility, pH dependence, and mechanical or thermal behavior [31]. The *bottom-up* approach starts from the molecular or atomic state of matter, employing methods that benefit the assembly or aggregation of material until the nanostructure is achieved. A common feature in the methods developed by this approach is that precursors or raw materials should be in molecular dispersion, that is, previously dissolved, to permit the addition of an antisolvent, or create an environment where nanoparticle formation is promoted; the use of surfactants or stabilizers plays an essential role in these methods [32]. Technologies such as microemulsion, single or multi-emulsion (double)—with subsequent evaporation, extraction, or diffusion solvent, as well as interfacial emulsion polymerization and nanoprecipitation—are included in this approach [33,34]. The *top-down* approach refers to the production of nanostructures from macrometric materials by means of size reduction through various techniques such as milling media, fragmentation, cutting or engraving, machining film processes, photolithography, printing, or homogenization at high or ultra-pressure. These manufacturing methods require large amounts of energy and generate large amounts of heat, posing a great danger to labile materials [31,32].

In this work, recombinant melanin nanoparticles (RMNPs) were obtained utilizing three different manufacturing methods (nanocrystallization, double emulsion–solvent evaporation, and high-pressure homogenization), by manipulating a preformed recombinant melanin (PRM) as a new raw polymeric material, with the aim of evaluating and exploring its capacity to be utilized to generate nanoparticles when these manufacturing approaches are applied, as well as of evaluating the effect of the process parameters on the physical properties of the nanoparticles. Methods were optimized to generate RMNPs with a target size of ~250 nm using statistical techniques. The nanoparticles were characterized by their physical, chemical, physicochemical, and in vitro biocompatibility properties.

2. Materials and Methods

2.1. Materials and Cell Lines

Preformed recombinant melanin was donated by the Institute of Biotechnology—UNAM. Briefly, melanin was obtained using a platform of recombinant *Escherichia coli*, expressing the tyrosinase gene from *Rhizobium etli* by Chávez-Béjar et al. [35]. Sodium hydroxide, ammonium hydroxide and hydrochloric acid were purchased from JT Baker (Monterrey, Mexico) (>99% pure; analytical reagent). Solvents such as propanol, hexane, benzene, dichloromethane, glyceryl triacetate, n-butanol, octanol, ethyl acetate, ethanol, acetone, isopropyl alcohol, and isobutyl alcohol were supplied by Fermont (Monterrey, Mexico) and Sigma-Aldrich (Saint Louis, MO, USA) (>98% pure; reagent grade). Polyvinyl alcohol (PVAL, Mowiol 4-88) and polysorbate 80 (Tween 80), used as stabilizing agents, were supplied by Sigma-Aldrich (Saint Louis, MO, USA). Distilled water from a Milli-Q filtration system was used throughout the experiments (Millipore, Billerica, MA, USA).

Two types of certified cell lines were used: human embryonic kidney (HEK293) (ATCC CRL-1573) and human epidermal keratinocytes from normal neonatal foreskin (HEKn) (ATCC PCS-200-010). HEK-293 cells were cultivated using Dulbecco's Modified Eagle's Medium/Nutrient Mixture F-12 (DMEM/F12-GIBCO) supplemented with 10% FBS (GIBCO), 3.574 g/L of (4-(2-hydroxyethyl)-1-piperazinaetanosulfonic [HEPES]), and 2.44 g/L of sodium bicarbonate. HEKn cells were grown using Dermal Cell Basal Medium (DCBM) (ATCC PCS-200-030) supplemented with a specific Keratinocyte Growth Kit (ATCC PCS-200-040) in T-75 culture flasks in an atmosphere of 5% CO_2 at 37 °C, according to the instructions provided by the manufacturers. Percentage of cell viability was determined using two different standard methods such as Trypan Blue staining (Sigma-Aldrich; Saint Louis, MO, USA) and MTS (CellTiter 96®Aqueous One Solution Cell Proliferation Assay; PROMEGA, Madison, WI, USA).

2.2. Characterization of Preformed Recombinant Melanin (PRM)

PRM was previously generated by Chávez-Béjar et al. [35,36], through a bioprocess in which plasmid *pTrcMutmelA* with the *MutmelA* gene of *Rhizobium etli* that encodes for the enzyme tyrosinase was used to transform *Escherichia coli W3110* into *E. coli W3110/pTrcMutmelA*, a strain capable of producing eumelanin from L-tyrosine as substrate. The donated melanin was a dark brown-colored material, with an irregular shape and an amorphous granular aspect (Figure S1A). One hundred g of this preformed melanin was placed in a mortar and pulverized for 10 min until a fine powder was obtained (Figure S1B). This powder was used as the raw material in all further studies and characterized in terms of chemical identity, apparent solubility, and capacity to be sterilized by wet steam to define the process conditions in the melanin-nanoparticle manufacturing methods.

2.2.1. Melanin Identity by FTIR and UV–Vis Spectrophotometry

The IR and UV–Vis spectra of the PRM were obtained. In brief, for IR studies, 2 mg of pulverized PRM was placed onto the ATR implement of the FTIR spectrophotometer (ABB-MB3000 FTIR, ABB Group, QC, Canada). Then, 150 scans per sample were measured, in the range of a wavenumber of 500–4000 cm^{-1}, with a resolution of 16. The reported IR spectrum represents the average of the scans. To record the UV–Vis spectrum, a PRM solution was prepared from 25 μg/mL in 1 N NaOH, with the scanning of the light absorption in the UV–Visible region of the electromagnetic spectrum, in a wavelength range of 200–800 nm, utilizing a double-beam Cary 60 UV–Vis spectrophotometer (Agilent, Santa Clara, CA, USA), employing 1 N NaOH solution as blank. The maximal absorption wavelength for PRM was set from the UV spectrum. PRM identity was verified by comparing the obtained spectra with the melanin IR and UV spectra previously reported in the literature.

2.2.2. Apparent Solubility

The apparent solubility of PRM in different solvents was carried out to determine the subsequent approach and strategy to generate recombinant melanin nanoparticles (RMNP). Briefly, 10 mg of PRM raw material was added to a series of 10 mL test tubes, which were labeled with the test solvent. Then, increasing amounts of 100 μL of different dissolution media, organic solvents (of different polarities) or aqueous media (at different pHs), were added until complete dissolution was observed, or a maximal volume of 10 mL was reached. All samples were shaken vigorously with the aid of a vortex. When 10 mL of solvent was used and no dissolution was observed, the result was taken as "practically insoluble in the medium". Tests were performed in triplicate, reporting an average in mg/mL.

2.2.3. Autoclavable Resistance

To assess the resistance of the PRM raw material to the wet steam sterilization process, three samples of 2 g of powder, placed in a 10 mL test tube, were sterilized in a Hinotek autoclave (YXQ-LS-SII Vertical type; Hinotek, Ningbo, China), for 20 min at 121 °C and at 15 psi of pressure. Afterward, to evaluate the chemical stability, the IR spectra of the samples were obtained, employing the methodology mentioned previously.

2.3. *Obtaining Recombinant Melanin Nanoparticles (RMNPs) by a Bottom-Up Approach*

Since the manufacture of nanoparticles under this approach requires manipulation of the material at a molecular level, the results of apparent solubility were considered to select the manufacturing methods. Due to melanin being only soluble in alkaline aqueous media (pH > 12), the manufacturing methods of nanocrystallization and double emulsion–solvent evaporation were selected, where the dissolution medium or the aqueous phase was a 1N of NaOH solution.

2.3.1. Nanocrystallization Method (NC)

The preparation of RMNPs by the nanocrystallization method (nanoprecipitation) was performed using an adaptation of the methodology proposed by Chen et al. [31,37]. Nanoprecipitation is based on generating a non-solvent environment for the material that is to be nanosized. Here, an alkaline PRM solution (at a concentration of 1 mg/mL in 1N NaOH) was neutralized with an acidic solution (HCl) of PVAL (stabilizing agent), utilizing a dosing unit (Dosimat 665; Metrohm AG, Herisau, Switzerland). The PRM alkaline solution was poured into the acid solution, at a rate of 14 μL/s and constant stirring at 600 rpm. To evaluate the effect of the manufacturing-process parameters on the characteristics of the RMNPs, a 2^K experimental design (k = 3) was established with the following factors and levels: X_1: volume of the acidic aqueous phase (10 mL or 30 mL); X_2: concentration of HCl (0.5 N or 1 N), and X_3: the concentration of stabilizing agent (PVAL at 1% or 5% *w*/*v*). Table 1 presents the design matrix and the total experimental runs. The volume of the alkaline PRM solution (10 mL) and the stirring time after neutralization (2 h)

(Dual-Range Mixer IKA RW20 digital, IKA-Werke, Satufen, Germany) remained constant in all experiments. The formed RMNPs were obtained in aqueous suspension; hence, the nanoparticles were washed three times by centrifugation at 3000× g, 4 °C, and 60 min (Eppendorf 5804R centrifuge, 15-amp version, Thermo Fisher Scientific, Waltham, MA, USA), then the nanoparticles were resuspended in each wash with 5 mL of distilled water. Average particle size (Y_1) and Z-potential (Y_2) were measured as response variables in all runs.

Table 1. Experimental designs to produce and optimize RMNPs, which were used for each manufacturing process.

	Bottom-Up **Approach**											***Top-Down*** **Approach**			
	Nanocrystallization (NC)					**Double Emulsion–Solvent Evaporation (DE)**					**High-Pressure Homogenization (HP)**				
	Factor-Levels			**Response Variables**		**Factor-Levels**			**Response Variables**		**Factor-Levels**		**Response Variables**		
Run	**X_1**	**X_2**	**X_3**	**Y_1**	**Y_2**	**X_1**	**X_2**	**X_3**	**Y_1**	**Y_2**	**X_1**	**X_2**	**Y_1**	**Y_2**	
1	30	0.5	1	590.9	−2.3	10	6.66:3.33	5	167.5	−31.2	18,000	8	317.2	−38.6	
2	10	0.5	1	958.6	−33.3	10	10:00	10	874.3	−25.5	23,000	10	340.6	−41.6	
3	30	0.5	5	358.9	−7.6	20	13.33:6.66	5	513.8	−31.0	12,000	5	505.1	−35.3	
4	10	1	5	270.4	−12.4	20	20:00	5	232.9	−39.8	23,000	5	408.9	−37.8	
5	10	0.5	5	160.0	−19.1	20	13.33:6.66	10	839.7	−27.8	12,000	10	402.8	−38.5	
6	30	1	5	687.0	−1.0	20	20:00	10	455.8	−18.9					
7	30	1	1	568.9	−30.1	10	10:00	5	496.8	−23.3					
8	10	1	1	549.7	−2.8	10	6.66:3.33	10	484.1	−23.7					
	X_1: Volume of HCl solution (mL) X_2: HCl concentration (N) X_3: PVAL concentration (%w/v)					X_1: Volume of O phase (mix of DCM-ACE) (mL) X_2: DCM:ACE ratio (mL) X_3: Homogenization time (min)					X_1: Pressure (psi) X_2: Number of cycles				
	Y_1: Particle size of RMNPs (nm) Y_2: Z-potential (mV)														

2.3.2. Double Emulsion–Solvent Evaporation Method (DE)

Due to the nonsolubility of melanin in organic solvents, the double emulsion–solvent evaporation method proposed by Liu et al. [38] was adapted to produce the RMNPs. Briefly, a W1/O first emulsion was prepared by homogenization at high shear, whereby the first aqueous phase (W1), composed of an alkaline solution of PRM (1 mg/mL, in 1 N NaOH) and a stabilizing agent (Tween 80), was dispersed into an organic phase (O) composed of dichloromethane or a mixture of dichloromethane acetone. This emulsion was then homogenized under gentle stirring in a second aqueous phase (W_2), comprising an acidic solution of HCl and a second stabilizing agent (PVAL). A neutralized $W_1/O/W_2$ emulsion type was formed. Evaporation of the O phase in conjunction with the phenomenon of neutralization caused the generation of the RMNPs as an aqueous suspension.

The manufacture of RMNPs by this method was optimized following a 2^k experimental design (k = 3), which was established with the following factors and levels: X_1: volume of O phase (mix of dichloromethane [DCM]-acetone [ACE]) (10 mL or 20 mL); X_2: DCM-ACE ratio in the O phase (at proportions of 1:0 or 2:1), and X_3: the homogenization time in the first emulsion (5 min or 10 min). Table 1 depicts the design matrix and the experimental runs. The first W_1/O emulsion was generated using an UltraTurrax homogenizer (IKA T18 Digital, IKA-Werken, Staufen, Germany) at 12,000 rpm according to the predefined homogenization times in the experimental design. Then, the first emulsion was poured into 40 mL of an acidic PVAL solution (HCl at 0.25 N), stirring at 600 rpm (Dual-Range Mixer IKA RW20 digital, IKA-Werken, Staufen, Germany) for 30 min. At the end of the stirring time of the second emulsion, the O phase was evaporated entirely under reduced pressure, using a rotary evaporator (IKA RV 10) at 90 rpm and 30 °C. During the experiments, the concentration of the stabilizing agents in the first and the second emulsions (Tween 80 at 5% v/v, and PVAL at 1% w/v, respectively) and the volume of the aqueous phase W1 (10 mL) in the first emulsion remained constant. In each experimental run, the RMNPs were recovered by centrifugation at 3000× g, 4 °C, and 60 min (Eppendorf 5804R centrifuge, 15-amp version) and washed three times to eliminate stabilizing agent residues, resuspending each wash with 5 mL of distilled water. Average particle size (Y_1) and Z-potential (Y_2) were measured as response variables in all runs.

2.4. Obtaining Recombinant Melanin Nanoparticles (RMNPs) by a Top-Down Approach

To produce RMNPs using this approach, the material should be reduced to nanoscale size by milling, collision, attrition, and sometimes cavitation. Considering the crystalline structure of PRM and the possibility of reducing its size by pulverization with a mortar, RMNPs were manufactured with the following technique:

High-Pressure Homogenization Method (HP)

A high-pressure homogenizer (Microfluidizer LM 10, Microfluidics International Corporation, Newton, MA, USA) was used. Particle size reduction was evaluated through a 2^k experimental design (k = 2), with one central point. Factors and levels considered in the study included X_1: applied pressure (12,000 or 23,000 psi), and X_2: number of cycles (5 cycles in the first step or 10 cycles in the final stage). One cycle corresponds to the complete passage of the samples from the hopper through the equipment's impact micro-camera until its exit. A volume of 40 mL of an aqueous suspension of PRM was treated at the predefined pressure and cycles. The average particle size (Y_1) and Z-potential (Y_2) were immediately measured in the final aqueous suspension. Design matrix and experimental runs are included in Table 1.

2.5. RMNP Optimization and Recovery

Once the RMNPs were obtained by the three evaluated technologies, the optimal manufacturing conditions were defined by response surface methodology (RSM) and desirability function for multiple responses. The combination of levels was estimated for the factors through which RMNPs with an average particle size of 250 ± 50 nm and high Z-potential were reached; these values were fixed as optimization objectives. The average particle-size target was established to mimic the nanometric natural size of melanin granules in keratinocytes, and a high Z-potential was desired to assure good stability in suspension (expecting ± 30 mV). The predicted levels for each method are presented in Table 2. Three batches of RMNPs were manufactured under the optimized conditions to verify the predictive model: the prepared nanoparticles were identified as RMNP-NC, RMNP-DE, and RMNP-HP.

As the final step in the manufacturing process, the RMNP suspensions were washed three times with 5 mL of distilled water, using centrifugation cycles (at 4 °C, 15,557× g, and 60 min). To prevent particle aggregation after centrifugation, all RMNP samples were sonicated using an ultrasonic probe (SONICS Vibra Cell VCX 130, Sonics and Materials Inc, Newtown, CT, USA) at 10 kHz, for 30 min. Finally, these aqueous suspensions of RMNPs were lyophilized. Briefly, all samples were frozen for 3 h on dry ice, and placed on a bulk rack in the condenser at −80 °C in the lyophilizer (BenchTop Pro with Omnitronics, SP Scientific VirTis, Warminster, PA, USA). Vacuum was applied to reach 200 mTorr for 48 h. These dried nanoparticles, that is, RMNP-NC, RMNP-DE, and RMNP-HP, were employed to conduct subsequent physical characterization studies and biocompatibility assays.

Table 2. RMNPs prepared under optimized conditions: model verification; n = 3 batches per process. Optimization targets: Y_1 = 250 nm and Y_2 = maximized response.

Method	Factors	Estimated Levels	$\hat{Y}_1$ (nm)	D	Y_1 (nm)	PEE (%)	Y_2 (mV)
NC	X_1: Volume of HCl solution X_2: HCl concentration X_3: PVAL concentration	10 mL 0.56 N 4.7 % *w*/*v*	250.2	0.86	245.9 ± 31.5	1.72	−20.2 ± 1.56
DE	X_1: Volume of organic phase X_2: DCM:ACE ratio X_3: Homogenization time	19.7 mL 19.7:0 ratio (mL) 5 min	252.7	0.89	253.1 ± 30.6	0.16	−39.2 ± 0.56
HP	X_1: Applied pressure X_2: Number of cycles	18,000 psi 10 cycles	~321.17	0.99	302.2 ± 69.9	5.91	−38.6 ± 2.25

NC: Nanocrystallization
DE: Double emulsion–solvent evaporation
HP: High-pressure homogenization (HP)

$\hat{Y}_1$: Predicted particle size
Y_1: Observed particle size
Y_2: Z-Potential
D: Desirability
PEE: Percentage of estimation error

2.6. *Physical Characterization of RMNP*

2.6.1. Particle-Size Analysis

The average particle size of the RMNPs and their distribution, for all experiments, was determined utilizing the dynamic light scattering technique (DLS) of Zetasizer equipment (Nano ZS90, Malvern Panalytical, Malvern Worcestershire, UK). Operational conditions included were distilled water as dispersing medium (refractive index of 1.33), with an equilibration time of 120 s, a diffraction angle of 90° at 25 °C, with a resolution of 16 readings, and in triplicate. Samples were diluted at 1:10 using the same dispersing medium, placing 1 mL of suspension in polystyrene cuvettes.

2.6.2. Zeta-Potential Measurement

Zeta-potential measurement was performed using the electrophoretic light scattering technique under similar conditions to those of the average particle size, using capillary polystyrene cuvettes (Zetasizer Nano ZS90; Malvern Panalytical, Malvern Worcestershire, UK). Measurements were performed on samples that were diluted at a 1:5 ratio using distilled water as dispersing medium.

2.6.3. Morphological Analysis

The morphology of RMNPs was described using transmission electron microscopy (TEM). A volume of 25 µL of each RMNP suspension was placed on a copper grill and was dried at room temperature, adding uranyl acetate as a contrast solution. The micrographs were recorded with an electron microscope (Zeiss EM900 80 kV, Carl Zeiss AG, Oberkochen, Germany) in conjunction with a Dual Vision 300 W CCD camera (Gatan, Inc, Warrendale, PA, USA).

2.6.4. Evaluation of Solid State

This characterization was carried out using the lyophilized RMNPs (RMNP-NC, RMNP-DE, and RMNP-HP).

ATR-FTIR Studies

A mass of approximately 2 mg of dried nanoparticles was placed on a horizontal ATR SeZn attachment in the FTIR spectrophotometer (ABB-MB3000 FTIR; ABB Group, QC, Canada). IR spectra were obtained in the vibrational wavenumber range from 500 to 4000 cm^{-1}, with a resolution of 16 and 150 scans.

Powder X-ray Diffraction (PXRD) Analysis

PXRD analyses were performed for PRM and for all of the prepared RMNPs in the transmission mode on a Bruker D-8 Advance diffractometer (Bruker, Billerica, MA, USA) equipped with a LynxEye detector (λCu-Kα1 = 1.5406 Å, monochromator: germanium) (Bruker, Billerica, MA, USA). The equipment was operated at 40 kV and 40 mA, and data were collected at room temperature in the range of 2θ = 5–40°.

Calorimetric Studies by Differential Scanning Calorimetry (DSC)

Thermal analysis was performed on PRM raw material and dried RMNPs using the DSC technique, with a Q20 Calorimeter (TA Instruments, New Castle, DE, USA). Briefly, a mass of 5–10 mg of each sample was placed in a Tzero aluminum hermetic pan (TA Instruments, New Castle, DE, USA) that was sealed with a press machine. Experimental assays were carried out under the following operating conditions: thermal equilibrium at 15 °C; temperature range from 20 to 300 °C; a heating rate of 10 °C/min, and an inert environment with nitrogen at a flow of 50 mL/min.

2.7. Biocompatibility Evaluation: In Vitro Cytotoxic Assay

Based on physical characterization, RMNP-DE were chosen as ideal for cytotoxic evaluation due to their morphology and reproducibility. Evaluations were made on HEK293 and HEKn cell lines. MTS methodology was used to measure cell activity, and cell viability in the presence of RMNP, and the cell counting was performed with the Trypan Blue technique. Briefly, 20,000 cells per well were seeded in 96-well plates, using DMEM/F12 medium supplemented with 10% FBS at 37 °C and 5% CO_2, which were incubated to promote cell adherence. After 24 h of incubation, different RMNP-DE concentrations (10, 50, and 100 μg/mL) were added to different wells and were then incubated for an additional 72 h; a well with no treatment was set as control. Lastly, the culture medium was eliminated carefully, and 90 μL of DMEM/F12 and 10 μL of MTS were added and incubated for 4 h. The absorbance of the supernatant was measured at 490 nm utilizing an Epoch microplate spectrophotometer (Biotek, Agilent Technologies, Santa Clara, CA, USA).

2.8. Statistical Analysis

Multiple regression analyses were performed using response surface methodology, with an analysis of variance (ANOVA) to assess significant effects and model fit. One-way ANOVA was also employed to compare groups, with the Dunnett post hoc test. Statistically significant differences were considered when $p < 0.05$. All statistical analyses were performed using Statgraphics statistical software (Version Centurion XVIII).

3. Results and Discussion

3.1. PRM Characterization

The recombinant melanin raw material produced in *E. coli W3110/pTrcmelA* was characterized by identity, autoclavable resistance and apparent solubility, because it is a material with a different origin from that of conventional melanin. PRM has a macroscopic physical appearance as a dark-colored irregular and granular material that, after being pulverized, produced a resulting particle size of 7.5 ± 2.9 μm (Figure S1). This initial grinding made it possible to obtain a homogeneous raw material, with an increased surface area, and one that enhanced the dissolution rate during the apparent solubility test.

The UV–Vis and IR spectra of PRM are revealed in Figure 1. The UV–Vis spectrum of PRM previously dissolved in an alkaline solution evidenced that the maximal UV-light absorbance occurs at a wavelength of 222 nm (Figure 1A), which is in agreement with the reports of Tan et al. [39], Dong and Yao [40], and Madhusudhan et al. [41]. Regarding the FTIR studies, the IR spectrum of the PRM demonstrated absorption peaks related to the known functional groups of melanin (Figure 1B); a signal at 3200 cm^{-1} is related to a broad band of the -NH and -OH groups stretching; peaks at 2920 cm^{-1} and 1450 cm^{-1} correspond to the -CH stretching and bending vibration, respectively; a strong signal attributed to the -C=O bond in -COOH groups was observed at 1595 cm^{-1}; vibration of the -C=C- bonds in the aromatic groups is associated with a peak at 1500 cm^{-1}; and finally, a peak at 2385 cm^{-1} can be recognized as the protonated amine of 5,6-dihydroxyindole-2-carboxylic acid (characteristic monomer in melanin). The peaks of these spectra correspond to those reported in the literature [39–43]. These findings allowed the identity of the melanin to be verified as eumelanin. In addition, as depicted in Figure 1B, PRM after the wet-steam sterilization process does not present any differences in its chemical structure, in that no changes in the IR spectrum were observed. This sterilization resistance is relevant when biocompatibility studies are executed or if biomedical applications are considered for this material.

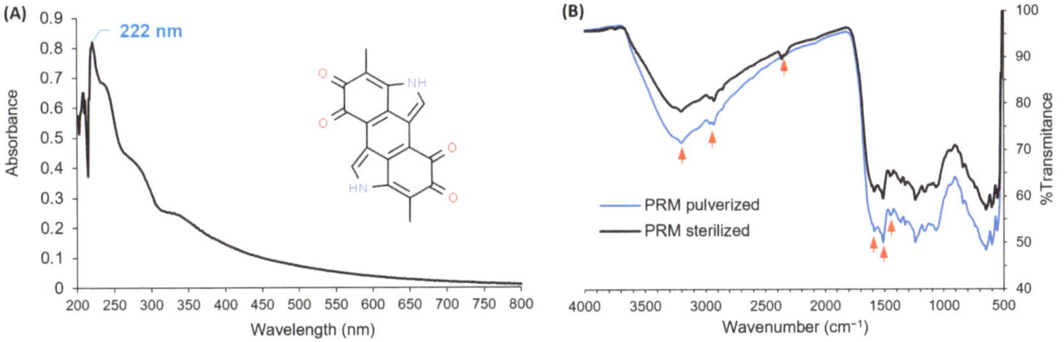

Figure 1. Spectra of PRM: (**A**) UV–Vis spectrum, (**B**) IR spectrum; the red arrows show the absorption peaks related to the known functional groups of melanin.

The results of PRM solubility in different solvents are summarized in Table 3. The insoluble nature of melanin was observed in a wide range of solvents with different polarities, except for alkaline aqueous media (high pH, >12, either 1N NaOH or 1N NH_4OH); however, the results classify it as slightly soluble according to United States Pharmacopeia (USP) criteria. Ye et al. and Guo et al. also reported this solubility behavior for natural and conventional melanin [44,45], which were materials obtained from a source other than a recombinant origin.

The determination of the apparent solubility of PRM was an essential aspect because it provided a starting point for designing a manufacturing strategy to obtain recombinant melanin nanoparticles (RMNPs). This was particularly apparent in the *bottom-up* approach, where materials must be manipulated at a molecular level. Based on these solubility results, two methods widely utilized in the development of nanocrystals were adapted: nanoprecipitation and emulsification–solvent evaporation. For the latter, a double emulsion alternative was chosen, because the nanostructure shape and size could be controlled from the first emulsion, it being possible to neutralize this first aqueous phase in a second emulsion because the organic solvent is evaporated, creating an anti-solvent effect for the PRM, which leads to the formation of the nanoparticles.

Table 3. Apparent solubility of PRM; practically insoluble: < 0.1 mg/mL, and slightly soluble: between 1 and 10 mg/mL.

Test Solvent	Apparent Solubility
Distilled water (pH~6)	Practically insoluble
NaOH 1N (pH > 12)	Slightly soluble (1.10 mg/mL)
NH_4OH 1N (pH > 12)	Slightly soluble (0.98 mg/mL)
HCl 1N (pH 1.2)	Practically insoluble
Ethanol	Practically insoluble
Isopropyl alcohol	Practically insoluble
n-Butanol	Practically insoluble
Acetone	Practically insoluble
Glyceryl triacetate	Practically insoluble
Octanol	Practically insoluble
Ethyl acetate	Practically insoluble
Dichloromethane	Practically insoluble
Benzene	Practically insoluble
Hexane	Practically insoluble

On the other hand, the aqueous-insoluble character of the PRM also facilitated the selection of the dispersing medium to generate nanoparticles in the *top-down* approach. For this project, it was desirable to achieve an aqueous suspension of nanoparticles; thus, distilled water was chosen as the vehicle in the manufacture of RMNPs by the high-pressure homogenization process.

3.2. Melanin Nanoparticles by Bottom-Up and Top-Down Approaches

Three nanoparticle manufacturing methods, including nanocrystallization (NC), double emulsion–solvent evaporation (DE), and high-pressure homogenization (HP), were adapted to prepare nanomaterials from preformed recombinant melanin (PRM). Table 1 presents the design matrix generated and executed for each method; average particle size and Z-potential are reported for all experimental runs.

For the manufacturing processes with a *bottom-up* approach, certain similarities could be observed in the behavior of the average particle size. The NC method generated RMNPs with a size ranging from 160.0 to 958.6 nm, and with Z-potential between −1 and −30 mV. Meanwhile, with the DE method, the RMNPs were obtained with a size distribution range of 167.5–874.3 nm but, unlike the previous method, a more homogeneous Z-potential was observed in the different runs, with values between −18 and −40 mV, which were ideal for good physical stability for the nanoparticles in aqueous suspension. For both methods, experimental run #4 produced RMNPs sufficiently close to the expected target size of 250 ± 50 nm (Table 1).

On the other hand, the top-down approach was applied to reduce the particle size of the pulverized PRM by the application of the HP method to prepare melanin nanoparticles (RMNPs). As shown in Table 1, this technique led to average particle sizes ranging from 317.2 to 505.1 nm, with homogeneous Z-potential values of between −35.3 and −41.6 mV. However, it was not possible to produce nanostructures smaller than 300 nm, even though operating conditions at the highest working pressure and the greatest number of cycles in the high-pressure homogenizer were utilized. For this reason, the manufacture of the optimized RMNPs employing the HP method was fixed at operational conditions under which the smallest particle size was generated, meaning experimental run #1.

Regarding the statistical analysis of the main effects of the factors, this was focused on particle size, in that the main objective was to produce nanoparticles of a specific size (Figure S2). However, during the optimization process, both variables were considered (particle size and Z-potential), carrying out predictions with the fitted models, searching the levels for the studied factors in which the target particle size and a maximal Z-potential could be reached. High Z-potential favors the physical stability of RMNPs in aqueous suspension. Hence, response surface methodology was utilized to define the most appropriate regression model. The response surfaces for the particle size of the RMNPs and the fitted mathematical models for all of the applied methods are presented in Figure 2.

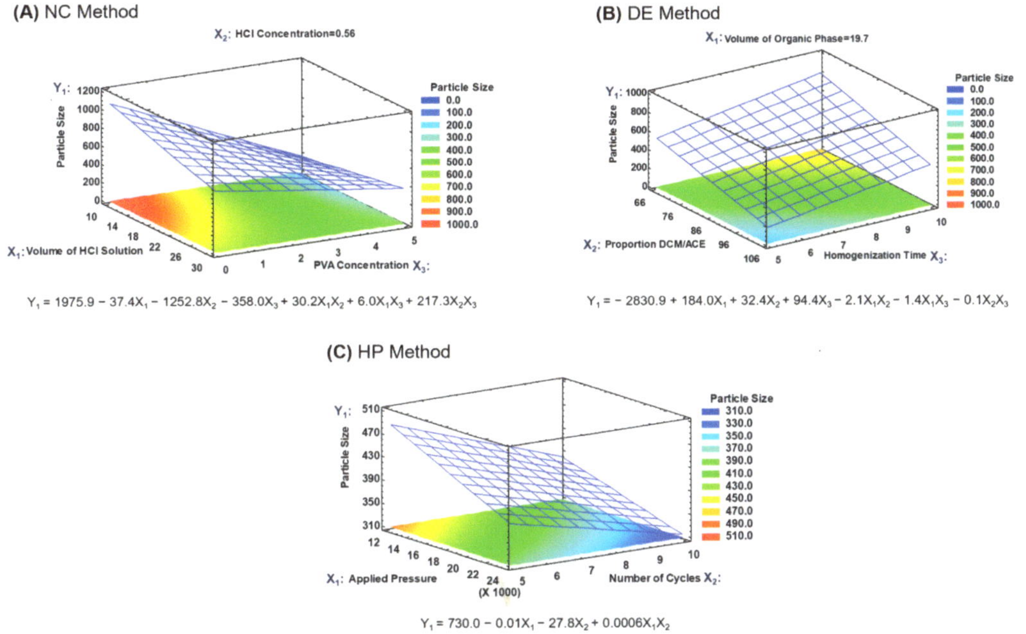

Figure 2. Response surfaces for the particle size of the recombinant melanin nanoparticles (RMNPs) for the three applied manufacturing methods. (**A**) Nanocrystallization (NC), (**B**) double emulsion–solvent evaporation (DE), and (**C**) high-pressure homogenization (HP).

To prepare RMNPs, the stabilizing agent concentration (X_3) was the factor that mainly affected the particle size in the NC method, generating a decrease in the average size as the PVAL concentration increased. This effect is due to the stabilizing effect of the PVAL on the solid–liquid interface during the precipitation process, in that its steric stabilization mechanism exerts an influence on the crystal growth during the formation of the nanoparticle [46,47]. Regarding the neutralization process, it is suggested that chemical equilibrium in some cases was reached, and in others, there was a change in pH sufficiently close to neutrality that it created an anti-solvent environment for melanin, leading to the precipitation of the dissolved material [31]. The volume of the acid medium and the concentration of the neutralizing agent, factors X_1 and X_2, respectively, were not significant, but a slight tendency was observed toward the generation of a larger average size when the volume of the acidic medium was increased. Thus, based on these findings, the response surface for the particle size was obtained with the factors X_1 and X_3 (Figure 2A), with the factor X_2 remaining at a low level; as can be observed, a second-order model was fitted with a determination coefficient (R^2) of 0.944.

Following the same strategy of analysis, during the preparation of RMNPs by the DE method, a significant effect on the particle size was observed when the homogenization time (factor X_3) was increased in the first emulsion. Because homogenization of the phases is conducted in a high-shear process, the reduction in the droplet size of the discontinuous W_1 phase (aqueous alkaline solution that contained the previously dissolved PRM) revealed time dependence; the latter is relevant because the droplet size in the first emulsion plays an essential role in reaching a desired particle size [48,49]. Regarding factors X_1 and X_2, the volume of the O phase and its composition (DCM:ACE proportion), respectively, did not exert a significant impact on the particle size. Moreover, the neutralization of aqueous phase W_1 with the W_2 acidic aqueous phase occurred as the O phase was evaporated from the $W_1/O/W_2$ emulsified system, generating an anti-solvent environment for the previously dissolved RPM, leading to the formation of the RMNPs. Figure 2B depicts the response surface for particle size for the RMNPs produced by the DE method, where factors X_2 and X_3 were employed to construct the graph and explore the surface, with factor X_1 remaining at a low level, in that its effect was practically null; a second-order model with a R^2 of 0.946 was found in a similar manner.

The experimental design of two factors was executed to produce RMNPs by a top-down approach which evidenced non-statistically significant differences in particle size at the pressure and the number of cycles set in the high-pressure homogenization process. Nevertheless, a tendency was detected for the particle size to be reduced as the applied pressure (factor X_1) or the number of cycles (factor X_2) increased. Additionally, a common reduction limit was observed in the particle size, independent of the high pressure or the number of times that the PRM suspension passed through the impact chamber in the equipment [50]. The response surface for the particle size when the HP method was applied is presented in Figure 2C; a second-order model exhibited the best fit, with an R^2 of 0.647. Thus, based on these findings and on the operational conditions tested in this manufacturing method, the RMNPs obtained under these experimental conditions provided the smallest particle size observed (run #1).

3.3. Optimization of RMNP Particle Size and Model Verification

Using the fitted regression model of each previously analyzed surface, the optimization process was carried out for each manufacturing method applied to produce RMNPs. From the DOE analysis menu in the Statgraphics software, the response optimization tool was used to make predictions for the levels of the factors in which it was possible to achieve the established optimization objectives, that is, a particle size of 250 nm and a Z-potential with the highest value possible. Optimization by multiple responses was performed using the desirability function, giving the greater weight for the particle size in the equation, mainly to achieve this objective. Once the optimized levels for each factor in the models were theoretically estimated, the expected or predicted particle size was determined for the estimated conditions. To verify that the models adequately predicted the behavior of the particle size, three batches of RMNPs were prepared under the optimized conditions for each manufacturing method, measuring the response variables, and comparing the predicted values versus the experimental values. The prediction error was determined by the difference in values between the expected and the experimental responses. The results of the estimates, optimization conditions, desirability, and model verification are summarized in Table 2. The RMNPs obtained under the estimated conditions were recognized as the optimized nanoparticles.

3.4. Physical Characterization of the Optimized RMNPs

Morphological studies to verify the particle size and to determine the shape of the RMNPs were carried out using the transmission electron microscopy (TEM) technique. The obtained micrographs are presented in Figure 3. It can be observed how the RMNPs produced under the *bottom-up* approach (Methods: NC and DE) exhibited a similar structure. These were nanoparticles with a spherical, homogeneous, and solid shape (Figure 3A: RMNP-NC, and 3B: RMNP-DE). It is worth mentioning that the particle size observed in TEM corresponded to those reported by DLS. This shape and particle size distribution have been reported, when obtaining melanin nanoparticles by *bottom-up* methods, using the conventional tyrosine oxide-reduction process [26]. In contrast, for the RMNP-HP obtained by the top-down approach with the HP method, the nanoparticles presented a morphology that described structures with an equivalent particle size similar to those measured by DLS, but with an irregular shape (Figure 3C: RMNP-HP); this characteristic may be explained as being due to the size-reduction mechanism that materials undergo with the high-pressure homogenization technique, which occurs by impact, thus provoking interparticle interactions and shocks against the walls during confinement within the impact chamber in the homogenizer [50,51].

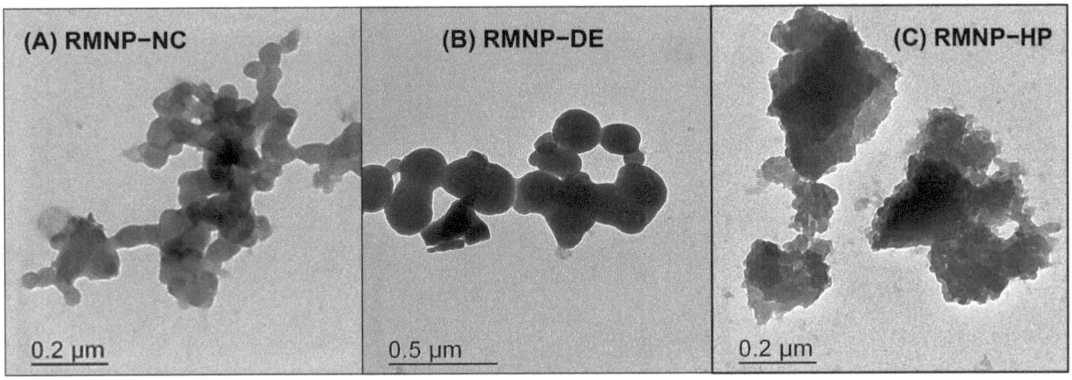

Figure 3. TEM micrographs of the recombinant melanin nanoparticles (RMNPs) obtained by the following methods: (**A**) NC = nanocrystallization; (**B**) DE = double emulsion–solvent evaporation, and (**C**) HP = high-pressure homogenization.

IR spectra were recorded for the recombinant melanin nanoparticles (RMNPs) obtained by means of optimal manufacturing conditions, for the *bottom-up* and *top-down* methods. The spectra evidenced the characteristic peaks that correspond to the functional groups previously mentioned for preformed recombinant melanin (PRM), which were also present in the RMNP spectra. As observed in Figure 4, in comparison to the IR spectrum of PRM, at 3200 cm^{-1} the peak of the -NH and -OH groups was observed; at 2920 cm^{-1}, the peak that corresponds to the -CH bond; at 1680 cm^{-1}, there appeared a strong signal attributed to the -C=O bond of -COOH groups; at 1520 cm^{-1} the vibration of the aromatic group bonds C=C was observed again, and finally, the peak related to the protonated amine of the DHICA at 2385 cm^{-1} was detected. New peaks that could evidence the formation of new bonds were not appreciable, while small displacements in some peaks could be explained by the reordered crystalline state that melanin acquires when nanoparticles are formed [52]. In addition, these results show that the chemical integrity of the PRM was maintained after applying the different methods to produce the RMNPs.

In addition, PXRD analysis was performed on the melanin samples, preformed recombinant melanin (PRM) as raw material, and recombinant melanin nanoparticles (RMNPs). The broad peaks observed in the PXRD analysis depicted an amorphous state for PRM and treated melanin (RMNP). As observed in Figure 5, a broad band presented a 2θ value at 21.5° in PRM that was attributed to an amorphous solid state. This state for melanin has been previously reported, even for melanin from other sources, including from another biological origin [53,54]. This result also confirms that the structure and common amorphous solid state of PRM was preserved due to the stacking of molecular planes in the melanin structure that contains a range of distinct macromolecules or oligomeric material (chemical disorder model) [55]. Similarly, the RMNPs exhibited non-crystalline signals, only a broad band with 2θ value at 62.4°, suggesting a lack of sheet stacking and no order in aggregation. However, a low formation of protomolecules that aggregate by hydrogen bonding, or aggregates generated by a smaller number of protomolecules, could be attributed to the nanoparticle preparation method, the dissolution process, or both [55].

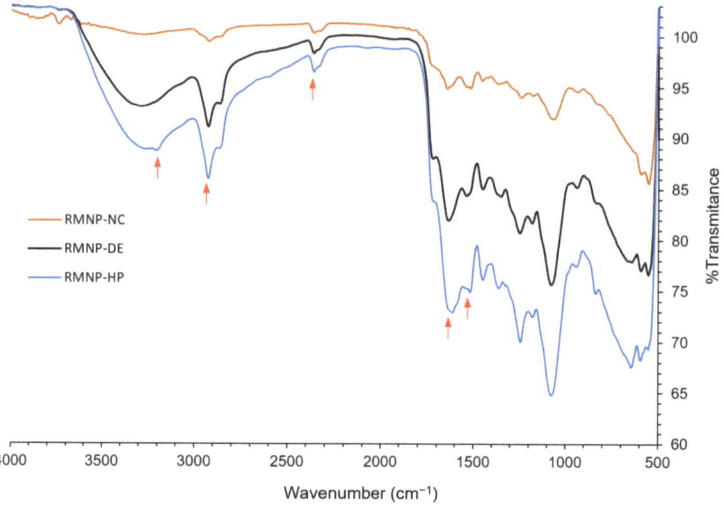

Figure 4. IR spectra of the recombinant melanin nanoparticles (RMNPs) prepared by the following methods: NC = nanocrystallization, DE = double emulsion–solvent evaporation, and HP = high-pressure homogenization. The red arrows show the absorption peaks related to the known functional groups of melanin.

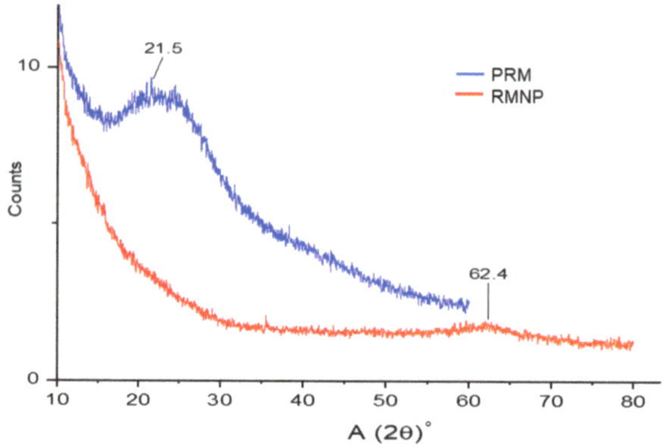

Figure 5. PXRD diffractograms of recombinant melanin (PRM) as raw material and the recombinant melanin nanoparticles (RMNPs).

The thermal behavior of the PRM and of the RMNPs is shown in the thermograms obtained by DSC, which are presented in Figure 6. For the PRM, a raw material without treatment, an endothermic peak onset temperature at 181.80 °C was recognized as its melting point or degradation peak [54], which was apparent without any significant displacement in the RMNP-NC (183.13 °C). However, a displacement of approximately 10 °C was observed in the melting point for RMNP-HP (onset temperature at 192.19 °C), which could be explained by an increase in the arrangement of melanin in the nanostructure during the manufacturing process; this new order in the solid state may be caused by the high energy added to the material when the size reduction is occurring due to pressure and impact forces. On the other hand, RMNP-DE demonstrated a displacement of 20 °C (onset temperature at 202.58 °C) in comparison to PRM, which could be explained by the order state that melanin molecules acquire during the crystallization process; this was because a slow evaporation rate was needed for the organic phase in comparison to the other two methods applied [56]. As previously mentioned, these movements at the melting point could be associated with the different degree of ordering of the melanin polymer chains due to the processes applied.

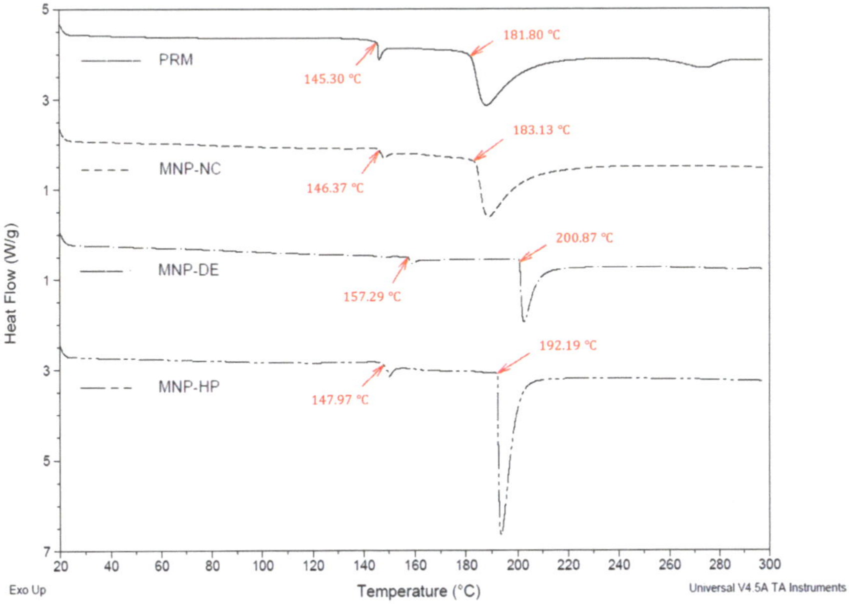

Figure 6. Thermograms by DSC for preformed recombinant melanin (PRM) and the recombinant melanin nanoparticles (RMNPs) obtained by the following methods: NC = nanocrystallization; DE = double emulsion–solvent evaporation, and HP = high-pressure homogenization.

Additionally, an endothermic event at 145.30 °C was observed in the thermograms (Figure 6), which corresponded to a first-order solid–solid transition in the solid state of the material (PRM). The latter was also observed in a similar manner for the RMNP-NC (146.37 °C), but was displaced in RMNP-DE (157.29 °C) and RMNP-HP (147.97 °C) due to changes in the arrangement of the melanin units inside the nanoparticle. The difference in degrees between the melting point and the vertex of the fusion endotherm was less than 2 °C for RMNP-DE and RMNP-HP, indicative of a higher purity for the recovered nanomaterial due to a lower amount of residue from the other materials that were added during the production process; the type of solvents used and the washing favored this condition. However, for PRM and RMNP-NC, there was a difference of 6–7 °C, which could be associated with an impurity content of at least 0.5–1% [57].

Considering the desired size of the particle, its distribution, and variability, the more reproducible manufacturing method, the lower prediction error in the modeling behavior, a solid, spherical, and compact shape for the RMNPs, the best purity after the recovery process, a Z-potential greater than ± 30 mV to favor physical stability in suspension with no aggregation phenomena, and a good chemical stability in the solid state, we decided to use the RMNP-DE for further studies. In this study, the first aim in the application of the RMNPs was related to treating or solving skin complications, such as tissue regeneration, diseases involving pigment deficiencies, drug delivery, or sun protection. Since the RMNP-DE have an expected use in health and cosmetics, biocompatibility assays with this nanomaterial were carried out using two non-carcinogenic-derived skin cell lines.

3.5. Biocompatibility Studies

Once the physical and chemical characterization of the different RMNPs had been carried out, those obtained by the DE method were used for the biocompatibility analyses. These analyses were performed using two non-carcinogenic established cell lines (HEK293 and HEKn); both cell lines were selected by considering the potential applications of the RMNPs, for biomedical and cosmetic devices. Figure 7 depicts the effect of RMNPs of biotechnological origin on cell viability measured by two different techniques at 72 h after exposure to the nanoparticles. As is already known, the Trypan Blue technique is utilized to determine the % of cell viability in HEKn cells considering membrane integrity, and the MTS technique is useful for measuring mitochondrial activity in HEK293, which is directly correlated with the % of cell viability. The tested RMNP concentrations were 10, 50, and 100 μg/mL. As observed, in both cell lines and using both methodologies, the average viability did not decrease from 95%, implying that RMNPs are not toxic and allow cell-culture development for in vitro studies. In contrast, in the findings reported by Blinova et al. [5], a marked inhibition in cell proliferation was exhibited at concentrations of 50 and 100 μg/mL of melanin from black yeast fungi.

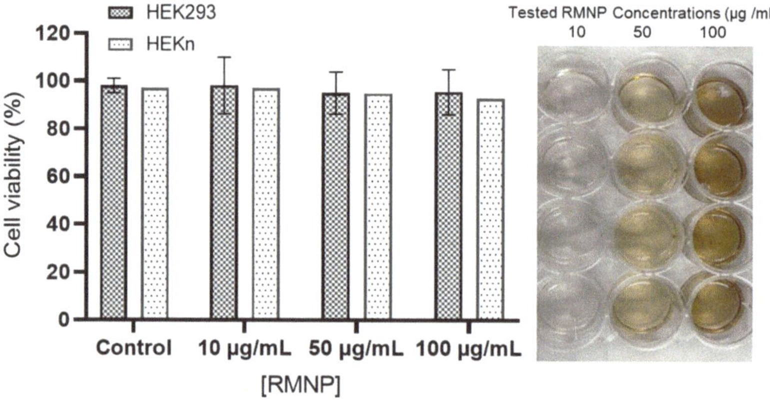

Figure 7. Percentage of cell viability of HEK293 and HEKn cultures using different concentrations of RMNP-DE. Three different concentrations were tested (10, 50 and 100 μg/mL). Each of the bars represents the mean of n = 8.

Considering that the RMNPs were steam-heat sterilized for the development of these experiments, it is important to note that the sterilization process was also effective, and no contamination was present in the cultures. This is an important fact as the cell culture media was formulated without antibiotics.

4. Conclusions

Melanin nanoparticles from recombinant preformed melanin were obtained applying *top-down* and *bottom-up* nanotechnology approaches, adapting three conventional manufacturing methods and starting from an insoluble raw material to obtain the nanostructures. The identity and apparent solubility of this melanin from a biotechnological origin were also characterized. Moreover, by applying experimental designs, statistical techniques, and response surface methodology, it was possible to optimize each of the three methods used for the preparation of RMNPs, in terms of the control of the particle size: nanocrystallization (NC), double emulsion–solvent evaporation (DE) and high-pressure homogenization (HP). The obtained RMNPs in all of the employed manufacturing methods presented properties that demonstrated their physical and chemical stability. In particular, the DE method evidenced the production of RMNPs in a reproducible way, generating nanostructures with a spherical and consolidated solid shape, with an average particle size of 250 ± 50 nm and an ideal Z-potential (>30 mV) to assure good physical stability and dispersion in aqueous suspension, without the formation of aggregates. The nanoparticles generated in this study showed the technological characteristics necessary for application in different areas, including medicine and cosmetics, in that they also proved to be sterilizable and biocompatible in HEK293 and HEKn cell lines. This nanomaterial could be utilized to formulate products with applications in radiation receptors in cancer therapy, as sunscreen, or in makeup to solve aesthetic skin problems (such as burned skin and vitiligo), and they may potentially be employed as contrast agents for the diagnosis of diseases or as drug delivery systems.

Supplementary Materials: The following supporting information can be downloaded at: https://www.mdpi.com/article/10.3390/polym15102381/s1, Figure S1: Preformed recombinant melanin (PRM); (A) granular raw material, (B) pulverized raw material; Figure S2: Pareto charts of main effects on particle size for the three manufacturing methods employed to produce RMNPs; (A) nanocrystallization (NC), (B) double emulsion–solvent evaporation (DE), and (C) high-pressure homogenization (HP).

Author Contributions: Conceptualization, S.A.-A.; methodology, S.A.-A., J.E.C.-A. and M.M.-G.; validation, S.A.-A., C.A.T.-M. and A.M.-A.; formal analysis, S.A.-A. and J.E.C.-A.; investigation, J.E.C.-A., M.B.-H., C.A.T.-M. and M.M.-G.; resources, G.G., D.L.G.-G. and A.M.-A.; data curation, S.A.-A.; writing—original draft preparation, S.A.-A.; writing—review and editing G.G., D.L.G.-G. and A.M.-A.; visualization, S.A.-A.; supervision, A.M.-A.; project administration, S.A.-A. and A.M.-A.; funding acquisition, A.M.-A. All authors have read and agreed to the published version of the manuscript.

Funding: This research was funded by CONACYT-Infrastructure 2014 Projects (226271), CONACyT-PEI 2014 (221546) and PRODEP-SEP Project No. DSA/103.5/15/6986.

Institutional Review Board Statement: Not applicable.

Informed Consent Statement: Not applicable.

Data Availability Statement: The data presented in this study are available on request from the corresponding author.

Acknowledgments: The authors are grateful to Efrén Hernández, from the Faculty of Pharmacy-UAEM, for the loan of the high-pressure homogenizer. PXRD analyses were carried out at the National Laboratory of Nano and Biomaterials, CINVESTAV-IPN; funded by projects FOMIX-Yucatán 2008-108160, CONACYT LAB-2009-01-123913, 292692, 294643, 188345 and 204822. Our thanks go to Patricia Quintana for access to LANNBIO, to M.C. Daniel Aguilar for obtaining the diffractograms, and to M.C. Mario Herrera Salvador for corrective maintenance work on the diffractometer.

Conflicts of Interest: The authors declare no conflict of interest.

References

1. Cordero, R.J.; Casadevall, A. Melanin. *Curr. Biol.* **2020**, *30*, R142–R143. [CrossRef] [PubMed]
2. Lea, A.J. Solubility of Melanins. *Nature* **1952**, *170*, 709. [CrossRef] [PubMed]
3. Riley, P.A. Melanin. *Int. J. Biochem. Cell Biol.* **1997**, *29*, 1235–1239. [CrossRef] [PubMed]
4. Serpentini, C.-L.; Gauchet, C.; de Montauzon, D.; Comtat, M.; Ginestar, J.; Paillous, N. First Electrochemical Investigation of the Redox Properties of DOPA–Melanins by Means of a Carbon Paste Electrode. *Electrochim. Acta* **2000**, *45*, 1663–1668. [CrossRef]
5. Różanowska, M.; Sarna, T.; Land, E.J.; Truscott, T. Free Radical Scavenging Properties of Melanin. *Free Radic. Biol. Med.* **1999**, *26*, 518–525. [CrossRef]
6. Blinova, M.; Yudintseva, N.; Kalmykova, N.; Kuzminykh, E.; Yurlova, N.; Ovchinnikova, O.; Potokin, I. Effect of Melanins from Black Yeast Fungi on Proliferation and Differentiation of Cultivated Human Keratinocytes and Fibroblasts. *Cell Biol. Int.* **2003**, *27*, 135–146. [CrossRef]
7. Ivins, B.E.; Holmes, R.K. Isolation and Characterization of Melanin-Producing (Mel) Mutants of *Vibrio cholerae*. *Infect. Immun.* **1980**, *27*, 721–729. [CrossRef]
8. Görner, H. New Trends in Photobiology. *J. Photochem. Photobiol. B Biol.* **1994**, *26*, 117–139. [CrossRef]
9. Mohagheghpour, N.; Waleh, N.; Garger, S.J.; Dousman, L.; Grill, L.K.; Tusé, D. Synthetic Melanin Suppresses Production of Proinflammatory Cytokines. *Cell. Immunol.* **2000**, *199*, 25–36. [CrossRef]
10. Hill, H.Z. The Photobiology of Melanin. *Photochem. Photobiol.* **1997**, *65*, 471. [CrossRef]
11. Maranduca, M.; Branisteanu, D.; Serban, D.; Branisteanu, D.; Stoleriu, G.; Manolache, N.; Serban, I. Synthesis and Physiological Implications of Melanic Pigments (Review). *Oncol. Lett.* **2019**, *17*, 4183–4187. [CrossRef]

12. D'Alba, L.; Shawkey, M.D. Melanosomes: Biogenesis, Properties, and Evolution of an Ancient Organelle. *Physiol. Rev.* **2019**, *99*, 1–19. [CrossRef]
13. Roy, S.; Rhim, J.-W. Preparation of carrageenan-based functional nanocomposite films incorporated with melanin nanoparticles. *Colloids Surfaces B Biointerfaces* **2019**, *176*, 317–324. [CrossRef]
14. Solano, F. Melanin and Melanin-Related Polymers as Materials with Biomedical and Biotechnological Applications—Cuttlefish Ink and Mussel Foot Proteins as Inspired Biomolecules. *Int. J. Mol. Sci.* **2017**, *18*, 1561. [CrossRef]
15. Vahidzadeh, E.; Kalra, A.P.; Shankar, K. Melanin-Based Electronics: From Proton Conductors to Photovoltaics and beyond. *Biosens. Bioelectron.* **2018**, *122*, 127–139. [CrossRef]
16. Eom, T.; Jeon, J.; Lee, S.; Woo, K.; Heo, J.E.; Martin, D.C.; Wie, J.J.; Shim, B.S. Naturally Derived Melanin Nanoparticle Composites with High Electrical Conductivity and Biodegradability. *Part. Part. Syst. Charact.* **2019**, *36*, 1900166. [CrossRef]
17. Cuzzubbo, S.; Carpentier, A.F. Applications of Melanin and Melanin-like Nanoparticles in Cancer Therapy: A Review of Recent Advances. *Cancers* **2021**, *13*, 1463. [CrossRef]
18. Zhou, Z.; Yan, Y.; Wang, L.; Zhang, Q.; Cheng, Y. Melanin-like Nanoparticles Decorated with an Autophagy-Inducing Peptide for Efficient Targeted Photothermal Therapy. *Biomaterials* **2019**, *203*, 63–72. [CrossRef]
19. Mavridi-Printezi, A.; Guernelli, M.; Menichetti, A.; Montalti, M. Bio-Applications of Multifunctional Melanin Nanoparticles: From Nanomedicine to Nanocosmetics. *Nanomaterials* **2020**, *10*, 2276. [CrossRef]
20. Yang, Z.; Zhang, J.; Liu, H.; Hu, J.; Wang, X.; Bai, W.; Zhang, W.; Yang, Y.; Gu, Z.; Li, Y. A Bioinspired Strategy toward UV Absorption Enhancement of Melanin-like Polymers for Sun Protection. *CCS Chem.* **2023**, 1–14. [CrossRef]
21. Menter, J.M. Melanin from a Physicochemical Point of View. *Polym. Int.* **2016**, *65*, 1300–1305. [CrossRef]
22. Ozlu, B.; Kabay, G.; Bocek, I.; Yilmaz, M.; Piskin, A.K.; Shim, B.S.; Mutlu, M. Controlled Release of Doxorubicin from Polyethylene Glycol Functionalized Melanin Nanoparticles for Breast Cancer Therapy: Part I. Production and Drug Release Performance of the Melanin Nanoparticles. *Int. J. Pharm.* **2019**, *570*, 118613. [CrossRef] [PubMed]
23. Zhang, Z.; Gupte, M.J.; Ma, P.X. Biomaterials and Stem Cells for Tissue Engineering. *Expert Opin. Biol. Ther.* **2013**, *13*, 527–540. [CrossRef] [PubMed]
24. Liu, J.-J.; Wang, Z.; Nie, L.-M.; Zhu, Y.-Y.; Li, G.; Lin, L.-L.; Chen, M.; Zhang, G.-J. RGD-Functionalised Melanin Nanoparticles for Intraoperative Photoacoustic Imaging-Guided Breast Cancer Surgery. *Eur. J. Nucl. Med. Mol. Imaging* **2022**, *49*, 847–860. [CrossRef] [PubMed]
25. Hong, Z.-Y.; Feng, H.-Y.; Bu, L.-H. Melanin-based nanomaterials: The Promising Nanoplatforms for Cancer Diagnosis and Therapy. *Nanomed. Nanotechnol. Biol. Med.* **2020**, *28*, 102211. [CrossRef]
26. Liopo, A.; Su, R.; Oraevsky, A.A. Melanin Nanoparticles as a Novel Contrast Agent for Optoacoustic Tomography. *Photoacoustics* **2015**, *3*, 35–43. [CrossRef]
27. Rageh, M.M.; El-Gebaly, R.H.; Abou-Shady, H.; Amin, D.G. Melanin Nanoparticles (MNPs) Provide Protection against Whole-Body γ-Irradiation in Mice via Restoration of Hematopoietic Tissues. *Mol. Cell. Biochem.* **2015**, *399*, 59–69. [CrossRef]
28. Keane, T.J.; Badylak, S.F. Biomaterials for Tissue Engineering Applications. *Semin. Pediatr. Surg.* **2014**, *23*, 112–118. [CrossRef]
29. Mostert, A.B.; Powell, B.J.; Pratt, F.L.; Hanson, G.R.; Sarna, T.; Gentle, I.R.; Meredith, P. Role of Semiconductivity and Ion Transport in the Electrical Conduction of Melanin. *Proc. Natl. Acad. Sci. USA* **2012**, *109*, 8943–8947. [CrossRef]
30. Eom, T.; Woo, K.; Shim, B.S. Melanin: A Naturally Existing Multifunctional Material. *Appl. Chem. Eng.* **2016**, *27*, 115–122. [CrossRef]
31. Chen, H.; Khemtong, C.; Yang, X.; Chang, X.; Gao, J. Nanonization Strategies for Poorly Water-Soluble Drugs. *Drug Discov. Today* **2011**, *16*, 354–360. [CrossRef]
32. Iqbal, P.; Preece, J.A.; Mendes, P.M. Nanotechnology: The "Top-Down" and "Bottom-Up" Approaches. In *Supramolecular Chemistry*; John Wiley & Sons, Ltd.: Chichester, UK, 2012.
33. Iqbal, M.; Zafar, N.; Fessi, H.; Elaissari, A. Double Emulsion Solvent Evaporation Techniques Used for Drug Encapsulation. *Int. J. Pharm.* **2015**, *496*, 173–190. [CrossRef]
34. Mahesh, K.V.; Singh, S.K.; Gulati, M. A Comparative Study of Top-Down and Bottom-Up Approaches for the Preparation of Nanosuspensions of Glipizide. *Powder Technol.* **2014**, *256*, 436–449. [CrossRef]
35. Chávez-Béjar, M.; Balderas-Hernandez, V.; Gutiérrez-Alejandre, A.; Martinez, A.; Bolívar, F.; Gosset, G. Metabolic Engineering of *Escherichia coli* to Optimize Melanin Synthesis from Glucose. *Microb. Cell Factories* **2013**, *12*, 108. [CrossRef]
36. Lagunas-Muñoz, V.; Cabrera-Valladares, N.; Bolívar, F.; Gosset, G.; Martínez, A. Optimum Melanin Production Using Recombinant *Escherichia coli*. *J. Appl. Microbiol.* **2006**, *101*, 1002–1008. [CrossRef]
37. Chen, H.; Wan, J.; Wang, Y.; Mou, D.; Liu, H.; Xu, H.; Yang, X. A facile Nanoaggregation Strategy for Oral Delivery of Hydrophobic Drugs by Utilizing Acid–Base Neutralization Reactions. *Nanotechnology* **2008**, *19*, 375104. [CrossRef]
38. Liu, J.; Qiu, Z.; Wang, S.; Zhou, L.; Zhang, S. A Modified Double-Emulsion Method for the Preparation of Daunorubicin-Loaded Polymeric Nanoparticle with Enhanced In Vitro Anti-Tumor Activity. *Biomed. Mater.* **2010**, *5*, 065002. [CrossRef]
39. Tan, M.; Gan, D.; Wei, L.; Pan, Y.; Tang, S.; Wang, H. Isolation and Characterization of Pigment from *Cinnamomum burmannii'* Peel. *Food Res. Int.* **2011**, *44*, 2289–2294. [CrossRef]
40. Dong, C.; Yao, Y. Isolation, Characterization of Melanin Derived from *Ophiocordyceps sinensis*, an Entomogenous Fungus Endemic to the Tibetan Plateau. *J. Biosci. Bioeng.* **2012**, *113*, 474–479. [CrossRef]

41. Madhusudhan, D.N.; Mazhari, B.B.Z.; Dastager, S.G.; Agsar, D. Production and Cytotoxicity of Extracellular Insoluble and Droplets of Soluble Melanin by *Streptomyces lusitanus* DMZ-3. *BioMed Res. Int.* **2014**, *2014*, 306895. [CrossRef]
42. Centeno, S.A.; Shamir, J. Surface Enhanced Raman Scattering (SERS) and FTIR Characterization of the Sepia Melanin Pigment Used in Works of Art. *J. Mol. Struct.* **2008**, *873*, 149–159. [CrossRef]
43. Tu, Y.; Sun, Y.; Tian, Y.; Xie, M.; Chen, J. Physicochemical Characterisation and Antioxidant Activity of Melanin from the Muscles of Taihe Black-Bone Silky Fowl (*Gallus gallus* Domesticus Brisson). *Food Chem.* **2009**, *114*, 1345–1350. [CrossRef]
44. Ye, Z.; Lu, Y.; Zong, S.; Yang, L.; Shaikh, F.; Li, J.; Ye, M. Structure, Molecular Modification and Anti-Tumor Activity of Melanin from *Lachnum singerianum*. *Process. Biochem.* **2018**, *76*, 203–212. [CrossRef]
45. Guo, X.; Chen, S.; Hu, Y.; Li, G.; Liao, N.; Ye, X.; Liu, D.; Xue, C. Preparation of Water-Soluble Melanin from Squid Ink Using Ultrasound-Assisted Degradation and Its Anti-Oxidant Activity. *J. Food Sci. Technol.* **2014**, *51*, 3680–3690. [CrossRef] [PubMed]
46. Tran, P.; Nguyen, H.T.; Fox, K.; Tran, N. In Vitro Cytotoxicity of Iron Oxide Nanoparticles: Effects of Chitosan and Polyvinyl Alcohol as Stabilizing Agents. *Mater. Res. Express* **2018**, *5*, 035051. [CrossRef]
47. Pota, G.; Zanfardino, A.; Di Napoli, M.; Cavasso, D.; Varcamonti, M.; D'errico, G.; Pezzella, A.; Luciani, G.; Vitiello, G. Bioinspired Antibacterial PVA/Melanin-TiO_2 Hybrid Nanoparticles: The Role of Poly-Vinyl-Alcohol on Their Self-Assembly and Biocide Activity. *Colloids Surfaces B Biointerfaces* **2021**, *202*, 111671. [CrossRef]
48. Iqbal, M.; Valour, J.-P.; Fessi, H.; Elaissari, A. Preparation of Biodegradable PCL Particles via Double Emulsion Evaporation Method Using Ultrasound Technique. *Colloid Polym. Sci.* **2015**, *293*, 861–873. [CrossRef]
49. Chong, D.; Liu, X.; Ma, H.; Huang, G.; Han, Y.L.; Cui, X.; Yan, J.; Xu, F. Advances in Fabricating Double-Emulsion Droplets and Their Biomedical Applications. *Microfluid. Nanofluidics* **2015**, *19*, 1071–1090. [CrossRef]
50. Homayouni, A.; Sohrabi, M.; Amini, M.; Varshosaz, J.; Nokhodchi, A. Effect of High Pressure Homogenization on Physicochemical Properties of Curcumin Nanoparticles Prepared by Antisolvent Crystallization Using HPMC or PVP. *Mater. Sci. Eng. C* **2019**, *98*, 185–196. [CrossRef]
51. Nna-Mvondo, D.; Anderson, C.M. Infrared Spectra, Optical Constants, and Temperature Dependences of Amorphous and Crystalline Benzene Ices Relevant to Titan. *Astrophys. J.* **2022**, *925*, 123. [CrossRef]
52. Romero, G.B.; Keck, C.M.; Müller, R.H. Simple Low-Cost Miniaturization Approach for Pharmaceutical Nanocrystals Production. *Int. J. Pharm.* **2016**, *501*, 236–244. [CrossRef]
53. Araujo, M.; Xavier, J.R.; Nunes, C.D.; Vaz, P.D.; Humanes, M. Marine Sponge Melanin: A New Source of an Old Biopolymer. *Struct. Chem.* **2012**, *23*, 115–122. [CrossRef]
54. Kiran, G.S.; Jackson, S.A.; Priyadharsini, S.; Dobson, A.D.; Selvin, J. Synthesis of Nm-PHB (Nanomelanin-Polyhydroxy Butyrate) Nanocomposite Film and Its Protective Effect against Biofilm-Forming Multi Drug Resistant *Staphylococcus aureus*. *Sci. Rep.* **2017**, *7*, 9167. [CrossRef]
55. Littrell, K.C.; Gallas, J.M.; Zajac, G.W.; Thiyagarajan, P. Structural Studies of Bleached Melanin by Synchrotron Small-Angle X-Ray Scattering. *Photochem. Photobiol.* **2003**, *77*, 115. [CrossRef]
56. Panith, N.; Assavanig, A.; Lertsiri, S.; Bergkvist, M.; Surarit, R.; Niamsiri, N. Development of Tunable Biodegradable Polyhydroxyalkanoates Microspheres for Controlled Delivery of Tetracycline for Treating Periodontal Disease. *J. Appl. Polym. Sci.* **2016**, *133*, 44128–44140. [CrossRef]
57. Attia, A.K.; Abdel-Moety, M.M.; Abdel-Hamid, S.G. Thermal Analysis Study of Antihypertensive Drug Doxazosin Mesilate. *Arab. J. Chem.* **2017**, *10*, S334–S338. [CrossRef]

polymers

MDPI

Article

Bio-Based Electrospun Fibers from Chitosan Schiff Base and Polylactide and Their Cu^{2+} and Fe^{3+} Complexes: Preparation and Antibacterial and Anticancer Activities

Milena Ignatova [1,*], Ina Anastasova [1], Nevena Manolova [1], Iliya Rashkov [1,*], Nadya Markova [2], Rositsa Kukeva [3], Radostina Stoyanova [3], Ani Georgieva [4] and Reneta Toshkova [4]

1 Laboratory of Bioactive Polymers, Institute of Polymers, Bulgarian Academy of Sciences, Acad. G. Bonchev St., Bl. 103A, BG-1113 Sofia, Bulgaria
2 Institute of Microbiology, Bulgarian Academy of Sciences, Acad. G. Bonchev St., Bl. 26, BG-1113 Sofia, Bulgaria
3 Institute of General and Inorganic Chemistry, Bulgarian Academy of Sciences, Acad. G. Bonchev St., Bl. 11, BG-1113 Sofia, Bulgaria
4 Institute of Experimental Morphology, Pathology and Anthropology with Museum, Bulgarian Academy of Sciences, Acad. G. Bonchev St., Bl. 25, BG-1113 Sofia, Bulgaria
* Correspondence: ignatova@polymer.bas.bg (M.I.); rashkov@polymer.bas.bg (I.R.); Tel.: +359-(0)2-9792239 (M.I.); Fax: +359-(0)2-8700309 (M.I.)

Citation: Ignatova, M.; Anastasova, I.; Manolova, N.; Rashkov, I.; Markova, N.; Kukeva, R.; Stoyanova, R.; Georgieva, A.; Toshkova, R. Bio-Based Electrospun Fibers from Chitosan Schiff Base and Polylactide and Their Cu^{2+} and Fe^{3+} Complexes: Preparation and Antibacterial and Anticancer Activities. *Polymers* **2022**, *14*, 5002. https://doi.org/10.3390/polym14225002

Academic Editor: Antonio M. Borrero-López

Received: 25 October 2022
Accepted: 16 November 2022
Published: 18 November 2022

Abstract: The Schiff base derivative (Ch-8Q) of chitosan (Ch) and 8-hydroxyquinoline-2-carboxaldehyde (8QCHO) was prepared and fibrous mats were obtained by the electrospinning of Ch-8Q/polylactide (PLA) blend solutions in trifluoroacetic acid (TFA). Complexes of the mats were prepared by immersing them in a solution of $CuCl_2$ or $FeCl_3$. Electron paramagnetic resonance (EPR) analysis was performed to examine the complexation of Cu^{2+}(Fe^{3+}) in the Ch-8Q/PLA mats complexes. The morphology of the novel materials and their surface chemical composition were studied by scanning electron microscopy (SEM), attenuated total reflection Fourier transform infrared spectroscopy (ATR-FTIR) and X-ray photoelectron spectroscopy (XPS). The performed microbiological screening demonstrated that in contrast to the neat PLA mats, the Ch-8Q-containing mats and their complexes were able to kill all *S. aureus* bacteria within 3 h of contact. These fibrous materials had efficiency in suppressing the adhesion of pathogenic bacteria *S. aureus*. In addition, Ch-8Q/PLA mats and their complexes exerted good anticancer efficacy in vitro against human cervical HeLa cells and human breast MCF-7 cells. The Ch-8Q-containing fibrous materials had no cytotoxicity against non-cancer BALB/c 3T3 mouse fibroblast cells. These properties render the prepared materials promising as wound dressings as well as for application in local cancer treatment.

Keywords: Schiff base; chitosan; electrospinning; polylactide; Cu^{2+} complexes; Fe^{3+} complexes; antibacterial activity; anticancer activity

1. Introduction

Electrospinning is an attractive and low-cost technique for the fabrication of continuous nanoscale polymer fibers. Electrospun fibrous materials exhibit intriguing characteristics such as high surface area-to-volume and aspect ratios and high porosity with very small pore size [1]. Due to these features, the materials are suitable candidates for diverse biomedical applications, such as wound dressings, drug delivery systems, tissue engineering scaffolds, etc. [2,3]. Electrospinning enables the incorporation of drugs of various natures into the polymer fibers. Systems for sustained drug release based on electrospun materials lead to an enhancement in the therapeutic effect of the drugs and a reduction in their side effects [4–6]. In recent years bio-based polymers such as the natural polysaccharide chitosan (Ch) and its derivatives, have been considered some of the most promising polymers obtained from renewable sources suitable as drug carriers [7–9]. Ch is biocompatible, biodegradable, non-toxic and non-immunogenic; in addition, it bears functional groups that enable its facile chemical modification [10,11]. It also possesses valuable biological properties, among which of particular importance are its haemostatic activity, its ability to influence the function of macrophages, and its good antibacterial, antioxidant and anticancer properties [12,13]. However, Ch is difficult to electrospin due to its limited solubility (in aqueous medium at pH lower than 6 and in harsh solvents) and its polyelectrolyte nature, which makes the necessary entanglement of the chains difficult in many solvents [11,14]. Ch-based fibrous materials have been successfully obtained by electrospinning of Ch alone using TFA as a solvent [15–19] or a mixture of TFA/CH_2Cl_2 [15,16,20,21] or in concentrated acetic acid [22,23]. The ability of Ch to be electrospun can be improved by mixing it with water-soluble non-ionogenic polymers with a flexible chain [15,24,25]. In order to improve the mechanical properties of Ch-containing fibrous materials suitable for biomedical applications, the incorporation of aliphatic polyesters into the fibers is of interest. In the present study, poly(L-lactide-*co*-D,L-lactide) (PLA) was selected as the aliphatic polyester because it is a biodegradable polymer with a degradation rate suitable for most musculoskeletal applications. In addition, PLA is a biocompatible polymer of low toxicity and has a good profile of its mechanical properties. These features indicated that PLA-based fibrous materials can serve as suitable candidates for diverse biomedical applications, e.g., implants, drug delivery systems and wound healing materials [26–28]. Fibrous materials from Ch or its derivatives and polyesters (e.g., PLA and its copolymers [16,26,29–31], poly(ethylene terephthalate) [32,33] and poly(ε-caprolactone) [34]) have been fabricated using TFA as a solvent. Previously, we have reported the preparation of fibrous materials containing Ch or its quaternized derivative by the electrospinning of mixed Ch (quaternized Ch)/PLA solutions in a common solvent TFA/CH_2Cl_2 [26]. The prepared materials have been shown to possess high antibacterial activity. We have demonstrated that mats based on quaternized Ch/PLA containing the anticancer drug doxorubicin hydrochloride exert good cytotoxicity in vitro against HeLa, MCF-7 and Graffi tumor cells, and high efficacy in vivo against myeloid Graffi tumor [35–37].

8-Hydroxyquinoline and its derivatives are very attractive for biomedical applications due to their beneficial biological properties—antimicrobial, anticancer, antiinflammatory and antioxidant [38–40]. In addition, they also possess low toxicity. These compounds have the ability to form complexes with biologically important transition metal ions—Cu^{2+}, Fe^{2+}, Fe^{3+}, etc.—which accounts for their biological action [39,40]. Previously, we have demonstrated that the incorporation of 8-hydroxyquinoline derivatives into electrospun materials from natural [24,41–44] and synthetic polymers [45–47] imparts to them antimicrobial and anticancer activity. We have also shown that 5-amino-8-hydroxyquinoline-modified fibrous materials based on copolymers of styrene and maleic anhydride exhibit high antibacterial and antifungal activity [48]. The possibility of obtaining electrospun materials containing a Schiff base from Jeffamine ED® and 8-hydroxyquinoline-2-carboxaldehyde (8QCHO) or its complex with Cu^{2+}, which possess antioxidant and anticancer properties, has also been reported [49]. However, until now, there have been no reports on the preparation of a Schiff base derivative of Ch and 8QCHO (Ch-8Q). An enhancement of the biological activity of the 8-hydroxyquinoline derivatives was observed by their coordination to metal ions of biological significance, such as Cu and Fe ions [38,50]. For this reason, in the present study, we have chosen these ions for the preparation of complexes.

The present study reports the successful fabrication of novel fibrous materials from PLA and Ch-8Q by one-pot electrospinning. Electron paramagnetic resonance (EPR) spectroscopy was used to study the complexation of Cu^{2+} (Fe^{3+}) in the complexes of Ch-8Q/PLA mats. The morphology of the mats was examined by scanning electron microscopy (SEM), and their thermal properties were evaluated by differential scanning calorimetry (DSC). The in vitro antibacterial activity of the novel fibrous materials against the Gram-positive bacteria *S. aureus* was assessed. The cytotoxicity of the mats and their complexes against human HeLa and MCF-7 cancer cells as well as non-cancer BALB/c 3T3 mouse fibroblast cells was also studied.

2. Materials and Methods

2.1. Materials

Poly(L-lactide-*co*-D,L-lactide) (PLA; Resomer® LR 708 (mass average molar mass ($\overline{M}_w$) 911,000 g/mol, $\overline{M}_w/\overline{M}_n$ = 2.46; molar ratio L-lactide:D,L-lactide = 69:31) kindly donated by Boerhinger-Ingelheim Chemicals Inc. (Ingelheim am Rhein, Germany) was used in this study. 8-hydroxyquinoline-2-carboxaldehyde (8QCHO) (Aldrich, St. Louis, MO, USA), $CuCl_2$ anhydrous (Acros Organics, Geel, Belgium), $FeCl_3$ anhydrous (Acros Organics, Geel, Belgium) were used without further purification. Chitosan with an average viscometric molar mass of 380,000 g/mol (Ch) and a degree of deacetylation of 80% was purchased from Aldrich (St. Louis, MO, USA). Glacial acetic acid (Merck, Billerica, MA, USA), absolute ethanol (Merck, Billerica, MA, USA) and trifluoroacetic acid (TFA, Aldrich, St. Louis, MO, USA) of analytical grade purity were used. Dulbecco's Modified Eagle's Medium (DMEM) (Sigma-Aldrich, Schnelldorf, Germany), fetal bovine serum (FBS) (Gibso/BRL, Grand Island, NY, USA), glutamine, penicillin and streptomycin (LONZA, Cologne, Germany) were also used in the study. 3-(4,5-Dimethylthiazol-2-yl)-2,5-diphenyltetrazolium bromide (MTT), acridine orange (AO) and ethidium bromide (EtBr) were obtained from Sigma-Aldrich, Schnelldorf, Germany. 4′,6-diamidino-2-phenylindole (DAPI) (AppliChem, Darmstadt, Germany) was used without further purification. The disposable consumables were purchased from Orange Scientific, Braine-l'Alleud, Belgium. Human permanent cancer cells (HeLa, MCF-7) and normal mouse fibroblasts (Balb/c3T3) were purchased from the American Type Culture Collection (ATCC) (Manassas, VA, USA). *Staphylococcus aureus* (*S. aureus*) 3359 were obtained from the National Bank for Industrial Microorganisms and Cell Cultures (NBIMCC), Sofia, Bulgaria.

2.2. Preparation of Schiff Base Derivative from Ch and 8QCHO

The synthesis of Ch-8Q is presented in Scheme 1.

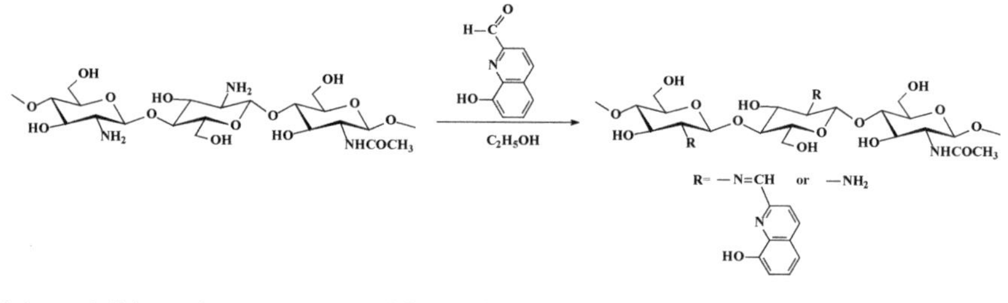

Scheme 1. Schematic representation of the synthesis of Ch-8Q.

To prepare the Schiff base (Ch-8Q) (Scheme 1), Ch (2.5 g, 14.8 mmol) was dissolved in a 125 mL dilute aqueous solution of acetic acid (1% *w*/*v*) for 24 h at 25 °C. Absolute ethanol (13 mL) was added under continuous stirring for 2 h. Then, an ethanolic solution of 8QCHO (2.1 g, 12.1 mmol, 55 mL) was added dropwise to the solution. The mixture was kept under stirring for 24 h at 50 °C. The orange precipitate that formed was filtered and washed with absolute ethanol several times. The unreacted aldehyde was extracted in a Soxhlet apparatus with absolute ethanol for two days. The Schiff base Ch-8Q was dried under reduced pressure at 40 °C for 24 h. Yield—90%. ATR-FTIR (film), ν/cm^{-1}: 3410 (ν(N-H), (ν(O-H)), 2886, 2866 (ν(C-H)), 1645 (ν(CH=N)), 1568 (amide II), 1506, 1468 (ν(C=C)), 1373 (δ(C-H)), and 1064 (ν(C-O-C)). The structure of Ch-8Q was also analyzed by ^1H NMR spectroscopy (Bruker Avance II+ 600, D_2O/DCl (2%), 333 K). The average degree of substitution was determined from the intensity ratio of the signal at 10.1 ppm for the proton of the imino groups to that at 3.67 ppm for H-2 from the 2-amino-2-deoxy-D-glucopyranose units of Ch. The degree of substitution was 73% for Ch-8Q.

2.3. Preparation of Ch-8Q/PLA Fibrous Materials by Electrospinning

Ch-8Q/PLA mixed solutions with Ch-8Q/PLA weight ratio (50:50 and 30:70) were prepared in TFA at a total polymer concentration of 5 wt.%. For comparison, PLA fibers and Ch/PLA fibers (weight ratio of 30:70) were obtained by the electrospinning of their spinning solutions in TFA at a concentration of 5 wt.%.

The system for electrospinning included a plastic syringe equipped with a 22-gauge stainless-steel needle, a syringe infusion pump (NE-300 Just InfusionTM Syringe Pump, New Era Pump Systems Inc., Farmingdale, NY, USA), a high-voltage power supply, and a grounded metal drum collector of diameter 56 mm. The distance between the needle tip and the collector was 13 cm. The applied voltage and flow rates were 30 kV and 1.0 mL/h, respectively. The rotating speed was maintained at 1300 rpm. The humidity was 45% and the temperature was 20 °C. The collected Ch-8Q/PLA fibers were kept under ammonia vapor in a desiccator for 1 h. Then, the mats were dried under reduced pressure at 40 °C for 24 h to remove excess ammonia.

2.4. Preparation of the Cu^{2+}(Fe^{3+}) Complexes of Ch-8Q/PLA Mats, of Ch-8Q, of 8QCHO and of Jeff-8Q

Cu^{2+} or Fe^{3+} complexes of Ch-8Q/PLA mats were obtained by immersing the mats into a 0.1 M absolute ethanol solution of $CuCl_2$ or $FeCl_3$ for 40 min at 25 °C. The mats were then taken out, purified of non-coordinated salt by washing with absolute ethanol several times and then freeze-dried. For comparison, complexes of Ch/PLA mats were also obtained using this procedure.

Cu^{2+} and Fe^{3+} complexes of Ch-8Q were prepared according to a procedure described in detail in the Supplementary Material (see the preparation of the Cu^{2+}(Fe^{3+}) complexes of Ch-8Q). Cu^{2+} complexes of 8QCHO and Jeff-8Q were obtained by the procedure described in our previous report [49]. Fe^{3+} complexes of 8QCHO and Jeff-8Q were synthesized by the procedure described in the Supplementary Material (see the preparation of the Fe^{3+} complexes of 8QCHO and Jeff-8Q).

2.5. Characterization

Electrospun mats were vacuum-coated with a gold layer using a fine coater Jeol (JFC-1200) and the morphology of the fibrous materials was observed by scanning electron microscopy (SEM, Jeol JSM-5510 (Tokyo, Japan)). The average fiber diameters were calculated over 40 fibers from each SEM image with ImageJ software (V.1.53e, Wayne Rasband, National Institute of Health, Bethesda, MD, USA). Chemical analysis was performed with ATR-FTIR spectroscopy using an IRAffinity-1 spectrophotometer (Shimadzu Co., Kyoto, Japan) equipped with a MIRacleTM ATR (diamond crystal; depth of penetration of the IR beam into the sample was ~2 μm) accessory (PIKE Technologies, Madison, WI, USA). XPS analyses were carried out in the ultrahigh-vacuum (UHV) chamber of an ESCALABMkII (VG Scientific) spectrometer with Mg Kα excitation. DSC analyses were conducted with a DSC TA instrument (DSC Q2000, New Castle, DE, USA) from 0 °C to 400 °C at a 10 °C/min heating rate under nitrogen.

The contact angles were measured by the static sessile drop method using an Easy Drop Krüss GmbH apparatus (DSA 10-MK2 model, Hamburg, Germany) as the average value of at least twenty 10 μL droplets of deionized water placed on the mat surface. The values of the water contact angle were calculated by computer analysis of the acquired images of the droplets.

The EPR spectra of the complexes were acquired as the first derivative of the absorption signal using a Bruker EMXplus EPR spectrometer (E7001039, Karlsruhe, Germany) in the X-band (9.4 GHz). The SpinCount™ software module (Bruker) was used for quantitative EPR calculations. The recorded temperature was changed from 100 to 295 K. The spectra were simulated by the program WIN-EPR SimFonia (Bruker).

2.6. Assessment of the Antibacterial Activity

The Ch-8Q/PLA mats and their complexes, PLA and Ch/PLA mats were screened for in vitro antibacterial activity against bacteria *S. aureus* 749 using the viable cell-counting method. Each of the fibrous mats was sterilized by UV irradiation for 30 min and then was added into a suspension of the bacteria (5 mL containing 10^5 cells/mL). Then the suspension was stirred and incubated at 37 °C. At predetermined time points, 1 mL samples were removed from each tube and serially diluted 10-fold with sterile phosphate-buffered saline (PBS). Finally, the surviving bacteria were counted using the spread-plate method. Triplicate counting was done for each experiment. The number of surviving cells was determined as colony-forming unit (CFU).

The adhesion of bacteria *S. aureus* 749 on the surface of the Ch-8Q/PLA mats and their complexes, and PLA and Ch/PLA mats was assessed using SEM observation with a Jeol JSM-5510 SEM spectrometer (Jeol Ltd., Tokyo, Japan). Mats were incubated in 3 mL of *S. aureus* culture (containing about 10^7 cells/mL) at 37 °C for 24 h. After incubation, the mats were washed with PBS and then immersed in 2.5 vol.% glutaraldehyde solution in PBS at 4 °C for 5 h. Finally, the mats were washed with PBS, freeze-dried, coated with gold with Jeol JFC-1200 fine coater and studied by a Jeol JSM-5510 SEM spectrometer (Tokyo, Japan).

2.7. MTT Cytotoxicity Assay

HeLa cells, MCF-7 cells, or mouse BALB/c 3T3 fibroblasts, were maintained in the logarithmic phase at 37 °C in a humidified atmosphere of 5% CO_2 in air using Dulbecco's Modified Eagle Medium (DMEM) (Sigma-Aldrich, Schnelldorf, Germany), containing 10% fetal bovine serum (FBS) (Gibco/BRL, Grand Island, NY, USA), antibiotics (50 units/mL penicillin and 50 µg/mL streptomycin) (LONZA, Cologne, Germany) and 2 mM l-glutamine. Cells were trypsinized by 0.25% Trypsin-EDTA and counted with a hemocytometer. Cell proliferation assay was performed according to the method of Mossman [51]. Briefly, the tested cells (1×10^5 cells/well) seeded in a 96-well microplate were cultivated overnight under standard conditions (37 °C, 5% CO_2 and 95% humidity) to form a monolayer. After 24 h, the medium was removed and replaced with fresh medium containing different types of fibrous mats (PLA, Ch/PLA, Ch-8Q/PLA, Cu^{2+} complex of Ch-8Q/PLA and Fe^{3+} complex of Ch-8Q/PLA) pre-sterilized with UV irradiation for 30 min. All Ch-8Q-containing mats and their Cu^{2+} and Fe^{3+} complexes were studied at a concentration of 8Q residues 60 µg/mL of culture medium. Cells incubated in culture medium only and in the presence of solutions of Jeff-8Q and its complexes as well as solutions of 8QCHO and its complexes (concentration of 8Q residues was 60 µg/mL of culture medium) were used as negative and positive controls, respectively. Each experimental variant was assayed by five measurements. At the end of 24 h and 72 h incubation, the cells were washed twice with PBS (pH 7.4), and further incubated with 100 µL of MTT solution (Sigma Chemical) at 37 °C for 3 h. Subsequently, the supernatants were aspirated and 100 µL of the lysing solution (DMSO:ethanol = 1:1) was added to each well to dissolve the obtained formazan. The absorbance was measured at 570 nm using a microplate reader (TECAN, Sunrise™, Grodig/Salzburg, Austria). The cell viability was calculated as follows:

$$\text{Cell viability } (\%) = \left[\frac{A_{570}(\text{experimental})}{A_{570}(\text{control})}\right] \times 100 \tag{1}$$

where A_{570}(experimental) was the absorbance at 570 nm of the experimental variants and A_{570}(control) was the absorbance at 570 nm of the respective negative control.

2.8. Fluorescent Microscopic Imaging

2.8.1. Double Staining with AO and EtBr

HeLa cells, MCF-7 cells, or mouse BALB/c 3T3 fibroblasts (1×10^5 cells/well) were cultivated overnight under standard conditions on glass coverslips placed on the bottom of a 24-well tissue culture plate. Following 24 h co-incubation of the cells with the different types of fibrous mats (PLA, Ch/PLA, Ch-8Q/PLA, Cu^{2+} complex of Ch-8Q/PLA and Fe^{3+} complex of Ch-8Q/PLA), the coverslips were rinsed twice with PBS (pH 7.4) and stained with fluorescent dyes—AO (5 μg/mL) and EtBr (5 μg/mL) in the ratio of 1:1—to visualize the cytomorphological alterations. It is known that AO can permeate both live and dead cells and shows strong yellow-green fluorescence after intercalation in DNA. EtBr is a fluorochrome that passes through damaged membranes and shows orange fluorescence as a result of the intercalation in DNA. EtBr stains late apoptotic cells and dead cells.

Cells maintained in the culture medium only, without treatment, served as a negative control. Cells treated with solutions of Jeff-8Q and its Cu^{2+}(Fe^{3+}) complexes as well as solutions of 8QCHO and its Cu^{2+}(Fe^{3+}) complexes were used as a positive control. The cells were examined under a fluorescence microscope (Leica DM 5000B; Wetzlar, Germany).

2.8.2. DAPI Staining

The nuclear morphology of the treated and control-untreated cells was further analyzed through DAPI staining according to a standard procedure [52]. For this purpose, the HeLa cells, MCF-7 cells, or mouse BALB/c 3T3 fibroblasts were processed as described in 2.8.1. After 24 h incubation, the glass coverslips were removed and washed with PBS (pH 7.4). Then the cells were fixed with 3% paraformaldehyde at room temperature and stained with a DAPI solution for 15 min at room temperature in the dark. The nuclear morphology of the stained cells was examined under a fluorescence microscope (Leika DM 5000B, Wetzlar, Germany).

2.9. *Statistical Analysis*

Statistical analysis was performed by one-way analysis of variance (ANOVA) followed by Bonferroni's post hoc comparison test (GraphPad Prism software package, version 5 (GraphPad Sofware Inc., San Diego, CA, USA)). *** $p < 0.001$ was considered statistically significant.

3. Results and Discussion

3.1. *Morphology*

The morphology and average diameter of the fibrous materials obtained by electrospinning are affected by various parameters such as the composition of the spinning solution, the concentration of the solution and the applied field strength. For the preparation of fibrous materials from Ch-8Q and PLA, a suitable co-solvent was chosen—TFA—that allowed the preparation of mixed solutions and their successful electrospinning. Defect-free and cylindrical fibers of PLA and Ch/PLA were also obtained.

Figure 1 shows the SEM micrographs of the prepared electrospun materials. Fibers prepared from a PLA solution with a concentration of 5 wt.% had an average diameter of 360 ± 90 nm (Figure 1a). A decrease in the average fiber diameter was observed when adding Ch-8Q or Ch to the PLA solution (Figure 1b–d).

For the Ch-8Q/PLA system at a weight ratio of Ch-8Q:PLA = 30:70, the average fiber diameter was 187 ± 128 nm, while at a weight ratio of Ch-8Q:PLA = 50:50, it was 124 ± 45 nm (Figure 1b,c). The average diameter of Ch/PLA fibers (weight ratio Ch:PLA = 30:70) was 238 ± 105 nm. The dynamic viscosities of the solutions of PLA, Ch/PLA (Ch:PLA = 30:70 *w/w*), Ch-8Q/PLA (Ch-8Q:PLA = 30:70 *w/w*) and Ch-8Q/PLA (Ch-8Q:PLA = 50:50 *w/w*) were 4200, 1900, 1700 and 940 cP, respectively. It has been found for other systems that the lower solution viscosity led to a decrease in fiber diameter or an increase in the amount of spindle-like defects [24,43,46,47]. In our study, decreasing the solution viscosity also resulted in the formation of fibers with smaller diameters. Moreover, Ch-8Q/PLA and Ch/PLA fibers are characterized by a broader diameter distribution compared to that of PLA fibers, which explains the elevated standard deviation of the average diameters of the Ch-8Q/PLA and Ch/PLA fibers. Some fiber splitting and branching of the main fibers with the appearance of very thin fibers was observed in the cases of the Ch-8Q/PLA and Ch/PLA mats (Figure 1b–d). The obtained results are consistent with those observed by other authors for Ch and Ch/PLA mats electrospun in TFA [15,29] and are most likely due to the fact that the jet elongation and solvent evaporation alter the shape and charge per unit area of the jet, which may cause some change in the balance between the surface tension and the electrical forces, which may, in turn, allow jet splitting and fiber branching to occur [19,53].

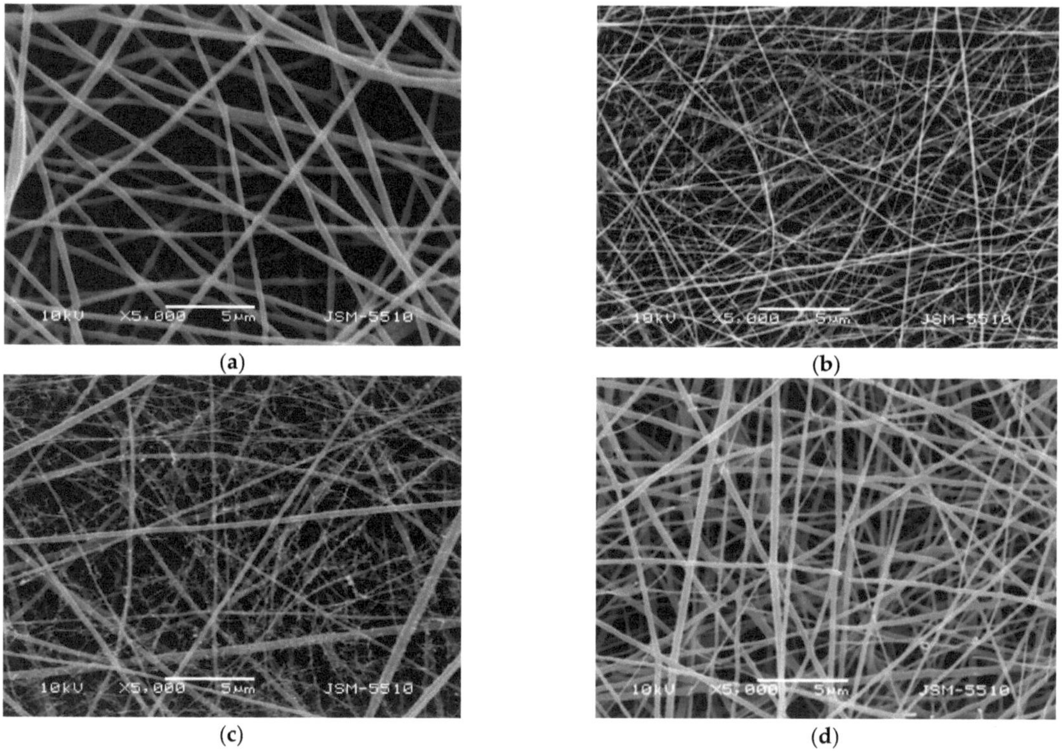

Figure 1. SEM micrographs of electrospun mats of (**a**) PLA, (**b**) Ch-8Q/PLA (50:50 *w*/*w*), (**c**) Ch-8Q/PLA (30:70 *w*/*w*) and (**d**) Ch/PLA (30:70 *w*/*w*); magnification ×5000.

Electrospun mats obtained at a weight ratio of Ch-8Q:PLA = 50:50 proved too brittle to manipulate. Thus, in the present study, mats from Ch-8Q and PLA in a weight ratio of 30:70 were selected for further biological experiments as the optimal formulation.

Experiments were conducted to obtain Cu^{2+} and Fe^{3+} complexes of Ch-8Q/PLA mats by immersing the fibrous materials in an ethanol solution of $CuCl_2$ and $FeCl_3$ for 40 min. As seen from the SEM micrographs, after this treatment, the mats retained their fibrous structure and the average diameter of the fibers remained unchanged—208 ± 126 nm and 189 ± 130 nm for Cu^{2+} and Fe^{3+} complexes of the Ch-8Q/PLA mats, respectively (Figure 2a,b).

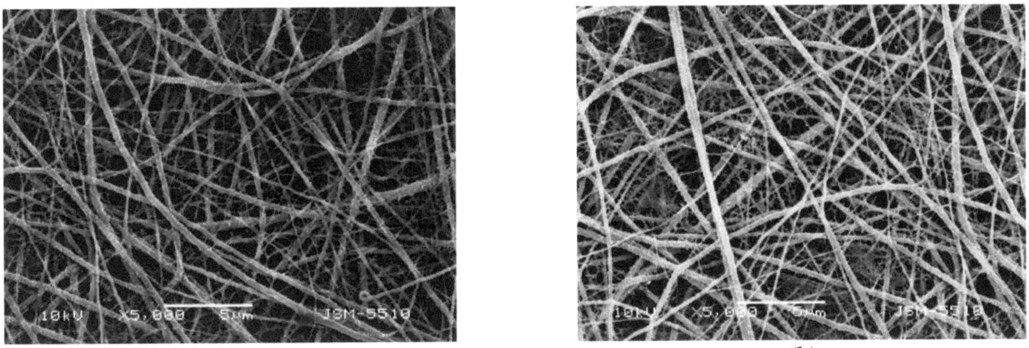

(**a**) (**b**)

Figure 2. SEM micrographs of electrospun Ch-8Q/PLA (30:70 *w*/*w*) mats after immersion in an ethanol solution of (**a**) $CuCl_2$ and (**b**) $FeCl_3$ for 40 min.; magnification ×5000.

3.2. ATR-FTIR Spectra of Fibrous Materials

ATR-FTIR spectroscopy was used to characterize the Ch-8Q/PLA mats and their Cu^{2+} and Fe^{3+} complexes. In the ATR-FTIR spectrum of the Ch/PLA mats, absorption characteristic bands were detected for both PLA (1751 cm^{-1}—C=O stretching vibration; 1084 cm^{-1}—C-O stretching vibration) and Ch (1676 cm^{-1}— stretching vibrations for protonated amino (-NH_3^+) groups; 1558 cm^{-1}—amide II; 3348 cm^{-1}—O-H and N-H stretching vibrations) (Figure 3 (a)). In the spectrum of the Ch-8Q/PLA mats, in addition to the PLA bands, characteristic bands were recorded at 1664, 1645 and 1506 cm^{-1}, respectively, for amide I from the polysaccharide structure of Ch-8Q, for C=N stretching vibrations from the azomethine group of the Schiff base Ch-8Q and for stretching vibrations from the ring of 8Q moieties in Ch-8Q, respectively (Figure 3 (b)). In the ATR-FTIR spectrum of Cu^{2+} and Fe^{3+} complexes of Ch-8Q/PLA mats, a shift of the absorption characteristic band for C=N stretching vibrations from the azomethine group to 1622 cm^{-1}, compared to the spectrum of Ch-8Q/PLA mats (1645 cm^{-1}) was detected. This shift is most likely due to the coordination of Cu^{2+} and Fe^{3+} to the azomethine nitrogen of Ch-8Q. This is consistent with the literature data for other metal complexes of Schiff bases [54,55]. In addition, in the spectra of the complexes (Figure 3 (c,d)), the band attributed to the C=N stretching vibrations of the 8Q residues in Ch-8Q, which was recorded in the spectrum of 8QCHO at 1591 cm^{-1} (Supplementary Material, Figure S1a), was shifted towards higher wavenumbers by 2 cm^{-1} to 1593 cm^{-1} and by 6 cm^{-1} to 1597 cm^{-1} for Cu^{2+} and Fe^{3+} complexes of the Ch-8Q/PLA mats, respectively. This is most likely due to the fact that the lone pair of electrons on the nitrogen of the 8Q residues participates in bond formation with the metal ion [56].

3.3. Thermal Behavior of the Fibrous Mats

The thermal properties of Ch-8Q/PLA mats and their Cu^{2+} and Fe^{3+} complexes were studied by DSC (Figure 4). In the DSC thermograms of the Ch-8Q/PLA mats and their complexes and of the Ch/PLA mats, a weakly intense broad endothermic peak between 25 and 100 °C was observed, which might be attributed to desorption of water or TFA from Ch and Ch-8Q (Figure 4 (b–e)). In the thermogram, the glass transition temperature (T_g), cold crystallization temperature (T_{cc}) and melting temperature (T_m) for the PLA mat were detected at 62 °C, 89 °C and 153 °C, respectively (Figure 4 (a)). The thermograms of Ch/PLA mats, as well as Ch-8Q/PLA and their Cu^{2+} and Fe^{3+} complexes, showed an absence of T_{cc} for PLA (Figure 4 (b–e)). This indicated that mixing with Ch, Ch-8Q and complex formation with the metal ions most likely affects the crystallization and the rate of crystallization of PLA.

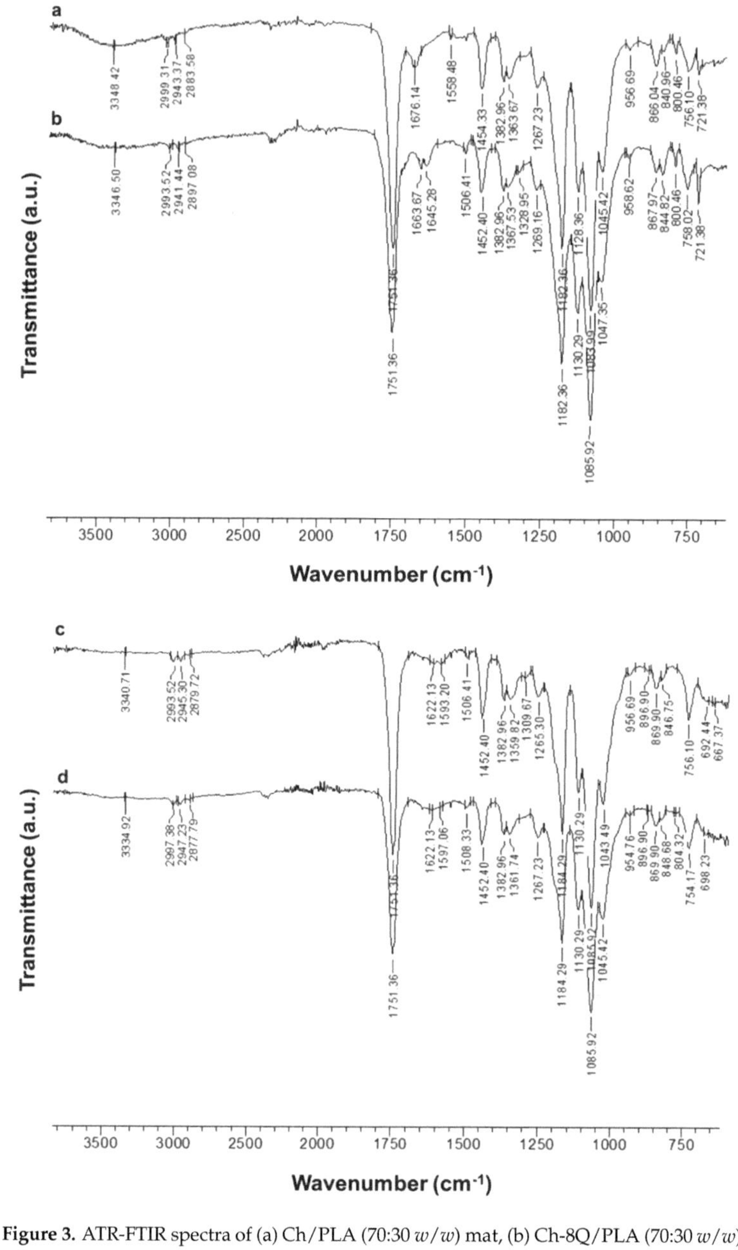

Figure 3. ATR-FTIR spectra of (a) Ch/PLA (70:30 *w*/*w*) mat, (b) Ch-8Q/PLA (70:30 *w*/*w*) mat, (c) Cu^{2+} complex of Ch-8Q/PLA mat and (d) Fe^{3+} complex of Ch-8Q/PLA mat.

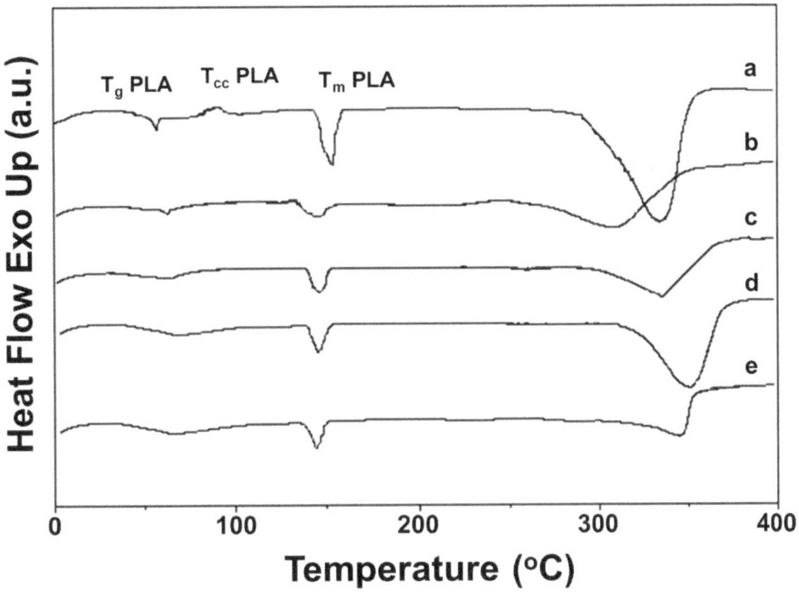

Figure 4. DSC thermograms of mats from (a) PLA, (b) Ch/PLA, (c) Ch-8Q/PLA, (d) Cu^{2+} complex of Ch-8Q/PLA and (e) Fe^{3+} complex of Ch-8Q/PLA.

Furthermore, the T_g for PLA did not change upon the incorporation of Ch and Ch-8Q into the mats, as well as upon coordination of Cu^2 or Fe^{3+} with the Ch-8Q-containing mats (61 °C for Ch/PLA mats as well as for Ch-N=C-8Q/PLA mats and their complexes) (Figure 4 (b–e)). As seen from Figure 4 (b–e), T_m for PLA was observed at a lower temperature—at 148 °C, 147 °C, 145 °C and 145 °C for Ch/PLA, Ch-8Q/PLA mats, and Cu^{2+} and Fe^{3+} complexes of Ch-8Q/PLA mats, respectively. Similarly to Ch, a melting peak for Ch-8Q was not recorded (Figure 4 (b,c)). In the DSC thermograms of the mats at a temperature above 280 °C, the appearance of endothermic peaks was detected (Figure 4 (a–e)), which were most likely due to the thermal degradation of the polymer constituents of the mats. These observations require further profound study to elucidate the changes that occur in the thermal stability of the mats upon the incorporation of Ch-8Q instead of Ch, as well as upon the coordination of Cu^{2+} or Fe^{3+} with the Ch-8Q/PLA mats.

3.4. Water Contact Angle Measurements

It is known that the adhesion of cells and their proliferation are highly dependent on the wettability of the surface of electrospun materials [57]. Therefore, in the present study, the wettability of the surface of the obtained fibrous materials that would come in contact with bacterial and cancer cells was measured. The PLA mat had a hydrophobic surface (the value of the water contact angle was 121.4 ± 2.0°) (Figure 5a). The water droplet retained its spherical shape on the surface of the mat. Hydrophilization of the mats was observed when Ch was present in the fibers (Figure 5b). The measured values of the water contact angle for Ch/PLA were 50.2 ± 5.3°. The Ch-8Q/PLA mats were hydrophobic (the water contact angle value was 112.9 ± 5.6°) (Figure 5c). Coordination of Cu^{2+} and Fe^{3+} with the Ch-8Q/PLA mats led to a decrease in the water contact angle to about 78.0° (Figure 5d,e).

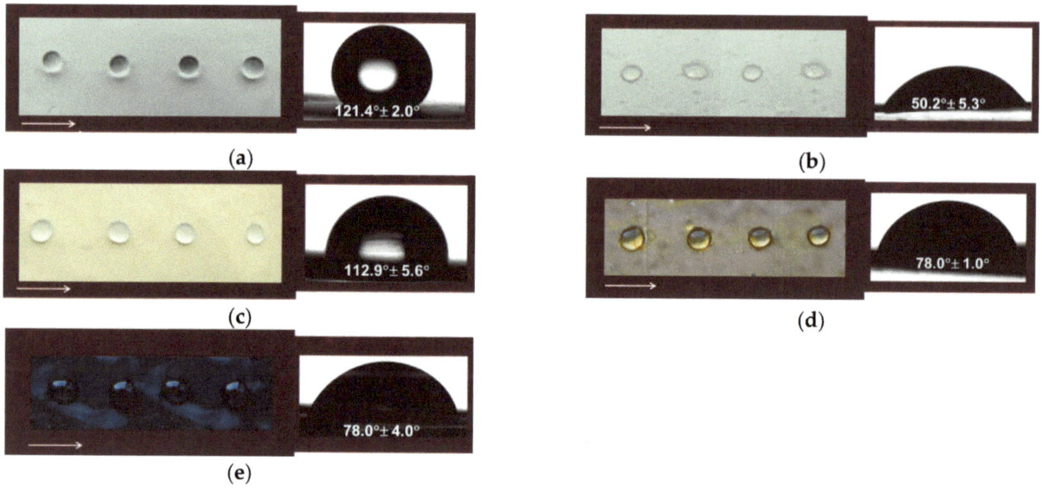

Figure 5. Digital images of water droplets (10 μL) deposited on the surfaces of fibrous mats from (**a**) PLA, (**b**) Ch/PLA, (**c**) Ch-8Q/PLA, (**d**) Cu^{2+} complex of Ch-8Q/PLA and (**e**) Fe^{3+} complex of Ch-8Q/PLA. The direction of the collector rotation is indicated by an arrow.

3.5. XPS Analysis

XPS analysis of the surface also confirmed the structure of the Ch-8Q/PLA mats and their Cu^{2+} and Fe^{3+} complexes. Peaks at 285.0 eV (-C-H or -C-C- of PLA and Ch-8Q, as well as -C-NH_2 of Ch-8Q), at 285.6 eV (-C-N and -C-OH of the 8Q residues), at 286.8 eV (-C-O, -C-OH and -C-N-C=O of Ch-8Q, -C-O of PLA and -C=N of Ch-8Q), at 288.4 eV (-O-C-O- and -N-C=O of Ch-8Q), at 289.1 eV (-O-C=O of PLA) and at 290.5 eV ($\pi \rightarrow \pi^*$ shake-up satellite characteristic of the 8Q ring of Ch-8Q) were detected in the high-resolution C_{1s} spectrum of Ch-8Q/PLA mats (Supplementary Material, Figure S2a). The O_{1s} spectrum showed four components at 530.8 eV assigned to -N-C=O of Ch-8Q, at 532.2 eV for -C=O of PLA, at 532.9 eV for -C-OH of Ch-8Q and of the 8Q residues and at 533.5 eV for -O-C-O and -C-O of PLA (Supplementary Material, Figure S2b). The N_{1s} signal consisted of four peaks at 398.8 eV, assigned to -N=C of Ch-8Q, at 399.6 eV to -N-C=O and -C-NH_2 of Ch-8Q, at 400.8 eV characteristic of the –N-C of the 8Q residues in Ch-8Q and at 401.8 eV for the -NH_3^+ groups of Ch (Supplementary Material Figure S2c). The theoretically calculated peak-area ratio for the corresponding carbon atoms was [C-C/C-H/C-NH_2]/[C-N/C-OH]/[C-O/C-OH/C-N-C=O/C-O/C=N]/[O-C-O/N-C=O]/[O-C=O]/[$\pi \rightarrow \pi^*$] = 34.1/2.3/36.1/3.5/23.3/0.7. The experimentally determined ratio was 35.5/2.3/35.3/3.4/22.8/0.7. The peak for the carbon atoms from the C-C/C-H/C-NH_2 bonds had the largest area. This is consistent with the determined surface hydrophobicity of the Ch-8Q/PLA mat (the water contact angle was 112.9° ± 5.6°).

In the detailed C_{1s} spectra of the complexes of the Ch-8Q/PLA mats, compared with those of the Ch-8Q/PLA mats, the appearance of a new component was observed at 287.2 eV, characteristic of -C-O—Cu/-C-N—Cu or -C-O—Fe/-C-N—Fe (Figure 6a,f). The intensity of the peak at 286.8 eV also decreased. In the expanded O_{1s} spectra of the complexes of the mats, a new -O—Cu or -O—Fe signal from the 8Q residues in Ch-8Q appeared at 531.2 eV (Figure 6b,g). In the N_{1s} spectra of the complexes of the Ch-8Q/PLA mats, a new peak at 400.1 eV was detected, attributed to -C-N—Cu or -C-N—Fe of the 8Q residues in Ch-8Q (Figure 6c,h). The complex formation between Cu^{2+} and the Ch-8Q/PLA mats was also indicated by the appearance of a peak composed of two components—$Cu_{2p1/2}$ and $Cu_{2p3/2}$ (Figure 6d). $Cu_{2p3/2}$ consisted of a main peak at 933.7 eV and two satellites at 940.9 eV and 944.6 eV. The main peak was characterized by a binding energy close to that observed by other authors for Cu^{2+} complexes (933.1 eV) [58]. In the XPS spectrum of the Fe^{3+} complex of the Ch-8Q/PLA mats, a new $Fe_{2p3/2}$ component from the Fe_{2p} region with a main peak at 711.9 eV and a satellite at 718.2 eV was also detected (Figure 6i). The binding energy of the main peak was close to that reported by other authors for Fe^{3+} complexes [59]. These results confirmed the coordination of Cu^{2+} or Fe^{3+} on the surface of the Ch-8Q/PLA mats. The appearance of a peak in the Cl2p region (at 198.4 eV (Cl2p3/2) and 200.0 eV (Cl2p1/2)) (Figure 6e,j) indicated the presence of Cl ions on the surface layer of complexes of the Ch-8Q/PLA mats.

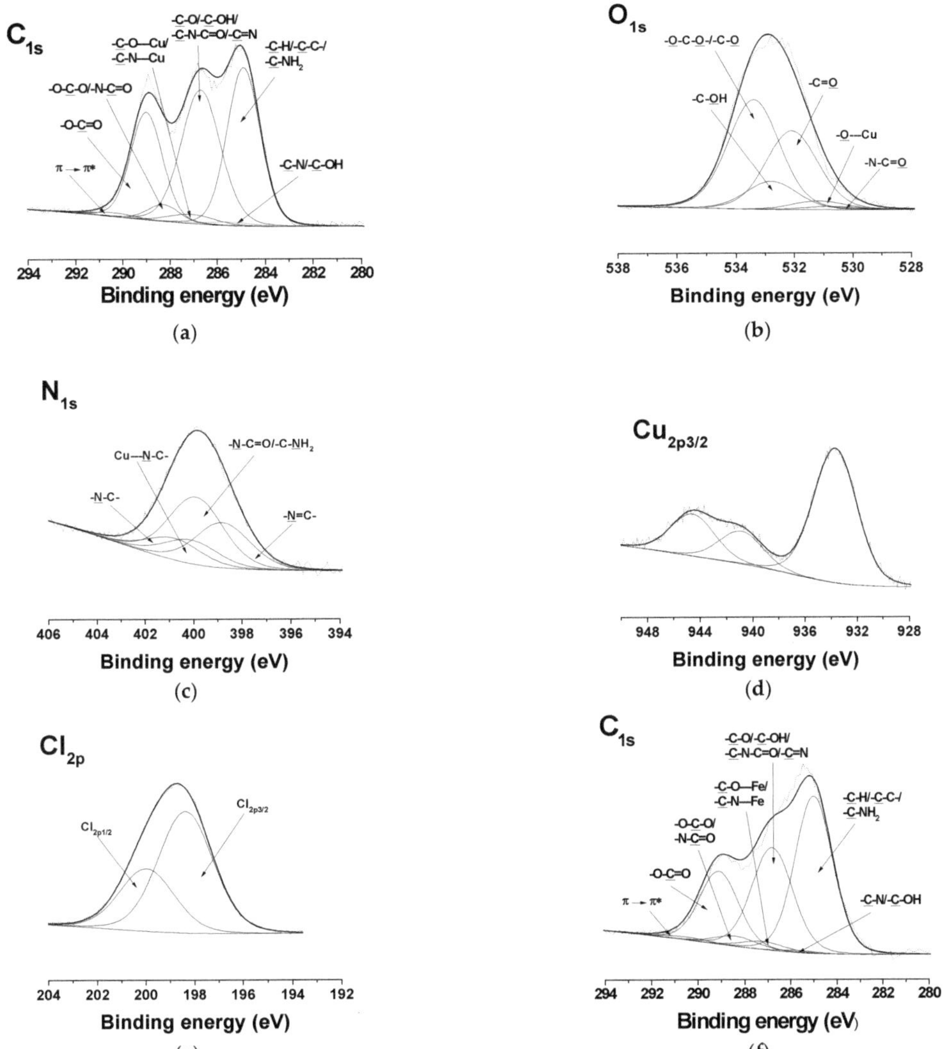

Figure 6. *Cont.*

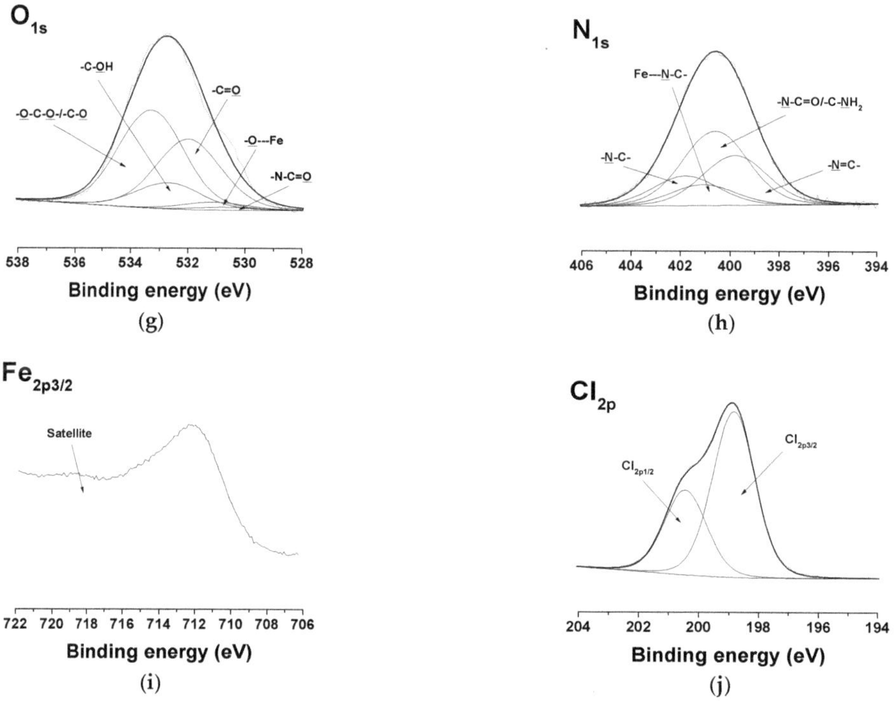

Figure 6. XPS peak fittings for Cu^{2+} complex of Ch-8Q/PLA mat [C_{1s} (**a**), O_{1s} (**b**), N_{1s} (**c**), Cu2p 3/2 (**d**) and Cl2p (**e**)] and Fe^{3+} complex of Ch-8Q/PLA mat [C_{1s} (**f**), O_{1s} (**g**), N_{1s} (**h**), Fe2p 3/2 (**i**) and Cl2p (**j**)].

3.6. EPR Spectroscopy Analysis of Cu^{2+} and Fe^{3+} Complexes of the Fibrous Materials

EPR analysis was used in order to reveal the coordination mechanism of Cu^{2+} and Fe^{3+} in the complexes of Ch-8Q/PLA fibrous materials. As EPR standards, we studied Cu^{2+} and Fe^{3+} complexes of Ch/PLA mats as well as complexes Cu^{2+}-Ch-8Q and Fe^{3+}-8QCHO in the solid state. The complexes were analyzed in the temperature range of 100 to 295 K and their spectra are shown in Figures 7 and 8.

The EPR spectrum of the Cu^{2+} complex of the Ch-8Q/PLA mats contained one nearly symmetric signal with slightly resolved $g_{\perp}$ and $g_{\parallel}$ ($g_{\perp}$ = 2.10, g_{II}~2.26). The signal retained its shape and position on cooling, while its intensity increased according to the Currie–Weiss law ($\theta = -147 \pm 6$ K). These EPR parameters are typical for Cu^{2+} ions that are magnetically coupled.

Since the Schiff base Ch-8Q (Scheme 1) used to obtain the Ch-8Q-containing mats is a derivative of Ch and 8QCHO, the possibilities for coordination of Cu^{2+} ions with 8Q residues as well as with Ch moieties could not be excluded. Thus, the question arises as to how the Cu^{2+} ions are coordinated in the complexes of Ch-8Q/PLA mats. To understand the Cu^{2+} coordination, two reference materials were used: Cu^{2+}-Ch/PLA mat and Cu^{2+}-Ch-8Q in the solid state. The EPR spectrum of the Cu^{2+}-Ch/PLA mats complex consisted of one symmetric signal with g = 2.133. The g-value was constant between 100 and 295 K, and the linewidth (ΔH_{pp}) varied from 22.8 mT at 295 K to 19.7 mT at 100 K. The temperature dependence of the reciprocal value of signal intensity followed the Currie–Weiss law ($\theta = -28 \pm 7$ K). All these EPR parameters indicated that the EPR signal of the Ch/PLA mats originated from exchanged coupled Cu^{2+} ions. The g-value of the Ch/PLA mats was close to that of g_{av} previously determined by Pawlicka et al. [60] for membranes of

Ch coordinated with Cu^{2+} ions (g_{av} = 2.123). Therefore, it could be concluded that Ch is coordinated around the Cu^{2+} ions in Cu^{2+}-Ch/PLA mats.

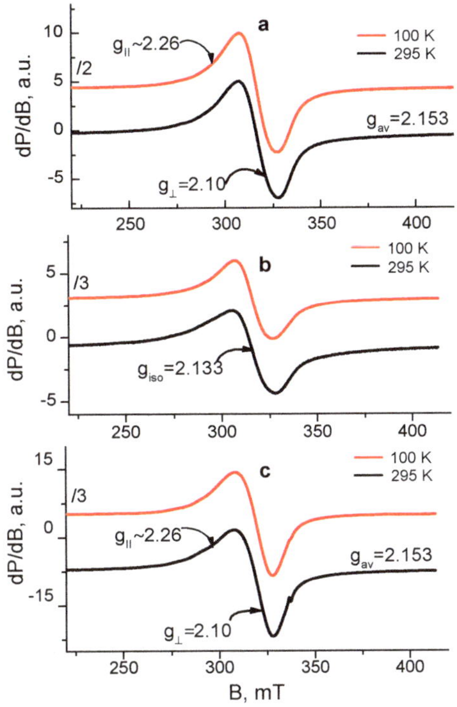

Figure 7. EPR spectra at 100 and 295 K of (**a**) Cu^{2+}-Ch-8Q/PLA mat, (**b**) Cu^{2+}-Ch/PLA mat and (**c**) Cu^{2+}-Ch-8Q in the solid state.

The spectrum of the Cu^{2+}-Ch-8Q complex in the solid state showed a slightly asymmetric signal with low-intensity and not fully resolved hyperfine lines. The EPR parameters at 295 K were: g_{II}~2.26, $g_{\perp} \approx 2.10$, g_{av} = 2.153. The signal asymmetry decreased on cooling from 295 to 100 K, but the temperature dependence of the signal intensity followed the Currie–Weiss law ($\theta = -52 \pm 4$ K). It is worth noting that the g_{iso} of the Cu^{2+}-8Q complex [50] is close to the g_{av} value of the Cu^{2+}-Ch-8Q complex. In addition, the EPR parameters of the Cu^{2+}-Ch-8Q complex deviated from that of the Cu^{2+}-Ch/PLA mat. Therefore, the signal of the solid-state Cu^{2+}-Ch-8Q complex was assigned to Cu^{2+} ions that were coordinated by 8Q residues.

The comparison between the EPR parameters of the Cu^{2+}-Ch-8Q/PLA mat and the two above-mentioned references allowed the outlining of several EPR features. Because of the difference in the g-value for the Ch/PLA mat and Ch-8Q/PLA mat as well as the close g-value of the Cu^{2+}-Ch-8Q/PLA mat and solid Cu^{2+}-Ch-8Q complex, it suggested a similar coordination of Cu^{2+} in both materials, i.e., it appeared that Cu^{2+} ions in Ch-8Q/PLA mats are preferentially coordinated with 8Q residues in Ch-8Q. This means that Cu^{2+} ions were coordinated with O and N atoms from the 8Q moieties.

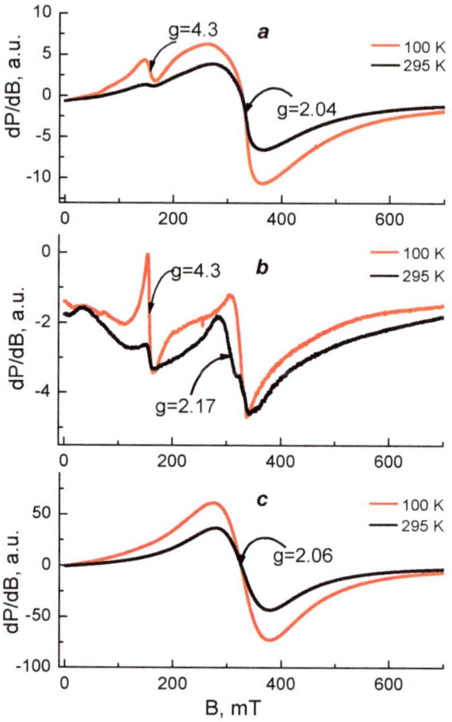

Figure 8. EPR spectra at 100 and 295 K of (**a**) Fe^{3+}-Ch-8Q/PLA mat, (**b**) Fe^{3+}-Ch/PLA mat and (**c**) Fe^{3+}-8QCHO in the solid state.

Figure 8 shows the EPR spectra of the studied Fe^{3+} complexes. The EPR spectrum of the Fe^{3+}-8QCHO complex consisted of a broad, symmetric signal with g = 2.06 and a linewidth $\Delta H_{pp} \approx 100$ mT at 295 K. A similar signal with g = 2.04 and $\Delta H_{pp} \approx 93$ mT at 295 K was observed also in the central part of the Fe^{3+}-Ch-8Q/PLA mat spectrum. These signals were assigned to exchanged coupled Fe^{3+} ions and their similarity could be related to similar coordination of Fe^{3+} ions in solid Fe^{3+}-8QCHO complex and Fe^{3+}-Ch-8Q/PLA mat complex. These findings showed that Fe^{3+} ions in the complex of the Ch-8Q/PLA mat were preferentially coordinated to 8Q residues of Ch-8Q.

In order to confirm this conclusion the EPR spectra of the Fe^{3+}-Ch-8Q/PLA mat and Fe^{3+}-Ch/PLA mat were compared. Their spectra were characterized by two different signals having g-values of g = 4.3 and g = 2.0. In the EPR spectrum of the Fe^{3+}-Ch/PLA mat the signal with $g \approx 4.3$ dominated, while the signal with $g \approx 2.0$ predominated in the Fe^{3+}-Ch-8Q/PLA mat spectrum. Furthermore, the intensity of the signal with $g \approx 2.0$ for the Fe^{3+}-Ch-8Q/PLA mat was more than 50 times higher than that for Fe^{3+}-Ch/PLA mat, thus indicating a higher concentration of coordinated Fe^{3+}. Therefore Fe^{3+} was coordinated in Fe^{3+}-Ch-8Q/PLA mat preferably by 8Q moieties of Ch-8Q.

3.7. Evaluation of the Antibacterial Activity

The antibacterial activity of the Ch-8Q/PLA mats and their Cu^{2+} and Fe^{3+} complexes against the Gram-positive bacteria *S. aureus* was estimated by counting the viable bacteria that remained after incubation of the mats in an *S. aureus* suspension for given time periods. In the present study, *S. aureus* bacteria were selected because they are one of the most common pathogenic bacteria responsible for secondary infections of wounds. For the sake of comparison, the antibacterial activity of PLA and Ch/PLA mats was also examined.

The *S. aureus* control was found to grow normally for the given time periods during the experiment and log(CFU/mL) reached 13.4 in 24 h. As can be seen in Figure 9, the neat PLA mats did not suppress the growth of *S. aureus* bacteria. In this case, the number of survived cells in 24 h was 13.0 log units. In contrast, Ch/PLA mats with a Ch content of 1000 μg/mL, for a contact time of 24 h led to a 7.2 log decrease in the *S. aureus* titer (Figure 9). A difference in the antibacterial activity was observed for the Ch-8Q/PLA mats and their Cu^{2+} and Fe^{3+} complexes at the same content of Ch-8Q—1000 μg/mL. For Ch-8Q/PLA mats, the *S. aureus* titer decreased by 2 log units for a contact time of 2 h, whereas for the same contact time for Cu^{2+} and Fe^{3+} complexes of Ch-8Q/PLA mats, a reduction of the *S. aureus* titer by 1 and 0.9 log units was detected, respectively. In the cases of Ch-8Q/PLA mats and their complexes, viable *S. aureus* cells were absent after 3 h of contact. The obtained results showed that the incorporation of Ch-8Q into the mats, as well as the complexation of the mats with Cu^{2+} and Fe^{3+}, imparted to the mats a higher antibacterial activity compared to that of the Ch/PLA mats. The Ch-8Q-containing mats and their complexes that manifest their activity by contact between the bioactive agent and the bacteria are perspective dressings for infected wounds.

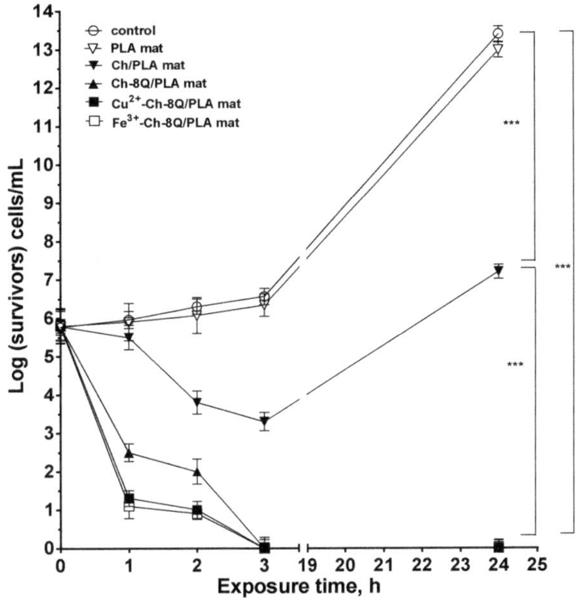

Figure 9. Logarithmic plot of the viable bacterial cell *S. aureus* number versus the exposure time for control (*S. aureus*), PLA mat, Ch/PLA mat, Ch-8Q/PLA mat, Cu^{2+} complex of Ch-8Q/PLA mat and Fe^{3+} complex of Ch-8Q/PLA mat. Data represent the mean ± standard deviation ($n = 3$). *** $p < 0.001$, statistical significance.

The adhesion of *S. aureus* cells on the surface of the fibrous materials was monitored by SEM. SEM micrographs of *S. aureus* cells adhered to the surface of the mats after contact with the *S. aureus* suspension for 24 h at 37 °C are presented in Figure 10. The *S. aureus* cells adhered very well to the surface of the hydrophobic PLA mat with a tendency to form a biofilm (Figure 10a). In the case of the Ch/PLA mats, the number of adhered bacterial cells decreased. The reduced adhesion of bacterial cells was most likely due to the presence of Ch, which had an antibacterial effect (Figure 10b).

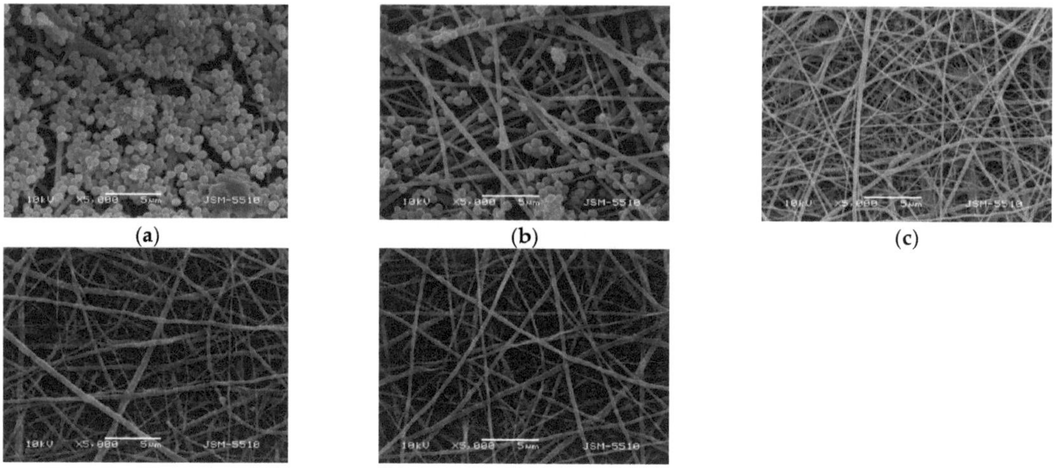

(a) (b) (c) (d) (e)

Figure 10. SEM micrographs of mats that have been incubated in *S. aureus* cell culture (10^7 cells/mL) for 24 h at 37 °C, (**a**) PLA, (**b**) Ch/PLA, (**c**) Ch-8Q/PLA, (**d**) Cu^{2+} complex of Ch-8Q/PLA and (**e**) Fe^{3+} complex of Ch-8Q/PLA; magnification ×5000.

The incorporation of Ch-8Q in the fibers, as well as the complex formation with Cu^{2+} and Fe^{3+}, led to a complete suppression of the growth of *S. aureus* bacteria on the surface of the mats (Figure 10d,e). These results show that pathogenic bacteria were killed on contact with the mats, which was attributed to the high bactericidal activity of Ch-8Q incorporated in the mats and their Cu^{2+} and Fe^{3+} complexes. The obtained Ch-8Q-containing fibrous materials and their complexes might be suitable candidates as materials having a surface that can kill the pathogenic bacteria *S. aureus*.

3.8. Cytotoxicity of Fibrous Mats against HeLa and MCF-7 Cells and BALB/c 3T3 Fibroblasts

It has been reported that 8QCHO exerts good antiproliferative activity against various human cancer cell lines, such as Hs578t, SaoS2, K562, MDA231, SKHep1, T-47D and Hep3B [61]. In the present study, the viability of MCF-7 and HeLa cancer cells cultured in the presence of Ch-8Q/PLA fibrous materials and their complexes was evaluated by the MTT assay. The cytotoxic effect of these materials against non-cancer BALB/c 3T3 mouse fibroblast cells in vitro was also assessed. In this assay, 8QCHO and its Cu^{2+} (Fe^{3+}) complexes, as well as Jeff-8Q and its Cu^{2+}(Fe^{3+}) complexes, were used as positive controls, and untreated MCF-7, HeLa, or BALB/c 3T3 cells were used as negative controls. The effect on cancer cell viability was less pronounced when they were treated with PLA and Ch/PLA mats (Figure 11a–d). In contrast, the viability of MCF-7 and HeLa cells treated with the Ch-8Q-containing fibrous materials and their complexes decreased significantly (Figure 11a–d). The observed antiproliferative effect increased on increasing the duration of the incubation period. The highest cytotoxicity of the Ch-8Q/PLA mats and their complexes was detected after 72 h of incubation (Figure 11b,d). The antiproliferative activity of these mats was more pronounced against HeLa cancer cells than MCF-7 cells. At the 72nd h of incubation, Cu^{2+} and Fe^{3+} complexes of Ch-8Q/PLA mats exhibited higher cytotoxicity (2.6 ± 1.8% and 0.8 ± 0.7% viable cells for Cu^{2+} and Fe^{3+} complexes of the mats, respectively) against HeLa cells compared to the Ch-8Q-containing mats (27.4 ± 2.1% viable cells). In the case of Cu^{2+} complexes of Ch-8Q/PLA mats, a stronger decrease in the proliferative activity of MCF-7 cancer cells (1.2 ± 1.3% viable cells) than that caused by Ch-8Q/PLA mats (48.6 ± 8.5% viable cells) and their Fe^{3+} complexes (34.0 ± 6.6% viable cells) was found. The percentage of viable HeLa and MCF-7 cells for the Ch-8Q-containing mats was close to that for free Jeff-8Q (approx. 37.9% viable HeLa cells and approx. 44.1%

viable MCF-7 cells). About 4.8 ± 1.4 and 7.2 ± 1.8% of HeLa cells and about 2.1 ± 3.1 and 37.5 ± 3.7% of MCF-7 cells remained viable after 72h of incubation in the presence of Cu^{2+} and Fe^{3+} complexes of Jeff-8Q, respectively. These complexes had cytotoxicity close to that of the complexes of Ch-8Q/PLA mats. As seen in Figure 11b,d, 8QCHO and its complexes exhibited higher antiproliferative effects against both types of cancer cells than that of Ch-8Q-containing fibrous materials and its complexes. It should be noted that the Ch-8Q/PLA fibrous mats did not show any statistically significant antiproliferative activity against non-cancer BALB/c 3T3 cells—viable cells were 90.0 ± 12.2% after 72 h of incubation. PLA and Ch/PLA mats also exhibited low cytotoxicity against BALB/c 3T3 cells (Figure 11e,f). In the case of the complexes of Ch-8Q/PLA mats, the decrease in viability of BALB/c 3T3 cells was less pronounced compared to that for both types of cancer cells (Figure 11e,f). After 72 h of incubation, the percentage of viable BALB/c 3T3 cells was 35.3 ± 4.0% and 5.4 ± 3.7% for Fe^{3+} and Cu^{2+} complexes of the Ch-8Q/PLA mats, respectively. Therefore, Ch-8Q-containing mats and their complexes exerted high anticancer activity against HeLa and MCF-7 cells, while being less toxic against normal mouse BALB/c 3T3 fibroblasts.

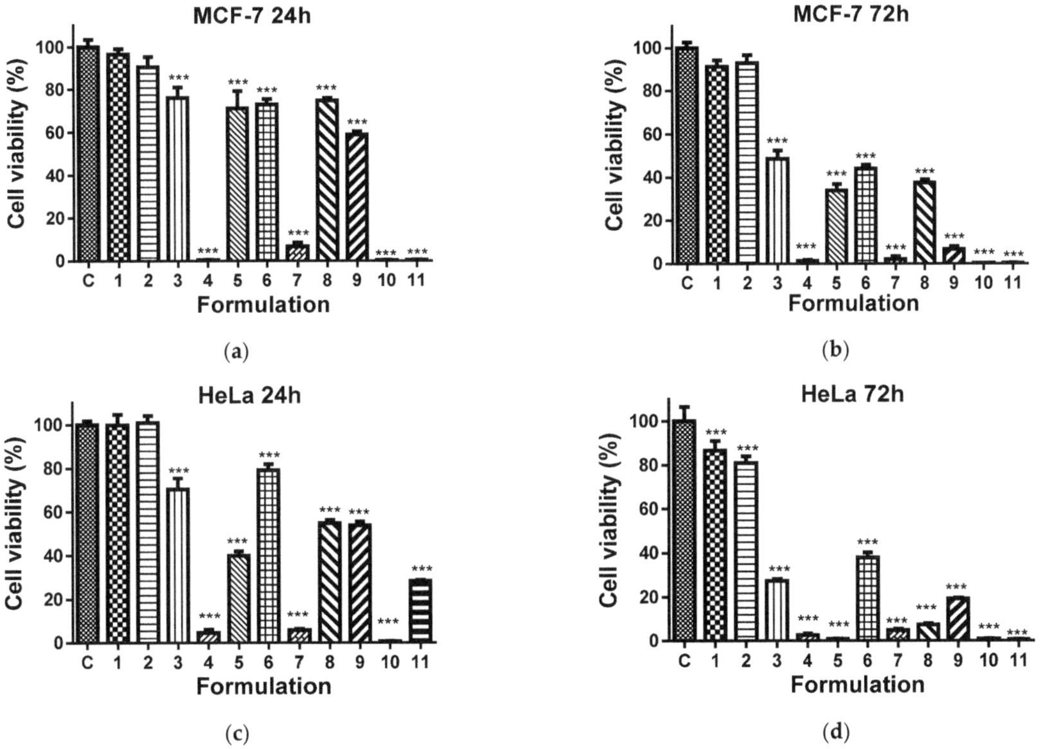

Figure 11. *Cont.*

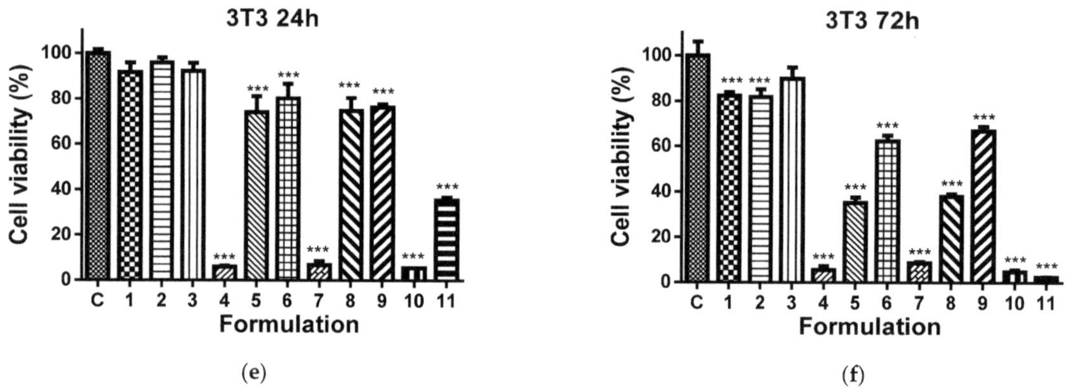

Figure 11. Effect of the different formulations on MCF-7 (**a**,**b**), HeLa (**c**,**d**), or BALB/c 3T3 (**e**,**f**) cells after 24 (**a**,**c**,**e**) and 72 h (**b**,**d**,**f**) of incubation. C—Untreated cells (control); 1—PLA mat; 2—Ch/PLA mat; 3—Ch-8Q/PLA mat; 4—Cu^{2+} complex of Ch-8Q/PLA mat; 5—Fe^{3+} complex of Ch-8Q/PLA mat; 6—aqueous solution of Jeff-8Q; 7—aqueous solution of Cu^{2+} complex of Jeff-8Q; 8—aqueous solution of Fe^{3+} complex of Jeff-8Q; 9—solution of 8QCHO; 10—solution of Cu^{2+} complex of 8QCHO and 11—solution of Fe^{3+} complex of 8QCHO. All 8Q-containing formulations were studied at a concentration of 8Q residues 60 μg/mL of culture medium. *** $p < 0.001$.

3.9. Analysis of Cell Death by Staining Methods

In order to determine whether the antiproliferative effect of fibrous Ch-8Q-containing mats and their Cu^{2+} and Fe^{3+} complexes was related to the induction of apoptosis, a fluorescence assay was applied to detect cell death by intravital double staining with the fluorescent dyes AO and EtBr. The morphological features of HeLa and MCF-7 cancer cells cultured for 24 h in the presence of the various fibrous mats were studied. Untreated cancer cells (negative control) had homogeneous pale green nuclei and bright yellow-green nucleoli (Figure 12a and Supplementary Material Figure S4a).

No change was observed in the staining of the nuclei and cytoplasm in HeLa and MCF-7 cells after their treatment with PLA and Ch/PLA mats (Figure 12b,c and Supplementary Material Figure S4b,c). In these cases, the cell morphology remained normal. In contrast, when cells were cultured in the presence of the Ch-8Q/PLA mats and their complexes, cell rounding and cell shrinkage, cell membrane blebbing, cellular and nuclear volume reduction (pyknosis), condensation and aggregation of nuclear chromatin, the appearance of apoptotic bodies and nuclear fragmentation occurred (Figure 12d–f and Supplementary Material Figure S4d–f). These are typical morphological signs of early or late apoptosis. Similar morphological changes were observed when both types of cancer cells were placed in contact with free Jeff-8Q and its complexes or with free 8QCHO and its complexes (Figure 12g–l and Supplementary Material, Figure S4g–l). The most significant morphological changes were detected in HeLa and MCF-7 cancer cells after their treatment with Cu^{2+} complexes of Ch-8Q/PLA mats, as well as with Cu^{2+} complexes of Jeff-8Q or 8QCHO (Figure 12e,h,k and Supplementary Material Figure S4e,h,k). A significant number of nuclei and cytoplasm of cells that were stained red-orange were observed, as well as a decrease in the number of cells and the presence of dead destructured cells with pyknotic nuclei (criteria for late apoptosis) (Figure 12e,h,k and Supplementary Material Figure S4e,h,k).

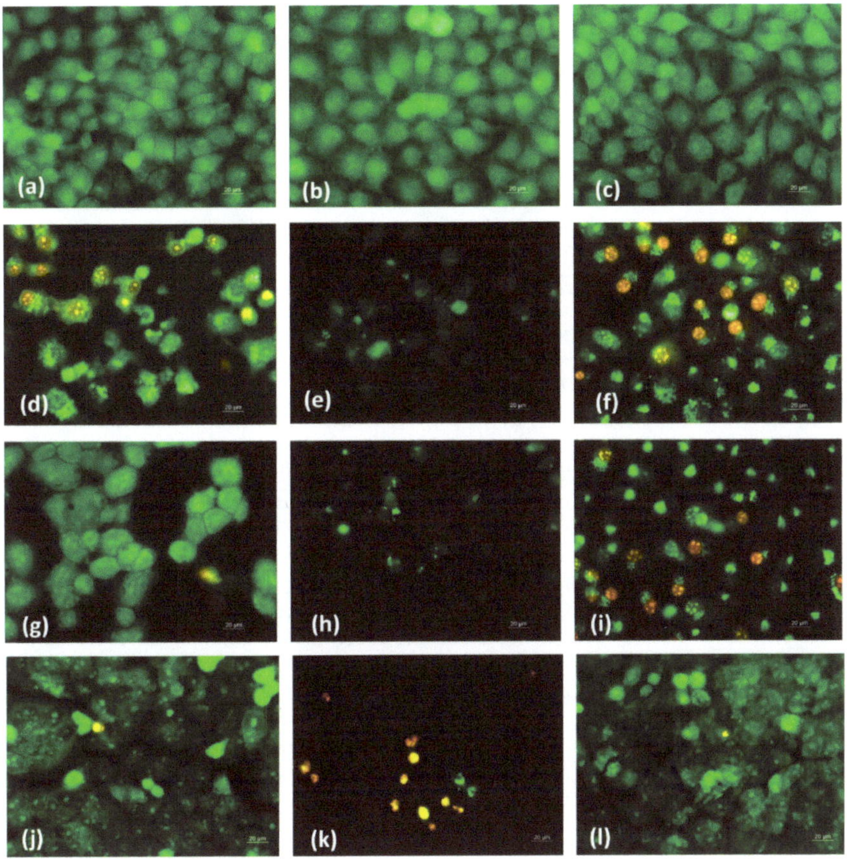

Figure 12. Fluorescence micrographs of AO and EtBr double-stained HeLa cancer cells incubated with different formulations for 24 h. Cells after incubation with (**a**) untreated HeLa cells, (**b**) PLA mat, (**c**) Ch/PLA mat, (**d**) Ch-8Q/PLA mat, (**e**) Cu^{2+} complex of Ch-8Q/PLA mat, (**f**) Fe^{3+} complex of Ch-8Q/PLA mat, (**g**) aqueous solution of Jeff-8Q, (**h**) aqueous solution of Cu^{2+} complex of Jeff-8Q, (**i**) aqueous solution of Fe^{3+} complex of Jeff-8Q, (**j**), solution of 8QCHO, (**k**) solution of Cu^{2+} complex of 8QCHO and (**l**) solution of Fe^{3+} complex of 8QCHO, scale bar = 20 μm. All 8Q-containing formulations were studied at a concentration of 8Q residues 60 μg/mL of culture medium.

The morphological changes in the nuclei of HeLa and MCF-7 cancer cells were analyzed after staining the cells with DAPI. Control untreated HeLa and MCF-7 cells possessed intact nuclei, slightly oval, in shape, of approximately equal size, with smooth edges and uniformly distributed chromatin (Supplementary Material Figures S3a and S5a). The morphology of the nuclei of HeLa and MCF-7 cancer cells treated with PLA and Ch/PLA mats was close to that of the control (Supplementary Material Figures S3b,c and S5b,c). Cancer cells that had been in contact with the Ch-8Q/PLA mats and their complexes, with solutions of Jeff-8Q and its complexes, and with solutions of 8QCHO and its complexes were characterized by changes in the morphology of the nuclei, typical of apoptosis, such as chromatin condensation, nuclei pyknosis, nuclei fragmentation and an increase in the number of apoptotic bodies (Supplementary Material Figures S3d–l and S5d–l). The strongest damage to the nuclei of HeLa and MCF-7 cells was observed when they were treated with the Cu^{2+} complex of Ch-8Q/PLA mats or with Cu^{2+} complexes of Jeff-8Q and 8QCHO (Supplementary Material Figures S3e,h,k and S5e,h,k).

The obtained results were consistent with the data obtained from the MTT test and revealed that Cu^{2+} and Fe^{3+} complexes of Ch-8Q/PLA mats exerted a high antiproliferative effect against HeLa and MCF-7 cancer cells. Ch-8Q-containing mats displayed weaker cytotoxicity toward cancer cells. The observations from the performed fluorescence microscopy analyses showed that these mats induce the death of cancer cells via apoptosis.

When Balb/c 3T3 mouse fibroblasts were cultured in the presence of Cu^{2+} and Fe^{3+} complexes of Ch-8Q/PLA mats, morphological changes of the cells and nuclei that are characteristic of early and late apoptosis were detected (Supplementary Material, Figures S6e,f and S7e,f). These changes were significantly greater in the case of Balb/c 3T3 cells treated with the Cu^{2+} complex of the Ch-8Q-containing mats (Supplementary Material, Figures S6e and S7e). The results obtained from the fluorescence methods indicated that the Ch-8Q-containing mats did not exhibit toxicity against Balb/c 3T3 cells (Supplementary Material, Figures S6d and S7d).

4. Conclusions

In the present study, the Schiff base derivative (Ch-8Q) of Ch and 8QCHO was synthesized and novel fibrous materials were successfully fabricated from Ch-8Q and PLA by one-pot electrospinning of their blend solution. Complexes of the mats were easily obtained by treating them with $CuCl_2$ or $FeCl_3$ solution. Based on ATR-FTIR, XPS and EPR spectroscopic analyses, it was concluded that Cu^{2+} and Fe^{3+} ions in both studied complexes of the Ch-8Q/PLA mats were preferably surrounded by 8Q-residue of Ch-8Q fibrous materials. The incorporation of Ch-8Q in the fibrous mats and complexation with Cu^{2+}(Fe^{3+}) imparted significant biocidal activity against *S. aureus* bacteria. These mats demonstrated the ability to kill all *S. aureus* bacterial cells within a contact time of 3h. Furthermore, in contrast to the Ch-containing mats, which only reduce the adhesion of pathogenic bacteria *S. aureus*, Ch-8Q-containing materials and their complexes inhibit bacterial adhesion. These fibrous mats exhibited high anticancer effects against human cervical HeLa and human breast MCF-7 carcinoma cell lines. Their in vitro anticancer activity depends on the incubation period. Fluorescence microscopy analyses indicated that the induction of apoptosis was one of the major mechanisms of the anticancer efficacy of the new materials. The cytotoxic effect of the mats was higher in cancer cells than in non-cancer BALB/c 3T3 mouse fibroblasts. Moreover, Ch-8Q/PLA mats displayed no cytotoxicity to the non-cancer cells. The obtained materials could find potential as wound dressing materials and in application in local treatment of cervical and breast cancer.

Supplementary Materials: The following supporting information can be downloaded at: https://www.mdpi.com/article/10.3390/polym14225002/s1, Figure S1: ATR-FTIR spectra of 8QCHO and Ch-8Q, Figure S2: XPS peak fittings for Ch-8Q/PLA mat, Figure S3: Fluorescence microscopic images of HeLa cancer cells stained with DAPI after treatment with PLA mat, Ch/PLA mat, Ch-8Q/PLA mat and its complexes, Jeff-8Q and its complexes and 8QCHO and its complexes, Figure S4: Fluorescence micrographs of AO and EtBr double-stained MCF-7 cancer cells incubated with different formulations for 24 h, Figure S5: Fluorescence microscopic images of MCF-7 cancer cells stained with DAPI after treatment with PLA mat, Ch/PLA mat, Ch-8Q/PLA mat and its complexes, Jeff-8Q and its complexes and 8QCHO and its complexes, Figure S6: Fluorescence micrographs of AO and EtBr double-stained BALB/c 3T3 cells incubated with different formulations for 24 h, Figure S7: Fluorescence microscopic images of BALB/c 3T3 cells stained with DAPI after treatment with PLA mat, Ch/PLA mat, Ch-8Q/PLA mat and its complexes, Jeff-8Q and its complexes and 8QCHO and its complexes.

Author Contributions: M.I., N.M. (Nevena Manolova) and I.R. conceived the original concept; I.A. and M.I. prepared and characterized the fibrous materials and their metal complexes. N.M. (Nadya Markova) performed the microbiological tests of the prepared materials; A.G. and R.T. conducted the estimation of in vitro anticancer activity; R.S. and R.K. performed the analyses of coordination of Cu^{2+} (Fe^{3+}) in complexes of the materials by EPR spectroscopy and discussed the obtained results; M.I. administrated the project; M.I., N.M. (Nevena Manolova) and I.R. wrote and revised the manuscript. All authors have read and agreed to the published version of the manuscript.

Funding: This research was funded by the Bulgarian National Science Fund, Grant KP-06-N39/13/2019. Research equipment of Distributed Research Infrastructure INFRAMAT, part of the Bulgarian National Roadmap for Research Infrastructures, supported by the Bulgarian Ministry of Education and Science was used in this investigation.

Institutional Review Board Statement: Not applicable.

Data Availability Statement: The data presented in this study are available on request from the corresponding author.

Acknowledgments: Financial support from the Bulgarian National Science Fund (Grant KP-06-N39/13/2019) is kindly acknowledged.

Conflicts of Interest: The authors declare no conflict of interest.

References

1. Fadil, F.; Affandi, N.D.N.; Misnon, M.I.; Bonnia, N.N.; Harun, A.M.; Alam, M.K. Review on electrospun nanofiber-applied products. *Polymers* **2021**, *13*, 2087. [CrossRef] [PubMed]
2. Reddy, V.S.; Tian, Y.; Zhang, C.; Ye, Z.; Roy, K.; Chinnappan, A.; Ramakrishna, S.; Liu, W.; Ghosh, R. A review on electrospun nanofibers based advanced applications: From health care to energy devices. *Polymers* **2021**, *13*, 3746. [CrossRef] [PubMed]
3. Liu, H.; Gough, C.; Deng, Q.; Gu, Z.; Wang, F.; Hu, X. Recent advances in electrospun sustainable composites for biomedical, environmental, energy, and packaging applications. *Int. J. Mol. Sci.* **2020**, *21*, 4019. [CrossRef] [PubMed]
4. Luraghi, A.; Peri, F.; Moroni, L. Electrospinning for drug delivery applications: A review. *J. Control. Release* **2021**, *334*, 463–484. [CrossRef]
5. Torres-Martínez, E.J.; Bravo, J.M.C.; Medina, A.S.; González, G.L.P.; Gómez, L.J.V. A summary of electrospun nanofibers as drug delivery system: Drugs loaded and biopolymers used as matrices. *Curr. Drug Deliv.* **2018**, *15*, 1360–1374. [CrossRef]
6. Balaji, A.; Vellayappan, M.V.; John, A.A.; Subramanian, A.P.; Jaganathan, S.K.; Supriyanto, E.; Razak, S.I.A. An insight on electrospun-nanofibers-inspired modern drug delivery system in the treatment of deadly cancers. *RSC Adv.* **2015**, *5*, 57984–58004. [CrossRef]
7. Li, J.; Cai, C.; Li, J.; Sun, T.; Wang, L.; Wu, H.; Yu, G. Chitosan-based nanomaterials for drug delivery. *Molecules* **2018**, *23*, 2661. [CrossRef]
8. Al-Jbour, N.D.; Beg, M.D.; Gimbun, J.; Moshiul Alam, A.K.M. An overview of chitosan nanofibers and their applications in the drug delivery process. *Curr. Drug Deliv.* **2019**, *6*, 272–294. [CrossRef]
9. Ali, A.; Ahmed, S. A review on chitosan and its nanocomposites in drug delivery. *Int. J. Biol. Macromol.* **2018**, *109*, 273–286. [CrossRef]
10. Chang, X.X.; Mubarak, N.M.; Mazari, S.A.; Jatoi, A.S.; Ahmad, A.; Khalid, M.; Walvekar, R.; Abdullah, E.C.; Karri, R.R.; Siddiqui, M.T.H.; et al. A review on the properties and applications of chitosan, cellulose and deep eutectic solvent in green chemistry. *J. Ind. Eng. Chem.* **2021**, *104*, 362–380. [CrossRef]
11. Paneva, D.; Ignatova, M.; Manolova, N.; Rashkov, I. Novel chitosan–containing micro- and nanofibrous materials by electrospinning: Preparation and biomedical application. In *Nanofibers: Fabrication, Performance, and Applications*; Chang, W.N., Ed.; Nova Science Publishers: New York, NY, USA, 2009; pp. 73–151.
12. Liu, X.F.; Guan, Y.L.; Yang, D.Z.; Li, Z.; De Yao, K. Antibacterial action of chitosan and carboxymethylated chitosan. *J. Appl. Polym. Sci.* **2000**, *79*, 1324–1335.
13. Qin, C.; Du, Y.; Xiao, L.; Li, Z.; Gao, X. Enzymic preparation of water-soluble chitosan and their antitumor activity. *Int. J. Biol. Macromol.* **2002**, *31*, 111–117. [CrossRef]
14. Martínez-Camacho, A.P.; Cortez-Rocha, M.O.; Castillo-Ortega, M.M.; Burgos-Hernández, A.; Ezquerra-Brauer, J.M.; Plascencia-Jatomea, M. Antimicrobial activity of chitosan nanofibers obtained by electrospinning. *Polym. Int.* **2011**, *60*, 1663–1669. [CrossRef]
15. Ohkawa, K.; Cha, D.; Kim, H.; Nishida, A.; Yamamoto, H. Electrospinning of chitosan. *Macromol. Rapid Commun.* **2004**, *25*, 1600–1605. [CrossRef]
16. Torres-Giner, S.; Ocio, M.J.; Lagaron, J.M. Development of active antimicrobial fiber-based chitosan polysaccharide nanostructures using electrospinning. *Eng. Life Sci.* **2008**, *8*, 303–314. [CrossRef]
17. Correia, D.M.; Amparo Gámiz-González, M.; Botelho, G.; Vidaurre, A.; Gomez Ribelles, J.L.; Lanceros-Mendez, S.; Sencadas, V. Effect of neutralization and cross-linking on the thermal degradation of chitosan electrospun membranes. *J. Therm. Anal. Calorim.* **2014**, *117*, 123–130. [CrossRef]
18. Su, H.; Liu, K.; Karydis, A.; Abebe, D.G.; Wu, C.; Anderson, K.M.; Ghadri, N.; Adatrow, P.; Fujiwara, T.; Bumgardner, J.D. In vitro and in vivo evaluations of a novel post-electrospinning treatment to improve the fibrous structure of chitosan membranes for guided bone regeneration. *J. Biomed. Mater. Res.* **2017**, *12*, 015003. [CrossRef]
19. Ohkawa, K.; Minato, K.-I.; Kumagai, G.; Hayashi, S.; Yamamoto, H. Chitosan nanofiber. *Biomacromolecules* **2006**, *9*, 3291–3294. [CrossRef]

20. Sangsanoh, P.; Supaphol, P. Stability improvement of electrospun chitosan nanofibrous membranes in neutral or weak basic aqueous solutions. *Biomacromolecules* **2006**, *7*, 2710–2727. [CrossRef]
21. Nirmala, R.; Il, B.W.; Navamathavan, R.; El-Newehy, M.H.; Kim, H.Y. Preparation and characterizations of anisotropic chitosan nanofibers via electrospinning. *Macromol. Res.* **2011**, *19*, 345–350. [CrossRef]
22. De Vrieze, S.; Westbroek, P.; Van Camp, T.; Van Langenhove, L. Electrospinning of chitosan nanofibrous structures: Feasibility study. *J. Mater. Sci.* **2007**, *42*, 8029–8034. [CrossRef]
23. Geng, X.; Kwon, O.H.; Jang, J. Electrospinning of chitosan dissolved in concentrated acetic acid solution. *Biomaterials* **2005**, *26*, 5427–5432. [CrossRef] [PubMed]
24. Spasova, M.; Manolova, N.; Paneva, D.; Rashkov, I. Preparation of chitosan-containing nanofibres by electrospinning of chitosan/poly(ethylene oxide) blend solutions. *e-Polymers* **2004**, *4*, 056. [CrossRef]
25. Duan, B.; Dong, C.; Yuan, X.; Yao, K. Electrospinning of chitosan solutions in acetic acid with poly (ethylene oxide). *J. Biomater. Sci. Polym. Ed.* **2004**, *15*, 797–811. [CrossRef]
26. Ignatova, M.; Manolova, N.; Markova, N.; Rashkov, I. Electrospun non-woven nanofibrous hybrid mats based on chitosan and PLA for wound-dressing applications. *Macromol. Biosci.* **2009**, *9*, 102–111. [CrossRef] [PubMed]
27. Maleki, H.; Azimi, B.; Ismaeilimoghadam, S.; Danti, S. Poly(lactic acid)-based electrospun fibrous structures for biomedical applications. *Appl. Sci.* **2022**, *12*, 3192. [CrossRef]
28. Toncheva, A.; Spasova, M.; Paneva, D.; Manolova, N.; Rashkov, I. Polylactide (PLA)-based electrospun fibrous materials containing ionic drugs as wound dressing materials: A review. *Int. J. Polym. Mater. Polym. Biomater.* **2014**, *63*, 657–671. [CrossRef]
29. Xu, J.; Zhang, J.; Gao, W.; Liang, H.; Wang, H.; Li, J. Preparation of chitosan/PLA blend micro/nanofibers by electrospinning. *Mater. Lett.* **2009**, *63*, 658–660. [CrossRef]
30. Siqueira, N.M.; Garcia, K.C.; Bussamara, R.; Both, M.H.; Vainstein, R.; Soares, M.D. Poly(lactic acid)/chitosan fiber mats: Investigation of effects of the support on lipase immobilization. *Int. J. Biol. Macromol.* **2015**, *72*, 998–1004. [CrossRef]
31. Tighzert, W.; Habi, A.; Ajji, A.; Sadoun, T.; Boukraa-Oulad Daoud, F. Fabrication and characterization of nanofibers based on poly(lactic acid)/chitosan blends by electrospinning and their functionalization with phospholipase A1. *Fibers Polym.* **2017**, *18*, 514–524. [CrossRef]
32. Jung, K.-H.; Huh, M.-W.; Meng, W.; Yuan, J.; Hyun, S.H.; Bae, J.-S.; Hudson, S.M.; Kang, I.-K. Preparation and antibacterial activity of PET/chitosan nanofibrous mats using an electrospinning technique. *J. Appl. Polym. Sci.* **2007**, *105*, 2816–2823. [CrossRef]
33. Sadeghi, D.; Karbasi, S.; Razavi, S.; Mohammadi, S.; Shokrgozar, M.A.; Bonakdar, S. Electrospun poly(hydroxybutyrate)/chitosan blend fibrous scaffolds for cartilage tissue engineering. *J. Appl. Polym. Sci.* **2016**, *133*, 44171. [CrossRef]
34. Dinan, B.; Bhattarai, N.; Li, Z.; Zhang, M. Characterization of chitosan based hybrid nanofiber scaffolds for tissue engineering. *J. Undergrad. Res. Bioeng.* **2007**, *7*, 33–37.
35. Ignatova, M.G.; Manolova, N.E.; Toshkova, R.A.; Rashkov, I.B.; Gardeva, E.G.; Yossifova, L.S.; Alexandrov, M.T. Electrospun nanofibrous mats containing quaternized chitosan and polylactide with in vitro antitumor activity against HeLa cells. *Biomacromolecules* **2010**, *11*, 1633–1645. [CrossRef]
36. Ignatova, M.; Yossifova, L.; Gardeva, E.; Manolova, N.; Toshkova, R.; Rashkov, I.; Alexandrov, M. Antiproliferative activity of nanofibers containing quaternized chitosan and/or doxorubicin against MCF-7 human breast carcinoma cell line by apoptosis. *J. Bioact. Compat. Polym.* **2011**, *26*, 539–551. [CrossRef]
37. Toshkova, R.; Manolova, N.; Gardeva, E.; Ignatova, M.; Yossifova, L.; Rashkov, I.; Alexandrov, M. Antitumor activity of quaternized chitosan-based electrospun implants against Graffi myeloid tumor. *Int. J. Pharm.* **2010**, *400*, 221–233. [CrossRef] [PubMed]
38. Gupta, R.; Luxami, V.; Paul, K. Insights of 8-hydroxyquinolines: A novel target in medicinal chemistry. *Bioorg. Chem.* **2021**, *108*, 104633. [CrossRef] [PubMed]
39. Song, Y.; Xu, H.; Chen, W.; Zhan, P.; Liu, X. 8-Hydroxyquinoline: A privileged structure with a broad-ranging pharmacological potential. *MedChemComm* **2014**, *6*, 61–74. [CrossRef]
40. Prachayasittikul, V.; Prachayasittikul, V.; Prachayasittikul, S.; Ruchirawat, S. 8-Hydroxyquinolines: A review of their metal chelating properties and medicinal applications. *Drug Des. Dev. Ther.* **2013**, *7*, 1157–1178. [CrossRef] [PubMed]
41. Mincheva, R.; Manolova, N.; Paneva, D.; Rashkov, I. Preparation of polyelectrolyte-containing nanofibers by electrospinning in the presence of a non-ionogenic water-soluble polymer. *J. Bioact. Compat. Polym.* **2005**, *20*, 419–435. [CrossRef]
42. Spasova, M.; Manolova, N.; Rashkov, I.; Naydenov, M. Electrospun 5-chloro-8-hydroxyquinoline-loaded cellulose acetate/polyethylene glycol antifungal membranes against Esca. *Polymers* **2019**, *11*, 1617. [CrossRef] [PubMed]
43. Spasova, M.; Manolova, N.; Rashkov, I.; Tsekova, P.; Georgieva, A.; Toshkova, R.; Markova, N. Cellulose acetate-based electrospun materials with a variety of biological potentials: Antibacterial, antifungal and anticancer. *Polymers* **2021**, *13*, 1631. [CrossRef] [PubMed]
44. Ignatova, M.; Manolova, N.; Rashkov, I.; Markova, N.; Kukeva, R.; Stoyanova, R.; Georgieva, A.; Toshkova, R. 8-Hydroxyquinoline-5-sulfonic acid-containing poly(vinyl alcohol)/chitosan electrospun materials and their Cu^{2+} and Fe^{3+} complexes: Preparation, antibacterial, antifungal and antitumor activities. *Polymers* **2021**, *13*, 2690. [CrossRef] [PubMed]
45. Spasova, M.; Manolova, N.; Markova, N.; Rashkov, I. Superhydrophobic PVDF and PVDF-HFP nanofibrous mats with antibacterial and anti-biofouling properties. *Appl. Surf. Sci.* **2016**, *363*, 363–371. [CrossRef]

46. Toncheva, A.; Mincheva, R.; Kancheva, M.; Manolova, N.; Rashkov, I.; Dubois, P.; Markova, N. Antibacterial PLA/PEG electrospun fibers: Comparative study between grafting and blending PEG. *Eur. Polym. J.* **2016**, *75*, 223–233. [CrossRef]
47. Stoyanova, N.; Paneva, D.; Mincheva, R.; Toncheva, A.; Manolova, N.; Dubois, P.; Rashkov, I. Poly (l-lactide) and poly (butylene succinate) immiscible blends: From electrospinning to biologically active materials. *Mater. Sci. Eng. C* **2014**, *41*, 119–126. [CrossRef]
48. Ignatova, M.; Stoilova, O.; Manolova, N.; Markova, N.; Rashkov, I. Electrospun mats from styrene/maleic anhydride copolymers: Modification with amines and assessment of antimicrobial activity. *Macromol. Biosci.* **2010**, *10*, 944–954. [CrossRef]
49. Ignatova, M.; Stoyanova, N.; Manolova, N.; Rashkov, I.; Kukeva, R.; Stoyanova, R.; Toshkova, R.; Georgieva, A. Electrospun materials from polylactide and Schiff base derivative of Jeffamine ED® and 8-hydroxyquinoline-2-carboxaldehyde and its complex with Cu^{2+}: Preparation, antioxidant and antitumor activities. *Mater. Sci. Eng. C* **2020**, *116*, 111185. [CrossRef]
50. Barilli, A.; Atzeri, C.; Bassanetti, I.; Ingoglia, F.; Dall'Asta, V.; Bussolati, O.; Maffini, M.; Mucchino, C.; Marchiò, L. Oxidative stress induced by copper and iron complexes with 8-hydroxyquinoline derivatives causes paraptotic death of HeLa cancer cells. *Mol. Pharm.* **2014**, *11*, 1151–1163. [CrossRef]
51. Mossmann, T. Rapid colorimetric assay for cellular growth and survival: Application to proliferation and cytotoxicity assays. *J. Immunol. Methods* **1983**, *65*, 55–63. [CrossRef]
52. Tsai, S.C.; Lu, C.C.; Lee, C.Y.; Lin, Y.C.; Chung, J.G.; Kuo, S.C.; Amagaya, S.; Chen, F.N.; Chen, F.N.; Chen, M.Y.; et al. AKT serine/threonine protein kinase modulates bufalin-triggered intrinsic pathway of apoptosis in CAL 27 humanoral cancer cells. *Int. J. Oncol.* **2012**, *41*, 1683–1692. [CrossRef] [PubMed]
53. Koombhongse, S.; Liu, W.; Reneker, D.H. Flat polymer ribbons and other shapes by electrospinning. *J. Polym. Sci. Part B Polym. Phys.* **2001**, *39*, 2598–2606. [CrossRef]
54. El-Dissouky, A.; Mohamad, G.B. Synthesis and characterization of copper(II) metal(II) binuclear complexes of N,N'-bis(8-hydroxyquinoline-7-carboxaldene)-1,3-diaminopropane. *Inorg. Chim. Acta* **1990**, *168*, 241–248. [CrossRef]
55. Khandar, A.A.; Nejati, K. Synthesis and characterization of a series of copper(II) complexes with azo-linked salicylaldimine Schiff base ligands.: Crystal structure of $Cu5PHAZOSALTN{\cdot}CHCl_3$. *Polyhedron* **2000**, *19*, 607–613. [CrossRef]
56. Gubendran, A.; Kesavan, M.P.; Ayyanaar, S.; Raja, J.D.; Athappan, P.; Rajesh, J. Synthesis and characterization of water-soluble copper(II), cobalt(II) and zinc(II) complexes derived from 8-hydroxyquinoline-5-sulfonic acid: DNA binding and cleavage studies. *Appl. Organomet. Chem.* **2017**, *31*, e3708. [CrossRef]
57. Kim, C.H.; Khil, M.S.; Kim, H.Y.; Lee, H.U.; Jahng, K.Y. An improved hydrophilicity via electrospinning for enhanced cell attachment and proliferation. *J. Biomed. Mater. Res. B Appl. Biomater.* **2006**, *78*, 283–290. [CrossRef] [PubMed]
58. Fujiwara, M.; Matsushita, T.; Ikeda, S. X-ray photoelectron spectroscopy of copper (II) complexes with donor sets of O_4, N_2O_4, N_2O_2, N_4, N_2S_2, and S_4. *Anal. Sci.* **1993**, *9*, 289–291. [CrossRef]
59. Rashid, S.; Shen, C.; Chen, X.; Li, S.; Chen, Y.; Wen, Y.; Liu, J. Enhanced catalytic ability of chitosan–Cu–Fe bimetal complex for the removal of dyes in aqueous solution. *RSC Adv.* **2015**, *5*, 90731–90741. [CrossRef]
60. Pawlicka, A.; Mattos, R.I.; Tambelli, C.E.; Silva, I.D.A.; Magon, C.J.; Donoso, J.P. Magnetic resonance study of chitosan bio-membranes with proton conductivity properties. *J. Membr. Sci.* **2013**, *429*, 190–196. [CrossRef]
61. Chan, S.H.; Chui, C.H.; Chan, S.W.; Kok, S.H.; Chan, D.; Tsoi, M.Y.; Leung, P.H.; Lam, A.K.; Chan, A.S.; Lam, K.H.; et al. Synthesis of 8-hydroxyquinoline derivatives as novel antitumor agents. *ACS Med. Chem. Lett.* **2012**, *4*, 170–174. [CrossRef]

polymers

MDPI

Article

Synthesis and Characterisation of Poly(3-hydroxybutyrate-*co*-3-hydroxyvalerate)-*b*-poly(3-hydroxybutyrate-*co*-3-hydroxyvalerate) Multi-Block Copolymers Produced Using Diisocyanate Chemistry

Jingjing Mai, Steven Pratt, Bronwyn Laycock * and Clement Matthew Chan *

School of Chemical Engineering, The University of Queensland, Brisbane, QLD 4072, Australia; jingjing.mai@uq.edu.au (J.M.); s.pratt@uq.edu.au (S.P.)
* Correspondence: b.laycock@uq.edu.au (B.L.); c.chan@uq.edu.au (C.M.C.)

Abstract: Bacterially derived polyhydroxyalkanoates (PHAs) are attractive alternatives to commodity petroleum-derived plastics. The most common forms of the short chain length (scl-) PHAs, including poly(3-hydroxybutyrate) (P3HB) and poly(3-hydroxybutyrate-*co*-3-hydroxyvalerate) (PHBV), are currently limited in application because they are relatively stiff and brittle. The synthesis of PHA-*b*-PHA block copolymers could enhance the physical properties of PHAs. Therefore, this work explores the synthesis of PHBV-*b*-PHBV using relatively high molecular weight hydroxy-functionalised PHBV starting materials, coupled using facile diisocyanate chemistry, delivering industrially relevant high-molecular-weight block copolymeric products. A two-step synthesis approach was compared with a one-step approach, both of which resulted in successful block copolymer production. However, the two-step synthesis was shown to be less effective in building molecular weight. Both synthetic approaches were affected by additional isocyanate reactions resulting in the formation of by-products such as allophanate and likely biuret groups, which delivered partial cross-linking and higher molecular weights in the resulting multi-block products, identified for the first time as likely and significant by-products in such reactions, affecting the product performance.

Keywords: polyhydroxyalkanoates; isocyanate chemistry; block copolymers; allophanates; cross-linking

Citation: Mai, J.; Pratt, S.; Laycock, B.; Chan, C.M. Synthesis and Characterisation of Poly(3-hydroxybutyrate-*co*-3-hydroxyvalerate)-*b*-poly(3-hydroxybutyrate-*co*-3-hydroxyvalerate) Multi-Block Copolymers Produced Using Diisocyanate Chemistry. *Polymers* **2023**, *15*, 3257. https://doi.org/10.3390/polym15153257

Academic Editors: Antonio M. Borrero-López, Concepción Valencia-Barragán, Esperanza Cortés Triviño, Adrián Tenorio-Alfonso and Clara Delgado-Sánchez

Received: 29 June 2023
Revised: 23 July 2023
Accepted: 26 July 2023
Published: 31 July 2023

1. Introduction

Polyhydroxyalkanoates (PHAs) are a family of natural polyesters that are synthesised by microorganisms as an intracellular material to store carbon and energy [1]. Their biocompatibility and biodegradability make them excellent candidates for biomedical applications [2–4]. They are also thermoplastic and/or elastomeric (depending on their copolymer composition), which makes them attractive substitutes for petroleum-based synthetic polymers such as polypropylene (PP), polyethylene (PE) and polyethylene terephthalate (PET) [5,6].

However, the most common of the PHAs, the homopolymer poly(3-hydroxybutyrate) (P3HB) and copolymer poly(3-hydroxybutyrate-*co*-3-hydroxyvalerate) (PHBV), are constrained in their use by limitations in their material properties. P3HB and PHBV with low (1 mol%) 3-hydroxyvalerate (3HV) content, in particular, have very narrow processing windows and are relatively brittle [7,8]. The synthesis of block copolymers based on these starting materials as one block, but coupled to another block of PHA that is incompatible, is one approach for the potential step-change improvement in PHA properties [9–12].

The production of blocks of copolymers rather than blends is a highly attractive option because of their versatility in developing unique microstructures that deliver very different material properties for the as-formed products, as well as the flexibility to fabricate variations in polymer chain aggregation [8,9,11,13,14]. Furthermore, the molecular weight, to a large extent, is responsible for the end-use properties of biopolymers [15,16]. The mechanical properties of PHBV random copolymer, for example, deteriorate when the weight average molecular weight (M_w) is lower than 100 kDa [17,18]. So, the target of this work is to produce PHA-*b*-PHA block copolymers of high molecular weight (M_w> 100 kDa).

Attempts to produce PHA-*b*-PHA block copolymers have previously been made using a biological approach. Two strategies have been adopted: one using direct pulse feeding of carbon feedstock to unmodified organisms and deliberately switching on and off the feeding over time and/or alternating the feedstock to produce different blocks [8,9]. It was concluded that such a biological approach lacks control over the timing and extent of block formation and in all probability produces both PHA/PHA blends as well as blocks and is thus a complex mixture for which structure–property relationships of PHA-*b*-PHA of different block lengths and compositions cannot readily be assessed [10,11,14,19–23]. The other strategy is to use genetically modified organisms and to manipulate the switching on and off of the metabolic pathways to deliver more controlled PHA-*b*-PHA blocks [10,11,22,23]. However, this strategy needs very tight control over the organisms present, with expensive processing and genetic manipulation, and also requires advanced biological manipulation of the synthetic strategy.

By contrast, the chemical synthesis of P3HB-/PHBV-based block copolymers using natural random copolymers as starting materials is a reasonably straightforward synthetic process that has the potential to deliver relatively controllable material structures and narrowly defined compositions [24–26]. Various chemical methods can be applied for the synthesis of PHA-based block copolymers, such as living radical polymerization, click reaction, transesterification, isocyanate coupling reaction and ring opening polymerisation, with Samui and Kamai providing a recent review [27]. However, only limited chemical methods have been used to synthesise PHA-*b*-PHA block copolymers, due to the absence of functional groups such as vinyl groups in most of the natural PHAs [28].

Further, in general, the typical PHA-based block copolymer synthesis reported in the literature has used relatively low molecular weight hydroxy-functionalised PHA oligomers (i.e., M_n < 4 kDa) derived from high molecular weight random copolymers as the starting materials. However, for some applications such as packaging, high molecular weight final products are needed, to achieve good mechanical properties. When starting with such low molecular weight materials, it is therefore necessary to produce multi-block copolymers in order to obtain relatively high molecular weight polymers with good mechanical properties [29–31]. However, this strategy is less controllable with respect to the chemical structure and architecture of the final products and introduces multiple urethane linkages, likely influencing the final product properties. In addition, one of the dominant side reactions is the formation of allophanates, which has been reported in polycaprolactone (PCL)-*b*-polyethylene glycol (PEG) and PHA-*b*-PEG systems [32,33]. This resulted in blocks with higher molecular weights and partially cross-linked products which could lead to a shift in properties and the formation of microgels [32,33]. Despite being reported, the allophanate side reactions were not characterised further in the past literature and none was reported in PHA-*b*-PHA systems.

Further, since much of the potential property improvement in block copolymers is obtained through local microphase separation of the different blocks [9,34–36], longer block sequences may be desirable. Yet, the one example to our knowledge of the synthesis of a PHA-*b*-PHA copolymer using the diisocyanate approach is where a low molecular weight block copoly(ester–urethane) of M_n = 10.6 kDa was synthesised using a one-step process from telechelic hydroxylated poly[(*R*)-3-hydroxyoctanoate] (PHO-diol, M_n = 2.4 kDa) and telechelic hydroxylated poly[(*R*)-3-hydroxybutyrate] (P3HB-diol, M_n = 2.6 kDa) with L-lysine methyl ester diisocyanate (LDI) as the junction unit. The synthesised block copolymer delivered good thermoplastic properties, with a melting temperature (T_m) of 146 °C and a glass transition temperature (T_g) of −6 °C. The mechanical properties were however those of a soft, low-strength material, consistent with its low molecular weight [37].

Therefore, the aim of this work was to synthesise novel, high molecular-weight PHBV-*b*-PHBV block copolymers, using PHA macroinitiators of relatively high molecular weights (e.g., M_n > 20 kDa).

The main approach to the chemical synthesis of PHA-based block copolymers is the use of isocyanate chemistry to form urethane linkages from hydroxy-terminated PHA blocks. This is a simple and efficient synthesis strategy, being quite rapid and clean, and being a well-established process [38]. PHA-based block copolymers can be produced using either a one-step or two-step process, where a one-step process adopts a single synthesis stage, with all reactants combined together at the start, while a two-step process adopts the alternate strategy of reacting a central hydroxy-terminated block with at least two times the molar equivalent of diisocyanate to produce a central isocyanate-terminated block, which is then subsequently reacted with an alternate hydroxy-terminated block in a second stage. There are now many examples of these strategies being adopted in the literature, leading to a wide range of block copolymeric products, often of significant molecular weight, with distinctly different properties to their starting materials [4,24,34,37,39–41]. Overall, PHA-based block copolymers joined by urethane links typically exhibit high thermal and mechanical properties and good processing ability over their counterpart random and blend copolymers [34,42]. However, this strategy has not been applied to PHBV-*b*-PHBV synthesis, and it is a particularly relevant one for targeting high molecular weight block copolymers where the proportion of functional end groups relative to the length of the main chain is low, needing efficient coupling.

In this first instance, the methods for synthesis using isocyanate chemistry were established using 1 mol% 3HV blocks. The effectiveness of both the one-step and two-step processes for the production of PHBV-*b*-PHBV multi-block copolymers were compared by tracing the reaction chemistry and the molecular weight of the products throughout. The extent of possible side reactions, i.e., the formation of allophanates, was also characterised. Our findings provide some insight into the methodology for the synthetic production of PHA-*b*-PHA block copolymers, resulting in block copolymers of industrially relevant molecular weight.

2. Materials and Methods

2.1. Materials

A commercial PHBV random copolymer of 1 mol% 3HV content was purchased from TianAn Biopolymer (Ningbo, China) (M_n = 194 kg/mol, M_w = 455 kg/mol and $Đ$ = 2.3, by GPC, 1 mol% 3HV, by ^{1}H-NMR, called Random_1HV#3 in this work). Hydroxy-functionalised PHBV copolymers of low (1 mol%) 3HV content (Random_1HV#1, Random_1HV#2, Random_1HV#4 and Random_1HV#5) were produced, as described in our previous work [43]. Details of these materials are provided in the supplementary information (Table S1), noting that these products are mixtures containing mono-hydroxy terminated PHBV and di-hydroxy-terminated (telechelic) PHBV, with some carboxylic acid group functionality as well. HPLC grade chloroform, 1,2-dichloroethane (anhydrous, 99.8%), tin(II) 2-ethylhexanoate (stannous octoate, 92.5–100.0%) and hexamethylene diisocyanate (puriss., ≥99.0% (GC)) (HDI) were purchased from Sigma-Aldrich (St. Louis, MI, USA) and used as received. Deuterated chloroform (99.8%) was purchased from Novachem (Calgary, AB, Canada) and used as received. Argon (Ultra High Purity) was purchased from Supagas Pty Ltd. (Branxholm, Australia) and used as received.

2.2. Block Copolymer Synthesis

Based on our previous work and the literature [4,44,45], three different experiments using hydroxy-terminated PHBV macromer starting materials and either a one-step or a two-step process were carried out. The experimental conditions, including the details of the hydroxy-terminated PHBV macromer used in these experiments and whether or not the experiment was multi-step, are shown in Table 1. A summary of the experimental approach is provided in Figure 1. Experiment #1 was a two-step synthesis. In the first step, the hydroxy-functionalised PHBV (Random_1HV#4) solution was added dropwise to hexamethylene diisocyanate (HDI) at an [NCO]/[OH] ratio of 2.2 and temperature of 65 °C. The reaction of the first step lasted for 5 h. In the second step, double portions of hydroxy-functionalised PHBV (Random_1HV#4) solution were quickly added into the reaction system and the reaction was run for 48 h. Only the reaction products in the second step of the two-step synthesis were analysed.

Table 1. Details of reaction conditions and materials used in block copolymer syntheses based on 1 mol% 3HV content PHBV.

Exp't No.	Hydroxy-functionalised PHBV Starting Materials	$\overline{M_n}$ (kDa)	$\overline{M_w}$ (kDa)	Đ	3HV (mol%)	1 °OH and 2 °OH Used in Reaction (mmoles/g PHA)	Description	Experimental Conditions
1	Random_1HV#4	28	51	1.8	1	0.084	First and second steps of the two-step synthesis	[NCO]/[OH] for step 1 = 2.2; Temp = 65 °C; Time (step 1) = 5 h; Theoretical [NCO]/[OH] for step 2 = 0.5 *; Temp = 65 °C; Time (step 2) = 48 h*Overall [NCO]/[OH] = 2.2:3*
2A	Random_1HV#4	28	51	1.8	1	0.084	One-step synthesis (first step of the two-step synthesis)	[NCO]/[OH] for step 1 = 2.2; Temp = 65 °C: Time = 18.5 h
2B	Random_1HV#5	25	44	1.7	1	0.103	One-step synthesis extended	[NCO]/[OH] = 2.2; Temp = 65 °C: Time = 24 h

* Based on initial NCO addition in step 1.

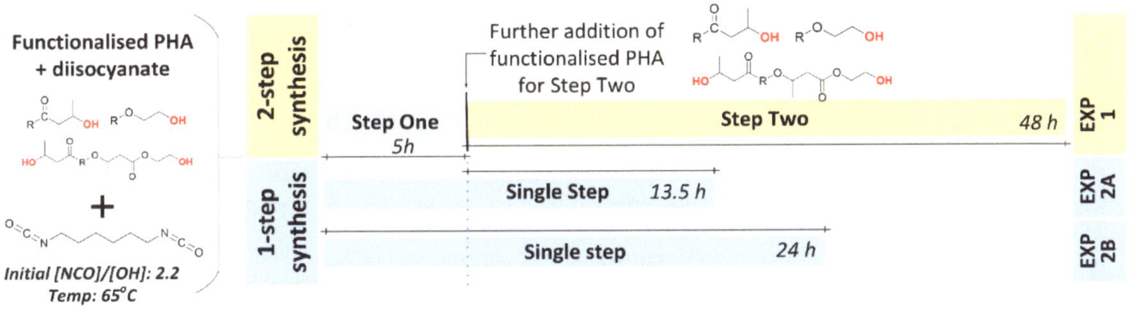

Figure 1. Graphical overview of PHBV-*b*-PHBV block copolymer syntheses, where R represents PHBV main chain. The yellow colour indicates the 2-step synthesis while the blue colour indicates the one-step synthesis.

Experiment 2 was a one-step synthesis. Experiment #2A was essentially a repeat of the first step of Experiment #1, and so the first 5 hours of it were used to understand the reaction kinetics of the first step in a two-step synthesis. Experiment #2B was the duplicate experiment of Experiment #2A with extended time and samples taken throughout so as to follow the reaction kinetics of the one-step synthesis.

2.2.1. Drying of Glassware, Solvents and PHBV Reagents

In all these experiments, glassware was dried in an oven at 200 °C overnight and then cooled in a desiccator over dried silica gel before use. The solvent 1,2-dichloroethane was dried before use [46] and distilled by azeotropic distillation using a short-path distillation apparatus. In this process, a two-necked round-bottomed flask (250 mL) equipped with a reflux condenser and gas inlet was connected to a bubbler and the glassware was flushed with argon and carefully flame dried. Then, 100 mL 1,2-dichloroethane and 3 g calcium hydride were quickly added into the flask, which was heated in an oil bath at 60 °C under an argon flux and magnetically stirred for 2 h. The reflux condenser was then replaced with a short-path distillation head, which was connected to a one-necked receiving flask (100 mL) and a bubbler. The pre-dried 1,2-dichloroethane was then dried further by azeotropic distillation, with the first 10% of azeotrope collected in the receiving flask being discarded to remove the trace water from the solvent. The remaining distillate was collected and the flask was flushed with argon before being sealed with a rubber septum and parafilm as a dried solvent for the subsequent isocyanate reactions.

The hydroxy-terminated PHBV was firstly dried under vacuum (−90 kPa) in an oven at 80 °C overnight then cooled in a desiccator over dried silica gel before use. A two-necked round-bottomed flask (250 mL) equipped with a rubber septum (sealed with parafilm) and a reflux condenser was connected to a bubbler and the glassware was flushed with argon by injecting a needle through the rubber septum. The glassware was carefully flame dried before dried hydroxy-terminated PHBV was quickly added into the flask under an argon stream. Then, dried 1,2-dichloroethene (5 mL/gPHA) was quickly added into the flask through the rubber septum using a 50 mL glass syringe, again under an argon stream. The flask was then heated in an oil bath at 140 °C under an argon flux and magnetically stirred until the PHBV was fully dissolved. Then, the solution was dried by azeotropic distillation until 10% of the azeotrope was collected in the receiving flask (50 mL). The oil bath was then cooled down to 65 °C.

2.2.2. Experiment #1: Two-Step Synthesis of PHBV-*b*-PHBV Multi-Block Copolymer

A three-necked round-bottomed flask (100 mL) equipped with a reflux condenser and gas inlet was connected to a bubbler and the glassware was flushed with argon and carefully flame dried. In the first step of the block copolymer synthesis, 3 g dried hydroxy-terminated PHBV random copolymer (Random_1HV#4) was dissolved in 15 mL dried 1,2-dichloroethane and the PHBV solution was dried by azeotropic distillation. After the PHBV solution was prepared, the reaction apparatus was flame dried again. Then, 2 mL 1,2-dichloroethane, 0.012 mL stannous octoate as a catalyst (0.8 μL per mL 1,2-dichloroethene) and 0.044 mL hexamethylene diisocyanate (HDI) (based on desired ratio relative to the concentration of OH groups) were quickly added into the flask through the rubber septum using 500 μL glass syringes. The solution was magnetically stirred and heated in an oil bath at 65 °C. Then, the PHBV solution was immediately transferred into the three-necked flask dropwise through the rubber septum using 20 mL glass syringes. The reaction was conducted for 5 h.

In the second step, 6 g of the hydroxy-functionalised PHBV random copolymer was dissolved in 30 mL dried 1,2-dichloroethane and the PHBV solution was dried by azeotropic distillation. Then, the dried PHBV solution was quickly transferred into the three-necked flask through the rubber septum using 20 mL glass syringes. The reaction was conducted for 48 h. In this second step, the [NCO]/[OH] ratio was 0.5, in accordance with normal practice, in order to leave hydroxyl end groups present post synthesis. Samples were collected by glass syringes through the rubber septum at different reaction times and quenched in a mixture of diethyl ether and methanol (20/1, *v*/*v*) for further characterisation. Finally, the quenched solids were recovered by vacuum filtration through a Buchner funnel fitted with a dried, pre-weighed quantitative filter paper (Whatman^{TM}, Maidstone, UK, 10312209) and rinsed with a large volume of methanol. Then, the product was dried in a vacuum oven with a negative pressure of −90 kPa at 60 °C overnight.

2.2.3. Experiment #2A: One-Step Synthesis of PHBV-*b*-PHBV Multi-Block Copolymers

A three-necked round-bottomed flask (100 mL) equipped with a reflux condenser and gas inlet was connected to a bubbler and the glassware was flushed with argon and carefully flame dried. Then, 5 mL 1,2-dichloroethane, 0.032 mL stannous octoate as a catalyst (0.8 μL per mL 1,2-dichloroethene) and 0.118 mL hexamethylene diisocyanate (HDI) (based on desired ratio relative to concentration of OH groups) were quickly added into the flask, through the rubber septum, using 5 mL and 500 μL glass syringes, respectively. The solution was magnetically stirred and heated in an oil bath at 65 °C. Then, the dried hydroxy-functionalised PHBV solution (containing 8 g PHBV) was immediately transferred into the three-necked flask dropwise through the rubber septum using a 20 mL glass syringe.

The reaction was conducted for 18.5 h. Samples were collected by glass syringes through the rubber septum at different reaction times and quenched in a mixture of diethyl ether and methanol (20/1, v/v) for further characterisation. Finally, the quenched solids were recovered by vacuum filtration through a Buchner funnel fitted with a dried, pre-weighed quantitative filter paper (Whatman^{TM}, Maidstone, UK, 10312209) and rinsed with a large volume of methanol. Then, the product was dried in a vacuum oven with a negative pressure of −90 kPa at 60 °C overnight.

2.2.4. Experiment #2B: One-Step Synthesis of PHBV-*b*-PHBV Multi-Block Copolymers with Longer Reaction Time

Experiment #2B was a duplicate experiment of Experiment #2A, using a hydroxy-functionalised PHBV (Random_1HV#5) of the same 3HV content and similar molecular weight, with the reaction being run for a longer time (24 h). In this experiment, 8 g hydroxy-terminated PHBV (Random_1HV#5) was reacted with 0.145 mL hexamethylene diisocyanate (HDI) (based on the desired NCO ratio relative to concentration of OH groups, see Table 1). The protocol was otherwise as described in Section 2.2.3.

2.3. *Preparation of PHBV Films by Solvent Casting*

To prepare solvent cast films, 1 g PHBV was dissolved in 10 mL HPLC chloroform (100 g/L) at 80 °C in an oil bath under reflux and cooled down after being well dissolved. Then, the solution was placed in a Petri dish. After most of the solvent was slowly evaporated, the film was dried under a vacuum to constant weight at room temperature. Then, the film was left in an open environment for aging for at least two weeks before testing.

2.4. *Solubility Assessment for As-Produced Block Copolymeric Products*

The as-produced products were dissolved in HPLC-grade chloroform (25 mg/mL) at 80 °C in an oil bath under reflux. After the PHBV was well dissolved, the solution was cooled and filtered through a Buchner funnel using pre-weighed polytetrafluoroethylene filter papers (0.22 μm, Whatman^{TM}, Maidstone, UK, WM4-022090). Then, the filtered solution was transferred into pre-weighed Petri dishes followed by rinsing of the receiving flask three times with 20 mL chloroform each time, which was then added to the petri dish. Then, the filtered solution was allowed to evaporate in the fume hood. The mass of the solid residue was measured. The filter paper and the remaining undissolved solid were transferred into a separate Petri dish and the mass of the remaining solid was weighed after evaporation of the solvent (when it had reached constant weight).

2.5. *Material Characterization*

2.5.1. Gel Permeation Chromatography (GPC)

The number average molar mass (M_n), weight average molar mass (M_w) and dispersity ($Đ$) of the PHBV copolymers were determined by GPC, using an Agilent 1260 Infinity Multi Detector Suite system (Cheshire, UK). Samples were dissolved in HPLC grade chloroform (2.5 mg/mL) followed by filtration using polytetrafluoroethylene syringe filters (0.22 μm, Kinesis, ESF-PT-13-022). An HPLC solvent delivery system was used in conjunction with an auto-injector. A column set consisting of a guard column (Agilent PLgel 10 μm 7.5 mm × 50 mm) followed 3 × Agilent PLgel 10 μm MIXED-B columns (7.5 mm × 300 mm) in series. The columns were kept at 30 °C. A refractometer, at 30 °C, was used to detect the signals. A chloroform flow rate of 1 mL/min was used for the analysis. Narrowly distributed molecular weight polystyrene standards and Agilent PS-H EasiVial calibration standards (PL2010-0201). The apparatus was calibrated with polystyrene standards. The Mark–Houwink–Sakurada (MHS) relation and specific MHS parameters (K and a) were used to correct the molar mass. $K = 7.7 \times 10^{-3}$ mL/g and $\alpha = 0.82$ were used in this study [47].

2.5.2. Nuclear Magnetic Resonance Spectroscopy (NMR)

Quantitative ^{1}H high-resolution one-dimensional NMR spectra were acquired at 298 K in deuterated chloroform ($CDCl_3$) (50 mg/mL) on Bruker Advance 700 and 500 spectrometers. The number of scans was set to 512. The relative peak intensities of ^{1}H-NMR spectra were determined using PeakFit 4.12 software [9]. Chemical shifts were referenced to the residual proton peak of $CDCl_3$ at 7.26 ppm. The list of chemical shifts of the as-identified peaks from ^{1}H-NMR throughout the study are presented in Table S2.

Solid-state ^{13}C-NMR spectra were performed at 298 K on a Bruker Advance III spectrometer with a 300 MHz magnet equipped with a 4 mm double air bearing, magic angle spinning probe. The powdered samples were placed in a zirconia rotor with a Kel-F cap and rotated at 5 kHz. Next, ^{13}C spectra were recorded with a CPMAS pulse sequence. The ramped cross-polarization time was 1 ms, and decoupling was carried out using a tppm 15 sequence with 100 kHz. The relaxation delay was 3 s and the acquisition time was 49 ms. A total of 2048 scans were collected. Adamantane was used as a reference. The list of chemical shifts of the as-identified peaks from ^{13}C-NMR throughout the study are presented in Table S3.

2.5.3. Differential Scanning Calorimetry (DSC)

DSC analysis was performed using a TA instrument Q2000. Samples of 2–4 mg in sealed aluminium pans were analysed under nitrogen flow (50 mL/min). A five-step procedure was applied as follows: (1) Equilibrate at 25 °C, then heat up from 25 °C to 190 °C with a 10 °C/min ramp and keep isothermal for 0.1 min to erase the thermal history; (2) cool down to −70 °C with a 10 °C/min ramp and keep isothermal for 5 min; (3) heat up from −70 °C to 190 °C with a 10 °C/min ramp; (4) cool down to −70 °C with a 100 °C/min ramp; and (5) heat up to 25 °C with a 20 °C/min ramp. The melting temperature (T_m) and enthalpy of melting (ΔH_m) were determined from the first heating cycle while the crystallisation temperature (T_c) was determined from the first cooling scan. The glass transition temperature (T_g) was determined from the final heating cycle. Data were analysed using TA Universal Analysis software (UA 4.5.0.5).

3. Results and Discussion

In this work, the synthesis of PHBV-*b*-PHBV block copolymers was achieved using diisocyanate chemistry, adopting either a two-step or a one-step approach. The starting materials were relatively high molecular weight hydroxy-terminated PHBV macromers of (1 mol%) 3HV content, and the diisocyanate was deliberately selected to be the less rigid aliphatic hexamethylene diisocyanate (HDI), to limit the risk of introducing rigidity into the resulting chain.

The loading of isocyanate to the reactive end group is an important variable. Excess of isocyanates, over the stoichiometric requirements, i.e., [NCO]/[OH] ratio of 2.2, was used in Experiment 2A.

3.1. Molecular Weight and Functional Groups of the Reaction Products

An increase in molecular weight is a key indicator for the success of the block copolymer synthesis with results shown in Figure 2. It should be noted that further reaction of the isocyanate with the urethane reaction products to produce allophanate and/or biuret bonds is possible, causing branching and chemical cross-linking [48,49]. Therefore, it is likely that after all the hydroxyl groups were reacted with isocyanate groups (Figure 3a,b), the further increase in molecular weight was mainly dependent on the formation of allophanate linkages by the reaction of urethane groups with isocyanate groups (Figure 3c), which has been illustrated in our previous study [44].

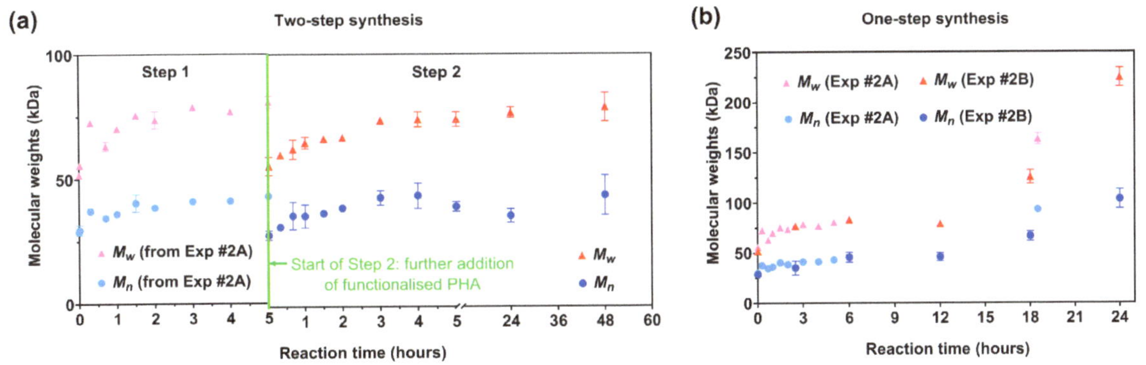

Figure 2. Molecular weights of the soluble fraction of reaction products in (**a**) a two-step block copolymer synthesis (Experiment #1); (**b**) a one-step synthesis of block copolymer (Experiment #2A and 2B). Values are shown as mean with standard deviation; the error bar may not be visible due to the marker. Note: different scales for molecular weight in (**a**,**b**).

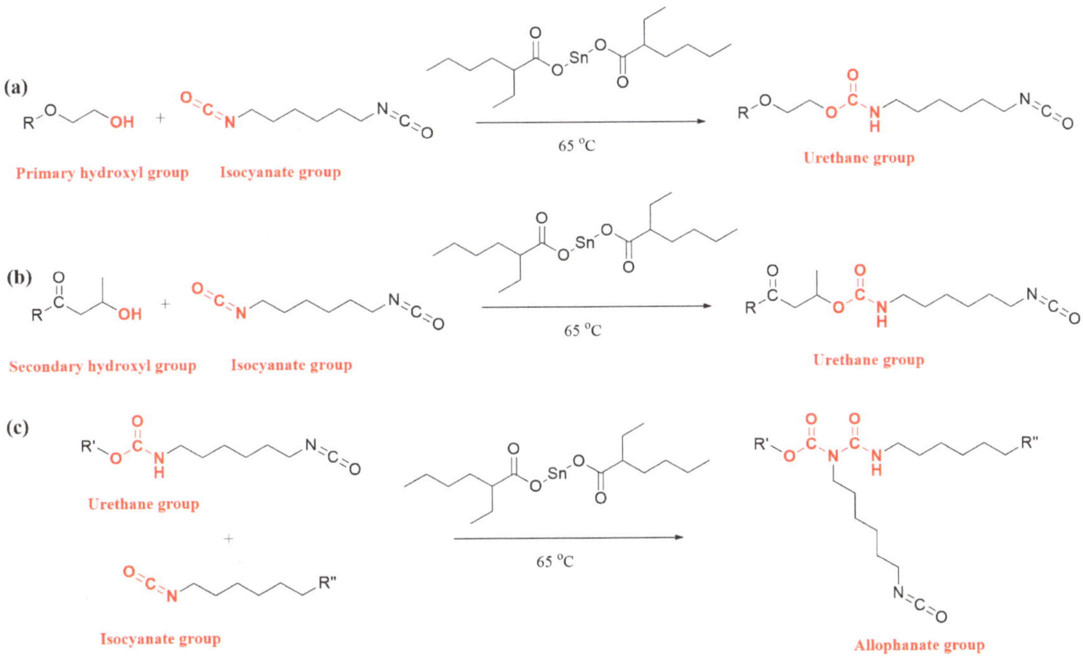

Figure 3. Chemical reaction of (**a**) primary hydroxyl groups with isocyanate groups to form urethane groups; (**b**) secondary hydroxyl groups with isocyanate groups to form urethane groups; and (**c**) urethane groups with isocyanate groups to form allophanate groups. R and R′ = PHBV random copolymer and R″ = NCO or NHC(O)OR′. The functional groups of interest are identified in red.

The proportion of insoluble components in the final products was assessed as an indicator of the extent of cross-linking (Table 2). A gelatinous insoluble component was observed, at 2.7–19.5 wt.%, indicating some by-product formation. In addition, the one-step products (Block_1HV#1 and Block_1HV#2) were characterised using solid-state NMR and compared with their counterpart random copolymers of the same 3HV contents and similar molecular weights (Random_1HV#1 and Random_1HV#2). As shown in Figure 4, peaks at around 28 ppm were only observed in the one-step products, which were attributed to the methylene group of an allophanate functionality (labelled as Al_2, 4C, -O-CO-(R)N-CO-NH-CH_2-CH_2-$\underline{C}H_2$-$\underline{C}H_2$-$\underline{C}H_2$-$\underline{C}H_2$-) [44,50,51], indicating the formation of allophanate groups (therefore cross-linking) in the block copolymer synthesis. This was also evidenced by the broad peaks at around 142–160 ppm, which were attributed to the carbonyl peaks of the allophanate (labelled as C_1, 1C, -O-CO-(R)N-$\underline{C}$O-NH-CH_2-) [50,51]. The detailed chemical shifts of the other peaks are shown in the supplementary material (Table S3). Thus, the gel as produced was likely primarily due to the formation of these allophanate cross-linking groups.

Table 2. Solubility of final reaction products of block copolymer synthesis.

Experiment No.	Final Reaction Products	Insoluble Component (wt.%)
1	Product of two-step synthesis	2.7
2A	Block_1HV#1	4.4
2B	Block_1HV#2	19.5

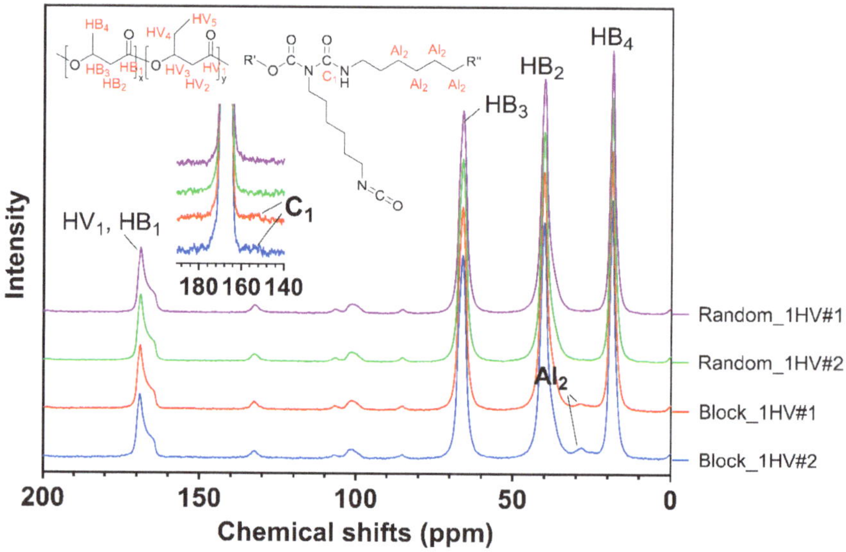

Figure 4. Solid-state ^{13}C-NMR results of PHBV materials of 1 mol% 3HV, where R′ = PHBV random copolymer and R″ = NCO or NHC(O)OR′. The carbons identified in the spectra are indicated in red in the chemical structures above.

3.1.1. Two-Step Synthesis (Experiment #1)

The molecular weight was quantified through only the final (second) step of the two-step reaction for Experiment #1 (shown in Figure 2a); the initial 5 h of reaction was inferred from the first 5 h of Experiment #2A. It is worth noting that the molecular weights of the mixture at time zero in the second step (i.e., after adding the second hydroxy-functionalised block, Random_1HV#4, M_n of 28 kDa, M_w of 51 kDa) are smaller than the product after 5 h of reaction for Experiment #2A (M_n of 43 kDa, M_w of 81 kDa). This is due to the fact that the addition of hydroxy-functionalised PHBV at the start of the second step decreased the average molecular weight of the mixture. Regardless, excess isocyanate was still expected to be present at the start of this second stage of the two-step block copolymer synthesis. This was evidently the case with the molecular weight of the reaction products in the second step increasing gradually, with M_n increasing from 28 to 43 kDa and M_w increasing from 56 to 74 kDa within 3 h of the addition of the second aliquot of hydroxy-functionalised PHBV (Figure 2a).

The [NCO]/[OH] ratio over the whole synthesis was 2.2:3. But this meant that in the second stage of this two-step synthesis, the [NCO]/[OH] ratio would be at best less than 1, even if no reaction had occurred in the initial 5 h stage. If this reaction had run as planned, the molecular weight should in the end have been triple that of the time zero material. However, the molecular weight of the final reaction products only increased around 1.5 times after 3 h in this second stage. And after that, no further increase in molecular weights was observed even after an extended 48 h, i.e., the addition of the second dose of hydroxy-functionalised PHBV effectively consumed all remaining isocyanate but without further blocking copolymer synthesis. As shown in Figure 5b, the ^{1}H-NMR spectra showed that neither the primary (Et_1 and Et_2) nor the secondary hydroxyl groups (2 °OH) were fully consumed after 48 h, while urethane peaks associated with the reaction of the primary and secondary hydroxyl groups (Ur_1 (1H, -O-CH_2-CH_2-O-CO-N<u>H</u>-) and Ur_2 (1H, -CO-CH_2-CH(CH_3)-O-CO-N<u>H</u>-), respectively) were observed in all reaction products [44], confirming that there was insufficient isocyanate present at the end to complete the reaction. There are many possible reasons for this, not least of which is that a large proportion of isocyanate has been consumed in the first step, with isocyanate reacting with both hydroxyl groups and urethane groups. It is also noted that trace water or other impurities present in the second batch of PHBV copolymers or introduced during their addition reacted with any unreacted diisocyanate or isocyanate functionalised PHBV that was present, removing the potential for further reaction. The reaction product was thus likely a blend of some block material with unreacted hydroxy-terminated PHBV macromonomer.

Although this experiment failed to build up molecular weight as expected, there was still 2.7 wt.% of the insoluble component in the final products, which is likely due to the formation of cross-linked materials caused by excess isocyanate in the first step.

3.1.2. One-Step Synthesis (Experiment #2A and 2B)

In contrast to the two-step synthesis, the molecular weights through time of the reaction products from Experiment #2A showed a large increase between the 5 h reaction product and the 18.5 h reaction product, with the M_n increasing from an initial 28 kDa to reach 93 kDa and M_w increasing from 51 kDa to 163 kDa after 18.5 h (Figure 2b). Both the M_n and the M_w of the final reaction products were greater than triple that of the starting hydroxy-functionalised PHBV, indicating the formation of a multi-block copolymer. It is clear that under these conditions, and with the excess isocyanate present, the isocyanate reaction continued over time to produce a likely blend of different multi-block copolymers of PHBV, as would be expected.

The ^{1}H-NMR spectra of the reaction products of Experiment #2A are shown in Figure 5a and revealed that the primary hydroxyl groups were fully consumed within 40 min, while the secondary hydroxyl groups were fully consumed within 1.5 h. The urethane peaks associated with the reaction of the primary and secondary hydroxyl groups were observed in all reaction products, confirming the proposed isocyanate chemistry. The presence of trace allophonate groups (labelled as Al_1, 2H, -O-CO-(R)N-CO-NH-C<u>H</u>$_2$-CH_2-CH_2-CH_2-CH_2-CH_2-) was also confirmed [44], which was consistent with solid-state NMR results.

Overall, from the results of Experiment #1 and #2A, the idea of preparing PHBV capped with isocyanate groups at each end in the first step and then synthesising a tri-block copolymer was shown to be challenging under the experimental conditions used. This is different from what has been reported in the literature, where tri-block copolymers have been successfully synthesised using a two-step strategy at 50 °C with a [NCO]/[OH] ratio of 2.2 in the first step [45,52]. However, in our work, the formation of allophanate groups coincidently helped to build up blocks, which is a key finding and was applied in this work to synthesise cross-linked PHA-*b*-PHA block copolymers that might potentially improve the toughness of PHAs.

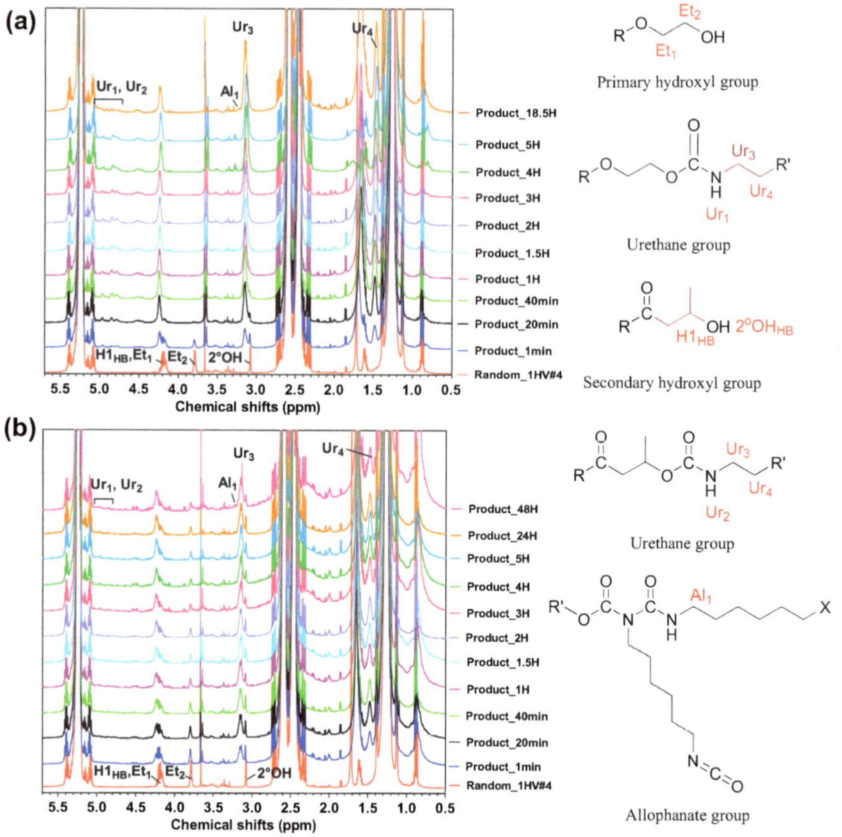

Figure 5. Detailed ^{1}H-NMR results of the reaction products of (**a**) the first step of the two-step synthesis (from the first 5 h of Experiment #2A) and (**b**) the second step of the two-step synthesis (Experiment #1), where time is post addition of second tranche of hydroxy-functionalised PHBV. Key functional groups identified in the ^{1}H-NMR spectra through labels are shown to the right, labelled in red. Et = ethylene glycol-derived end group, where Et_1 and Et_2 are the protons associated with that ethylene group; $H1_{,HB}$ is the tertiary proton associated with the butyric acid derived end group; 2 °OH = secondary hydroxyl proton associated with the butyric acid derived end group; Ur_1 = proton of primary hydroxyl-derived urethane group; Ur_2 = proton of secondary hydroxyl-derived urethane group; Ur_3 and Ur_4 = protons adjacent to the urethane groups that were derived from the diisocyanate; Al_1 = proton adjacent to allophanate group; R and R′ = PHBV random copolymer; and X = NCO or NHC(O)OR′.

Based on the above discussion, the one-step synthesis was shown to be feasible for building molecular weight, with evidence of this being primarily through the formation of urethane groups, and the synthesis could be manipulated by extending the reaction time. In support of this, the molecular weights of the reaction products from the one-step synthesis in Experiment #2B almost trebled over the initial 18 h, with M_n increasing from 24 to 67 kDa and M_w increasing from 43 to 126 kDa (Figure 2b). As the reaction continued to progress, the M_n and M_n of the final reaction products (after 24 h) increased around fivefold overall compared to the starting hydroxy-functionalised PHBV. Once again, it is likely that the formation of allophanate linkages—causing branching and chemical cross-linking—contributed greatly to the increase in molecular weight (with 19.5 wt.% insoluble gel, Table 2). It is therefore assumed that the synthesised final product—where the molecular weight increased fivefold—is likely a mixture of linear, branched and/or cross-linked block copolymers. The ^{1}H-NMR spectra were very similar to those of Experiment #2A, and, hence, are not discussed in detail here.

3.2. Thermal Properties of As-Synthesised Block Copolymers

The thermal properties of the as-synthesised final products are given in Table 3. The respective DSC thermograms are shown in the supplementary information (Figures S1–S5). Three random copolymers of 1 mol% 3HV (Random_1HV#1, Random_1HV#2 and Random_1HV#3) were compared with the two block copolymer products from Experiment #2A (Block_1HV#1) and #2B (Block_1HV#2). The product from Experiment #1 was not included as it was a mixture of block copolymers and the unreacted starting random copolymers. Comparing the initial cell-produced random copolymer of 1 mol% 3HV (Random_1HV#3, M_n of 194 kDa) with the hydroxy-functionalised transesterification products (Random_1HV#1, M_n of 86 kDa and Random_1HV#2, M_n of 117 kDa), the thermal properties of the PHBV random copolymers were similar, except for the slight increase in T_g for Random_1HV#3, which was due to the lower chain mobility of the longer polymer chains.

Table 3. Thermal properties of PHBV-*b*-PHBV block copolymers and random copolymers.

	PHBV Material	M_n (kDa)	M_w (kDa)	Đ	T_m (°C)	ΔH_m (J/g)	T_c (°C)	ΔH_c (J/g)	T_g (°C)
random	Random_1HV#1	86	186	2.2	157.6/ 174.0	102.3	98.3	87.1	2.6
	Random_1HV#2	117	273	2.3	160.0/ 175.0	102.5	103.4	89.2	2.6
	Random_1HV#3	194	455	2.3	159.5/ 170.0	106.0	104.7	87.3	4.0
block	Block_1HV#1	94	159	1.7	153.1/ 168.5	94.0	104.9	78.9	6.3
	Block_1HV#2	107	221	2.1	152.2/ 167.3	89.2	110.0	83.7	8.9

However, the thermal properties differed slightly more between random copolymers and the chemically synthesised block copolymers, with slightly lower T_m and ΔH_m values and a small increase in the T_g being observed for the latter. This is likely due to constrained chain mobility as a consequence of the formation of three-dimensional allophanate or potential biuret structures [53], as well as possible hydrogen bonding effects from the urethane groups (although they are a small component of the composition overall) and possible further constraints due to microphase separation between links.

By comparing the two synthesised block copolymers of different degrees of cross-linking (Block_1HV#1 and Block_1HV#2), it is also observed that with the increase in cross-linking, the T_m and ΔH_m values decreased slightly, while the T_g values increased from 6.3 °C (for Block_1HV#1) to 8.9 °C (for Block_1HV#2).

4. Conclusions and Outlook

Overall, a comparison between the one-step and the two-step synthesis of PHBV-*b*-PHBV block copolymers has been established, based on relatively high molecular weight hydroxy-functionalised copolymeric PHBV starting materials containing 1 mol% 3HV. In both strategies, block copolymers were successfully synthesised, as evidenced by the increase in molecular weight and the formation of urethane groups from NMR analysis. However, the one-step synthesis is a more efficient approach. Moreover, there was evidence of by-product formation, particularly allophonate and likely biuret groups, through the further reaction of isocyanates with the urethane groups initially formed. This resulted in cross-linking, particularly following extended reactions with high isocyanate loadings, and the formation of relatively high molecular weight products. It is a key finding of this work and helps to establish an approach for the synthesis of novel PHA-*b*-PHA through the relatively simple one-step strategy of block copolymer production. In addition, the fundamentals of the synthesis were investigated, which, to our knowledge, have not been reported for the synthesis of PHA-*b*-PHA block copolymers based on diisocyanate chemistry.

It is also suggested for future research that the effect of reaction conditions—such as temperature and amounts of catalyst—on the formation of the side reactions of the diisocyanate chemistry could be investigated. By reducing or even eliminating the side reactions—i.e., maximising the formation of isocyanate-terminated intermediate reaction products based on a [NCO]/[OH] ratio of 2.2—the synthesis of a target PHA-*b*-PHA-*b*-PHA tri-block copolymers would be achievable.

Supplementary Materials: The following supporting information can be downloaded at: https://www.mdpi.com/article/10.3390/polym15153257/s1, Figure S1: First heating scans of 1 mol% 3HV block and random copolymers and high 3HV block, blend and random copolymers; Figure S2: First cooling scans of 1 mol% 3HV block and random copolymers and high 3HV block, blend and random copolymers; Figure S3: Second heating scans of 1 mol% 3HV block and random copolymers and high 3HV block, blend and random copolymers; Figure S4: Second cooling scans of 1 mol% 3HV block and random copolymers and high 3HV block, blend and random copolymers; Figure S5: Last heating scans of 1 mol% 3HV block and random copolymers and high 3HV block, blend and random copolymers; Table S1: Details of hydroxy-functionalised PHBV random copolymers used; Table S2: Chemical shifts of ^{1}H-NMR spectra of reaction products of HDI with hydroxy-functionalised PHBV; Table S3: Chemical shifts of PHBV materials in solid-state ^{13}C-NMR spectra.

Author Contributions: J.M.: conceptualization, investigation, methodology, validation, data curation, formal analysis, visualization, and writing—original draft and editing. B.L.: supervision, project administration, methodology, and writing—review and editing. C.M.C.: supervision, writing—review and editing. S.P.: supervision, project administration, and writing—review and editing. All authors have read and agreed to the published version of the manuscript.

Funding: This work is funded by the Australian Research Council via Discovery Project DP200101144 and the ARC Training Centre for Bioplastics and Biocomposites IC210100023.

Institutional Review Board Statement: Not applicable.

Data Availability Statement: The raw/processed data required to reproduce these findings cannot be shared at this time due to technical or time limitations.

Acknowledgments: The authors acknowledge the facilities and the scientific and technical assistance of the UQ School of Chemical Engineering and the Centre of Advanced Imaging. J.M. acknowledge the Australian Government Research Training Program for scholarship during this study.

Conflicts of Interest: The authors declare that they have no known competing financial interest or personal relationships that could have appeared to influence the work reported in this paper.

References

1. Kellerhals, M.B.; Kessler, B.; Witholt, B.; Tchouboukov, A.; Brandl, H. Renewable Long-Chain Fatty Acids for Production of Biodegradable Medium-Chain-Length Polyhydroxyalkanoates (mcl-PHAs) at Laboratory and Pilot Plant Scales. *Macromolecules* **2000**, *33*, 4690–4698. [CrossRef]
2. Chen, G.Q.; Wu, Q. The application of polyhydroxyalkanoates as tissue engineering materials. *Biomaterials* **2005**, *26*, 6565–6578. [CrossRef]
3. Li, Z.B.; Loh, X.J. Water soluble polyhydroxyalkanoates: Future materials for therapeutic applications. *Chem. Soc. Rev.* **2015**, *44*, 2865–2879. [CrossRef]
4. Pan, J.; Li, G.; Chen, Z.; Chen, X.; Zhu, W.; Xu, K. Alternative block polyurethanes based on poly (3-hydroxybutyrate-co-4-hydroxybutyrate) and poly (ethylene glycol). *Biomaterials* **2009**, *30*, 2975–2984. [CrossRef]
5. Bejagam, K.K.; Iverson, C.N.; Marrone, B.L.; Pilania, G. Composition and Configuration Dependence of Glass-Transition Temperature in Binary Copolymers and Blends of Polyhydroxyalkanoate Biopolymers. *Macromolecules* **2021**, *54*, 5618–5628. [CrossRef]
6. Shah, D.T.; Tran, M.; Berger, P.A.; Aggarwal, P.; Asrar, J.; Madden, L.A.; Anderson, A.J. Synthesis and Properties of Hydroxy-Terminated Poly(hydroxyalkanoate)s. *Macromolecules* **2000**, *33*, 2875–2880. [CrossRef]
7. Akaraonye, E.; Keshavarz, T.; Roy, I. Production of polyhydroxyalkanoates: The future green materials of choice. *J. Chem. Technol. Biotechnol.* **2010**, *85*, 732–743.

8. Ferre-Guell, A.; Winterburn, J. Biosynthesis and Characterization of Polyhydroxyalkanoates with Controlled Composition and Microstructure. *Biomacromolecules* **2018**, *19*, 996–1005. [CrossRef]
9. Arcos-Hernández, M.V.; Laycock, B.; Donose, B.C.; Pratt, S.; Halley, P.; Al-Luaibi, S.; Werker, A.; Lant, P.A. Physicochemical and mechanical properties of mixed culture polyhydroxyalkanoate (PHBV). *Eur. Polym. J.* **2013**, *49*, 904–913. [CrossRef]
10. Li, S.Y.; Dong, C.L.; Wang, S.Y.; Ye, H.M.; Chen, G.Q. Microbial production of polyhydroxyalkanoate block copolymer by recombinant Pseudomonas putida. *Appl. Microbiol. Biotechnol.* **2011**, *90*, 659–669. [CrossRef]
11. Tripathi, L.; Wu, L.P.; Chen, J.C.; Chen, G.Q. Synthesis of Diblock copolymer poly-3-hydroxybutyrate -block-poly-3-hydroxyhexanoate PHB-b-PHHx by a beta-oxidation weakened Pseudomonas putida KT2442. *Microb. Cell. Fact.* **2012**, *11*, 11. [CrossRef]
12. Westlie, A.H.; Chen, E.Y.X. Catalyzed Chemical Synthesis of Unnatural Aromatic Polyhydroxyalkanoate and Aromatic–Aliphatic PHAs with Record-High Glass-Transition and Decomposition Temperatures. *Macromolecules* **2020**, *53*, 9906–9915. [CrossRef]
13. Lin, Y.; Böker, A.; He, J.; Sill, K.; Xiang, H.; Abetz, C.; Li, X.; Wang, J.; Emrick, T.; Long, S.; et al. Self-directed self-assembly of nanoparticle/copolymer mixtures. *Nature* **2005**, *434*, 55–59. [CrossRef]
14. Hu, D.; Chung, A.-L.; Wu, L.-P.; Zhang, X.; Wu, Q.; Chen, J.-C.; Chen, G.-Q. Biosynthesis and Characterization of Polyhydroxyalkanoate Block Copolymer P3HB-b-P4HB. *Biomacromolecules* **2011**, *12*, 3166–3173. [CrossRef]
15. Oliveira, F.C.; Dias, M.L.; Castilho, L.R.; Freire, D.M.G. Characterization of poly(3-hydroxybutyrate) produced by Cupriavidus necator in solid-state fermentation. *Bioresour. Technol.* **2007**, *98*, 633–638. [CrossRef]
16. Luo, S.; Grubb, D.T.; Netravali, A.N. The effect of molecular weight on the lamellar structure, thermal and mechanical properties of poly(hydroxybutyrate-co-hydroxyvalerates). *Polymer* **2002**, *43*, 4159–4166. [CrossRef]
17. Yu, J.; Chen, L.X.L. Cost-Effective Recovery and Purification of Polyhydroxyalkanoates by Selective Dissolution of Cell Mass. *Biotechnol. Prog.* **2006**, *22*, 547–553. [CrossRef]
18. Tanadchangsaeng, N.; Yu, J. Microbial synthesis of polyhydroxybutyrate from glycerol: Gluconeogenesis, molecular weight and material properties of biopolyester. *Biotechnol. Bioeng.* **2012**, *109*, 2808–2818. [CrossRef]
19. McChalicher, C.W.J.; Srienc, F. Investigating the structure–property relationship of bacterial PHA block copolymers. *J. Biotechnol.* **2007**, *132*, 296–302. [CrossRef]
20. Pederson, E.N.; McChalicher, C.W.J.; Srienc, F. Bacterial Synthesis of PHA Block Copolymers. *Biomacromolecules* **2006**, *7*, 1904–1911. [CrossRef]
21. Tripathi, L.; Wu, L.-P.; Meng, D.; Chen, J.; Chen, G.-Q. Biosynthesis and Characterization of Diblock Copolymer of P(3-Hydroxypropionate)-block-P(4-hydroxybutyrate) from Recombinant Escherichia coli. *Biomacromolecules* **2013**, *14*, 862–870. [CrossRef]
22. Kageyama, Y.; Tomita, H.; Isono, T.; Satoh, T.; Matsumoto, K. Artificial polyhydroxyalkanoate poly [2-hydroxybutyrate-block-3-hydroxybutyrate] elastomer-like material. *Sci. Rep.* **2021**, *11*, 22446. [CrossRef]
23. Nakaoki, T.; Yasui, J.; Komaeda, T. Biosynthesis of P3HBV-b-P3HB-b-P3HBV Triblock Copolymer by Ralstonia eutropha. *J. Polym. Environ.* **2019**, *27*, 2720–2727. [CrossRef]
24. Li, Z.B.; Yang, J.; Loh, X.J. Polyhydroxyalkanoates: Opening doors for a sustainable future. *NPG Asia Mater.* **2016**, *8*, 20. [CrossRef]
25. Meng, D.C.; Shen, R.; Yao, H.; Chen, J.C.; Wu, Q.; Chen, G.Q. Engineering the diversity of polyesters. *Curr. Opin. Biotechnol.* **2014**, *29*, 24–33. [CrossRef]
26. Müller, A.J.; Arnal, M.L.; Balsamo, V. Crystallization in Block Copolymers with More than One Crystallizable Block. In *Progress in Understanding of Polymer Crystallization*; Reiter, G., Strobl, G.R., Eds.; Springer: Berlin/Heidelberg, Germany, 2007; pp. 229–259.
27. Samui, A.B.; Kanai, T. Polyhydroxyalkanoates based copolymers. *Int. J. Biol. Macromol.* **2019**, *140*, 522–537. [CrossRef]
28. Adamus, G.; Sikorska, W.; Janeczek, H.; Kwiecień, M.; Sobota, M.; Kowalczuk, M. Novel block copolymers of atactic PHB with natural PHA for cardiovascular engineering: Synthesis and characterization. *Eur. Polym. J.* **2012**, *48*, 621–631. [CrossRef]
29. Li, J.; Li, X.; Ni, X.; Leong, K.W. Synthesis and Characterization of New Biodegradable Amphiphilic Poly(ethylene oxide)-b-poly[(R)-3-hydroxy butyrate]-b-poly(ethylene oxide) Triblock Copolymers. *Macromolecules* **2003**, *36*, 2661–2667. [CrossRef]
30. Li, J.; Li, X.; Ni, X.; Wang, X.; Li, H.; Leong, K.W. Self-assembled supramolecular hydrogels formed by biodegradable PEO–PHB–PEO triblock copolymers and α-cyclodextrin for controlled drug delivery. *Biomaterials* **2006**, *27*, 4132–4140. [CrossRef]
31. Loh, X.J.; Goh, S.H.; Li, J. New Biodegradable Thermogelling Copolymers Having Very Low Gelation Concentrations. *Biomacromolecules* **2007**, *8*, 585–593. [CrossRef]
32. Güney, A.; Gardiner, C.; McCormack, A.; Malda, J.; Grijpma, D.W. Thermoplastic PCL-b-PEG-b-PCL and HDI Polyurethanes for Extrusion-Based 3D-Printing of Tough Hydrogels. *Bioengineering* **2018**, *5*, 99. [CrossRef]
33. Li, Z.; Yang, X.; Wu, L.; Chen, Z.; Lin, Y.; Xu, K.; Chen, G.Q. Synthesis, characterization and biocompatibility of biodegradable elastomeric poly(ether-ester urethane)s Based on Poly(3-hydroxybutyrate-co-3-hydroxyhexanoate) and Poly(ethylene glycol) via melting polymerization. *J. Biomater. Sci. Polym. Ed.* **2009**, *20*, 1179–1202. [CrossRef]
34. Ou, W.; Qiu, H.; Chen, Z.; Xu, K. Biodegradable block poly(ester-urethane)s based on poly(3-hydroxybutyrate-co-4-hydroxybutyrate) copolymers. *Biomaterials* **2011**, *32*, 3178–3188. [CrossRef]
35. Ramier, J.; Renard, E.; Grande, D. Microwave-assisted synthesis and characterization of biodegradable block copolyesters based on poly(3-hydroxyalkanoate)s and poly(D,L-lactide). *J. Polym. Sci. Part A Polym. Chem.* **2012**, *50*, 1445–1455. [CrossRef]
36. Velankar, S.; Cooper, S.L. Microphase separation and rheological properties of polyurethane melts. 1. Effect of block length. *Macromolecules* **1998**, *31*, 9181–9192. [CrossRef]

37. Andrade, A.P.; Neuenschwander, P.; Hany, R.; Egli, T.; Witholt, B.; Li, Z. Synthesis and characterization of novel copoly (ester− urethane) containing blocks of poly-[(R)-3-hydroxyoctanoate] and poly-[(R)-3-hydroxybutyrate]. *Macromolecules* **2002**, *35*, 4946–4950. [CrossRef]
38. Shen, Y.; Deng, J.J.; Luo, X.; Zhang, X.; Zeng, X.; Feng, M.; Pan, S.R. Synthesis and Characterization of a Sterically Stabilized Polyelectrolyte Using Isophorone Diisocyanate as the Coupling Reagent. *J. Biomater. Sci.-Polym. Ed.* **2009**, *20*, 1217–1233. [CrossRef]
39. Chen, Z.F.; Cheng, S.T.; Xu, K.T. Block poly(ester-urethane)s based on poly(3-hydroxybutyrate-co-4-hydroxybutyrate) and poly(3-hydroxyhexanoate-co-3-hydroxyoctanoate). *Biomaterials* **2009**, *30*, 2219–2230. [CrossRef]
40. Liu, Q.; Zhu, M.; Chen, Y. Synthesis and characterization of multi-block copolymers containing poly [(3-hydroxybutyrate)-co-(3-hydroxyvalerate)] and poly(ethylene glycol). *Polym. Int.* **2010**, *59*, 842–850. [CrossRef]
41. Kai, D.; Loh, X.J. Polyhydroxyalkanoates: Chemical Modifications Toward Biomedical Applications. *ACS Sustain. Chem. Eng.* **2014**, *2*, 106–119. [CrossRef]
42. Li, X.; Loh, X.J.; Wang, K.; He, C.; Li, J. Poly(ester urethane)s Consisting of Poly[(R)-3-hydroxybutyrate] and Poly(ethylene glycol) as Candidate Biomaterials: Characterization and Mechanical Property Study. *Biomacromolecules* **2005**, *6*, 2740–2747. [CrossRef]
43. Mai, J.J.; Chan, C.M.; Colwell, J.; Pratt, S.; Laycock, B. Characterisation of end groups of hydroxy-functionalised scl-PHAs prepared by transesterification using ethylene glycol. *Polym. Degrad. Stab.* **2022**, *205*, 110123. [CrossRef]
44. Mai, J.; Chan, C.M.; Laycock, B.; Pratt, S. Understanding the Reaction of Hydroxy-Terminated Poly (3-hydroxybutyrate-co-3-hydroxyvalerate)(PHBV) Random Copolymers with a Monoisocyanate. *Macromolecules* **2023**, *56*, 2328–2338. [CrossRef]
45. Li, G.; Li, D.; Niu, Y.; He, T.; Chen, K.C.; Xu, K. Alternating block polyurethanes based on PCL and PEG as potential nerve regeneration materials. *J. Biomed. Mater. Res. Part A* **2014**, *102*, 685–697. [CrossRef]
46. Kinard, L.A.; Kasper, F.K.; Mikos, A.G. Synthesis of oligo(poly(ethylene glycol) fumarate). *Nat. Protoc.* **2012**, *7*, 1219–1227. [CrossRef]
47. Marchessault, R.; Okamura, K.; Su, C. Physical properties of poly (β-hydroxy butyrate). II. Conformational aspects in solution. *Macromolecules* **1970**, *3*, 735–740. [CrossRef]
48. Sekkar, V.; Bhagawan, S.S.; Prabhakaran, N.; Rama Rao, M.; Ninan, K.N. Polyurethanes based on hydroxyl terminated polybutadiene: Modelling of network parameters and correlation with mechanical properties. *Polymer* **2000**, *41*, 6773–6786. [CrossRef]
49. Petrović, Z.S.; Javni, I.; Divjaković, V. Structure and physical properties of segmented polyurethane elastomers containing chemical crosslinks in the hard segment. *J. Polym. Sci. Part B Polym. Phys.* **1998**, *36*, 221–235. [CrossRef]
50. Ishida, M.; Yoshinaga, K.; Horii, F. Solid-State 13C NMR Analyses of the Microphase-Separated Structure of Polyurethane Elastomer. *Macromolecules* **1996**, *29*, 8824–8829. [CrossRef]
51. Stern, T. Hierarchical fractal-structured allophanate-derived network formation in bulk polyurethane synthesis. *Polym. Adv. Technol.* **2018**, *29*, 746–757. [CrossRef]
52. Li, G.; Li, P.; Qiu, H.; Li, D.; Su, M.; Xu, K. Synthesis, characterizations and biocompatibility of alternating block polyurethanes based on P3/4HB and PPG-PEG-PPG. *J. Biomed. Mater. Res. Part A* **2011**, *98*, 88–99. [CrossRef]
53. Desai, S.; Thakore, I.M.; Sarawade, B.D.; Devi, S. Effect of polyols and diisocyanates on thermo-mechanical and morphological properties of polyurethanes. *Eur. Polym. J.* **2000**, *36*, 711–725. [CrossRef]

polymers

MDPI

Article

In Situ Tensile Testing under High-Speed Optical Recording to Determine Hierarchical Damage Kinetics in Polymer Layers of Flax Fibre Elements

Emmanuelle Richely [1], Johnny Beaugrand [1,*], Michel Coret [2], Christophe Binetruy [2], Pierre Ouagne [3], Alain Bourmaud [4] and Sofiane Guessasma [1,*]

1 INRAE, Research Unit BIA UR1268, 3, Impasse Yvette Cauchois, 44316 Nantes, France
2 Lab Therm & Energie Nantes, LTeN, École Centrale de Nantes, Nantes Université, CNRS, GeM, UMR 6183, 44321 Nantes, France; michel.coret@ec-nantes.fr (M.C.); christophe.binetruy@ec-nantes.fr (C.B.)
3 Laboratoire Génie de Production (LGP), Université de Toulouse, INP-ENIT, 65016 Tarbes, France; pierre.ouagne@enit.fr
4 Université de Bretagne Sud, IRDL UMR CNRS 6027, 56100 Lorient, France; alain.bourmaud@univ-ubs.fr
* Correspondence: johnny.beaugrand@inrae.fr (J.B.); sofiane.guessasma@inrae.fr (S.G.)

Abstract: This study aims at better understanding the damage and fracture kinetics in flax fibre elements at both the unitary and bundle scales, using an experimental setup allowing optical observation at high recording rate in the course of tensile loading. Defects and issues from flax unitary fibre extraction are quantitated using polarized light microscopy. Tensile loading is conducted according to a particular setup, adapted to fibres of 10 to 20 µm in diameter and 10 mm in length. Optical recording using a high-speed camera is performed during loading up to the failure at acquisition, with speed ranging from 108,000 to 270,000 frames per second. Crack initiation in polymer layers of fibre elements, propagation as well as damage mechanisms are captured. The results show different failure scenarios depending on the fibre element's nature. In particular, fractured fibres underline either a fully transverse failure propagation or a combination of transverse and longitudinal cracking with different balances. Image recordings with high time resolution of down to 3.7 µs suggest an unstable system and transverse crack speed higher than 4 m/s and a slower propagation for longitudinal crack deviation. Failure propagation monitoring and fracture mechanism studies in individual natural fibre or bundles, using tensile load with optical observation, showed contrasted behaviour and the importance of the structural scale exanimated. This study can help in tailoring the eco-design of flax-based composites, in terms of toughness and mechanical performances, for both replacement of synthetic fibre materials and innovative composites with advanced properties.

Keywords: flax fibre; tensile testing; crack propagation; fractography; high-speed optical imaging; microstructure; in situ tensile testing

Citation: Richely, E.; Beaugrand, J.; Coret, M.; Binetruy, C.; Ouagne, P.; Bourmaud, A.; Guessasma, S. In Situ Tensile Testing under High-Speed Optical Recording to Determine Hierarchical Damage Kinetics in Polymer Layers of Flax Fibre Elements. *Polymers* **2023**, *15*, 2794. https://doi.org/10.3390/polym15132794

Academic Editors: Antonio M. Borrero-López, Concepción Valencia-Barragán, Esperanza Cortés Triviño, Adrián Tenorio-Alfonso and Clara Delgado-Sánchez

Received: 28 May 2023
Revised: 15 June 2023
Accepted: 16 June 2023
Published: 23 June 2023

1. Introduction

Flax is currently viewed as one of the most promising plant fibres to replace synthetic materials, such as glass, in composite industry [1,2]. Thanks to the large specific mechanical performance of flax fibres, light and sustainable technical parts can challenge traditional counterparts [3]. Flax fibres are derived from the stems of the flax plant [4]. It has a hierarchical structure, with a large complexity composed of primary and secondary cell walls. The primary wall is the outermost layer of the fibre [5]. It is composed of cellulose microfibrils, hemicellulose and pectin [6]. The primary cell wall is the origin of the fibre flexibility and its ability to bend and twist [7]. The secondary cell wall is located beneath the primary cell wall, and is composed mainly of cellulose microfibrils and amorphous polymers, such as hemicellulose and lignin [8]. The secondary cell wall is responsible for the fibre's strength and stiffness, making it resistant to breaking and tearing [9,10]. As is the case for plant fibres, flax usage is threatened by property variability, which is triggered by its genuine phenotype and extraction processes [11]. The effect of this variability on the mechanical performance is amplified by global warming [12]. In fact, the variability of plant fibres within the context of valorisation in the composite industry is threatened by several factors, such as climate change, where the need for quality control is important to guarantee that biobased composite performance meets the specification thresholds. Thus, the understanding of the link between the structure and the performance of plant fibre elements, such as flax, is important to quantify the effect of variability and adopt strategies to attenuate its effect within the context of large-scale manufacturing. Several ways exist to act in that regard, with both leverages at fibre genetics [13–15] and extraction parameters. For instance, generation of new porosity and disruption in MFA (microfibril angle) have been reported due to plant fibre extraction machines [16,17]. The damage caused by the mechanical extraction influences the mechanical performance of genuine fibre elements, as well as the fibre-based composites [18–21]. This damage involves complex deformation mechanisms due to the interaction with the core structure of the fibre, such as the lumen, and the roughness [19,22–24]. These mechanisms promote stress localisation phenomena, which are difficult to capture if in situ testing or numerical models that use exact fibre elements ultrastructure are not considered. These defects gain a role at the composite level, through the intrinsic fibre properties or the interfacial properties. Both result in various failure scenarios, depending on the compatibility and the fibre surface treatments used to improve the load transfer between the matrix and the fibre [25]. When it is not possible to use elementary (also called unitary) fibres as reinforcement, bundles potentially close the gap. However, bundles add more complexity, because of the role of the middle lamella [26,27]. Here, at least three main scales need to be investigated to anticipate the composite performance, namely: unitary fibre, bundle and composite microstructure.

Many attempts have been made to understand the unitary fibre and bundle scale on the failure properties of flax [28,29]. By means of scanning electron microscopy and acoustic emission, Romhany et al. [28] showed that flax bundles exhibit a combination of axial splitting of the unitary fibres and radial cracking. Barbulée et al. [30] reported a frictional behaviour of unitary fibres under tension within the flax bundle and the delamination process, which compete with a transverse fibre rupture. Ahmed et al. [31] compared the effect of the unitary fibre and bundle scales on the mechanical behaviour of flax using scanning electron microscopy. A sequential rupture scenario has been suggested, which encompasses bundle rotation, unitary fibres segregation and ultimate failure. Fuentes et al. [24] reported deformation mechanisms of hemp bundles by using digital image correlation. The authors showed that interfacial behaviour explains the ultimate performance of bundles, as a significant shearing is witnessed for weak bundles, and transverse rupture of unitary fibres for strong ones.

In terms of the role of defects in rupture properties, Mott et al. [32] reported tensile results in wood fibres conducted under SEM observation, where defects acted as preferred sites for failure. Further, Baley [33] demonstrated, through SEM observations, crack initiation from kink bands in flax fibres. Aslan et al. [34] demonstrated the link between variability in properties and the defect distribution in unitary flax fibres. In particular, optical microscopy and SEM results showed a fracturing behaviour originating from both internal and surface defects.

Despite great efforts to quantify the outcome of fracture behaviour in unitary natural fibres and bundles, either by post-mortem or in situ SEM observations, the kinetics and ranking of deformation mechanisms are not easily understood. On top of that, the role of defects in defining rupture scenarios are not thoroughly described. High-speed optical recording has been envisaged as a way to capture the rupture events in natural fibre elements and related composites [25,27,35,36]. These observations highlight the increasing complexity in damage initiation and growth, depending on the scale. Transversely, longitudinal cracking and bridging phenomena are all reported as competing deformation mechanisms, where defects are identified as potential sites for crack initiation. Because of the multiple surface defects, combined with a rapid crack propagation, monitoring failure is challenging because of the difficulty to maintain a high frame rate while observing a large region of interest. Some attempts were made to address this difficulty by performing notches on the fibres prior to testing, to obtain a more localised rupture [35].

With recent progresses in terms of recording velocity by high-speed cameras, it is now possible to decipher the mechanisms involved during the failure in flax fibres without guiding the crack initiation. This study aims at providing a clear view of the deformation mechanisms in flax, at both the unitary fibre and bundle scales, with focuses on the failure scenarios, crack initiation and propagation paths.

This study addresses, firstly, the topography and chemical composition of flax fibre elements, as well as the quantification of the defect density and extent, from polarized light microscopy. Secondly, optical imaging at high acquisition rates is conducted during tensile loading to explore the damage growth in flax fibre elements. The damage mechanisms are discussed according to various cracking models.

2. Experimental Layout

2.1. Materials

The Bolchoï variety was selected as a high-fibre yield textile flax fibre for testing. Dew-retted flax stems were provided by Groupe Depestele/Teillage Vandecandelaère. These were cultivated in France and dew-retted in the field in 2017. After harvesting, the plants were stored, scutched and hackled. The fibre bundles obtained were then intentionally damaged by crushing them between rollers on a pilot scutching line (ENI-Tarbes). Therefore, unitary fibres extracted from scutched bundles with contrasted defect density were extracted manually, depending on the location, i.e., close or in-between the marks left by the rollers.

2.2. Biochemical Composition

For all biochemical analyses, a first step of homogenization was performed by cryo-grinding (SPEX 6700 freezer Mill) approximately 2 g of raw flax bundles into powder, taken from the middle part of the scutched technical flax fibres (about 60 cm long) from each batch. The results given are the mean values, expressed as the percentage of dry matter.

- Monosaccharides:

Flax powdered samples (approximately 5 mg per trial) were subjected to pre-hydrolysis in 12 M H_2SO_4 (Sigma Aldrich, Darmstadt, Germany) for 2 h at 25 °C. An additional hydrolysis step in 1.5 M H_2SO_4 was carried out for 2 h at 100 °C after the addition of inositol, which is used as an internal standard. The alditol acetate derivatives of the neutral sugars [37] were analysed by gas phase chromatography (Perkin Elmer, Shelton, CT, USA), equipped with an DB 225 capillary column (J&W Scientific, Folsorn, CA, USA) at 205 °C, with H_2 as the carrier gas and a flame ionisation detector. Standard monosaccharide solutions of known concentrations were used for calibration. The galacturonic and glucuronic acids were merged as uronic acids, and determined by the m-hydroxybiphenyl method [38], a colorimetric quantification. Three independent trials were performed for each analysis.

- Lignin:

The lignin content was quantified by spectrophotometry following the acetyl-bromide method [39]. Three independent trials were performed for each fibre batch, and the chemicals were laboratory-grade, provided by Sigma Aldrich (Darmstadt, Germany).

- Proteins:

The protein content was estimated by the C/N Dumas method, which allows the nitrogen, carbon and sulphur content to be measured by fast and complete combustion of the sample, using an elemental analyser (VarioMicro, Elementar, Lyon, France). The protein content is derived from the nitrogen content using a conversion factor of 6.25, following the general case. The determination of conversion factors is still controversial [40]. Two independent trials were performed.

2.3. Sample Preparation

Flax fibre elements, as unitary or in bundles, were glued onto cardboard frames prepared by laser cut, and adapted for an optimal control of the fibre alignment on the tensile bench. Unitary fibres were extracted manually from bundles. All samples used in this study were scanned by optical microscopy prior to tensile testing, in order to characterize their morphology and defect distribution. Optical images were obtained using a microscope (Leitz DMRB, Leica Microsystems, Nanterre, France) equipped with a Hamamatsu digital camera (C11440 ORCA-Flash4.0 LT, Hamamatsu city, Japan). Large image scanning was performed thanks to a motorized stage (Marzhauser, Wetzlar, Germany), allowing scanning along all of the fibres. Images were acquired using both bright light and linearly polarized light, and objectives of $\times 10$ and $\times 20$, leading to a maximum spatial resolution of 18,000 $\times$ 2000 pixels, with a pixel size of 328 to 656 nm. The diameters were measured at 3 to 6 locations along each fibre element by optical microscopy. The cross-sectional area was calculated from the mean of the diameters, assuming a circular cross-section. The edges of the paper frame were cut prior to the test.

2.4. Optical Image Processing

Optical images were processed to evaluate the density of defects within flax fibres. The process is illustrated in Figure 1. All acquired images were converted from RGB to grey level images. Image superposition is required to encapsulate the grey levels of the polarised image within the fibre element space. For such a purpose, a series of image processing operators, such as threshold, fill hole and max feature isolation, were used to convert the bright image into an outlined one, resulting in the image of fibre boundary extraction as shown in Figure 1. Next, the retrieved boundaries from the outlined image were added and superimposed to the polarised image. The grey level distribution within the fibre, delimited by the added boundaries, were analysed to derive the linear density of the defects and their intensity compared to the mechanical properties.

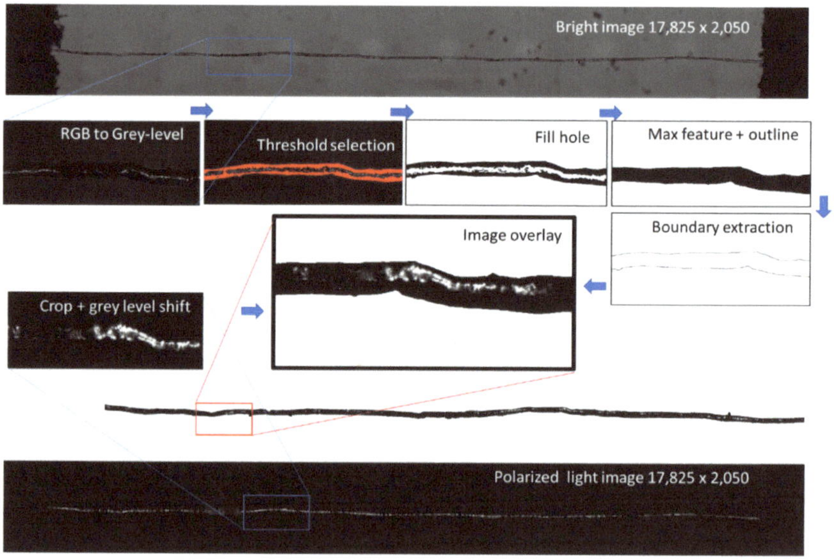

Figure 1. Image processing used to measure the density of defects from optical sources combining bright and linearly polarized lights.

2.5. Standard Tensile Experiments

- Unitary fibre scale:

Tensile tests were carried out on unitary fibres up to the fracture point in an MTS machine (MTS System, Créteil, France) at IRDL, Lorient (France), using a 2 N load cell and a strain rate of 1 mm/min, in accordance with the AFNOR NF T 25-501 standard. The controlled testing environment was set to 25 ± 1 °C and 48 ± 2% relative humidity. The fibres were manually extracted and glued to a paper frame with a gauge length of 10 mm. The edges of the paper frame were trimmed prior to testing. The compliance of the system was taken into account, in order to compensate for the influence of non-specimen extension in tensile testing, which leads to underestimation of Young's modulus and overestimation of strain [41]. Regarding the test reliability, 45 unitary fibres were tested to ensure confidence in measured properties. The stress–strain curves were plotted, taking into account the determination of the cross-sectional areas, which are based on 6 optical measurements along their length and assuming a circular approximation.

- Bundle scale:

Tensile testing experiments were carried out on 42 bundles in an MTS machine (MTS System, Créteil, France) at IRDL, Lorient (France), using a 50 N load cell, an acquisition frequency of 10 Hz and a strain rate of 1 mm/min. The controlled testing environment was set to 23 ± 1 °C and 43 ± 1% relative humidity. The fibre bundles were manually extracted and glued to a paper frame with a gauge length of 75 mm. Mean diameters were determined as the average of 6 diameter measurements along the bundles, taken by optical microscopy. The edges of the paper frame were trimmed before testing.

2.6. In Situ Tensile Experiments

Tensile testing experiments were carried out on unitary fibres using a Linkam machine TST350 (Linkam Scientific Instruments, Epsom, UK) with a 20 N load cell under high-speed camera recording. The tensile device was placed on a stage equipped with micrometre drives, allowing displacements with a precision in the μm range (Figure 2). As only the central zone could be imaged, the gauge length was adjusted from 15 to 5 mm, in order to capture the damage mechanisms without the need to localize the rupture. Moreover, laser markings were placed at 0.5 mm intervals on the 5 mm cardboard frame, to indicate specific locations of interest noted during the optical scanning. The testing environment was set to 24 ± 1 °C and 27 ± 3% relative humidity. Continuous tensile tests were conducted with a displacement rate of 1 mm/min, in accordance with the AFNOR NF T 25-501 standard. Moreover, the displacement was applied to both jaws. A minimum of 11 unitary fibres and 2 bundles were tested for the damage study.

High-speed camera recording, at speeds ranging from 100,000 to 270,000 frames per second (fps), was achieved using a Photron FASTCAM SA1.1 camera (Photron, Tokyo, Japan), equipped with a telecentric lens (TC16M009, Opto engineering, Mantova, Italy). The full frame dimensions were 1024 × 1024 pixels for frame rates of up to 5400 fps. The dimensions around the fibre elements were reduced to 1024 × 16 pixels, in order to achieve 270,000 fps. The pixel size in this setup is 5.1 μm. The acquisition was triggered manually as soon as the failure was observed, allowing the sequence to be recorded before triggering. The exposure time was reduced to minimize blurring effects. Additional lighting was applied to improve the image quality, using an LED source placed behind the tensile device (Figure 2).

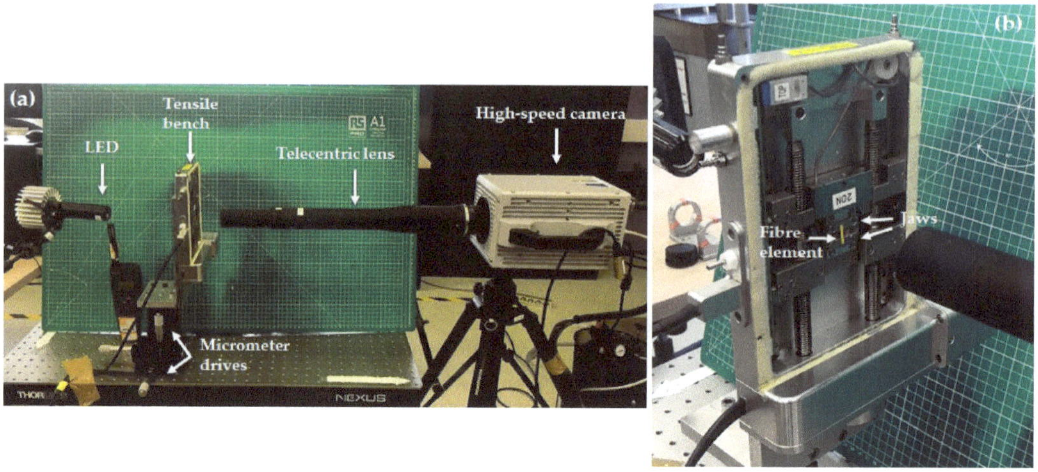

Figure 2. (**a**) Overview of the experimental set up of high-speed imaging in situ tensile testing experiments. (**b**) Magnified view of the specimen mount on the tensile machine.

2.7. Scanning Electron Microscopy Observations

Scanning electron microscopy (SEM) was performed close to the rupture point on the fibre elements after tensile testing, using a SEM Quattro S (ThermoFisher, Waltham, MA, USA). The fibre elements, still glued on the cardboard frame at both ends, were clamped and observed under low vacuum conditions at a pressure of 100 Pa, an acceleration voltage of 7 kV and a magnification between ×150 and ×15,000.

3. Results and Discussion

3.1. Biochemical Results

The monosaccharide composition of the Bolchoï variety, determined by gas phase chromatography (GPC), and expressed as a function of dry matter, shows that the main monosaccharide is glucose, which represents 78% of the dry matter. Glucose is often considered to represent the cellulose content of bast fibres, and values between 55 and 90% are reported in the literature for retted textile flax [6,42]. Regarding the other minor in-mass saccharides, the galactose content is 3.6%. Galactose is the main constituent of β(1–4)-galactan, the main encrusting component of unitary fibres, and is likely to be part of rhamnogalacturonan-I (RG-I) [43]. Both have been observed in the secondary cell wall at maturity. The galactose/rhamnose ratio has been reported to adjust the length and number of the RG-I chains, allowing the cellulose microfibril packing to be calculated [44]. In our study, this ratio is close to 4.8. The uronic acid content is 1.9 ± 0.1%, and reflects the presence of pectin in the middle lamella [45].

The lignin content is about 2.4% of the dry matter. These values are in agreement with those reported in the literature, ranging from 1 to 5% for flax [46]. A higher lignin rate has been described by several authors [47,48] as a possible response to an abiotic stress, such as lodging, temperature or lack of water. Since lignin is predominantly found in the middle lamella and cell junctions [49,50], a high degree of retting and a thin middle lamella might explain the relatively low lignin content in the flax studied.

The protein content was estimated from a nitrogen content assay, and shows the same trend as the lignin content, with a low value of around 1.4%. The differences suggest that the compound middle lamella (CML), composed of the middle lamella and adjacent primary cell walls and enriched in structural proteins [43], is less important for the flax studied. However, the origin of the proteins is diverse, and includes both those contained in the fibres (mentioned in the CML) and those introduced by the retting step. It should be noted that retted flax certainly owes its low protein content to the fact that it underwent extensive enzymatic digestion, since proteases are also secreted by the decomposing microorganisms. Moreover, some proteins may be involved in the plant's defence mechanisms against external stresses [51].

3.2. Overall Mechanical Behaviour

Figure 3 shows the typical stress-strain response of the studied flax fibre at both the unitary and bundle scales. At the unitary fibre scale, where the average flax fibre diameter is 16 ± 3 µm, the average Young's modulus reflecting the compilation of 45 testing results is 40 ± 10 GPa. In the case of plant fibres, due to the natural variability, especially the diameter, presence of kink-bands and size of lumen forms, standard deviations between 20 and 30% are often reported [34]. The observed low stiffness of the flax studied may be related to a high microfibril angle or the presence of a large number of defects. The other properties measured were the elongation at break (2.04 ± 1.05%) and the tensile strength (666 ± 232 MPa). When comparing these results with the literature data [52], the strength and stiffness are positioned in the lower range, while the elongation at break is in the middle range.

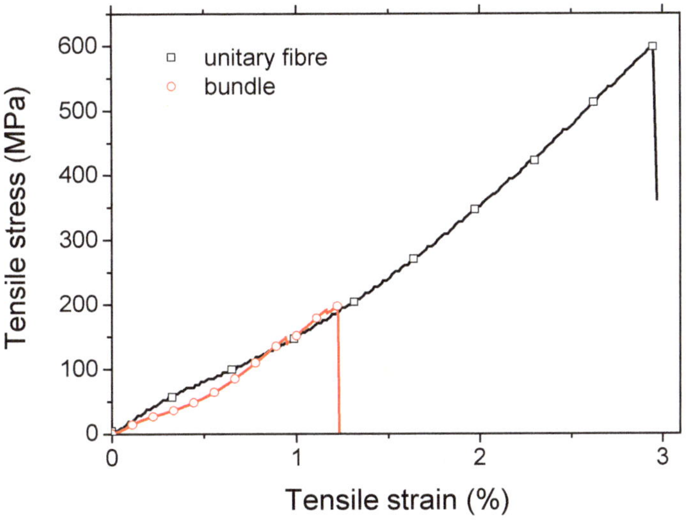

Figure 3. Typical tensile response of flax fibre elements at the unitary and bundle scale.

Tensile test results at the bundle scale show tensile strengths of 205 ± 85 and 193 ± 92 MPa. Taking into account the mean diameter of the bundles of 114 ± 21 µm, which is approximately 7 times the unitary fibre diameter, the high probability of critical flaws would explain a low strength value at the bundle scale, according to Griffith's theory [53]. These strength values could be attributed to a high degree of retting, resulting in a less cohesive middle lamella, or to a more technical aspect related to the influence of the gauge length in specimen [15]. Nevertheless, the values obtained are consistent with the literature results for 75 mm gauge length [28,54], and the decrease in strength as a function of the gauge length has already been extensively studied. Indeed, for gauge lengths above a threshold value of about 25 mm, the bundles can be considered as aligned short composites, driven by the weaker mechanical properties of the middle lamella. Load drops prior to the ultimate failure have also been observed as shown in Figure 3, possibly corresponding to the successive early failure of some fibres within the bundles [28,31,55]. The strain at break (1.24 ± 0.34%) appears to be smaller than the unitary fibre scale. This supports the idea of the role of a weaker phase driving the failure of the bundle, namely the middle lamella. The Young's modulus results (19.1 ± 6.2 GPa) are obtained with a variability of about 32% (i.e., the ratio between the standard deviation and the mean). This variability underlines the change in bundle diameter (18%), but also the non-linear behaviour, against which stiffness assessment is assessed in the last linear segment of the stress-strain curves before the load drops.

The relatively low score for tensile properties at both the unitary and bundle scales may be related to the circular cross-section approximation, which can be better captured by, for instance, automated laser scanning techniques. A correction factor similar to those determined for other types of plant fibres [56,57] could be applied to compensate for the underestimation of the fibre strength and moduli caused by the circular cross-section assumption.

3.3. Defect Analysis Results

Figure 4a illustrates the variation in the defect intensity along the fibre length, for which the average size is 19 ± 2 μm. The defect intensity is measured here as the average grey level within the transverse direction. Within the same double Y-axis plot, the variation in the fibre diameter is shown. Large peaks reaching 32 μm indicate the presence of remains of the middle lamella. Smaller peaks of about 12 μm suggest the presence of a twist, which is often encountered during the sample preparation. Attempts to correlate the intensity of the defects with the diameter of the fibres seems to be inconclusive. Further, it appears that the average defect intensity within the entire fibre does not fit the mechanical data, as a large scatter is observed with the tensile strength, for example. However, the measured scatter, which represents the standard deviation of the defect density divided by the average defect intensity, shows a trend with respect to the ultimate properties. The highest scatter appears to be associated with lower tensile stress values.

Figure 4b shows the compiled defect density results for 45 unitary fibres as a function of the average diameter. The diameter is measured along the entire length, as shown in Figure 1, by processing outline images from the bright optical micrographs. The defect density is plotted as the average grey level intensity from optical images acquired under linearly polarized light. Figure 4b shows that there is a large scatter in the results, but a global trend can still be read that supports the idea of increasing defect density with increasing fibre diameter. The scatter also increases as the diameter increases, which can be also related to the difficulty of capturing the defect within the fibre cross-section, unless it is rotated during acquisition.

Unfortunately, when these data are related to tensile behaviour, there is no general trend between the defect density and the tensile response. Figure 4c shows, however, a negative correlation between the tensile strength and defect density scatter. This scatter is measured as the ratio of the standard deviation to the average defect density along the length of the unitary fibres.

3.4. In Situ Mechanical Behaviour at Unitary Fibre Scale

Detailed characteristics and mechanical properties, obtained from in situ tensile testing of flax unitary fibres and bundles with high-speed camera recordings, are presented in Table 1. Failure sequences were recorded using a high-speed camera, with recording speeds ranging from 108,000 to 270,000 frames per second (fps). The initial, pre-failure and post-failure images were compared with the initial image obtained by bright and polarized light microscopy prior to testing. SEM images of the fracture surfaces after tensile testing are also presented.

The tensile strength and elongation at break of unitary fibres are close to the values obtained from regular tensile experiments for a gauge length of 10 mm, with a mean strength of 656 ± 164 MPa and a mean strain at break of 2.78 ± 1.18%, even though only 11 fibres were tested here.

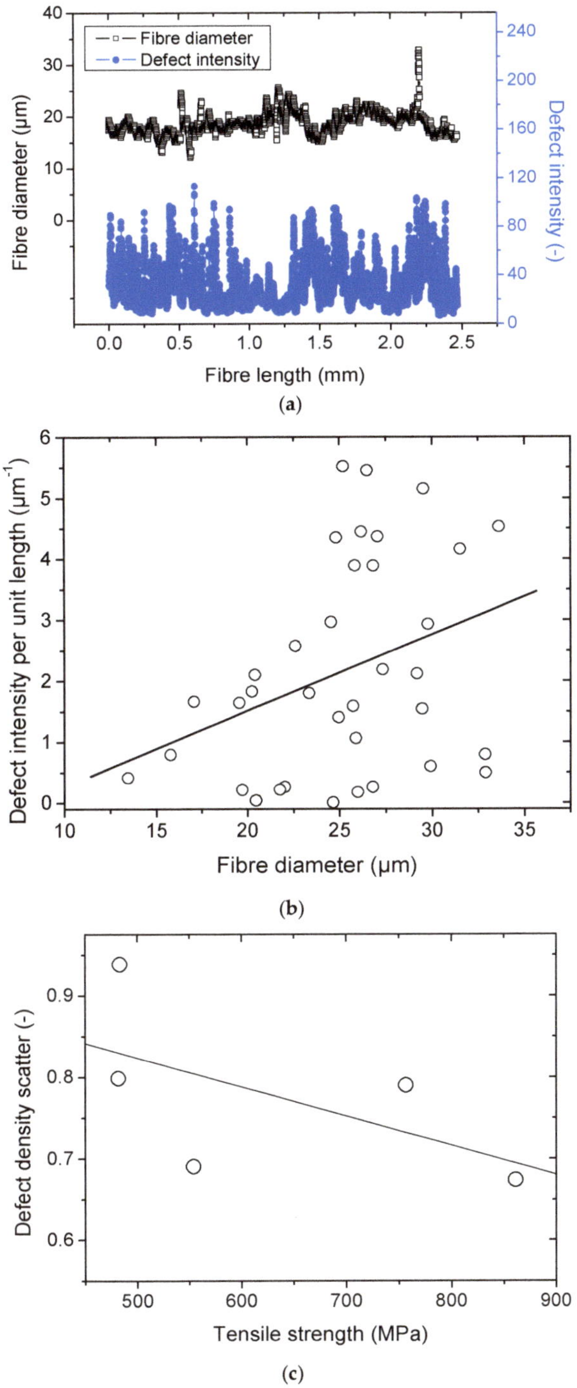

Figure 4. Defect analysis from optical imaging: (**a**) linear density of the defects measured according to the average grey level divided by the diameter, (**b**) correlation between the defect density and the fibre diameter, (**c**) relationship between tensile strength and defect density.

Table 1. Characteristics and mechanical properties of unitary fibres (a) and bundles (b) after high-speed camera in situ tensile testing. ROI: Region Of Interest.

Sample	Recording Speed (fps)	ROI Size (pix.)	Strength (MPa)	Elongation at Break (%)
b-1-41	100,000	704 × 48	403	1.80
a-1-10	108,000	1024 × 48	757	3.19
a-1-11			557	3.76
b-1-37			794	3.45
a-1-12	150,000	1024 × 32	482	3.60
a-1-13			875	1.79
a-1-14			554	2.78
a-1-22			483	3.97
a-1-16			852	1.59
a-1-18			861	3.57
a-1-20	270,000	1024 × 16	462	0.80
a-1-24			595	1.39
a-1-26			741.3	4.17
Mean values for unitary fibres	-	-	656 ± 164	2.78 ± 1.18

The complementary observations lead to different failure scenarios, depending on the fibre, as shown in Figures 5–9. In particular, the fracture fibre ends highlight a fully transverse failure propagation in Figure 6, or a combination of transverse and longitudinal propagation along different directions in other cases (Figures 5 and 7–9). Moreover, the short recording times, down to 3.7 µs, did not allow consecutive images of the failure mechanisms to be obtained in the case of almost complete transverse failure (Figures 5 and 6). This suggests a very unstable system and transverse crack propagation speeds higher than 1.5 and 4 m/s respectively. The longitudinal deviation could therefore lead to a slower overall failure, acting as an energy dissipating mechanism, as hypothesized by Beaugrand et al. [58]. However, observations at the surface only prevent crack propagation in the bulk from being captured. Moreover, it should also be remembered that only five fibres were fully described here.

In Figure 5, we can see that failure has occurred in a defected part of the fibre, which appears as a bright and curved area under polarized light prior to loading. However, the precise location of the failure initiation is difficult to evaluate, and therefore we can only assess the failure initiation in a globally defected area. The fracture surfaces reveal a large transverse failure with a small longitudinal deviation, which could be attributed to a sub-layer delamination or the presence of a central lumen [59]. Finally, longitudinal cracks close to the fracture surface can be underlined, with a zoom in on a crack bridging phenomenon.

The fracture surfaces in Figure 6 also show a predominant transverse failure, revealing a central dark area, which could correspond to the lumen. In addition, the possible involvement of surface mechanisms is evidenced in the high-speed camera images, with failure occurring in the vicinity of a surface impurity, indicated by blue arrows. The latter could be remnants of residual middle lamella or cortical tissue. Interestingly, the failure did not occur on the initially observed curved area, which is progressively straightened upon tensile testing.

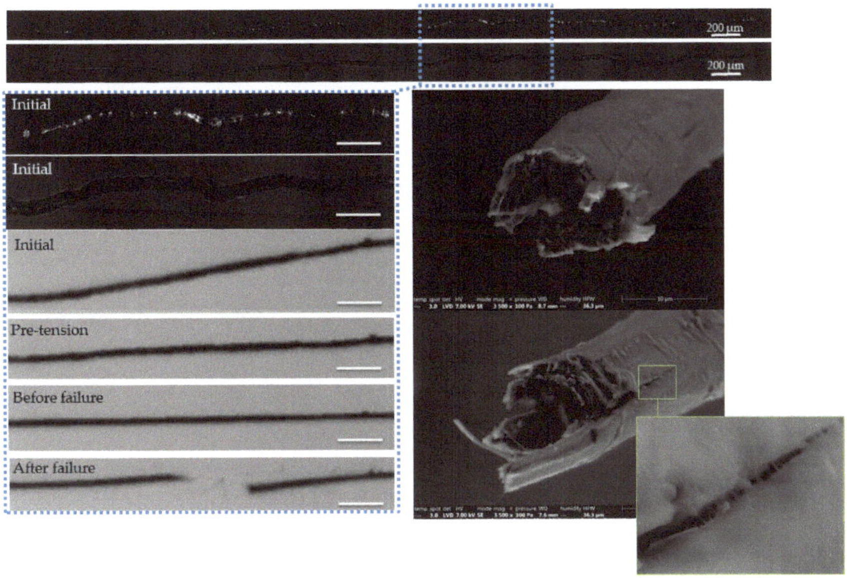

Figure 5. Images of the initial stage observed by optical microscopy under both bright and polarized light, and a zoom on the steps of failure of a flax unitary fibre a-1-11, observed during in situ tensile testing thanks to a high-speed camera. SEM images of the two fracture surfaces after complete failure. The scalebars indicate 100 µm, unless specified.

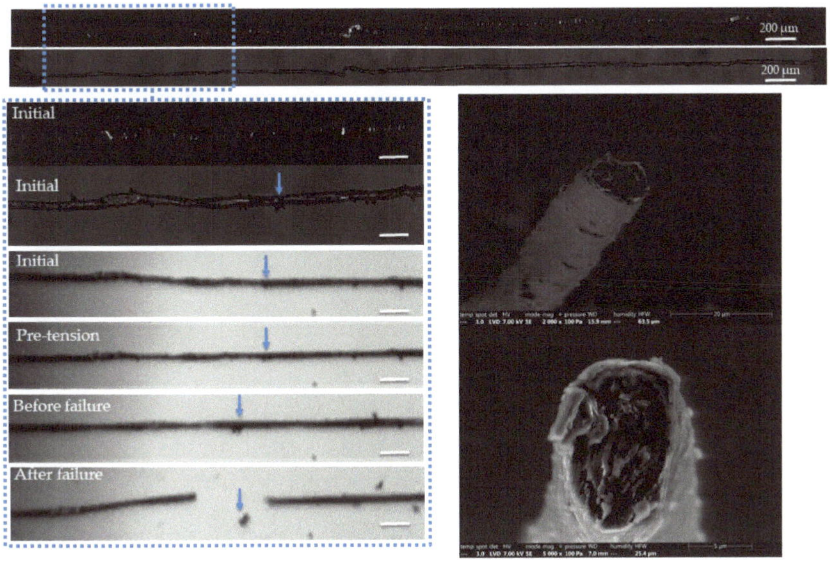

Figure 6. Images of the initial stage observed by optical microscopy under both bright and polarized light, and a zoom on the steps of failure of a flax unitary fibre a-1-22, observed during in situ tensile testing thanks to a high-speed camera. SEM images of the two fracture surfaces after complete failure. The scalebars represent 100 µm, unless specified, and the separation of a surface impurity is indicated by blue arrows.

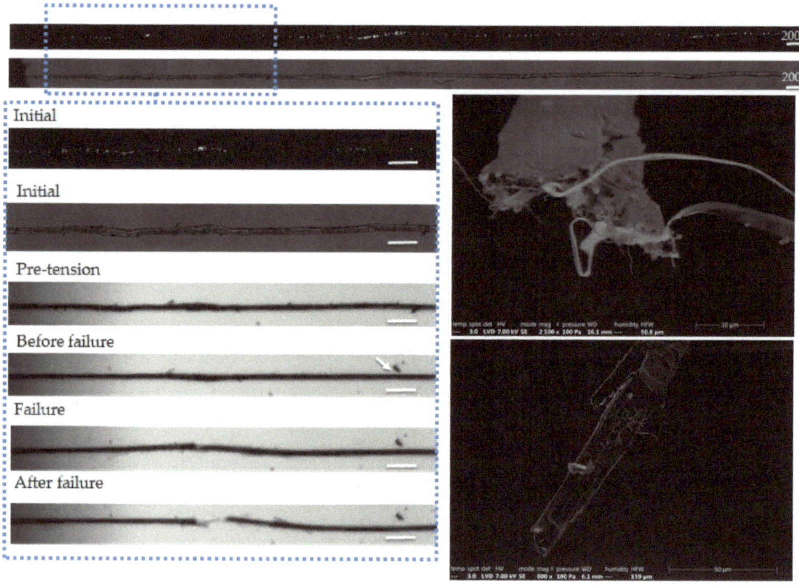

Figure 7. Images of the initial stage observed by optical microscopy under both bright and polarized light, and a zoom on the steps of failure of a flax unitary fibre a-1-14, observed during in situ tensile testing thanks to a high-speed camera. SEM images of the two fracture surfaces after complete failure. The scale bars represent 100 µm, and the separation of a surface impurity is indicated by a white arrow.

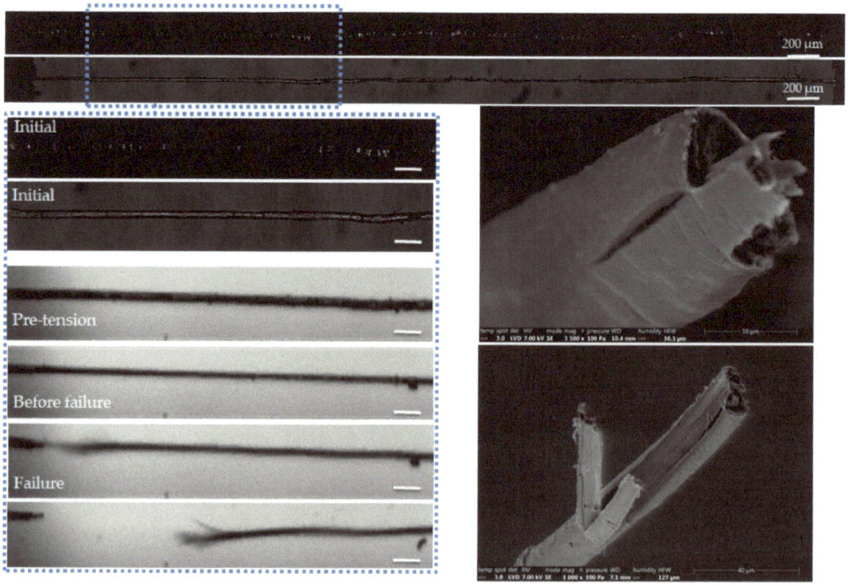

Figure 8. Images of the initial stage observed by optical microscopy under both bright and polarized light, and a zoom on the steps of failure of a flax unitary fibre a-1-18, observed during in situ tensile testing thanks to a high-speed camera. SEM images of the two fracture surfaces after complete failure. The scalebars represent 100 µm.

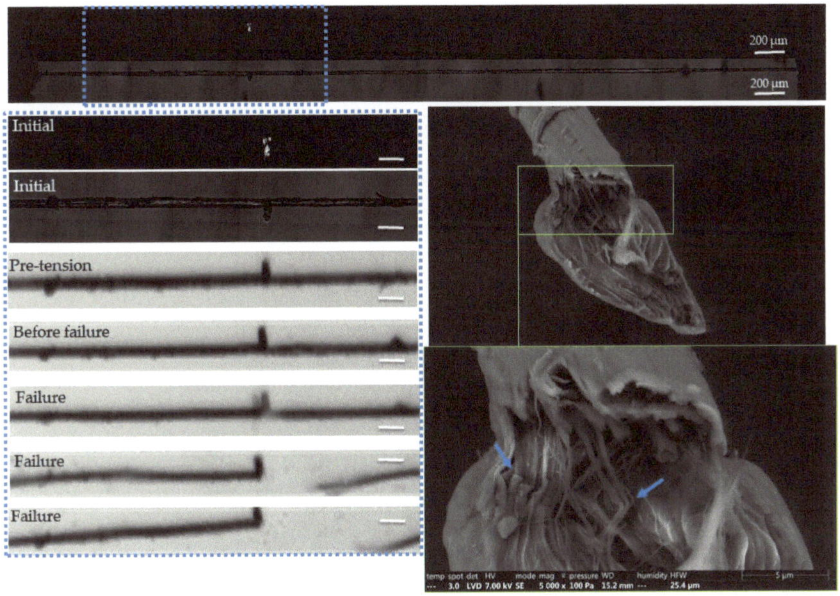

Figure 9. Images of the initial stage observed by optical microscopy under both bright and polarized light, and a zoom on the steps of failure of a flax unitary fibre a-1-20, observed during in situ tensile testing thanks to a high-speed camera. SEM images of the fracture surface of the left part of the fibre after complete failure. The scalebars represent 100 μm, and mesofibril misorientations highlighting a defected area are indicated by blue arrows.

Transverse cracking, followed by large longitudinal crack propagation, is highlighted in the SEM images of Figure 7. Moreover, it is also noted that the failure occurs in a zone of higher fibre diameter, which appears as a bright area under polarized light. This could be the result of pre-test processing damage, due to the compression of the cell walls between the rollers. Moreover, we can also see the removal of a surface impurity, indicated by a white arrow on the right part of the fibre. Regarding the fractured surfaces, a large longitudinal crack and disorganized cellulose macrofibrils can be seen in the top image.

The combination of transverse and longitudinal mechanisms is highlighted on the SEM images in Figure 8. Failure does not occur in the brightest area, and therefore in the zone of higher defect density along the fibre. The fracture surfaces reveal a significant crack at the surface of the upper part of the fibre, and a cell wall split longitudinally into three parts at the lower part of the fibre.

Finally, the fibre shown in Figure 9 has a failure close to a surface protrusion. At first sight, this could be attributed to a remaining part of adherent tissue, probably from the middle lamella or cortical tissue. However, the fracture surface shows the presence of cellulose macro- and microfibrils close to this protrusion, with a misalignment characteristic of a defect which could have caused the failure [60]. Unfortunately, with the observation of the other part of the fibre, it was not possible to confirm this hypothesis. Furthermore, the very low defect content and straight fibres observed in the initial stage before loading could be the result of a pre-tension applied during the sample preparation, explaining the low strain at break value of 0.8%.

In conclusion, the careful observation of the failure sequences in five different unitary fibres reveals different mechanisms involving pure transverse, and therefore more brittle, failure, or a combination of transverse failure and longitudinal splitting, in agreement with previous results reported in the literature [19,34,35]. Fractographic analysis reveals rough surfaces, typically encountered after tensile solicitation [31], and the presence of cracks close to the fracture surface. Moreover, some fibres have been observed to fail in a zone of high defect density, as shown by polarized light microscopy, but this scenario cannot be generalized to all samples. In agreement with the explanation given by Madsen et al. [19], the failure mechanisms may depend on microstructural parameters, such as the severity and spacing of defects, as well as the size of the lumen. Therefore, it is difficult to explain the differences between the fibres and the correlation with their mechanical properties based on the surface information provided in this study only. The use of X-ray tomography will be of interest in the second part of the study to access bulk information and, especially, defect severity and porosity-related mechanisms. Finally, the possible implication of surface contaminants in failure initiation is highlighted in two cases. As evidenced in Figure 9, the surface flaws may also coincide with zones of cellulose microfibril misalignment. As a perspective, bulk information will help to confirm and understand the underlying mechanisms.

3.5. In Situ Mechanical Behaviour at Bundle Scale

The investigation continued at the bundle scale, to better understand the contribution of the middle lamella to the failure mechanisms. Two bundles reflecting contrasted behaviour are shown in Figures 10 and 11. The first bundle shows a hierarchical failure sequence characterized by inter-fibre delamination, followed by the transverse failure of some unitary fibres (Figure 10). This is consistent with the observations of several authors [61], who point to fibre splitting prior to complete failure. Details of the rough interface are observed in the SEM images with the residual compound middle lamella and fibrillar structure, which may correspond to cellulose macrofibrils of the outer layers. In contrast, Figure 11 shows a predominant transverse failure of the unitary fibres without delamination. It occurred in a thicker zone, which could correspond to a defected area. However, polarized light microscopy did not allow the observation of defects in thick bundles, and volume information will help decipher the role of porosity and initial delamination in the different behaviours. Indeed, the second scenario lad to an early failure, characterized by a lower strength and strain at break (around 400 MPa and 1.8%, compared to 794 MPa and 3.5% reported for the first bundle). The value of 800 MPa for the strength is consistent with the values reported by Bos et al. [62] for such a short gauge length. As hypothesized by Fuentes et al. [24], the biochemical composition of the middle lamella could also explain the differences between the failure scenarios, which can be altered by moisture content [61,63–65]. However, it must be remembered that the gauge length is inferior to most fibre lengths. Therefore, the behaviours observed here may not reflect the loading of the middle lamella as for the longer gauge length, but rather resemble the behaviour of the unitary fibre scale.

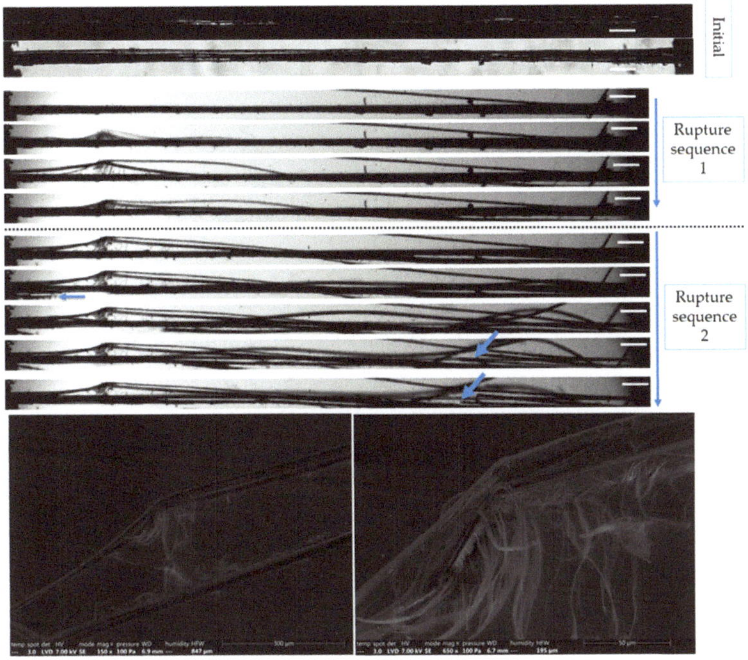

Figure 10. Top: Images of the initial stage observed by optical microscopy under both bright and polarized light, and a zoom on the two consecutive failure sequence of a flax bundle b-1-37, observed during in situ tensile testing thanks to a high-speed camera. Bottom: SEM images taken along the bundle after complete failure. The scale bars represent 200 μm, unless specified, and the blue arrows indicate transverse failures of unitary fibres.

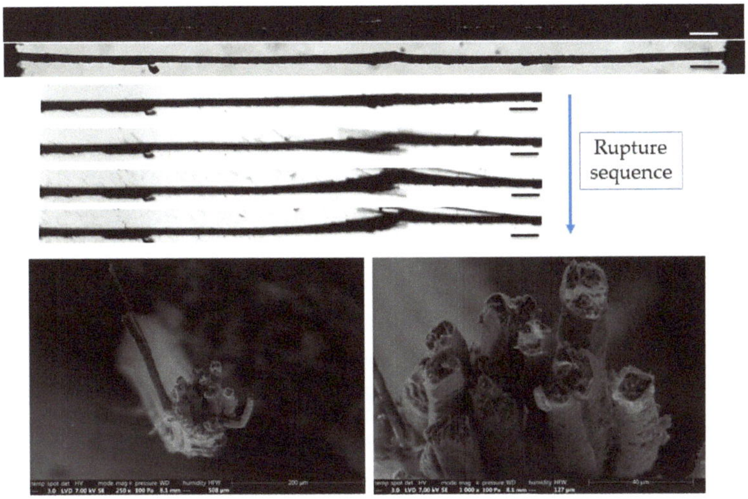

Figure 11. Top: Images of the initial stage observed by optical microscopy under both bright and polarized light, and a zoom on the failure sequence of a flax bundle b-1-41, observed during in situ tensile testing thanks to a high-speed camera. Bottom: SEM fracture surface of one extremity after complete failure. The scalebars represent 200 μm, unless specified.

4. Conclusions

High-speed camera observations during tensile testing revealed the fastest purely transverse failure of some unitary fibres, and a combination of transverse and longitudinal propagation in others, resulting in a delayed failure. Polarized light microscopy revealed a possible involvement of defects, and bright light observations highlighted the possible role of surface effects. However, the underlying mechanisms may be related to porosity, and need further investigation using more adequate tools, such as X-ray tomography. Indeed, surface observation can hardly reveal the extent of the lumen's role in driving or contributing to the failure at the unitary fibre scale, where transverse failure predominates.

The major limitation of the method is the accessibility to core deformation within the fibres, where, for instance, stress localisation close to lumen space cannot be captured. Another secondary limitation is lack of spatial and time resolutions to capture the cracking behaviour of unitary fibres along the transverse direction. The main advantage is the assessment of fast cracking involving surface defects, allowing revelation of intricate details under high magnification, and allowing the building of realistic scenarios of deformation sequences for a deeper understanding of crack deviation.

However, contrasted behaviours are observed at the bundle scale, linked to the additional complexity induced by the presence of numerous fibres connected by a middle lamella. Indeed, the high-speed camera showed a hierarchical failure by delamination between the fibres, followed by transverse failure of the unitary fibres in one case, and purely transverse failure of fibres close to a defected area in another case. The bulk effects can be again further explained by X-ray tomography experiments, which will be the subject of future work. The idea is to provide realistic scenarios of the transverse rupture for both compact and airy bundles, considering the role of the middle lamella.

Author Contributions: Conceptualization, J.B. and S.G.; Formal analysis, M.C., C.B. and S.G.; Investigation, E.R., M.C., P.O. and A.B.; Resources, J.B., S.G., A.B.; Writing—original draft, E.R. and S.G.; Writing—review & editing, E.R., J.B., M.C., C.B., P.O. and A.B.; Supervision, S.G.; Project administration, J.B.; Funding acquisition, A.B. All authors have read and agreed to the published version of the manuscript.

Funding: INTERREG VA FCE Program, FLOWER project, Grant Number 23.

Institutional Review Board Statement: Not applicable.

Data Availability Statement: Data Availability Statements are available on demand.

Acknowledgments: The authors would like to express their gratitude to Antoine Kervoelen from Université Bretagne Sud, Lorient, France for his technical assistance in preparing mechanical testing of fibre samples. The authors extend their gratitude to Depestele company, France for providing fibre samples. The authors thank Biogenouest (Western France life science and environment core facility network supported by the Conseil Régional des Pays de la Loire) and Angelina d'Orlando for supporting the SEM BIBS facility (INRAE, UR BIA, NANTES, France).

Conflicts of Interest: The authors declare no conflict of interest.

References

1. Rahman, M.Z. Mechanical and damping performances of flax fibre composites—A review. *Compos. Part C Open Access* **2021**, *4*, 100081. [CrossRef]
2. Mohanty, A.K.; Vivekanandhan, S.; Pin, J.-M.; Misra, M. Composites from renewable and sustainable resources: Challenges and innovations. *Science* **2018**, *362*, 536–542. [CrossRef] [PubMed]
3. Baley, C.; Bourmaud, A.; Davies, P. Eighty years of composites reinforced by flax fibres: A historical review. *Compos. Part A Appl. Sci. Manuf.* **2021**, *144*, 106333. [CrossRef]
4. Ray, F.; Evert, S.E.E. *Esau's Plant Anatomy: Meristems, Cells, and Tissues of the Plant Body: Their Structure, Function, and Development*, 3rd ed.; Wiley: Hoboken, NJ, USA, 2006; p. 624.
5. Gorshkova, T.; Brutch, N.; Chabbert, B.; Deyholos, M.; Hayashi, T.; Lev-Yadun, S.; Mellerowicz, E.J.; Morvan, C.; Neutelings, G.; Pilate, G. Plant Fiber Formation: State of the Art, Recent and Expected Progress, and Open Questions. *Crit. Rev. Plant Sci.* **2012**, *31*, 201–228. [CrossRef]

6. Morvan, C.; Andème-Onzighi, C.; Girault, R.; Himmelsbach, D.S.; Driouich, A.; Akin, D.E. Building flax fibres: More than one brick in the walls. *Plant Physiol. Biochem.* **2003**, *41*, 935–944. [CrossRef]
7. Park, Y.B.; Cosgrove, D.J. A revised architecture of primary cell walls based on biomechanical changes induced by substrate-specific endoglucanases. *Plant Physiol.* **2012**, *158*, 192880. [CrossRef]
8. Bourmaud, A.; Beaugrand, J.; Shah, D.U.; Placet, V.; Baley, C. Towards the design of high-performance plant fibre composites. *Prog. Mater. Sci.* **2018**, *97*, 347–408. [CrossRef]
9. Lucas, P.W.; Tan, H.T.W.; Cheng, P.Y. The toughness of secondary cell wall and woody tissue. *Philos. Trans. R. Soc. B-Biol. Sci.* **1997**, *352*, 341–352.
10. Gorshkova, T.A.; Gurjanov, O.P.; Mikshina, P.V.; Ibragimova, N.N.; Mokshina, N.E.; Salnikov, V.V.; Ageeva, M.V.; Amenitskii, S.I.; Chernova, T.E.; Chemikosova, S.B. Specific type of secondary cell wall formed by plant fibers. *Russ. J. Plant Physiol.* **2010**, *57*, 328–341. [CrossRef]
11. Thomason, J.L.; Carruthers, J.; Kelly, J.; Johnson, G. Fibre cross-section determination and variability in sisal and flax and its effects on fibre performance characterisation. *Compos. Sci. Technol.* **2011**, *71*, 1008–1015. [CrossRef]
12. Melelli, A.; Durand, S.; Alvarado, C.; Kervoëlen, A.; Foucat, L.; Grégoire, M.; Arnould, O.; Falourd, X.; Callebert, F.; Ouagne, P.; et al. Anticipating global warming effects: A comprehensive study of drought impact of both flax plants and fibres. *Ind. Crops Prod.* **2022**, *184*, 115011. [CrossRef]
13. Khadka, B.; Cloutier, S. Genetics of Abiotic Stress in Flax. In *The Flax Genome*; You, F.M., Fofana, B., Eds.; Springer International Publishing: Cham, Switzerland, 2023; pp. 101–120. [CrossRef]
14. Guo, D.; Jiang, H.; Ye, J.; Zhang, A.; Wang, Y.; Gao, Y.; Yan, Q.; Chen, J.; Duan, L.; Liu, H.; et al. Transcriptome combined with population level validation reveals genomic loci controlling plant height in flax (*Linum usitatissimum* L.). *Ind. Crops Prod.* **2021**, *172*, 113998. [CrossRef]
15. Chabi, M.; Goulas, E.; Leclercq, C.C.; de Waele, I.; Rihouey, C.; Cenci, U.; Day, A.; Blervacq, A.S.; Neutelings, G.; Duponchel, L.; et al. A Cell Wall Proteome and Targeted Cell Wall Analyses Provide Novel Information on Hemicellulose Metabolism in Flax. *Mol. Cell. Proteom.* **2017**, *16*, 1634–1651. [CrossRef]
16. Hernandez-Estrada, A.; Reza, M.; Hughes, M. The structure of dislocations in hemp (*Cannabis sativa* L.) fibres and implications for mechanical behaviour. *BioResources* **2020**, *15*, 2579–2595. [CrossRef]
17. Thygesen, L.G.; Gierlinger, N. The molecular structure within dislocations in *Cannabis sativa* fibres studied by polarised Raman microspectroscopy. *J. Struct. Biol.* **2013**, *182*, 219–225. [CrossRef]
18. Bourmaud, A.; Pinsard, L.; Guillou, E.; De Luycker, E.; Fazzini, M.; Perrin, J.; Weitkamp, T.; Ouagne, P. Elucidating the formation of structural defects in flax fibres through synchrotron X-ray phase-contrast microtomography. *Ind. Crops Prod.* **2022**, *184*, 115048. [CrossRef]
19. Madsen, B.; Aslan, M.; Lilholt, H. Fractographic observations of the microstructural characteristics of flax fibre composites. *Compos. Sci. Technol.* **2016**, *123*, 151–162. [CrossRef]
20. Hanninen, T.; Thygesen, A.; Mehmood, S.; Madsen, B.; Hughes, M. Mechanical processing of bast fibres: The occurrence of damage and its effect on fibre structure. *Ind. Crops Prod.* **2012**, *39*, 7–11. [CrossRef]
21. Madsen, B.; Thygesen, A.; Lilholt, H. Plant fibre composites—porosity and stiffness. *Compos. Sci. Technol.* **2009**, *69*, 1057–1069. [CrossRef]
22. Richely, E.; Nuez, L.; Pérez, J.; Rivard, C.; Baley, C.; Bourmaud, A.; Guessasma, S.; Beaugrand, J. Influence of defects on the tensile behaviour of flax fibres: Cellulose microfibrils evolution by synchrotron X-ray diffraction and finite element modelling. *Compos. Part C Open Access* **2022**, 100300.
23. Nilsson, T.; Gustafsson, P.J. Influence of dislocations and plasticity on the tensile behaviour of flax and hemp fibres. *Compos. Part A Appl. Sci. Manuf.* **2007**, *38*, 1722–1728. [CrossRef]
24. Fuentes, C.A.; Willekens, P.; Petit, J.; Thouminot, C.; Müssig, J.; Trindade, L.M.; Van Vuure, A.W. Effect of the middle lamella biochemical composition on the non-linear behaviour of technical fibres of hemp under tensile loading using strain mapping. *Compos. Part A Appl. Sci. Manuf.* **2017**, *101*, 529–542. [CrossRef]
25. Lawrence Sy, B.; Oguamanam, D.; Bougherara, H. Impact response of a new kevlar/flax/epoxy hybrid composite using infrared thermography and high-speed imaging. *Compos. Struct.* **2022**, *280*, 114885. [CrossRef]
26. Charlet, K.; Béakou, A. Mechanical properties of interfaces within a flax bundle—Part I: Experimental analysis. *Int. J. Adhes. Adhes.* **2011**, *31*, 875–881. [CrossRef]
27. Puech, L.; Ram Ramakrishnan, K.; Le Moigne, N.; Corn, S.; Slangen, P.R.; Le Duc, A.; Boudhani, H.; Bergeret, A. Investigating the impact behaviour of short hemp fibres reinforced polypropylene biocomposites through high speed imaging and finite element modelling. *Compos. Part A Appl. Sci. Manuf.* **2018**, *109*, 428–439. [CrossRef]
28. Romhany, G.; Karger-Kocsis, J.; Czigany, T. Tensile Fracture and Failure Behavior of Technical Flax Fibers. *J. Appl. Polym. Sci.* **2003**, *90*, 3638–3645.
29. Thygesen, A.; Madsen, B.; Bjerre, A.B.; Lilholt, H. Cellulosic Fibers: Effect of Processing on Fiber Bundle Strength. *J. Nat. Fibers* **2011**, *8*, 161–175. [CrossRef]
30. Barbulée, A.; Jernot, J.-P.; Bréard, J.; Gomina, M. Damage to flax fibre slivers under monotonic uniaxial tensile loading. *Compos. Part A Appl. Sci. Manuf.* **2014**, *64*, 107–114. [CrossRef]

31. Ahmed, S.; Ulven, C. Dynamic In-Situ Observation on the Failure Mechanism of Flax Fiber through Scanning Electron Microscopy. *Fibers* **2018**, *6*, 17. [CrossRef]
32. Mott, L.; Shaler, S.M.; Groom, L.; Liang, B. The tensile testing of individual wood fibers using environmental scanning electron microscopy and video image analysis. *Tappi J.* **1995**, *78*, 143–148.
33. Baley, C. Influence of kink bands on the tensile strength of flax fibers. *J. Mater. Sci.* **2004**, *39*, 331–334. [CrossRef]
34. Aslan, M.; Chinga-Carrasco, G.; Sørensen, B.F.; Madsen, B. Strength variability of single flax fibres. *J. Mater. Sci.* **2011**, *46*, 6344–6354. [CrossRef]
35. Beaugrand, J.; Guessasma, S. Scenarios of crack propagation in bast fibers: Combining experimental and finite element approaches. *Compos. Struct.* **2015**, *133*, 667–678. [CrossRef]
36. Silva, F.d.A.; Zhu, D.; Mobasher, B.; Soranakom, C.; Toledo Filho, R.D. High speed tensile behavior of sisal fiber cement composites. *Mater. Sci. Eng. A* **2010**, *527*, 544–552. [CrossRef]
37. Blakeney, A.B.; Harris, P.J.; Henry, R.J.; Stone, B.A. A simple and rapid preparation of alditol acetates for monosaccharide analysis. *Carbohydr. Res.* **1983**, *113*, 291–299. [CrossRef]
38. Thibault, J.F. Automatisation du dosage des substances pectiques par la methode au meta-hydroxydiphenyl. *Lebensm.-Wiss. Technol. Food Sci. Technol.* **1979**, *12*, 247–251.
39. Hatfield, R.; Fukushima, R.S. Can Lignin Be Accurately Measured? *Crop Sci.* **2005**, *45*, 832–839. [CrossRef]
40. Mariotti, F.; Tomé, D.; Mirand, P. Converting Nitrogen into Protein—Beyond 6.25 and Jones' Factors. *Crit. Rev. Food Sci. Nutr.* **2008**, *48*, 177–184.
41. Turek, D.E. On the tensile testing of high modulus polymers and the compliance correction. *Polym. Eng. Sci.* **1993**, *33*, 328–333.
42. Richely, E.; Bourmaud, A.; Placet, V.; Guessasma, S.; Beaugrand, J. A critical review of the ultrastructure, mechanics and modelling of flax fibres and their defects. *Prog. Mater. Sci.* **2021**, *124*, 100851. [CrossRef]
43. Roach, M.J.; Mokshina, N.Y.; Badhan, A.; Snegireva, A.V.; Hobson, N.; Deyholos, M.K.; Gorshkova, T.A. Development of cellulosic secondary walls in flax fibers requires beta-galactosidase. *Plant Physiol.* **2011**, *156*, 1351–1363. [CrossRef]
44. Chemikosova, S.B.; Pavlencheva, N.V.; Gur'yanov, O.P.; Gorshkova, T.A. The effect of soil drought on the phloem fiber development in long-fiber flax. *Russ. J. Plant Physiol.* **2006**, *53*, 656–662. [CrossRef]
45. Gautreau, M.; Durand, S.; Paturel, A.; Le Gall, S.; Foucat, L.; Falourd, X.; Novales, B.; Ralet, M.-C.; Chevallier, S.; Kervoelen, A.; et al. Impact of cell wall non-cellulosic and cellulosic polymers on the mechanical properties of flax fibre bundles. *Carbohydr. Polym.* **2022**, *291*, 119599. [CrossRef]
46. Faruk, O.; Bledzki, A.K.; Fink, H.-P.; Sain, M. Biocomposites reinforced with natural fibers: 2000–2010. *Prog. Polym. Sci.* **2012**, *37*, 1552–1596. [CrossRef]
47. Frei, M. Lignin: Characterization of a multifaceted crop component. *Sci. World J.* **2013**, *2013*, 436517. [CrossRef]
48. Cabane, M.; Afif, D.; Hawkins, S. Lignins and abiotic stresses. In *Advances in Botanical Research*; Elsevier: Amsterdam, The Netherlands, 2012; Volume 61, pp. 219–262.
49. Akin, D.E.; Gamble, G.R.; Morrison, W.H.; Rigsby, L. Chemical and Structural Analysis of Fibre and Core Tissues from Flax. *J. Sci. Food Agric.* **1996**, *72*, 155–165. [CrossRef]
50. Gorshkova, T. Composition and Distribution of Cell Wall Phenolic Compounds in Flax (*Linum usitatissimum* L.) Stem Tissues. *Ann. Bot.* **2000**, *85*, 477–486. [CrossRef]
51. Corbin, C.; Drouet, S.; Markulin, L.; Auguin, D.; Lainé, É.; Davin, L.B.; Cort, J.R.; Lewis, N.G.; Hano, C. A genome-wide analysis of the flax (*Linum usitatissimum* L.) dirigent protein family: From gene identification and evolution to differential regulation. *Plant Mol. Biol.* **2018**, *97*, 73–101.
52. Baley, C.; Bourmaud, A. Average tensile properties of French elementary flax fibers. *Mater. Lett.* **2014**, *122*, 159–161. [CrossRef]
53. Griffith, A.A. The phenomena of rupture and flow in solids. *Philos. Trans. R. Soc. Lond.* **1921**, *221*, 163–198.
54. Alix, S.; Lebrun, L.; Marais, S.; Philippe, E.; Bourmaud, A.; Baley, C.; Morvan, C. Pectinase treatments on technical fibres of flax: Effects on water sorption and mechanical properties. *Carbohydr. Polym.* **2012**, *87*, 177–185. [CrossRef] [PubMed]
55. Barbulee, A.; Gomina, M. Variability of the mechanical properties among flax fiber bundles and strands. *Procedia Eng.* **2017**, *200*, 487–493.
56. Garat, W.; Corn, S.; Le Moigne, N.; Beaugrand, J.; Bergeret, A. Analysis of the morphometric variations in natural fibres by automated laser scanning: Towards an efficient and reliable assessment of the cross-sectional area. *Compos. Part A Appl. Sci. Manuf.* **2018**, *108*, 114–123. [CrossRef]
57. Summerscales, J.; Virk, A.S.; Hall, W. Fibre area correction factors (FACF) for the extended rules-of-mixtures for natural fibre reinforced composites. *Mater. Today Proc.* **2020**, *31*, S318–S320. [CrossRef]
58. Beaugrand, J.; Guessasma, S.; Maigret, J.E. Damage mechanisms in defected natural fibers. *Sci. Rep.* **2017**, *7*, 14041. [CrossRef]
59. Richely, E.; Durand, S.; Melelli, A.; Kao, A.; Magueresse, A.; Dhakal, H.; Gorshkova, T.; Callebert, F.; Bourmaud, A.; Beaugrand, J.; et al. Novel Insight into the Intricate Shape of Flax Fibre Lumen. *Fibers* **2021**, *9*, 24. [CrossRef]
60. Melelli, A.; Jamme, F.; Legland, D.; Beaugrand, J.; Bourmaud, A. Microfibril angle of elementary flax fibres investigated with polarised second harmonic generation microscopy. *Ind. Crops Prod.* **2020**, *156*, 112847. [CrossRef]
61. Péron, M.; Célino, A.; Castro, M.; Jacquemin, F.; Le Duigou, A. Study of hygroscopic stresses in asymmetric biocomposite laminates. *Compos. Sci. Technol.* **2019**, *169*, 7–15. [CrossRef]

62. Bos, H. The Potential of Flax Fibres as Reinforcement for Composite Materials. Ph.D. Thesis, Technische Universiteit Eindhoven, Eindhoven, The Netherlands, 2004.
63. Lu, M.M.; Fuentes, C.A.; Van Vuure, A.W. Moisture sorption and swelling of flax fibre and flax fibre composites. *Compos. Part B Eng.* **2021**, *231*, 109538. [CrossRef]
64. Moudood, A.; Rahman, A.; Öchsner, A.; Islam, M.; Francucci, G. Flax fiber and its composites: An overview of water and moisture absorption impact on their performance. *J. Reinf. Plast. Compos.* **2018**, *38*, 323–339. [CrossRef]
65. El Hachem, Z.; Célino, A.; Challita, G.; Moya, M.-J.; Fréour, S. Hygroscopic multi-scale behavior of polypropylene matrix reinforced with flax fibers. *Ind. Crops Prod.* **2019**, *140*, 111634. [CrossRef]

polymers

MDPI

Article

Silk Fibroin-*g*-Polyaniline Platform for the Design of Biocompatible-Electroactive Substrate

Elsa Veronica Flores-Vela [1], Alain Salvador Conejo-Dávila [1,*], Claudia Alejandra Hernández-Escobar [1], Rocio Berenice Dominguez [2], David Chávez-Flores [3], Lillian V. Tapia-Lopez [1], Claudia Piñon-Balderrama [1], Anayansi Estrada-Monje [4], María Antonia Luna-Velasco [1], Velia Carolina Osuna [2] and Erasto Armando Zaragoza-Contreras [1,*]

1 Centro de Investigación en Materiales Avanzados, SC, Miguel de Cervantes No. 120, Complejo Industrial Chihuahua, Chihuahua 31136, Mexico
2 Consejo Nacional de Ciencia y Tecnología CONACYT-Centro de Investigación en Materiales Avanzados, SC, CIMAV, Miguel de Cervantes 120, Complejo Industrial Chihuahua, Chihuahua 31136, Mexico
3 Facultad de Ciencias Químicas, Universidad Autónoma de Chihuahua, Chihuahua 31125, Mexico
4 Centro de Innovación Aplicada en Tecnologías Competitivas, A.C. Calle Omega No. 201, Industrial Delta, León 37545, Mexico
* Correspondence: alain.conejo@cimav.edu.mx (A.S.C.-D.); armando.zaragoza@cimav.edu.mx (E.A.Z.-C.)

check for updates

Citation: Flores-Vela, E.V.; Conejo-Dávila, A.S.; Hernández-Escobar, C.A.; Dominguez, R.B.; Chávez-Flores, D.; Tapia-Lopez, L.V.; Piñon-Balderrama, C.; Estrada-Monje, A.; Luna-Velasco, M.A.; Osuna, V.C.; et al. Silk Fibroin-*g*-Polyaniline Platform for the Design of Biocompatible-Electroactive Substrate. *Polymers* **2022**, *14*, 4653. https://doi.org/10.3390/polym14214653

Academic Editors: Antonio M. Borrero-López, Concepción Valencia-Barragán, Esperanza Cortés Triviño, Adrián Tenorio-Alfonso and Clara Delgado-Sánchez

Received: 30 September 2022
Accepted: 28 October 2022
Published: 1 November 2022

Abstract: The structural modification of biopolymers is a current strategy to develop materials with biomedical applications. Silk fibroin is a natural fiber derived from a protein produced by the silkworm (*Bombyx mori*) with biocompatible characteristics and excellent mechanical properties. This research reports the structural modification of silk fibroin by incorporating polyaniline chain grafts through a one-pot process (esterification reaction/oxidative polymerization). The structural characterization was achieved by ^{1}H-NMR and FT-IR. The morphology was studied by scanning electron microscopy and complemented with thermogravimetric analysis to understand the effect of the thermal stability at each step of the modification. Different fibroin silk (Fib): polyaniline (PAni) mass ratios were evaluated. From this evaluation, it was found that a Fib to PAni ratio of at least 1 to 0.5 is required to produce electroactive polyaniline, as observed by UV-vis and CV. Notably, all the fibroin-*g*-PAni systems present low cytotoxicity, making them promising systems for developing biocompatible electrochemical sensors.

Keywords: silk fibroin; polyaniline; electroactive system; cytotoxicity

1. Introduction

Conducting polymers have contributed significantly to biomedical applications [1] because their properties change depending on the pH and temperature, the presence of an oxidizing or reducing agent, and the acidic or basic environment [2]. Polyaniline (PAni), in particular, has shown essential roles in biomedical applications, such as scaffolds for cell reproduction [3]. However, limited biocompatibility and the absence of biodegradability are drawbacks that restrict its use in applications in the field of biomedical materials and biosensors [4]. Consequently, the design of electroconductive-electroactive polymeric materials, especially copolymer systems [5] and graft copolymers [6], is a strategy for solving such disadvantages. The intrinsic properties are different according to the architecture, e.g., random copolymers, their conductive characteristics, and the dedoping point could be modified, depending on the specific requirements [7]. On the other hand, the graft copolymers and block copolymers architectures do not change the main chain conjugation; however, they can increase colloidal stability, surface adherence, and in general, the processability of the systems [8].

On the other hand, the silk fibroin (Fib) is the base protein of silkworm (*Bombyx mori*) cocoons, consisting of the following amino acid sequence: glycine-alanine-glycine-alanine-glycine-serine (GAGAGS). Fib is a biocompatible material with exceptional properties [9];

for example, high strength, slow degradation, and water-processable [10]. Furthermore, grafting polymers from Fib has been a strategy for developing biocompatible materials. For example, Boonpavanitchakul et al. reported that the development of Fib-*g*-PLA copolymers with an increase in processability due to Fib reduces the crystallinity of PLA [11]. In addition, Nong et al. found a methodology for coloring silk through graft polymerization of acrylamides utilizing Lacasse as a catalyzer [12]. Lastly, Zhou et al. studied the Fib-*g*-polyacrylic acid, a graft copolymer, as a scaffold for biomimetic mineralization of Ca/P solutions [13]. Therefore, Fib is a material with potential and value in biomedical applications [14].

The synergy of Fib with conductive polymers allows the design of biocompatible and biodegradable systems with electroactive and electroconductive properties, which have promising potential in the regeneration of electrically-active tissues such as nerve fibers [15]. Scaffolds for tissue engineering with electrical properties and high mechanical properties with nerve regeneration potential have also been studied [16]. The development of biocompatible electrodes is another field of interest, in which is possible to monitor electrophysiological tests, e.g., electrocardiograms or electroencephalograms [17]. In addition, Fib-PAni copolymers with electroconductive properties were used as resistive sensors for ammonium and acetaldehyde with remarkable reversibility [18]. Similarly, Hong et al. reported an outstanding electrical conductivity (0.6 S/cm) for a Fib copolymer in the form of yarn [19].

Moreover, electrochromic-electroactive properties have also been studied in electronic textiles based on Fib-PAni fibers obtained by electrospinning, allowing the design of materials with environmentally friendly and biocompatible characteristics [20]. Likewise, Li et al. reported filaments of a Fib-PAni copolymer obtained through the wet-spinning technique, using an ionic liquid obtained from shell wastes and formic acid, as the solvent. It was found that the electrical conductivity of the filaments depends on the PAni content [21].

In this work, the synthesis of Fib-*g*-PAni copolymers from a one-pot methodology is reported, consisting of consecutive Fischer esterification and oxidative polymerization. Different characterization techniques were employed to analyze in detail the esterification reaction between Fib and aminobenzoic acid, and grafting to polyaniline chains by oxidative polymerization. This copolymer is proposed as a biocompatible-electroactive platform for designing electrochemical sensors that could be in contact with living tissue. This first manuscript describes the method of synthesis and characterization of the copolymer, its electrochemical behavior, and a first approach to cytotoxicity studies.

2. Materials and Methods

2.1. Materials

Silkworm cocoons, sodium carbonate (Merk, St. Louis, MO, USA, >98%), nitric acid (Fermont, Monterrey, Nuevo Leon, Mexico), dimethylsulfoxide (DMSO) (Merk, St. Louis, MO, USA, >99%), 3-aminobenzoic acid (Merk, St. Louis, MO, USA, >98%), aniline (Merk, St. Louis, MO, USA, >99.5%), sulfuric acid (Merk, St. Louis, MO, USA, >98%), ammonium persulfate (APS) (Merk, St. Louis, MO, USA, 98%), triple distilled water.

2.2. Silk Degumming

When working with Fib, it is essential first to remove sericin because it covers the fibroin structure, precluding surface interactions with other compounds. The methodology commenced by cutting the cocoons into fragments of around 1 cm^2. The degumming process was carried out in a reflux system with magnetic stirring. The cut cocoons (1 g) and 50 mL of sodium carbonate aqueous solution (20 mM) were placed inside the reflux system. The mixture was heated to boiling point under continuous stirring for 2 h. Subsequently, the liquid was withdrawn and replaced with another 50 mL of the sodium carbonate solution, followed by another 2 h under the same conditions. Finally, the product was filtered and washed with tridistilled water. The Fib obtained was dried at room temperature [22].

2.3. Synthesis of the Fibroin-g-Polyaniline Copolymer

The fibroin-*g*-polyaniline (Fib-*g*-PAni) copolymer was prepared through a one-pot methodology. Fischer esterification and oxidative polymerization were carried out consecutively. Figure 1 illustrates the reaction scheme and the proposed model of Fib-*g*-PAni.

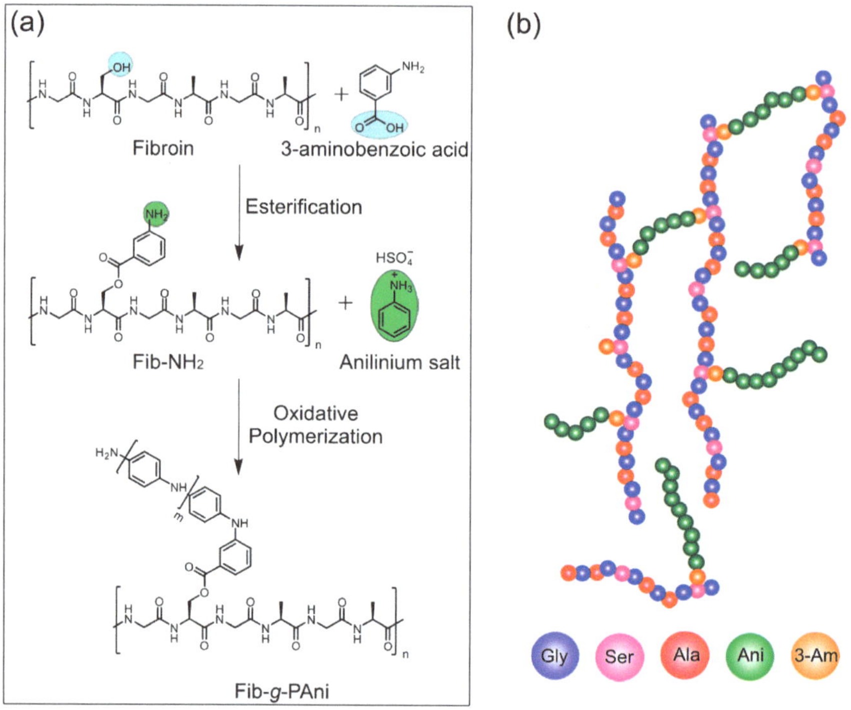

Figure 1. (**a**) Consecutive reactions for copolymer synthesis fibroin-*g*-polyaniline (Fib-*g*-PAni), and (**b**) proposed model of Fib-*g*-PAni.

2.3.1. Fischer Esterification

The organic reaction was performed in a reactor provided with temperature control and magnetic stirring. DMSO (10 mL), Fib (1 g), and nitric acid (1 mL) were then added and stirred for 1 h at laboratory temperature. Next, 3-aminobenzoic acid (150 mg, 1.1 mmol) was added dropwise, and then the reactor was heated to 70 °C, maintaining magnetic stirring for 6 h. The product was called Fib-NH_2.

It is worth saying that to carry out the characterization, Fib-NH_2 was purified by dialysis in water (2 days) to remove the DMSO. Afterward, the compound was purified through a chromatographic column using an ethanol: ethyl ether (1:9) mixture as the mobile phase. After purification, the Fib-NH_2 was an orange solid, and the reaction presented a yield of 35%.

2.3.2. Oxidative Polymerization

The Fib-NH_2 solution (without purification) was cooled to room temperature, and the anilinium salt (polyaniline precursor monomer) was added. Table 1 reports the experimental systems. APS was added in a molar ratio of 5:4, regarding the anilinium sulfate monomer. Both solutions were stirred separately and subsequently mixed. The polymerization was carried out at 60 °C for 6 h. After polymerization, the product was purified by three precipitation–centrifugation cycles. The precipitation was performed with 100 mL of sulfuric acid aqueous solution (H_2SO_4, 3 wt%) to prevent the dedoping effect. Finally, the product was washed with the same acid solution and dried at room temperature. The Fib-*g*-PAni was solid with a dark green color.

Table 1. Formulations employed for fibroin-*g*-PAni (Fib-*g*-PAni).

Sample	Fib (g)	3-Aminobenzoic Acid/g (mmol)	Anilinium Sulfate/g (mmol)	APS/g (mmol)	Yield/%
Fib-*g*-Pani 1:0.1	1	0.15 (1.1)	0.10 (0.523)	0.149 (0.654)	15%
Fib-*g*-Pani 1:0.25	1	0.15 (1.1)	0.25 (1.307)	0.373 (1.634)	45%
Fib-*g*-Pani 1:0.5	1	0.15 (1.1)	0.50 (2.615)	0.746 (3.269)	87%
Fib-*g*-Pani 1:1	1	0.15 (1.1)	1.00 (5.230)	1.492 (6.537)	95%

2.4. PAni Polymerization

The neat PAni homopolymer was synthesized in a flask equipped with a stirring bar. Firstly, 100 mL of H_2SO_4 (1 M) and aniline (10.737 mmol) were added. Once the anilinium salt was formed, the oxidizing initiator APS (13.42 mmol) was added. The solution was cooled at 4 °C for 24 h. The homopolymer, a dark green suspension, was purified by centrifugation at 5000 rpm. The solid was redispersed with H_2SO_4 (0.1 M) to prevent the dedoping process for three cycles. The yield of polymerization was 75%.

2.5. Cytotoxicity Assays on Fib-g-PAni

The cytotoxicity of the Fib-*g*-PAni 1:1, Fib-*g*-PAni 1:0.5, Fib-NH_2, and PAni was evaluated in NIH/3T3 fibroblast cells (ATCC® CRL-1658™) through the MTT (methyl thiazole tetrazolium) assay. Just prior to the assay, the powdered copolymer and instruments were sterilized under UV light for 25 min (5 cycles of 5 min). Tested compounds were then dispersed in a culture media to reach 0.1, 0.5, 2, and 4 mg/mL of the Fib-*g*-PAni copolymers. For Fib-NH_2, the contents were 0.05, 0.33, 1, and 2 mg/mL and for PAni 0.03, 0.25, 1, and 2 mg/mL, respectively. For the assay, cells were cultured in a DMEM-F12 HAM media supplemented with 5% fetal bovine serum and incubated at 37 °C with 90% humidified/5% CO_2/95% air. The media was replaced every 24–48 h until 80–90% confluence was reached for seeding treatments. NIH/3T3 cells were seeded in 96-well microplates at a density of 2×10^4 cells/well (100 µL supplemented media) and incubated under identical conditions for 24 h. Old media was then replaced with 100 µL of treatment dilutions. A dilution of 0.4% SDS was parallel assayed as a positive control and cells not treated were used as reference controls. All treatments were carried out in triplicate, and the microplate was incubated for 24 h. After incubation, all media was removed, and cells were gently washed with PBS at pH 7.4. A total of 100 µL MTT of culture media (0.5 mg/mL) was then added to each well. The microplate was incubated for another 4 h. The media/MTT was removed, and 100 µL of acidified isopropanol was added to each well to dissolve the MTT metabolized by the cells. Finally, the plate was read at an absorbance of 570 nm using a microplate reader (Varioskan Lux VLBLATD2, Thermo Scientific, Waltham, MA, USA). Cell viability relative to untreated cells (reference control) was determined as follows.

$$\text{Cell viability (\%)} = [(\text{O.D. Test})/(\text{O.D. Reference control})] \times 100 \quad (1)$$

where O.D. (Test) and O.D. (Reference) represent the mean absorbance of the treated cells and the absorbance of the reference cells, respectively.

2.6. Characterization

The structural characterization of the products was performed by ^{1}H-NMR spectroscopy, using the spectrophotometer (NMR Bruker Ascend 400 MHz, Billerica, MA, USA) at 400 MHz, and DMSO-d_6 as the solvent. To solubilize Fib in DMSO-d_6, adding 10 µL of nitric acid and sonication was necessary. The structural characterization was complemented with infrared spectroscopy (FT-IR) using a Fourier transform spectrophotometer (GX-FTIR, Perkin Elmer, Waltham, MA, USA). Spectra were obtained using the attenuated total reflectance (ATR) technique. The thermal stability of the copolymers was studied by thermogravimetric analysis (SDR Q600, TA Instruments, New Castle, DE, USA) with a sensitivity of 0.1 μg. The samples were heated from room temperature to 700 °C with a heating ramp of 10 °C min^{-1} in an air atmosphere.

Additionally, the absorption of all compounds in the UV and visible range was studied using a UV-VIS-NIR spectrometer (Evolution 220, Thermo Scientific, Waltham, MA, USA). In addition, a dispersion of the samples (1 mg) in 5 mL of distilled water and then sonicated 10 min, was utilized for analysis. The morphology of the Fib-*g*-PAni was observed using a field emission scanning electron microscope (JSM-7401F, Jeol Ltd., Akishima, Tokyo, Japan), operating in emission mode (FE-SEM).

The electroactivity properties were studied through a three-electrode system, employing as a reference electrode a Ag/AgCl electrode, a platinum plate (1 cm^2) as the counter electrode, a glassy carbon electrode acting as a working electrode, and a potentiostat (Emstat3+ blue, PalmSense BV, Houten, The Netherlands). The electrochemical properties were explored by cyclic voltammetry with a potential window from −0.5 to 1.0 V, and the experiments were realized with H_2SO_4 (1 M) as an electrolyte.

3. Results and Discussion

The structural modification of Fib was developed through a Fisher esterification reaction between the 3-aminobenzoic acid and the serine present in Fib. The typical molar content of serine is 12.2% in fibroin [23]. Subsequently, the PAni chains were grafted, employing oxidative polymerization varying the ratio of the anilinium salt monomer and Fib-NH_2, synthesizing four Fib-*g*-PAni copolymers.

3.1. Chemical Characterization

The structural characterization of Fib-*g*-PAni and its predecessor compounds was performed employing ^{1}H-NMR and FT-IR spectroscopy. Figure 2 displays the ^{1}H-NMR spectra of Fib, Fib-NH_2, and Fib-*g*-PAni 1:0.1 (this copolymer was selected because the analogs with higher PAni contents are not soluble). The three spectra present the characteristic signals of glycine and serine at 3.4 ppm, and also the signal at 2.10 ppm corresponding to alanine of the fibroin structure [24]. The Fib-NH_2 exhibits a new band at 8.3 ppm, corresponding to aromatic protons of the 3-aminobenzoate substituent product of Fisher esterification. Finally, the spectrum of Fib-*g*-PAni 1:0.1 displays typical signals between 7 and 8 ppm, corresponding to oligomers or PAni chains, and a new band appears around 3.1 ppm assigned to serine segments grafted with PAni chains.

The FT-IR spectroscopy complemented the results of ^{1}H-NMR. The degummed Fib presents the bands reported in the literature [18], see Figure 3. The spectrum shows signals at 1611 and 1504 cm^{-1} corresponding to the stretching vibration of the carbonyl group (C=O) and the bending vibration of the N–H bond of a secondary amine. Both signals are characteristic of the amide functional group in the Fib. In addition, the stretching vibration of the C–O bond of the alcohol group of serine and tyrosine is observed at 1234 cm^{-1} [25]. Compared with the Fib, the spectrum of Fib-NH_2 has two new peaks. Specifically, the stretching vibration corresponding to the ester group (C=O) at 1724 cm^{-1}, and the peak at 1595 cm^{-1} is assigned to the C=C bond of the aromatic ring of benzoate. The amine (N–H) vibration shifts to 1515 cm^{-1}. Therefore, Fischer esterification between Fib and 3-amino benzoic acid was confirmed.

he spectrum of Fib-*g*-PAni 1:0.5 shows the characteristic signals of the PAni. For example, the vibrations corresponding to the quinoid and benzenoid rings were observed at 1588 and 1494 cm^{-1}, respectively. A typical vibration of oxidative polymerization corresponding to C–N appears at 1304 cm^{-1}. As noted, the peak height of the benzenoid structure is greater than the band of the quinoid structure, so it can be deduced that the PAni is in the emeraldine oxidation state [26].

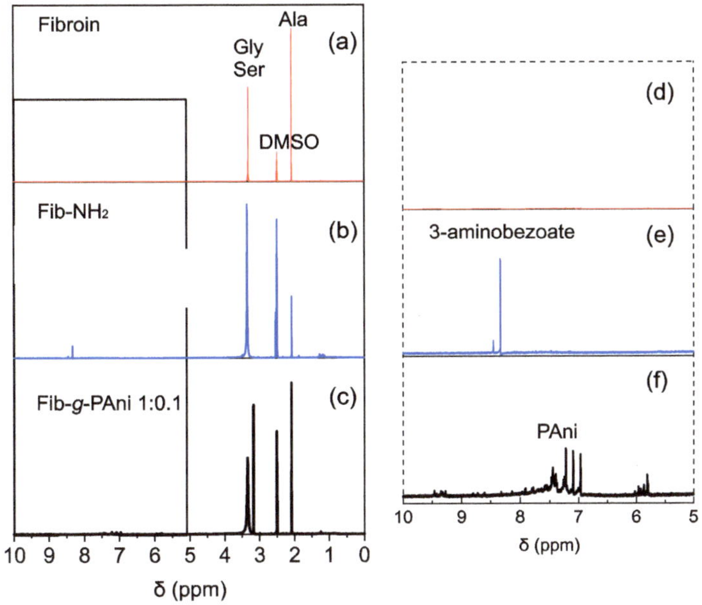

Figure 2. Proton NMR spectra (**a**–**c**) and zoom (**d**–**f**) from 5 to 10 ppm of these spectra. (**a**,**d**) Fibroin (Fib); (**b**,**e**) Fibroin esterified with aminobenzoate (Fib-NH_2); (**c**,**f**) Fibroin-*g*-polyaniline (Fib-*g*-PAni).

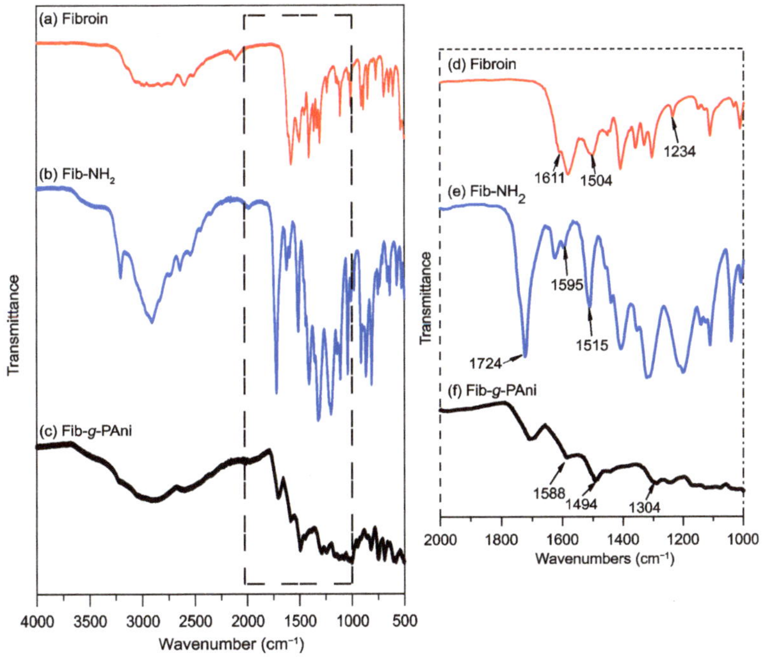

Figure 3. FT-IR spectra (**a**–**c**) and zoom (**d**–**f**) from 1000 to 2000 cm^{-1} of these spectra. (**a**,**d**) Fib, (**b**,**e**) Fib-NH_2, and (**c**,**f**) Fib-*g*-Pani 1:0.5.

3.2. Optical Activity

Absorption studies in the UV-Visible range were conducted to understand band changes due to Fisher esterification and the formation of benzenoid and quinoid bands (oxidative polymerization of PAni). The esterification reaction between aminobenzoic acid and Fib produces a band at 292 nm, assigned to the π–π^* transition of the aromatic ring of the aminobenzoate substituent. The esterification product was employed as grafting points for oxidative polymerization [27]. Nevertheless, the 292 nm band decreased after the polymerization process, due to the propagation of PAni chains and the change in the conjugation structure, see Figure 4a.

The Fib-*g*-PAni copolymers spectra with different mass ratios present different absorbances due to the formation of PAni chains, see Figure 4b. However, the Fib-*g*-PAni 1:0.5 and 1:1 spectra display the quinoid and benzenoid bands at 360 and 543 nm, respectively. The intensity and wavelength of absorption indicate the oxidation state of polyaniline (emeraldine base form) [28]. Therefore, a 2:1 ratio of Fib to aniline salt at least is required to produce PAni chains with a long enough degree of polymerization to produce an optical signal. At lower mass ratios only oligomers are produced [29]; nevertheless, they do not have the characteristic optical and electronic properties of PAni.

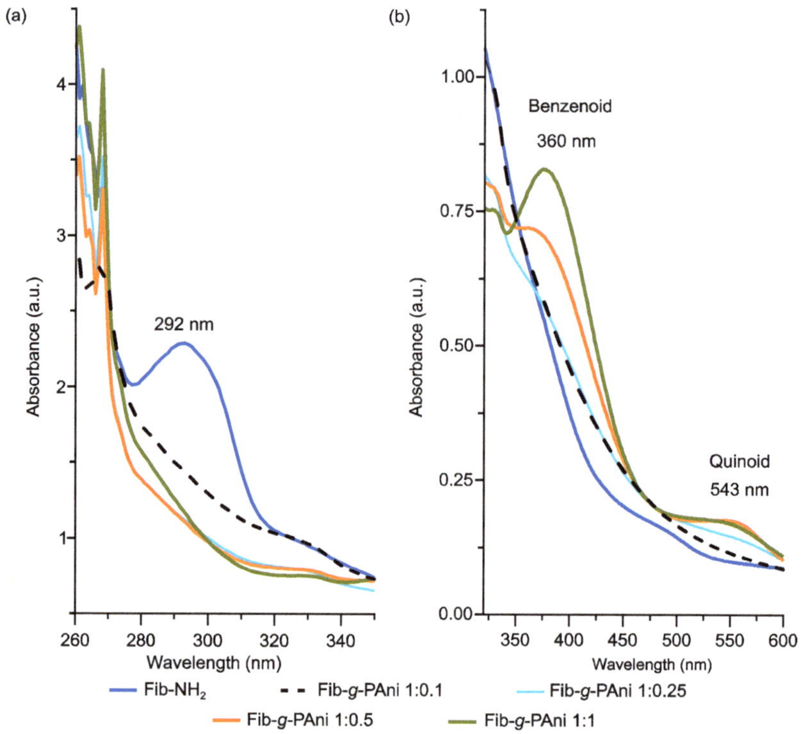

Figure 4. UV-Vis spectra of Fib-NH_2 and Fib-*g*-PAni copolymers (**a**) UV-Vis spectra from 260 to 350 nm and (**b**) UV-Vis spectra from 340 to 600 nm.

3.3. Thermal Stability

The thermal stability of the materials was determined by thermogravimetric analysis. Figure 5 displays the thermograms and first derivative (DTGA) of Fib, Fib-NH_2, and Fib-*g*-PAni 1:1 and Fib-*g*-PAni 1:0.5 copolymers. The Fib presents two stages of weight loss; the first is at 238 °C, and represents 70% of the sample weight. This stage is allocated to fragmented Fib chains produced by the degumming process [30]. The second step occurs at 530 °C, attributed to the Fib chain with a high molecular weight [31].

The thermal degradation of Fib-NH_2 occurs at two stages, similar to Fib. The first occurs at 202 °C and is assigned to fragments of Fib hydrolyzed during Fisher esterification, due to the acid-catalyst-produced hydrolysis (chain degradation). This stage represents 80% of the total weight loss. Furthermore, the second occurs at 638 °C, with 20% of weight loss, corresponding to the Fib which is less hydrolyzed, and esterified with the aniline analog.

Fib-*g*-PAni 1:0.5 traces present three stages. The first weight loss was observed at 179 °C, assigned to organic low molecular weight compounds such as solvent or byproduct derivatives of one-pot synthesis. The second occurs at 259 °C, with a 20% of sample weight loss, corresponding to Fib chains with low PAni grafting [32]. The third stage occurs at 532 °C, with 80% of weight loss. In contrast to its intermediates, Fib-*g*-PAni 1:0.5 presents the main weight loss at around 500 °C. Likewise, Fib-*g*-PAni 1:1 only presents one weight loss stage, at around 470 °C. These findings suggest the thermal degradation of PAni [33–35].

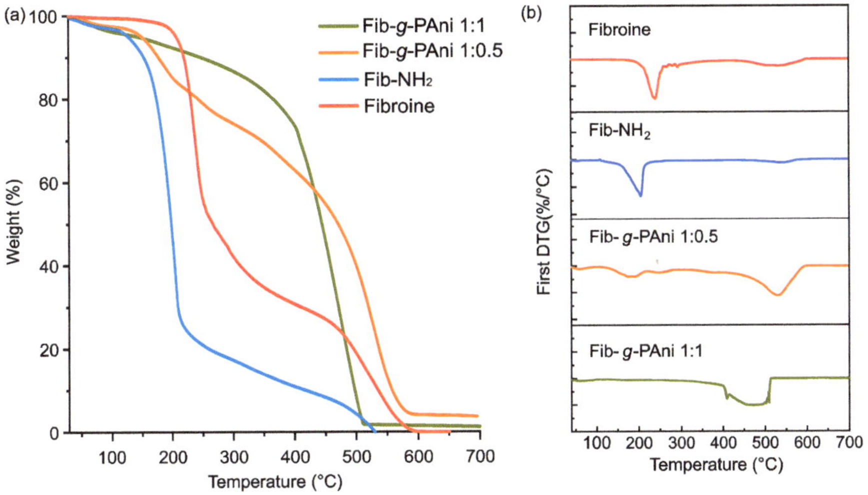

Figure 5. (**a**) Thermograms and (**b**) DTG of Fib, Fib-NH_2, Fib-*g*-PAni 1:0.5 and Fib-*g*-PAni 1:1.

3.4. Morphology

The morphological changes suffered by the Fib after each consecutive reaction were studied. Figure 6 illustrates the micrographs of degummed Fib, Fib-NH_2, Fib-*g*-PAni 1:0.5, and Fib-*g*-PAni 1:1. The edge of the fibroin exhibits the fibrous nature of the material, see Figure 6a. The magnification of the image (inner box) illustrates this structure more evidently [36]. A change in morphology is evident when comparing Fib and Fib-NH_2, see Figure 6b. In addition, the fibrous nature of the Fib disappears, suggesting that the incorporation of 3-aminobenzoic acid significantly alters the molecular arrangement [14].

Figure 6c,d correspond to the Fib-*g*-PAni samples with a mass ratio of 1:0.5 and 1:1, respectively. As noted, the PAni graft again produced important modifications on the Fib. The copolymers present particulate and globular structures for Fib-*g*-PAni 1:1 and Fib-*g*-PAni 1:0.5, respectively. The Fib-*g*-PAni 1:0.5 copolymer still retains a certain cohesion of the material; however, the aforementioned globular structure is perfectly observed under magnification. On the contrary, the Fib-*g*-PAni 1:1 copolymer has developed, to a greater extent, a particulate structure, and it is observed even in the magnification that the globular structures have, in turn, an internal structure.

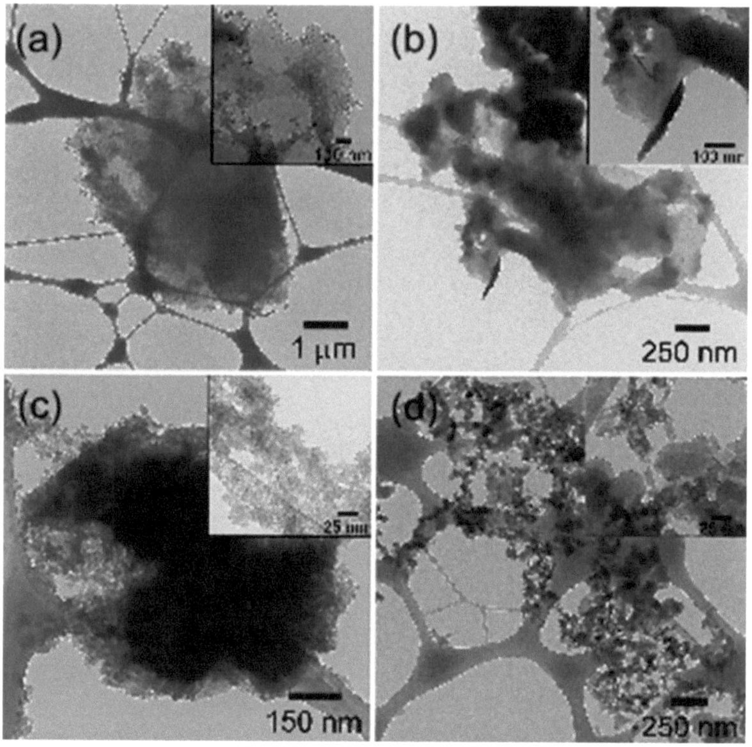

Figure 6. Micrographs of (**a**) Fib, (**b**) Fib-NH_2, (**c**) Fib-*g*-PAni 1:0.5 and (**d**) Fib-*g*-PAni 1:1.

3.5. Electroactivity

To corroborate the electrochemical activity of the graft copolymers, cyclic voltammetry tests were carried out. Figure 7a shows the voltammograms of the Fib-*g*-PAni copolymers run in 0.5 M sulfuric acid electrolyte. Both graphs display the oxidation and reduction signals that the polyanilines regularly exhibit [37]. For example, for Fib-*g*-PAni 1:1, the first peak corresponds to the leucoemeraldine-emeraldine transition at 0.31 V (vs. Ag/AgCl). With the mass ratio of 1:0.5, the signal appears at 0.2 V (vs. Ag/AgCl). This difference could be due to the interaction between PAni and Fib or the degree of polymerization of PAni [38]. The second signal, corresponding to the emeraldine–pernigraniline transition, appears at 0.73 V (vs. Ag/AgCl) in both cases. The location of the signals is evidence of the electroactivity of the graft copolymers, which corroborates the results of the UV-vis study.

It is important to note that there are differences in electroactive properties between a physical mixture and a graft copolymer architecture. Figure 7b compares a physical blend of PAni-Fib, and the copolymer graft (Fib-*g*-PAni 1:1). It is noted that the intensity of peaks in the physical blend is significantly higher than the grafted copolymer. Thus, its general behavior is similar to conventional PAni, which indicates that interaction between Fib and PAni is only physical, and that the PAni in the blend has a higher molecular weight.

It is worth noting that in the graft PAni copolymers, the conjugation present in the conductive chain is not interrupted, which does not modify the electrochemical prop-erties of PAni, which is different to other copolymer systems [8]. Furthermore, the electrochemical activity of PAni will depend on the polymerization process, dopant agent, and interaction with other polymers [38–40].

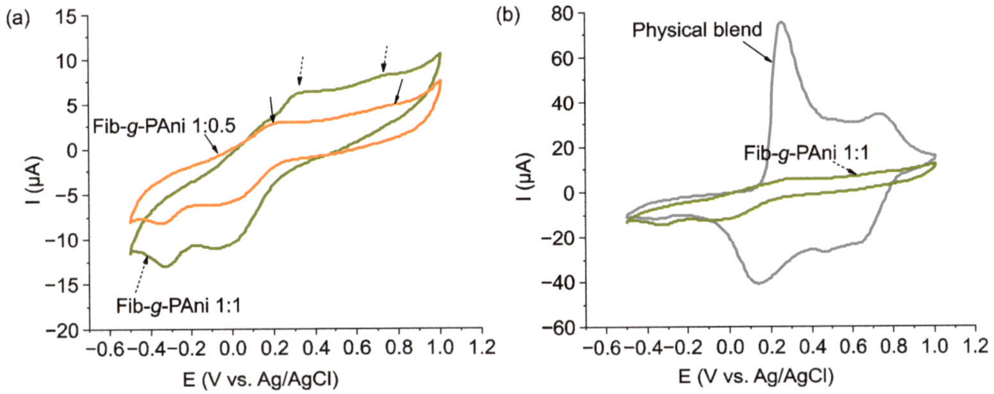

Figure 7. Voltamogramms of (**a**) Fib-*g*-PAni copolymers and (**b**) comparison between physical blends and graft copolymers.

3.6. Cytotoxicity

It is well known that Fib displays biocompatibility properties, slow biodegradability, and low cytotoxicity [41]. Nevertheless, these properties are not extrapolated to structural modification and composites derived from Fib. For this reason, the cytotoxicity of Fib-*g*-PAni (1:1 and 1:05), Fib-NH_2, and PAni were evaluated with NIH/3T3 fibroblast cells, measured at 24 h using the MTT assay [42], see Figure 8. The time for cell-tested viability was selected as the first approach. The use of fibroblasts is because they are the main active cells of connecting tissue, and are commonly used as models for viability/toxicity studies. In addition, the fibroblast presents an improved proliferation in the presence of Fib [43]. It is important to note that the concentrations used were established based on the content of PAni reported by previous research. For example, the PAni cytotoxicity results for fibroblast NIH/3T3 cells [44] and macrophage cells [45] are similar to previous works.

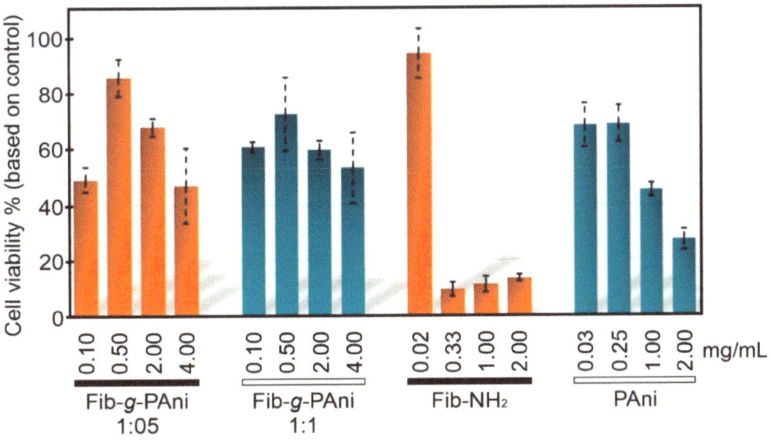

Figure 8. Cell viability of NIH/3T3 fibroblast cells treated with Fib-*g*-PAni (1:1 and 1:0.5), Fib-NH_2, and PAni, taking the viability of untreated cells as a reference control.

The PAni homopolymer presents content-dependent cytotoxicity to NIH/3T3 cells, noting a moderate to severe effect at the contents of 1 and 2 mg/mL, respectively. On the contrary, severe effects were noted at the PAni contents of 0.03 and 0.2 mg/mL. The differences in PAni toxicity could be associated with the composition and morphology of PAni. Fib-NH_2 presents a severe cytotoxicity event at 0.33 mg/mL, due to the substituted aniline, which could react with the cells.

Fib-*g*-PAni 1:1 and Fib-*g*-PAni 1:0.5 showed a similar pattern with content-dependent cytotoxicity, except at the lowest content (0.1 mg/mL), where moderate toxicity was noted. This result could be explained by the excellent dispersion of PAni under this condition, favoring a closer interaction with cells. From 0.5 to 2 mg/mL of the Fib-*g*-PAni copolymers, no toxicity or mild toxicity was noted, and at 4 mg/mL, moderate toxicity was observed. The low cytotoxic effect of the copolymers contrasted with the highly marked cytotoxic effect of Fib-NH_2 and PAni.

Generally, the Fib-*g*-PAni from 0.5 to 2 mg/mL could be considered biocompatible with cells. The Fib-*g*-PAni 1:1 presents higher cellular viability; these results are attributed to the fact that at lower contents, the interaction and probable internalization in cells is favored, which could be toxic for the cells; however, at higher contents, the molecules agglomerate, reducing the interaction with cells [46]. These results correspond to a comparable copolymer of Fib/PAni core/shell coaxial fiber-type, which exhibited good biocompatibility, cell attachment, and proliferation of L929 fibroblast cells [47]. Consequently, based on the results, the copolymers could be a promising material for biomedical applications.

4. Conclusions

Silk fibroin modification was performed through one-pot synthesis with 3-aminobenzoic acid esterification and subsequently with polyaniline grafting by oxidative polymerization. The one-pot methodology for grafting PAni chains onto fibroin structure provides a synthetic method that is easy to scale. The morphological study showed that each modification process changes the material structure, which was also reflected in the chemical structure and thermal stability. The silk fibroin:polyaniline mass ratio also showed a significant effect, mainly in optical properties and electroactivity. These differences are related to a higher degree of polymerization resulting from a higher polyaniline content, accentuating both properties. As a first approach, the cytotoxicity results for copolymers are promising; nevertheless, more experiments will be carried out, e.g., increasing the exposure time of fibroblasts to the copolymers, and increasing the concentration, among others. Based on the results of this research and the already known properties of fibroin, it can be expected that the graft copolymer is suitable as a biocompatible platform for developing electrochemical sensors that can be in contact with living tissue.

Author Contributions: Conceptualization, E.A.Z.-C. and A.S.C.-D.; methodology, E.V.F.-V., A.S.C.-D., M.A.L.-V. and C.A.H.-E.; software, R.B.D., C.P.-B. and D.C.-F.; validation, E.V.F.-V., A.S.C.-D., M.A.L.-V. and D.C.-F.; formal analysis, E.V.F.-V., V.C.O. and M.A.L.-V.; investigation, E.V.F.-V., A.S.C.-D., A.E.-M., C.A.H.-E. and L.V.T.-L.; resources, E.A.Z.-C.; data curation, A.S.C.-D., E.A.Z.-C., A.E.-M., R.B.D. and L.V.T.-L.; writing—original draft preparation, A.S.C.-D., E.A.Z.-C. and M.A.L.-V.; writing—review and editing, E.A.Z.-C. and V.C.O.; visualization, A.S.C.-D. and E.A.Z.-C.; supervision, E.A.Z.-C.; project administration, E.A.Z.-C.; funding acquisition, E.A.Z.-C. All authors have read and agreed to the published version of the manuscript.

Funding: This research was funded by the Centro de Investigación en Materiales Avanzados, SC (CIMAV), grant number CCDPI-09 and PI-22-03.

Institutional Review Board Statement: Not applicable.

Informed Consent Statement: Not applicable.

Data Availability Statement: Not applicable.

Acknowledgments: We wish to thank CONACYT for the scholarships awarded to Elsa Veronica Flores-Vela (1155684) and Alain Salvador Conejo-Dávila (627922). We also thank Daniel Lardizabal Gutiérrez, Raúl Armando Ochoa Gamboa, and Luis de la Torre Sáenz for their valuable support during the development of this research.

Conflicts of Interest: The authors declare no conflict of interest.

References

1. Kenry; Liu, B. Recent Advances in Biodegradable Conducting Polymers and Their Biomedical Applications. *Biomacromolecules* **2018**, *19*, 1783–1803. [CrossRef]
2. Nezakati, T.; Seifalian, A.; Tan, A.; Seifalian, A.M. Conductive Polymers: Opportunities and Challenges in Biomedical Applications. *Chem. Rev.* **2018**, *118*, 6766–6843. [CrossRef]
3. Zhang, J.; Qiu, K.; Sun, B.; Fang, J.; Zhang, K.; El-Hamshary, H.; Al-Deyab, S.S.; Mo, X. The aligned core–sheath nanofibers with electrical conductivity for neural tissue engineering. *J. Mater. Chem. B* **2014**, *2*, 7945–7954. [CrossRef]
4. Guo, B.; Glavas, L.; Albertsson, A.-C. Biodegradable and electrically conducting polymers for biomedical applications. *Prog. Polym. Sci.* **2013**, *38*, 1263–1286. [CrossRef]
5. AL-Oqla, F.M.; Sapuan, S.M.; Anwer, T.; Jawaid, M.; Hoque, M.E. Natural fiber reinforced conductive polymer composites as functional materials: A review. *Synth. Met.* **2015**, *206*, 42–54. [CrossRef]
6. Shahadat, M.; Khan, M.Z.; Rupani, P.F.; Embrandiri, A.; Sultana, S.; Ahammad, S.Z.; Wazed Ali, S.; Sreekrishnan, T.R. A critical review on the prospect of polyaniline-grafted biodegradable nanocomposite. *Adv. Colloid Interface Sci.* **2017**, *249*, 2–16. [CrossRef]
7. Ou, X.; Xu, X. A simple method to fabricate poly(aniline-co-pyrrole) with highly improved electrical conductivity via pre-polymerization. *RSC Adv.* **2016**, *6*, 13780–13785. [CrossRef]
8. Li, Y.; Zhou, M.; Xia, Z.; Gong, Q.; Liu, X.; Yang, Y.; Gao, Q. Facile preparation of polyaniline covalently grafted to isocyanate functionalized reduced graphene oxide nanocomposite for high performance flexible supercapacitors. *Colloids Surf. A Physicochem. Eng. Asp.* **2020**, *602*, 125172. [CrossRef]
9. Hardy, J.G.; Römer, L.M.; Scheibel, T.R. Polymeric materials based on silk proteins. *Polymer* **2008**, *49*, 4309–4327. [CrossRef]
10. Zare, E.N.; Makvandi, P.; Ashtari, B.; Rossi, F.; Motahari, A.; Perale, G. Progress in Conductive Polyaniline-Based Nanocomposites for Biomedical Applications: A Review. *J. Med. Chem.* **2020**, *63*, 1–22. [CrossRef]
11. Boonpavanitchakul, K.; Jarussophon, S.; Pimpha, N.; Kangwansupamonkon, W.; Magaraphan, R. Silk sericin as a bio-initiator for grafting from synthesis of polylactide via ring-opening polymerization. *Eur. Polym. J.* **2019**, *121*, 109265. [CrossRef]
12. Nong, Y.; Zhou, Z.; Yuan, J.; Wang, P.; Yu, Y.; Wang, Q.; Fan, X. Bio-Inspired Coloring and Functionalization of Silk Fabric via Laccase-Catalyzed Graft Polymerization of Arylamines. *Fibers Polym.* **2020**, *21*, 1927–1937. [CrossRef]
13. Zhou, B.; He, M.; Wang, P.; Fu, H.; Yu, Y.; Wang, Q.; Fan, X. Synthesis of silk fibroin-g-PAA composite using H_2O_2-HRP and characterization of the in situ biomimetic mineralization behavior. *Mater. Sci. Eng. C* **2017**, *81*, 291–302. [CrossRef]
14. Anantha-Iyengar, G.; Shanmugasundaram, K.; Nallal, M.; Lee, K.-P.; Whitcombe, M.J.; Lakshmi, D.; Sai-Anand, G. Functionalized conjugated polymers for sensing and molecular imprinting applications. *Prog. Polym. Sci.* **2019**, *88*, 1–129. [CrossRef]
15. Xu, M.; Cai, H.; Liu, Z.; Chen, F.; Chen, L.; Chen, X.; Cheng, X.; Dai, F.; Li, Z. Breathable, Degradable Piezoresistive Skin Sensor Based on a Sandwich Structure for High-Performance Pressure Detection. *Adv. Electron. Mater.* **2021**, *7*, 2100368. [CrossRef]
16. Wang, Y.; Yu, H.; Liu, H.; Fan, Y. Double coating of graphene oxide–polypyrrole on silk fibroin scaffolds for neural tissue engineering. *J. Bioact. Compat. Polym.* **2020**, *35*, 216–227. [CrossRef]
17. Gao, D.; Parida, K.; Lee, P.S. Emerging Soft Conductors for Bioelectronic Interfaces. *Adv. Funct. Mater.* **2020**, *30*, 1907184. [CrossRef]
18. Ahmad, S.; Sultan, A.; Mohammad, F. Electrically Conductive Polyaniline/Silk Fibroin Composite for Ammonia and Acetaldehyde Sensing. *Polym. Polym. Compos.* **2018**, *26*, 177–187. [CrossRef]
19. Hong, J.; Han, X.; Shi, H.; Jin, L.; Yao, J. Preparation of conductive silk fibroin yarns coated with polyaniline using an improved method based on in situ polymerization. *Synth. Met.* **2018**, *235*, 89–96. [CrossRef]
20. Chen, C.-Y.; Huang, S.Y.; Wan, H.-Y.; Chen, Y.-T.; Yu, S.-K.; Wu, H.-C.; Yang, T.-I. Electrospun Hydrophobic Polyaniline/Silk Fibroin Electrochromic Nanofibers with Low Electrical Resistance. *Polymers* **2020**, *12*, 2102. [CrossRef]
21. Li, X.; Ming, J.; Ning, X. Wet-spun conductive silk fibroin–polyaniline filaments prepared from a formic acid–shell solution. *J. Appl. Polym. Sci.* **2018**, *136*, 47127. [CrossRef]
22. Vepari, C.; Kaplan, D.L. Silk as a biomaterial. *Prog. Polym. Sci.* **2007**, *32*, 991–1007. [CrossRef] [PubMed]
23. Sashina, E.S.; Bochek, A.M.; Novoselov, N.P.; Kirichenko, D.A. Structure and solubility of natural silk fibroin. *Russ. J. Appl. Chem.* **2006**, *79*, 869–876. [CrossRef]
24. Ohgo, K.; Bagusat, F.; Asakura, T.; Scheler, U. Investigation of Structural Transition of Regenerated Silk Fibroin Aqueous Solution by Rheo-NMR Spectroscopy. *J. Am. Chem. Soc.* **2008**, *130*, 4182–4186. [CrossRef]
25. Zhang, Y.-Q.; Shen, W.-D.; Xiang, R.-L.; Zhuge, L.-J.; Gao, W.-J.; Wang, W.-B. Formation of silk fibroin nanoparticles in water-miscible organic solvent and their characterization. *J. Nanopart. Res.* **2007**, *9*, 885–900. [CrossRef]
26. Conejo-Dávila, A.S.; Hernández-Escobar, C.A.; Vega-Rios, A.; Rodríguez-Sánchez, I.; Estrada-Monje, A.; de León-Gómez, R.E.D.; Zaragoza-Contreras, E.A. Selective polymerization of a new bifunctional monomer via free radical polymerization and oxidative route. *Synth. Met.* **2020**, *259*, 116258. [CrossRef]
27. Shaw, A.A.; Wainschel, L.A.; Shetlar, M.D. The Photochemistry of p -Aminobenzoic Acid. *Photochem. Photobiol.* **1992**, *55*, 647–656. [CrossRef]
28. Conejo-Dávila, A.S.; Moya-Quevedo, M.A.; Chávez-Flores, D.; Vega-Rios, A.; Zaragoza-Contreras, E.A. Role of the Anilinium Ion on the Selective Polymerization of Anilinium 2-Acrylamide-2-methyl-1-propanesulfonate. *Polymers* **2021**, *13*, 2349. [CrossRef] [PubMed]

29. Manseki, K.; Yu, Y.; Yanagida, S. A phenyl-capped aniline tetramer for Z907/tert-butylpyridine-based dye-sensitized solar cells and molecular modelling of the device. *Chem. Commun.* **2013**, *49*, 1416. [CrossRef]
30. Ayub, Z.H.; Arai, M.; Hirabayashi, K. Quantitative structural analysis and physical properties of silk fibroin hydrogels. *Polymer* **1994**, *35*, 2197–2200. [CrossRef]
31. Baranowska-Korczyc, A.; Hudecki, A.; Kamińska, I.; Cieślak, M. Silk Powder from Cocoons and Woven Fabric as a Potential Bio-Modifier. *Materials* **2021**, *14*, 6919. [CrossRef] [PubMed]
32. Borah, R.; Kumar, A.; Das, M.K.; Ramteke, A. Surface functionalization-induced enhancement in surface properties and biocompatibility of polyaniline nanofibers. *RSC Adv.* **2015**, *5*, 48971–48982. [CrossRef]
33. Wang, X.; Deng, J.; Duan, X.; Liu, D.; Guo, J.; Liu, P. Crosslinked polyaniline nanorods with improved electrochemical performance as electrode material for supercapacitors. *J. Mater. Chem. A* **2014**, *2*, 12323–12329. [CrossRef]
34. Alves, W.F.; Venancio, E.C.; Leite, F.L.; Kanda, D.H.F.; Malmonge, L.F.; Malmonge, J.A.; Mattoso, L.H.C. Thermo-analyses of polyaniline and its derivatives. *Thermochim. Acta.* **2010**, *502*, 43–46. [CrossRef]
35. Chen, C.H. Thermal studies of polyaniline doped with dodecyl benzene sulfonic acid directly prepared via aqueous dispersions. *J. Polym. Res.* **2002**, *9*, 195–200. [CrossRef]
36. Rice, W.L.; Firdous, S.; Gupta, S.; Hunter, M.; Foo, C.W.P.; Wang, Y.; Kim, H.J.; Kaplan, D.L.; Georgakoudi, I. Non-invasive characterization of structure and morphology of silk fibroin biomaterials using non-linear microscopy. *Biomaterials* **2008**, *29*, 2015–2024. [CrossRef]
37. Yavarinasab, A.; Abedini, M.; Tahmooressi, H.; Janfaza, S.; Tasnim, N.; Hoorfar, M. Potentiodynamic Electrochemical Impedance Spectroscopy of Polyaniline-Modified Pencil Graphite Electrodes for Selective Detection of Biochemical Trace Elements. *Polymers* **2021**, *14*, 31. [CrossRef]
38. Korent, A.; Žagar Soderžnik, K.; Šturm, S.; Žužek Rožman, K. A Correlative Study of Polyaniline Electropolymerization and its Electrochromic Behavior. *J. Electrochem. Soc.* **2020**, *167*, 106504. [CrossRef]
39. Conejo-Dávila, A.S.; Casas-Soto, C.R.; Aparicio-Martínez, E.P.; Chávez-Flores, D.; Ramos-Sánchez, V.H.; Dominguez, R.B.; Osuna, V.C.; Estrada-Monje, A.; Vega-Rios, A.; Zaragoza-Contreras, E.A. Brush-like Polyaniline with Optical and Electroactive Properties at Neutral pH and High Temperature. *Int. J. Mol. Sci.* **2022**, *23*, 8085. [CrossRef]
40. Armando Zaragoza-Contreras, E.; Stockton-Leal, M.; Hernández-Escobar, C.A.; Hoshina, Y.; Guzmán-Lozano, J.F.; Kobayashi, T. Synthesis of core–shell composites using an inverse surfmer. *J. Colloid Interface Sci.* **2012**, *377*, 231–236. [CrossRef]
41. Zhao, Z.; Li, Y.; Xie, M.-B. Silk Fibroin-Based Nanoparticles for Drug Delivery. *Int. J. Mol. Sci.* **2015**, *16*, 4880–4903. [CrossRef]
42. Holland, C.; Numata, K.; Rnjak-Kovacina, J.; Seib, F.P. The Biomedical Use of Silk: Past, Present, Future. *Adv. Healthc. Mater.* **2019**, *8*, 1800465. [CrossRef]
43. Yamada, H.; Igarashi, Y.; Takasu, Y.; Saito, H.; Tsubouchi, K. Identification of fibroin-derived peptides enhancing the proliferation of cultured human skin fibroblasts. *Biomaterials* **2004**, *25*, 467–472. [CrossRef]
44. Kucekova, Z.; Humpolicek, P.; Kasparkova, V.; Perecko, T.; Lehocký, M.; Hauerlandová, I.; Sáha, P.; Stejskal, J. Colloidal polyaniline dispersions: Antibacterial activity, cytotoxicity and neutrophil oxidative burst. *Colloids Surf. B Biointerfaces* **2014**, *116*, 411–417. [CrossRef]
45. Li, Y.-S.; Chen, B.-F.; Li, X.-J.; Zhang, W.K.; Tang, H.-B. Cytotoxicity of Polyaniline Nanomaterial on Rat Celiac Macrophages In Vitro. *PLoS ONE* **2014**, *9*, e107361. [CrossRef]
46. Kaba, S.; Egorova, E. In vitro studies of the toxic effects of silver nanoparticles on HeLa and U937 cells. *Nanotechnol. Sci. Appl.* **2015**, *8*, 19. [CrossRef]
47. Xia, Y.; Lu, X.; Zhu, H. Natural silk fibroin/polyaniline (core/shell) coaxial fiber: Fabrication and application for cell proliferation. *Compos. Sci. Technol.* **2013**, *77*, 37–41. [CrossRef]

polymers

MDPI

Article

Lignin as a High-Value Bioaditive in 3D-DLP Printable Acrylic Resins and Polyaniline Conductive Composite

Goretti Arias-Ferreiro [1], Aurora Lasagabáster-Latorre [2], Ana Ares-Pernas [1], Pablo Ligero [3], Sandra María García-Garabal [4], María Sonia Dopico-García [1] and María-José Abad [1,*]

1 Grupo de Polímeros-CITENI, Campus Industrial de Ferrol, Universidade da Coruña, Campus de Esteiro, 15403 Ferrol, Spain
2 Departamento Química Orgánica I, Facultad de Óptica, Universidad Complutense de Madrid, Arcos de Jalón 118, 28037 Madrid, Spain
3 Enxeñería Química Ambiental Group, Centro de Investigacións Científicas Avanzadas (CICA), Universidade da Coruña, 15071 A Coruña, Spain
4 Grupo Mesturas, Universidade da Coruña, Campus da Zapateira s/n, 15071 A Coruña, Spain
* Correspondence: maria.jose.abad@udc.es

check for updates

Citation: Arias-Ferreiro, G.; Lasagabáster-Latorre, A.; Ares-Pernas, A.; Ligero, P.; García-Garabal, S.M.; Dopico-García, M.S.; Abad, M.-J. Lignin as a High-Value Bioaditive in 3D-DLP Printable Acrylic Resins and Polyaniline Conductive Composite. *Polymers* **2022**, *14*, 4164. https://doi.org/10.3390/polym14194164

Academic Editors: Antonio M. Borrero-López, Concepción Valencia-Barragán, Esperanza Cortés Triviño, Adrián Tenorio-Alfonso and Clara Delgado-Sánchez

Received: 15 September 2022
Accepted: 30 September 2022
Published: 4 October 2022

Abstract: With increasing environmental awareness, lignin will play a key role in the transition from the traditional materials industry towards sustainability and Industry 4.0, boosting the development of functional eco-friendly composites for future electronic devices. In this work, a detailed study of the effect of unmodified lignin on 3D printed light-curable acrylic composites was performed up to 4 wt.%. Lignin ratios below 3 wt.% could be easily and reproducibly printed on a digital light processing (DLP) printer, maintaining the flexibility and thermal stability of the pristine resin. These low lignin contents lead to 3D printed composites with smoother surfaces, improved hardness (Shore A increase ~5%), and higher wettability (contact angles decrease ~19.5%). Finally, 1 wt.% lignin was added into 3D printed acrylic resins containing 5 wt.% p-toluensulfonic doped polyaniline (pTSA-PANI). The lignin/pTSA-PANI/acrylic composite showed a clear improvement in the dispersion of the conductive filler, reducing the average surface roughness (R_a) by 61% and increasing the electrical conductivity by an order of magnitude (up to 10^{-6} S cm^{-1}) compared to lignin free PANI composites. Thus, incorporating organosolv lignin from wood industry wastes as raw material into 3D printed photocurable resins represents a simple, low-cost potential application for the design of novel high-valued, bio-based products.

Keywords: lignin; DLP; polyaniline; filler dispersibility; additive manufacturing; acrylic resin

1. Introduction

The development of 3D printing technologies requires new materials that are environmentally friendly and able to provide new features, specifically to increase the range of applications of vat polymerization-based techniques [1–3]. New 3D printing technologies are becoming important as production models in the frame of Industry 4.0 and the Internet of Things [4,5]. Stereolithography (SLA) and digital light processing (DLP) are vat polymerization printing techniques based on the use of UV-light to solidify a liquid resin layer-by-layer [6]. To achieve successful printing, the photo-rheological and mechanical properties of the multicomponent resins need to be adjusted [7]. These photosensitive resins are complex mixtures of a photoinitiator, monomers, oligomers, and other additives intended to tune their properties [6].

In this context, lignin can play an important role in the development of new resin formulations as the second most abundant natural polymer on the planet, low cost, and renewable [8–10]. Most lignin is obtained as a by-product of the wood industry and valorized energetically as a fuel. With increasing environmental awareness and the circular economy, the use of lignin has been reconsidered to look for its application in high-valued bio-based products [11–13]. Lignin is a reticular polyphenolic polymer with a three-dimensional structure of high structural complexity, amorphous, and very heterogeneous. Its structure and final properties are directly related to both its natural origin and the treatment followed for its extraction and purification, such as organosolv, soda, or kraft process [10,14]. Within polymers, lignin has a wide range of applications such as additive to enhance their mechanical behavior (hardness, elasticity and tensile strength), provide antioxidant properties, tune the surface wettability, improve biodegradability [14], or as compatibilizer in polymer composites [15].

In the field of electronics, lignin has been widely investigated [11,12,16–22] for the manufacture of energy storage devices [22], electromagnetic shielding [16,17], organic cathodes [11], as a natural binder or even as a dopant in conductive polymers for the production of electrochemical capacitors or supercapacitors [12,18–21]. Due to the redox activity of the quinone/hydroquinone moieties, lignin derivatives effectively enhanced the capacitance of intrinsically conductive polymers, including poly(3,4-ethylenedioxythiophene), polypyrrole, and polyaniline (PANI) [23] as well as carbon-based materials such as graphene [19]. Sulfomethylated lignin has been shown to effectively dope PANI and increase the conductivity and dispersibility of lignin/PANI composites in water [24].

In this context, future electronic devices could provide great benefits from the synergistic combination of AM techniques and green chemistry, delivering a wider range of 3D printed polymer composites with improved electrochemical performance and enhanced physicochemical properties. Nowadays, although lignin has been employed in a variety of 3D printing applications, its use in vat polymerization is still scarce [7,25].

In order to add lignin in SLA/DLP resins two strategies have been investigated: directly mixing unmodified lignin and functionalized bulk lignin with the photoreactive monomers formulation. The blending of unmodified kraft lignin with methacrylate resin improved the mechanical properties of the printed samples to a limited extent, as amounts higher than 1% lignin seriously hindered the crosslinking reaction [26]. Further, the ability of small organosolv lignin loadings as a compatibilizer for graphene nanoplatelets (G) has been positively proven in the mechanical properties enhancement of photocurable polyurethane–Lignin/G [27]. On the other hand, Sutton et al. [28] incorporated up to 15 wt.% of organosolv lignin functionalized with methacrylic anhydride in a commercial SLA resin, increasing the ductility of the materials due to the plasticizing effect of the lignin side chains [28].

In a previous work, the authors developed a formulation based on the conductive filler PANI-lignin in a photocurable acrylic matrix, which proved to be valid for the manufacture of flexible wearable electronics and sensors [23]. The present research is focused on the incorporation of unmodified organosolv lignin as an additive to the photocurable resin. The purpose is to deepen the composite behavior and explore other potential applications of lignin in the manufacture of functional materials, such as the dispersant of conductive fillers in composites for electronic devices. We present a study of the structure-property relationship of 3D printed composites based on organosolv lignin isolated from Betula alba bark and a home-made photocurable acrylic matrix. Lignin was used as a non-reactive filler without any further chemical modification.

In a preliminary stage, the resin curing properties were assessed through viscosity measurements, real-time Fourier transform infrared spectroscopy, and the Jacob's working curve approach. Then, the 3D printed lignin composites were thoroughly characterized by SEM, FTIR, TGA, tensile properties, Shore D hardness, and water contact angle. As way of example, the combination of lignin and para-toluene sulfonic acid doped polyaniline (pTSA-PANI) lead to an improvement of the filler dispersion within the acrylic matrix, resulting in more homogeneous samples with lower surface roughness and enhanced electrical conductivity.

2. Experimental Section/Methods

2.1. Materials and Samples Preparation

Aniline (ANI, 99.5%) and para-toluene sulfonic acid (pTSA, 98%) were supplied by Sigma-Aldrich (St. Louis, MO, USA) and ammonium persulfate (APS, 99%) was obtained from Acros (Geel, Belgium). Monomer, crosslinker and photoinitiator were ethylene glycol phenyl ether acrylate (EGPEA, molecular weight = 192.21 g/mol), 1,6-hexanediol diacrylate (HDODA, molecular weight = 226.27 g/mol), and diphenyl (2,4,6-trimethylbenzoyl) phosphine oxide (TPO, molecular weight = 348.37 g/mol), respectively; they were obtained from Sigma-Aldrich (St. Louis, MO, USA). Acetone, chloroform, glycerin, methanol and 2-propanol were purchased from Scharlau (Sentmenat, Spain).

Preparation of Lignin. Lignin was extracted from Betula Alba bark by organosolv fractionation using as a solvent acetic acid and hydrochloric acid, as described in a previous reference [23]. The obtained lignin was lyophilized before its use.

The characterization of the molecular weight distribution of lignin was performed by gel permeation chromatography (GPC). A Waters 2695 system (Waters, Mildford, MA, USA) was used, equipped with two linear columns (Styragel 4E and Styragel HR3, 4.6 × 300 mm) and an ultraviolet diode array detector (PDA, model 996 UV); the temperature of the column was 35 °C and the mobile phase tetrahydrofuran (THF) at 0.3 mL min^{-1}. A calibration curve obtained with polystyrene standards in the range 580 to 93,800 Da was used (220 nm). Samples of lignin were dissolved in the mobile phase with a concentration of 1000 mg/L by gently stirring the samples up to 24 h. The values obtained were Mw = 2115, Mn = 924 and polydispersity = 2.3 (average of 3 replicates).

PANI synthesis. PANI was synthesized adapting the emulsion polymerization method described by Dopico et al. for the synthesis of PANI doped with dodecyl benzenesulfonic acid (DBSA-PANI) in a mixture of water and chloroform, changing the dopant to pTSA [29]. Aniline hydrochloride (2 mL, 1.88 g) was dissolved in 200 mL chloroform. The oxidant ammonium peroxydisulfate (APS, 4.5 g) and para-toluene sulfonic acid (pTSA, 11.4 g) were dissolved in 100 mL water. The polymerization reaction was initiated by adding dropwise the aqueous solution to the aniline solution and allowed to proceed for 24 h under mechanic stirring (150 rpm) and low temperature (<6 °C). The resulting product was collected through precipitation with methanol and acetone and filtered under vacuum. pTSA doped polyaniline (pTSA-PANI) was obtained as a very dark green powder after drying during 24 h at 40 °C in a vacuum oven. The yield recovered was 2.39 g. The elemental analysis of pTSA-PANI was performed in duplicate on a ThermoFinnigan Flash EA1112 analyzer. The average result was: 8.34 wt.% N; 51.74% C; 4.74% H; 9.52% S.

Preparation and printing of the composites. Samples were formulated taking as reference the base resin of previous studies [30,31]. Fixed quantities of the HDODA crosslinker (15 wt.%) and the TPO photoinitiator (7 wt.%) were mixed with the EGPEA, as a major monomer. Increasing amounts of unmodified lignin were then added. The formulations of the lignin 3D printing composites analyzed in this investigation are presented in Table 1. In order to evaluate the effect of lignin as a dispersant of conductive charges in functional materials, two additional formulations were prepared using 5 wt.% pTSA-PANI with 1 wt.% lignin (LG1PANI5) and without lignin (LG0PANI5) as a control.

Table 1. Samples formulation, 3D printer exposure times, viscosity at 1 s^{-1}, maximum rate of polymerization, and ultimate degree of conversion at 9 mW cm^{-2} obtained from Real-Time ATR-FTIR spectroscopy.

Sample	Lignin (wt.%)	Exposure Time (s)	Bottom Exposure (s)	Viscosity (Pa.s) 1 s^{-1}	Induction Period (s)	Max. Rate of Polymerization (mol L^{-1} s^{-1})	Ultimate DBC_{∞} (%)
LG0	0	1	15	0.011	<8	62.0 ± 0.6	98.0 ± 0.2
LG05	0.5	2	15	0.019	<8	17.9 ± 2.4	97.9 ± 1.6
LG1	1	2.5	20	0.026	24 ± 7	5.8 ± 1.8	94.4 ± 0.1
LG2	2	5	30	0.056	32 ± 13	1.9 ± 0.5	95.3 ± 0.5
LG3	3	20	50	0.060	38 ± 4	1.9 ± 0.1	94.8 ± 2.0
LG35	3.5	30	75	0.064	-	-	-
LG4	4	35	90	0.072	- *	0.9 ± 0.2	79.0 ± 11.0

* Not possible accurate determination.

To promote homogeneous dispersion each sample formulation was sonicated during 30 min using a Digital Sonifier at 10% of intensity (Branson 450, Danbury, CT, USA) and further mechanically stirred in a Vortex mixer for 2 min at 1000 rpm (VELP Scientific, Schwabach, Germany) just prior printing. The 3D DLP printer was an Elegoo mars PRO (wavelength 405 nm; 9 mW.cm^{-2}); its settings were adjusted considering the printer technical requirements and the kinetics of polymerization of the formulations tested. The exposure time per layer was adjusted for each formulation (Table 1). Bottom exposure corresponds to the exposure time of the first 5 layers, which is always longer to ensure suitable adhesion of the sample to the metal platform. The layer thickness (z) was set at 0.025 mm. The printability worsens from 3% lignin composites which can be explained by their longer curing times and adhesion problems to the printing platform.

The uncured resin remaining in the printed samples was removed by soaking them in 2-propanol for 10 min. A post-curing process was performed using a lamp (Form Cure, Formlabs) for 5 min at 35 °C. The sample characterization was carried out on flexible dog-bone-shaped specimens according to ISO 527. The preparation of the samples for 3D printing and an example of the printed samples are shown in Figure 1.

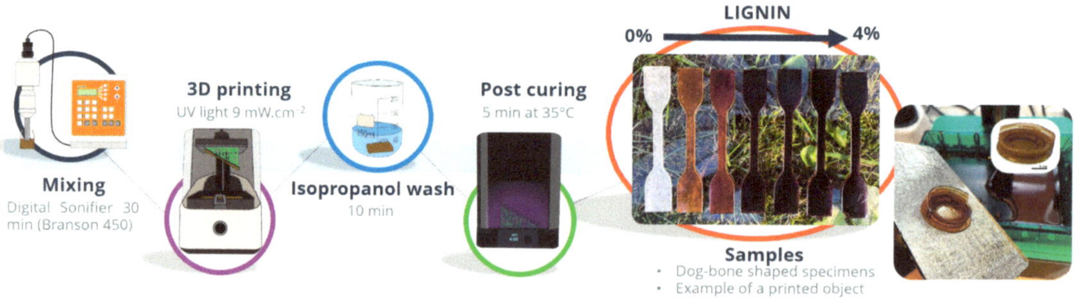

Figure 1. Diagram of sample preparation.

2.2. Sample Characterization

The viscosity of the acrylic composite formulations was measured using a controlled strain rheometer (ARES, TA Instruments, Newcastle, DE, USA) with parallel-plate geometry (25 mm diameter, 1 mm gap) at room temperature. Steady shear viscosity (η) was measured in the range of shear rates 0.3–100 s^{-1}.

The UV-Vis spectra of 4 different suspensions of lignin and pTSA-PANI in glycerin (spanning from 50 to 200 ppm) were recorded on a Jasco V-750 double-beam UV-Vis spectrophotometer (Jasco Analítica S.L., Madrid, Spain) between 330 and 800 nm with a sampling interval of 1 nm and 25 accumulations.

The kinetics of the UV-initiated radical polymerization of the acrylic composites with 0, 1, 2, 3, and 4 wt.% lignin were studied in situ by Real Time FTIR spectroscopy (Jasco 4700 spectrometer, Jasco Analítica S.L., Madrid, Spain) in Attenuated Reflectance Mode (ATR) (MK II Golden Gate^{TM} Diamond 45° ATR). The details of the full procedure are included in a previous reference [32]. The analyses were performed in triplicate with a 4 cm^{-1} resolution between 1800 and 550 cm^{-1} over 5 scans (6 s). A Visicure 405 nm spot lamp connected to an LED Spot-Curing System (BlueWave, Dymax Corp., Torrington, CT, USA), was used to cure the sample from its top side. Bruker OPUS® software version 5.5 (Bruker Española S.A, Madrid, Spain) was employed for the spectra analysis.

The Jacobs working curves of the liquid formulations were calculated from samples cured over the screen of the 3D printer, varying the exposure times; circular films of 18 mm in diameter and 1 mm thickness were printed, followed by a clean-up step with 2-propanol. The thickness of the samples was measured with a thickness DUALSCOPE® MP0R measuring instrument (Fisher) and plotted as a function of Exposure (E_{max}) (mJ.cm^{-2}). Each sample was performed in duplicate.

The morphology of lignin and pTSA-PANI was evaluated by Transmission Electron Microscopy (TEM) (JEOL JEM 1010 (80 KeV), after applying 10 mL of the aqueous powder dispersions to a copper grid, and by Scanning Electron Microscopy (SEM) (JEOL JSM-7200F Field Emission Scanning Electron Microscope at an accelerating voltage of 10 kV). Prior to SEM observation, the samples were sputter-coated with a thin palladium/platinum layer (Cressintong 208HR). The printed composites were also analyzed by SEM. In this last case, the specimens were previously broken under cryogenic conditions. Confocal microscopy was used to determine the surface roughness of the conductive printed composites with pTSA-PANI, by using a PLu 2300 Sensofar® optical imaging profiler. Images were captured using an EPI 10×-N objective, a depth resolution of 2 μm, and a lateral resolution of 1 nm. Roughness parameters such as R_a (average roughness), R_v (average maximum valley depth), and R_p (average maximum peak height) were obtained using SensoMaP 5.0.4 software. At least five measurements were performed for each sample in order to calculate the average values and standard deviations.

For lignin and pTSA-PANI powders, the FTIR spectra were performed in Potassium Bromide (KBr) pellets between 4000 and 400 cm^{-1}. The post-cured printed films were analyzed in ATR mode between 4000 and 550 cm^{-1} with a 4 cm^{-1} resolution over 64 scans. The degree of the acrylate double bond conversion (DBC%) based on Equation (2) was also calculated. For each sample, the average spectra of three replicates were examined.

Thermogravimetric Analysis (TGA) (TGA 4000—Perkin–Elmer, Waltham, MA, USA) of the cured films was performed under a nitrogen atmosphere at 50 mL min^{-1} using ceramic crucibles (60 μL) as composite holders. The heating rate was 10.0 ± 0.1 °C min^{-1} from 50 up to 700 °C.

Tensile stress–strain mechanical properties were characterized using an Instron 5569 universal testing machine (Instron Canton, Norwood, MA, USA). The analysis was performed using a cross-head speed of 5 mm min^{-1} until failure, at room temperature. At least five dog-bone-shaped specimens were tested following ISO 527 (dimensions 75 × 13 × 2 mm; width of narrow section 5 mm). Measurement of the hardness of the composites with a Shore "A" Durometer (Durotech M202) was carried out on the dog-bone-shaped specimens at a distance of ~6 mm from the edge of the material after 15 s of force application. The measurements were taken at 10 measuring points on each sample and the mean values and standard deviations were calculated according to ISO 868:2003 [33].

Surface wetting measurements were carried out with a Theta Lite Attention tensometer (Biolin Scientific, Gothenburg, Sweden) and the software program "One Attention". The static water contact angle (θ) formed by a single droplet was measured at least five times on dry samples using a 4 μL sessile drop of deionized water as test fluid at room temperature and the average values are reported. Images were recorded every 10 s.

The electrical conductivity (σ) of the samples was calculated from the electrical resistance data by the four-probe method (LORESTA-GP, Mitsubishi Chemical, MCP-T610,

Tokyo, Japan) at room temperature. For pTSA-PANI, square compression molded pellets of 2.5 cm × 2.5 cm × 0.5 mm were employed. The values reported are the mean of at least eight readings measured on three different samples.

3. Results

3.1. Prepolymerization Studies

Prior to 3D printing, the effect of lignin on viscosity, UV-spectroscopy, photopolymerization kinetics and the Jacob's working curves were evaluated, being these parameters significantly interesting for a DLP process and the quality of printed parts.

3.1.1. Viscosity

Viscosity is one of the parameters to be taken into account to achieve satisfactory prints. In general, low viscosities are desired to allow the proper coating of the last printed layer or the surface of the immersive platform [34]. On the contrary, too high a viscosity means longer exposure times, as well as limited adhesion of layers to each other and to the printer platform [28,31].

The viscosity values of all liquid formulations as a function of shear rate and Lignin amount at room temperature are depicted in Figure S1, whereas the viscosities at 1 s^{-1} are compiled in Table 1. It is obvious that the higher the lignin content, the greater the viscosity of the formulations at 1 s^{-1}, since the presence of lignin limits the mobility of the polymer chains [25,35]. Increased lignin content (0–4 wt. %) raised the viscosity of the polymer from 0.011 Pa s to 0.072 Pa s at 1 s^{-1}. Although the reported viscosities for current commercial resins are usually higher, in the range of 0.85−4.5 Pa [36], the viscosities of all the lignin formulations remain low enough to promote the resin layer uniformity in the 3D printer in use and are comparable to the viscosities of PANI, PANI-Lignin and PANI-MWCNT formulated with the same base acrylic resin and 3D printed successfully [31,32].

In contrast to the base formulation (LG0), which has Newtonian behavior in the range between 1 s^{-1} and 10^3 s^{-1}, all doped formulations show typical shear thinning behavior at 1 s^{-1} (Figure S1 and Table 1), in agreement with previous studies on resins incorporating nanofillers [30,35,37]. This shear thinning behavior can facilitate the spreading of homogeneous layers in vat polymerization techniques [38]. Thus, from the viscosities point of view, the studied formulations are viable for creating any type of design by 3D printing.

3.1.2. UV-Visible and Real Time-FTIR Spectroscopy

The UV–Vis absorption spectra between 330–800 nm of lignin is shown in Figure 2A. Lignin has absorption bands around 280–300 nm related to the phenolic hydroxyl and aromatic moieties [39]. The absorptivity value at 405 nm, ε, has been calculated from the slope of the regression line obtained when plotted, the experimental absorbance values, A, vs the concentrations of the filler in dispersion, using the Lambert–Beer law (Equation (1))

$$A = \varepsilon \times b \times c \quad (1)$$

where b is the length of the UV pathway (1 cm for the cuvettes used) and c is the concentrations of the tested fillers. The calculations have been performed on the basis of filler suspensions, thus the Lambert–Beer law is an approximation since no light scattering effects have been taken into account. The wavelength 405 nm selected is the critical wavelength on which the UV lamp of the printer has its highest radiation power. The calculated ε value is 1.45 g. L^{-1}. cm^{-1} (r^2 = 0.9967). This outcome suggests that UV absorption by lignin competes with the photoinitiator and hinders the curing process of the dispersions.

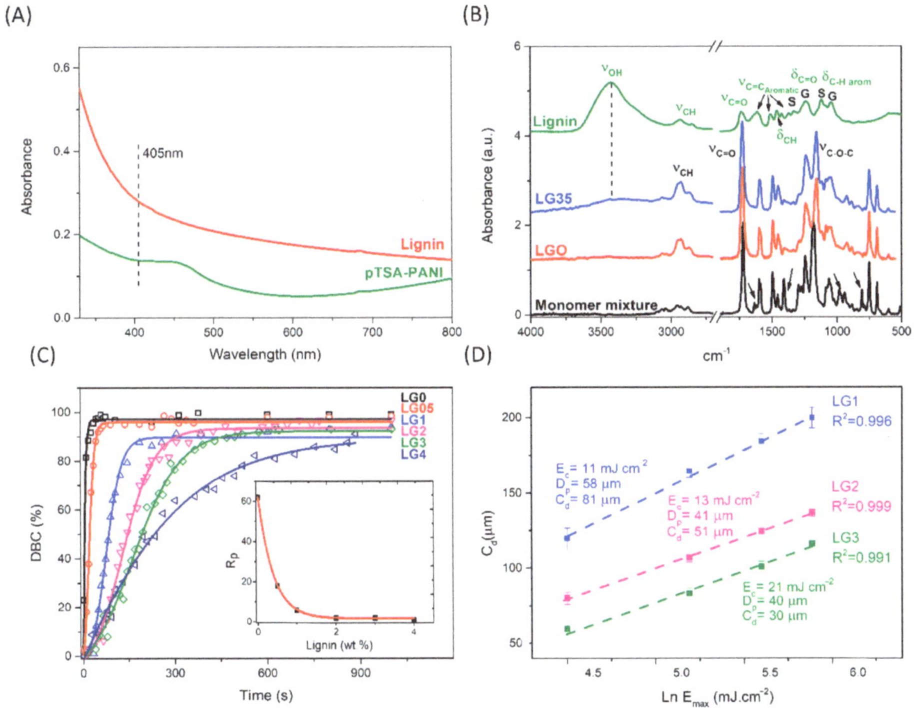

Figure 2. (**A**) UV-visible absorption spectra of lignin and pTSA-PANI in glycerin at 150 ppm. (**B**) ATR-FTIR spectra of the monomer mixture, pristine acrylic (LG0), LG35, and Lignin. (**C**) ATR conversion curves versus irradiation time for acrylic dispersions with 0.5, 1, 2, 3, and 4 wt.% lignin. The experimental data (symbols) were fitted with Boltzmann sigmoidal equation (lines) and (**D**) Jacobs working curves showing cure thickness as a function of the natural log of UV dosage for LG1, LG2, and LG3 formulations. Linear regressions are depicted.

The influence of lignin (0, 0.5, 1, 2, 3, and 4 wt.%) on the photocuring rate of the acrylate resin has been studied by in situ ATR-FTIR spectroscopy. The bands related with the C=C double bond of the acrylate groups gradually disappeared, specifically 1636 cm^{-1} ($\nu_{C=C}$, doublet), 1409 cm^{-1} (in plane deformation, scissoring, $\delta_{=CH2}$), 984 cm^{-1} and 810 cm^{-1} (out of plane deformation, $\delta_{=CH2}$) (indicated by arrows in Figure 2B). To follow the evolution of the polymerization reaction, the degree of double bond conversion (DBC%) was calculated according to Equation (2) [40]. The decrease of the band at 810 cm^{-1} was normalized to the carbonyl ester stretching band ($\nu_{C=O}$) of the resin at 1728 cm^{-1}, using this last one as the internal reference; t indicates the time irradiation [40]:

$$\text{DBC\%} = \frac{(A_{810}/A_{1728})_{t=0} - (A_{810}/A_{1728})_t}{(A_{810}/A_{1728})_{t=0}} \times 100\% \quad (2)$$

Figure 2C shows the conversion curves of DBC% versus irradiation time (s), were fitted with the Boltzmann sigmoidal model [41]. Three sections can be observed, corresponding to induction, propagation, and equilibrium or ultimate degree of conversion (DBC_{∞}%). The maximum rate of polymerization (R_P) has been calculated by Equation (3) for the conversion interval between 25–55% [32,40].

$$R_P = [M_0]\frac{(A_{810})_{t1} - (A_{810})_{t2}}{t_2 - t_1} \quad (3)$$

Being $[M_0]$ the initial concentration of acrylate double bonds (5.63 mol l^{-1} considering that HDODA is a bi-functional monomer and EGPEA is monofunctional), $(A_{810})_{t1}$ and $(A_{810})_{t2}$ are the areas of the at the irradiation times t_1 and t_2, corresponding to 25 and 55% conversion, respectively [32].

The maximum rate of polymerization (R_P), the induction period and ultimate degree of conversion are shown in Table 1. Longer retardation in polymerization initiation and slightly lower final conversions are seen for lignin contents ≥1 wt.%. Most importantly, as shown in Figure 2C inset, there is a pronounced exponential decrease in the rate of photopolymerization (r^2 = 0.997) when increasing the filler amount from 0.5 wt.% onwards. This is an expected behavior, partly due to the absorption of UV light, which competes with the light adsorption of the photoinitiator. The role of lignin UV blocker is due to the presence of UV-active functional groups such as C=O and aromatic rings [7,25]. The decrease in curing kinetics observed for lignin composites could also be related to the reported role of the phenolic groups of lignin as free radical scavengers and antioxidants [42–44]. On the contrary, some types of modified lignin have proved to have a certain photoinitiator ability when adequately combined with an amine co-initiator [39].

3.1.3. Jacobs Working Curves

In this section, the printability of the developed formulations by DLP is evaluated. Owing to the UV absorption of lignin and its reported effect as a free radical scavenger, the photopolymerization reaction is hindered, jeopardizing successful printing. To determine the optimal exposure times for each formulation, Jacobs working curves for LG1, LG2 and LG3, as well as printing tests, were carried out. Jacob working curves were calculated according to Equation (4) [45], where C_d is the cured depth (μm), D_p is the penetration depth (μm) of the light into the resin, E_{max} is the light irradiation dosage on the surface (mJ cm^{-2}) and E_c is the critical exposure required for polymerization (mJ cm^{-2}) [32]. As previously indicated, the light intensity of the 3D printer was 9 mW.cm^{-2} and the layer thickness (z) was 25 μm for all samples.

$$C_d = D_p \cdot \ln\left(\frac{E_{max}}{E_c}\right) \tag{4}$$

As shown in Figure 2D, the correlation coefficients (R^2) of the logarithmic regression lines were all higher than 0.99. Based on Equation (4), the photosensitive parameters related with the intrinsic properties of the resin, E_c and D_p, were calculated. For LG1, LG2 and LG3, the E_c values were 11 mJ.cm^{-2}, 13 mJ.cm^{-2} and 21 mJ.cm^{-2} respectively. Regarding the penetration depth of the resin, the D_p values were 58 μm, 41 μm and 40 μm. As expected, the general trend is that as the amount of lignin increases, the amount of energy needed to induce polymerization (E_c) increases, while penetration depth (D_p) decreases.

For lignin contents ≤2 wt.% the critical exposure (E_c) to induce polymerization remains actually unchanged, whereas it doubles its value for 3 wt.% lignin. Notwithstanding, the E_c values are in the same range or lower compared to those reported for commercial nonconductive resins [25,28,36]. The D_p of LG1 (58 μm) lies within the low range of commercial resins without conductive fillers tested by Bennet et al. (53–568 μm), whereas D_p values decreased around 31% for 2 and 3 wt.% lignin contents [36]. Low D_p values have the advantage of allowing accurate control of the polymerization process and minimal over-cure, although the shortcoming of longer building times [36]. Further, the curing parameters are similar to those calculated for PANI and PANI MWCNT composites fabricated with the same acrylic matrix and photoinitiator content [23,31,32].

To ensure that a resin is suitable for a specific printer, it is advisable to calculate the C_d corresponding to the light source employed from Equation (4) and compare it with the layer thickness (z) [46]. The layer thickness must be equal to or less than the curing depth. The exposure times indicated in Table 1 were set based on D_p and E_c values and by test prints; their corresponding E_{max} were calculated considering the printer energy used (9 mW.cm^{-2}). The C_d values for LG1, LG3, and LG3 were 81, 51, 30 μm, respectively. To ensure good adhesion between layers, the target is $C_d > z$, since the stiffness of a polymer below the gel point would hinder the printing process [31]. In this way, the obtained values for LG1 and LG2 (C_d = 81 and 51 μm, respectively; z = 25 μm) show that these formulations are suitable for 3D printing, whereas the printing of LG3 (C_d = 30 μm) has proved difficult and less reproducible.

3.2. Composites Characterization

3.2.1. Structural Characterization: Morphology and ATR

TEM and SEM imaging were performed to examine the morphology of lignin (Figures S2A,B and S3A). TEM and SEM images portray an irregular distribution of particles with a wide range of sizes that consist on aggregates of individual spheres. Figure S2B shows an isolated sphere of about 200 nm in diameter. These spherical particles are probably formed of disordered entangled chains that have shrunk into a "collapsed ball" [21].

The morphological differences between the 3D printed films of pure acrylate resin and the lignin compounds were further evaluated by SEM. Analysis of the composites films surfaces (Figure 3) showed polymer-rich surfaces as lignin is not observed in the images, indicating that the filler is embedded within the resin. Moreover, the surface of the composite films became smoother upon increasing the lignin loading, with fewer and shallower scratches visible on the surface, which points to an increase in surface hardness. The observation of a reduced surface roughness of polymer nanocomposites films with increased lignin content has been previously reported [15,47].

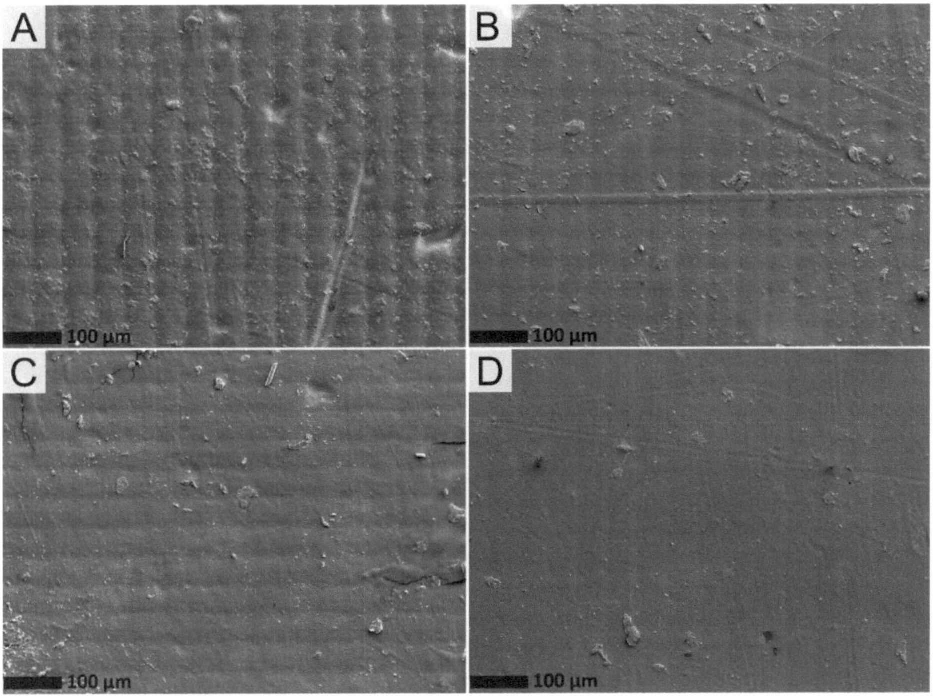

Figure 3. Representative SEM images of the films surfaces of (**A**) LG0, (**B**) LG1, (**C**) LG2, and (**D**) LG35 with a magnitude amplification of 500×.

Concerning the cryo-fractured cross-section shown in Figure 4, all the samples showed the layer stack gaps every ~25 μm, consistent with the layer thickness set at 0.025 mm. Nevertheless, there are some differences related with the fracture mechanism, the crack formation, and propagation. The LG0 sample shows a much smoother cross-section with several long cracks in accordance with the rubbery nature of the polymer (Figure 4); opposite to this, the rigid lignin microparticles augment the number of stress concentration spots, leading to an increasingly uneven and rougher fracture surface for composites with lignin contents $\geq$ 2%, as previously reported [26]. This behavior is enhanced by both the reduction on the UV-photopolymerization rate and the lignin aggregation and uneven dispersion in the matrix, which leads to increasing gaps and holes (Figure 4C,D insets). A poor dispersion of the filler will provide more concentrated stress locally, negatively affecting the mechanical properties of the material [27].

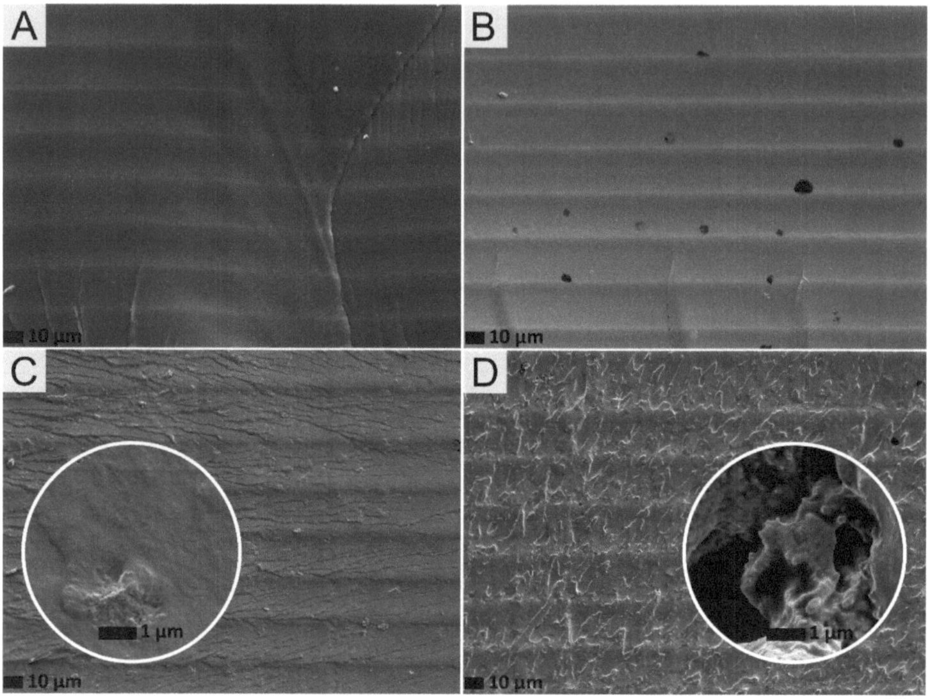

Figure 4. Representative SEM images of cryo-fractured cross-sections of (**A**) LG0, (**B**) LG1, (**C**) LG2, and (**D**) LG35 with magnitude amplification of 500× and 10,000× for the inset.

The KBr FT-IR spectrum of lignin is plotted in Figure 2B in comparison with the ATR spectra of the liquid monomers mixture, the pristine acrylic matrix, and the composite LG35. The spectrum of lignin showed typical lignin patterns after organosolv fractionation. The stretching vibrations of hydroxyl, CH alkane, carbonyl, and typical aromatic skeletal vibrations were observed at 3425 (ν_{OH}), 2936, and 2851 ($\nu_{C\text{-}H}$), 1734 ($\nu_{C=O}$), 1616, 1508, and 1425 ($\nu_{C\text{-}=C}$) cm^{-1}, respectively. An absorption at 1463 cm^{-1} related to C-H bending vibration ($-CH_2$, $-CH_3$) can be seen. Furthermore, the bands at 1239 and 1044 cm^{-1} may be ascribed to the C=O bending and aromatic CH in plane deformation of the guaiacyl (G) unit, whereas those centered at 1328 and 1117 cm^{-1} have been assigned to the same functional groups of the syringyl (S) unit. [16,48,49].

In relation with the pristine acrylic resin, the strongest bands in the spectrum centered at 1728 and 1156 cm^{-1}, are due to the stretching vibrations of the carbonyl bond ($\nu_{C=O}$) and the asymmetric stretching of the C–O–C bond ($\nu_{C\text{-}O\text{-}C}$), respectively [23]. The presence of lignin is not detected in the spectra of composites with lignin contents below 3 wt.% due to the overlapping of most lignin signals with the bands of the acrylic matrix. Nonetheless, for higher lignin contents, as in the spectrum of LG35 (Figure 2B), the filler can be perceived by a slight increase in the absorbance region between 3600–3200 cm^{-1}. Unlike previous reports, which used higher amounts of unmodified lignin (5 and 10 wt.%) [44], no bands shifts, or new bands, corresponding to the H-bonding between the acrylic matrix and lignin, have been detected in the ATR spectra, which is coherent with the observations made by SEM, suggesting that lignin is mostly embedded within the resin.

The effects of lignin on the degree of monomer conversion (%DBC) of the post-cured printed films were confirmed by ATR-FTIR. The almost complete disappearance of all the bands allotted to the C=C double bond of the acrylate groups revealed high degrees of monomer conversion on both sides of the films, irrespective of lignin loading. Despite the observed effect on the rate of polymerization, no differences with respect to pristine acrylic matrix have been found within experimental error (DBC% = 97.5 ± 0.6 and 96.6 ± 0.8 for LG0 and LG35, respectively). As discussed by previous authors, these results proved the need and efficiency of the UV post-curing step in completing the photoreaction [26].

3.2.2. Thermal and Mechanical Properties of Printed Composites

The thermal and mechanical behavior of lignin-containing materials is key to ensuring their processability and good final properties. Lignin can be used in polymers as a stiff filler with a purpose similar to the role played in plants providing cell wall rigidity. However, the reinforcing effect of lignin is highly influenced by several factors, such as the lignin source, content, polymer resin, and the printing techniques used [26] that can even cause undesired effects. In this way, lignin can worsen the mechanical properties of the polymer due to its naturally variable composition and irregular structure or introduce heat instability due to its phenolic moieties [44]. For these reasons, it is important to assess the final properties of the obtained lignin composites. At first, the thermal stability of lignin, pristine acrylic, and the printed composites were evaluated by TGA (Figure 5A and Table 2).

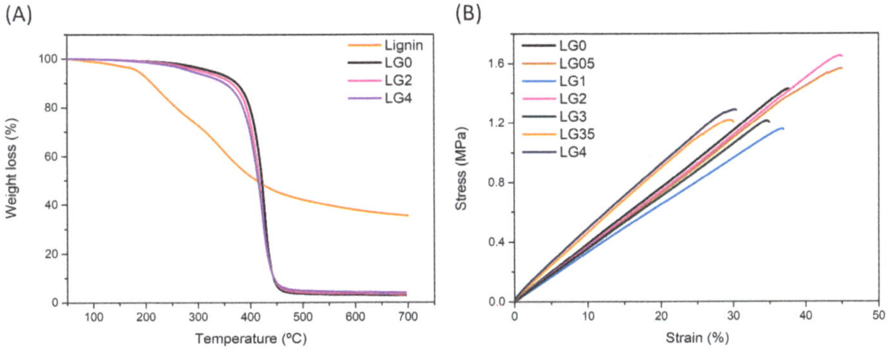

Figure 5. (**A**) TGA curves of Lignin and printed composites LG0, LG2, and (**B**) Stress-strain behavior of pure acrylate resin and lignin printed composites (representative curves).

Lignin shows three-step weight-loss behavior. The first loss step below 140 °C is attributed to the release of volatile components (2.5 ± 0.2%). Between 146 and 500 °C lignin loses 55.8 ± 0.5% of the initial mass in two thermal degradation processes with maximum degradation temperatures (T_{max}) at 223 and 346 °C. The degradation below 400 °C is mainly caused by the fragmentation of the weak inter-unit linkages (β-O-4) [50–52]. The loss of mass above 500 °C can be explained by the decomposition of aromatic rings and the cracking of C–C linkages [50,51,53], followed by the release of OCH_3 groups from aromatic rings [50,54]. After 700 °C, there was still a 35.1 ± 0.5% mass attributed to the formation of highly condensed aromatic structures [51,54].

Table 2. Characteristic temperatures measured from TGA thermograms. T_{10}: temperature at which 10% of the total mass is volatilized; T_{onset}: degradation temperature; DTG_{max}: maximum of thermal decomposition temperature; Residue: non-volatized weight fraction at 700 °C.

Sample	T_{10} (°C)	T_{onset} (°C)	DTG_{max} (°C)	Residue (wt.%)
Lignin	208.2 ± 3.7	176.7 ± 4.0	223 and 346	35.1 ± 0.2
LG0	374.5 ± 0.6	400.0 ± 0.1	426.5 ± 3.5	2.9 ± 0.4
LG05	369.2 ± 0.2	395.6 ± 1.6	420.5 ± 2.1	3.2 ± 0.2
LG1	366.9 ± 1.6	395.5 ± 0.3	421.5 ± 0.7	3.4 ± 0.1
LG2	363.6 ± 0.8	389.4 ± 3.8	419.5 ± 0.7	3.4 ± 0.1
LG3	359.0 ± 0.2	386.5 ± 3.5	419.0 ± 1.4	3.7 ± 0.1
LG35	356.9 ± 0.9	384.8 ± 1.9	420.0 ± 1.4	3.9 ± 0.1
LG4	353.2 ± 3.0	383.8 ± 4.4	419.5 ± 0.7	4.1 ± 0.1

Regarding pure acrylic resin, it shows two-step degradation behavior. The first small loss step is attributed to the release of volatiles (T_{max} = 188 ± 2.8. °C, 2.6 wt.%). The scission of the main polymeric backbone chain occurs within 350–450 °C [30,55], leaving a very small residue at 700 °C; therefore, the copolymer can be considered thermally stable up to 350 °C.

Concerning 3D printed lignin composites, they all exhibit degradation profiles similar to that of the pristine acrylic resin. No variation on volatiles evaporation is detected, within experimental error, confirming the high degrees of conversion observed by ATR and the efficiency of the post-curing step. Nevertheless, upon increasing the amount of lignin from 0.5 to 4 wt.% T_{10} and T_{onset} linearly decrease by about 6 and 4%, respectively (r^2 = 0.98 and 0.97). By contrast, although the maximum rate of weight loss temperature (DTG) shifted from 426.5 °C for the neat acrylic resin to 420.5 °C for lignin contents of 0.5% wt.%, the addition of higher amounts of lignin does not lead to further changes. Finally, a small increase in weight residue is observed upon increasing lignin content. The small changes in thermal behavior are coherent with the lower thermal stability of the lignin compared with the acrylic matrix [56] and with the absence of interactions between the matrix and lignin. At any rate, all composites are thermally stable up to 300 °C.

Lignin has been investigated as reinforcement in the field of 3D printing to enhance the mechanical properties of the printed materials, although the reinforcing effect of lignin greatly differs depending on the lignin source, content, polymer resin, the printing techniques, and the post-curing step [26,28,56], as previously indicated. To evaluate the mechanical properties of the composites, surface hardness (Shore A) and uniaxial tensile tests till rupture were performed according to ISO 868 and 527, respectively. The stress–strain curves are plotted in Figure 5B and the calculated modulus, stress at break, and elongation at break are shown in Table 3.

Table 3. Influence of lignin content on the mechanical properties of the printed films: Shore A hardness and tensile test parameters (E = Young's modulus, σ = stress at break, and ε = elongation at break).

Sample	Hardness Shore A (°Sh)	E (MPa)	σ (MPa)	ε (%)
LG0	78.3 ± 1.5	4.5 ± 0.3	1.58 ± 0.26	40.6 ± 5.1
LG05	78.8 ± 0.6	4.2 ± 0.2	1.60 ± 0.10	43.9 ± 2.2
LG1	78.0 ± 1.1	3.9 ± 0.3	1.30 ± 0.13	37.9 ± 2.1
LG2	77.7 ± 1.6	4.3 ± 0.1	1.71 ± 0.11	44.13 ± 2.4
LG3	80.9 ± 1.3	5.4 ± 0.4	1.49 ± 0.20	36.1 ± 5.8
LG35	81.6 ± 0.8	5.77 ± 0.4	1.02 ± 0.30	29.2 ± 9.0
LG4	82.2 ± 1.2	6.4 ± 0.6	1.29 ± 0.30	30.4 ± 8.0

The addition of low values of lignin does not modify the hardness of the acrylic resin, whereas a small increase (~5%) is detected for composites with lignin contents ≥3 wt.%. This small effect of unmodified lignin on the hardness of rubbery resins can be expected due to the lignin stiffness, which is higher than that of the rubbery matrix and the lack of interaction between lignin and the resin [57].

No variations in tensile parameters are detected with respect to LG0 samples at low lignin values (≤2 wt.%) within experimental error. For the tensile modulus results, a similar trend as hardness outcomes is appreciated. The small increase in tensile modulus for lignin contents ≥ 3 wt.% has been attributed by some authors to the rigid phenolic units in lignin [26]. At the same time, a reduction around 25–28% on the elongation is revealed. This diminution is explained by the stiffness increase together with the lack of strong interaction between lignin and the matrix. The stress is not transmitted from the polymer matrix to the filler and vice versa, so the material breaks more easily [27,57]. For the tensile strength, although the variations detected for lignin values of 3.5 and 4% fall within experimental error, the observed trend is in agreement with previous publications explaining that reduction at higher loadings may be due to imperfect curing caused by lignin aggregation and uneven dispersion within the acrylic matrix, producing gaps and holes in the composites [56], which can be perceived in the SEM pictures of Figure 4D inset.

3.2.3. Contact Angles (CA)

There are relatively few publications regarding the wettability of additively manufactured materials. Depending on the production process as well as on the chemical composition, specifically the effect of additives, materials can differ in their surface morphology and consequently in their CAs [58]. The influence of lignin on the average water contact angles of printed composites is displayed in Figure 6. The contact angle for the control sample (86.6° ± 0.6) showed a rather hydrophobic nature, which agrees with that reported in the literature for acrylic rubbers with contact angles ranging between 78–89° [59,60]; the contact angle decreases slightly for lignin contents of 0.5 wt.%. and stabilizes at 70° ± 2 for lignin contents ≥ 1 wt.%. The reason may be dual. On the one hand, lignin possesses polar and hydrophilic groups (carboxyl, phenols…), therefore, more hydrophilic groups are distributed on the surface, which reduce the interfacial tension and increase its hydrophilic character, as has been reported for different polymer matrices [61] of hydrophobic nature. On the other hand, surface roughness slightly decreases with lignin content, as shown in Figure 3. Opposite to this, lignin confers a hydrophobic character on composites based on more hydrophilic resins like polylactic (PLA) or polyhydrobutyrate (PHB [62]), indicating that surface properties can be widely tuned with lignin.

This decrease in contact angle with small lignin contents may contribute to improving the impression because the resin wets the surface of the platform and the anterior layer better. Further, it facilitates subsequent painting, if necessary, for finishing the part. It has been postulated that adequate surface wettability can aid in the production of 3D printed elements for applications involving interactions with fluids, such as antistatic coatings, electrochemical sensors, microfluidic devices, among others [60].

3.3. Use of Lignin as Dispersant of Conductive Fillers

With the aim of evaluating the potential application of lignin for the manufacture of functional materials, the filler dispersing effect of lignin in conductive matrices has been evaluated in this section. For this purpose, two photocurable acrylic formulations with 5 wt.% conductive filler pTSA-PANI have been developed. To one of these formulations, 1 wt.% lignin has been added (LG1PANI5) to assess the potential improvement of its properties compared with those of the lignin-free reference (LG0PANI5).

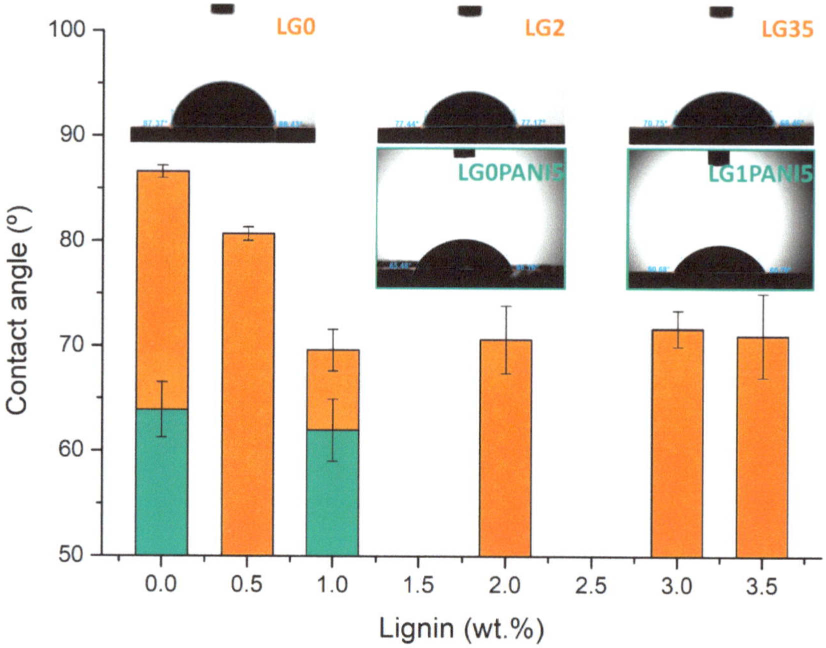

Figure 6. Water contact angle measurements on neat resin (L0), lignin composites of increasing lignin content (orange), and pTSA-PANIi composites (L0PANI5 and L1PANI5) (green).

3.3.1. pTSA-PANI Characterization

At first, the properties of pTSA-PANI were characterized by TEM, SEM, UV-vis spectroscopy, FTIR, elemental analysis and electrical conductivity. Regarding the morphology, neat pTSA-PANI consists of long fibrillar chains ($\approx$ 0.03–0.05 $\times$ 1.0–1.5 µm) partially surrounded by short nanosized granular structures with a great tendency to aggregate (Figure S2C,D). The aggregation of the fibrillar and granular structures leads to the formation of flat platelets interwoven by individual fibrillar chains (Figure S3B). The hollow-tube morphology described by Khalid et al. for PANI-PTSA is not perceived [63]. The key factor that determines the pTSA-PANI structure is the acid dopant:aniline ratio. As a general rule, the fibrous structure of pTSA-PANI, favorable for high conductivity, has been described for high acid dopant: aniline ratios, as is the present case; by contrast, lower acid dopant concentration leads to coral-like structures. Nevertheless, when the volume of the acid anion is large, as is the case for pTSA, the slower movement of molecules decreases the doping rate, which is responsible for the decrease of pTSA-PANI fiber structure [64] and the observed mixed morphology.

Both the UV-visible (Figure 2A) and FTIR spectra (Figure 7) confirm that the synthesized pTSA-PANI is in the protonated doped state. The UV-Vis spectra of pTSA-PANI dispersed in glycerin shows a band between 330–800 nm centered at 450 nm and the upward slope of a band located above 800 nm. These features are in accordance with previous literature that describes for pTSA-PANI 2 peaks in the same range at 433 and 800 nm, assigned to the shift from polaron to π^* band and from π to polaron band of the doped pTSA-PANI chains, respectively [65,66]. By contrast, Beygisangchin et al. reported the maxima of these bands at neatly lower wavelengths, 321 and 578 nm, respectively [66]. The absorptivity coefficient at 405 nm was calculated in a similar way to lignin, as 1.03 g L^{-1} cm^{-1} (r^2 = 0.9656). It was slightly lower than the value obtained for lignin, so a similar hindering of the curing process of the resin is expected.

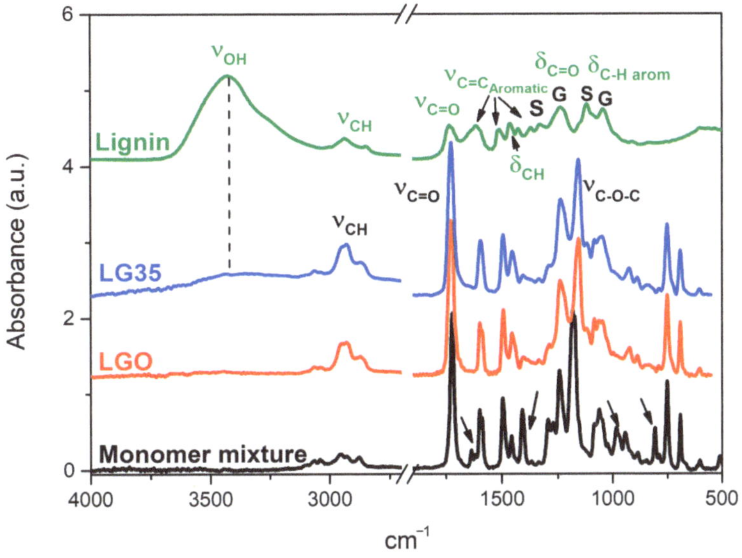

Figure 7. FTIR spectra of pTSA-PANI, pristine acrylic (LG0) and the composites LG0PANI5 and LG1PANI5.

All the characteristic peaks of -PANI plus those of the dopant, p-Toluensulfonic acid, are observed in the FTIR spectrum of d pTSA-PANI (described in detail in the Supplementary file) [63,67]. The oxidation state of the polymer was calculated through the intensity ratio of the Q/B absorption bands (I_Q/I_B) at 1561 cm^{-1} (Quinoid rings, Q) and 1467 cm^{-1} (Benzenoid rings, B); the emeraldine type structure corresponds to a value of 1.0. Oxidation mostly depends on the oxidant concentration and the pH of the reaction medium [68]. The I_Q/I_B ratio, 0.82 $\pm$ 0.03, is somewhat lower but still close to unity and similar to that obtained for PANI-DBSA synthesized under the same conditions [29]. The intrinsic oxidation of pTSA-PANI is an important feature as the quinoid imines are preferentially protonated in the protonic acid doping. The level of doping when sulfonic acids are employed can be calculated from the elemental analysis. Hence, the S/N bulk atomic ratio derived from the elemental analysis data, 0.50, agrees with a high doping level [69,70].

The electrical conductivity of pTSA-PANI filler was on average 7.6 $\pm$ 0.5 S. cm^{-1}, which is in the range of good semiconductors and is a result of the fibrous morphology, the high doping level, and the oxidation state. This conductivity is similar to the values of PANI-HCl previously obtained by this research group [31] and lies within the range of several reported pTSA-PANIs synthesized by oxidative polymerization with conductivities spanning from 1.46 to 34.8 S. cm^{-1} [63,66,71,72]; by contrast, the electrical conductivity is 2 orders of magnitude higher than pTSA-PANI obtained by redoping [73].

3.3.2. Characterization of Printed pTSA-PANI Composites

Figure 8A,B depicts the physical appearance of the two pTSA-PANI printed films. As evident from the photographs, LG1PANI5 shows a homogenous shiny surface in contrast to the uneven matte surface with pTSA-PANI lumps protruding from the lignin-free sample. The smoother morphology and more uniform distribution of pTSA-PANI aggregates in the presence of 1 wt.%. lignin is confirmed by confocal microscopy. As can be appreciated in Figure 8C,D, the relatively small pTSA-PANI clusters, evenly distributed on the surface of L1PANI5, contrast with the presence of large lumps in the lignin-free film; as a result, all the roughness parameters are clearly lower for the LG1PANI5 sample (Table 4). Its average surface roughness (R_a) decreases by 61% compared to that of LG0PANI5 film, suggesting a clear improvement in the dispersion of the fillers by incorporating lignin. Regarding the R_v and R_p parameters, a decrease of 60% is observed in R_v of LG1PANI5 compared with LG0PANI5, while for R_p, this decrease is 29%. This parameter presents a relatively high standard deviation which can be associated with the presence of scratches, cracks and irregularities unevenly distributed across and throughout the surface; the scratches are due to low surface hardness and the pTSA-PANI clusters to poor dispersion of the conductive filler within the acrylic matrix. These results agree with those reported by Yang et al. [24] concerning the surface morphologies of different PANI-lignin films, spin-coated on ITO substrates. They observed by AFM smoother and more uniform surfaces in comparison with pure PANI. These smoother surfaces were attributed to the better dispersibility [24].

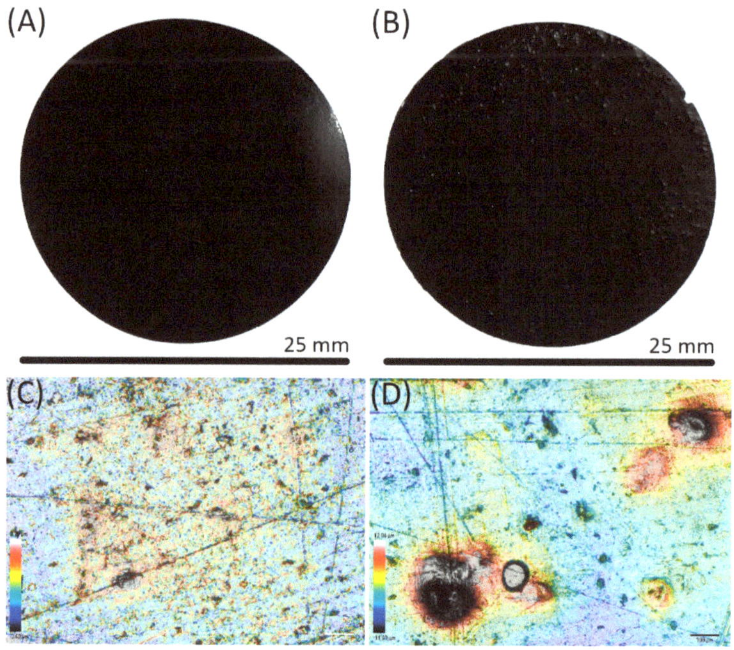

Figure 8. Digital pictures of printed films of (**A**) LG1PANI5 and (**B**) LG0PANI5 and representative confocal microscope images of (**C**) LG1PANI5 and (**D**) LG0PANI5.

Table 4. R_a (roughness average), R_v (average max roughness valley depth), and R_p (average max roughness peak height) of the pTSA-PANI acrylic composites.

Sample	R_a	R_v	R_p
PANI5LG0	3.6 ± 0.5	80.1 ± 9.9	72.5 ± 16.3
PANI5LG1	1.4 ± 0.2	32.2 ± 6.2	51.5 ± 26.5

R_a value indicates the absolute values of the profile heights over the evaluation length, R_p the height of the main peak structures. and R_v estimates the average depths of the fissures.

To further monitor the dispersion condition of PANI in the acrylic matrix, the morphologies LG0PANI and LG1PANI are assessed by SEM. The surface morphologies are consistent with the physical appearances and images were taken by confocal microscopy. In relation with the cryo-fractured cross-section of both films (Figure S5), the horizontal lines corresponding to the individual printing layers are perceived together with phase separation, as great PANI agglomerates lodged in cavities, completely detached from the matrix, are observed. These images prove the little interaction between the acrylic matrix and the conductive charge in the absence of lignin, whereas the addition of lignin improves the embedding of PANI in the matrix, as shown in more detail in Figure S6. Further, there are also some differences related to the fraction mechanism, as the crack's formation and propagation are greater in LG1PANI5 with respect LG0PANI5 or even LG2 and similar to LG3.

From another point of view, the contact angles were also determined in order to deepen the knowledge about surface properties and their ability to create new interfaces (Figure 6), LG0PTSA 5 and LG1PTSA5 3D printed film's contact angles were (63.90° ± 2.64) and (61.70° ± 2.90), respectively. No differences are found between them within experimental error (Student test, $p > 0.05$), although both surfaces are more hydrophilic than the neat acrylic resin (86.60° ± 0.60) and LG1 (69.92° ± 2.00). The decrease in contact angles compared to the neat resin is due to the more hydrophilic nature of PANI salts and the presence of amine groups in its structure, although the contact angles ultimately depend on the dopant used [74]. Contact angles of 69.9° ± 1.7 have been reported for polymer membranes coated with pTSA-PANI [75]. The higher hydrophilicity provided by PANI compared to lignin explains why no differences are observed between the two PANI composites. Further, the difference in surface roughness between LG0PTSA5 AND LG1PTSA5 is not reflected in the contact angles, probably due to the relatively high experimental error in the measurement. At any rate, these outcomes support the good wetting properties of both composites.

Regarding the ATR spectra of the composites, L0PANI5 and L1PANIT 5, in Figure 7 the specific bands of PANI and lignin are not clearly identified on the printed surface, as the spectra of both composites are practically superimposable to that of the pure resin, indicating that fillers are mostly embedded within the resin. Nevertheless, a detailed observation allows perceiving minor increases in the absorbance region between 3400–3200 cm^{-1}, corresponding to the stretching vibrations of the NH groups of PANI and the OH groups of lignin and a small shoulder at around 1688 cm^{-1}, which may be related to the stretching of the C=O groups of the acrylic resin H-bonded with PANI. Further, a small band is distinguished at 1640 cm^{-1} (C=C stretching region of aromatic rings present in the acid dopant (pTSA)). These changes are more noticeable in the composite without lignin (L0PANI5), where an additional band at 1540 cm^{-1} (Q band of PANI) is observed.

The influence of 5 wt.% pTSA-PANI without o with 1 wt.% lignin on the degree of monomer conversion (DBC_∞%) of the post-cured printed films was studied. The DBC_∞% values of the composites showed no differences between them, within experimental error, but were ~3–4% lower than the DBC_∞% of the pristine resin (DBC%: 97.5 ± 0.6, 93.8 ± 4.8 and 94.3 ± 2.8 for LG0, L0PANI5 and L1PANI5, respectively). These are expected results similar to those obtained in previous works for composites of the same acrylic matrix [23,31] filled with similar contents of PANI-HCl, and are explained by the relatively high absorptivity value of pTSA-PANI at 405 nm, which strongly competes with the light adsorption of the photoinitiator. Besides, the standard deviations of these values are greater in the composites than in the reference sample due to poor filler dispersion and is worse in the composite without 1wt.% lignin. This outcome agrees with the results of confocal microscopy showing the better inclusion of PANI aggregates within the resin in LG1PANI5.

Last but not least, the electrical conductivity of the neat acrylic resin is lower than 1.0×10^{-8} S cm^{-1}, which is the limit of detection of the instrument. The conductivity increased with the inclusion of 5 wt.% pTSA-PANI up to $(1.7 \pm 0.1) \times 10^{-7}$ S cm^{-1}. When 1 wt.% Lignin is added to the formulation with 5 wt.% pTSA-PANI (PANI5LG1), the conductivity increases by an order of magnitude to $(1.6 \pm 0.1) \times 10^{-6}$ S cm^{-1}, suggesting a small enhancement in the electrical properties of the ternary composites. These results agree with confocal microscopy images, since the presence of lignin suggests an improvement of the filler dispersion in the matrix, which translates into the existence of a continuous network of pTSA-PANI across the insulating acrylic matrix. Henceforth, these values are in the range of semiconductors. Similar conductivity on the order of 10^{-6} S cm^{-1} was previously achieved by conductive composites based on photocurable epoxy resins doped with 15 wt.% pTSA-PANI [76,77]. Regarding the incorporation of PANI as conductive filler in photocurable resins, very few studies exist, and the results obtained show medium–low conductivity values. Table 5 summarizes the main recent research on light-curable 3D printed resins with electrical properties based on PANI. Although PANI/MWCNT has been extensively studied in the field of electronics, only one previous research used this filler in a photocurable resin for SLA/DLP.

Table 5. Recent research on light-curable 3D printed resins with electrical properties based on PANI.

Printing Technique	Filler	Loading (wt.%)	Base Matrix	Conductivity (S cm^{-1})	Ref
DLP	graphene sheets/PANI	1.2/5	Polyacrylate resin	4.0×10^{-9}	[78]
DLP	PANI	5	Polyacrylate resin	1.0×10^{-10}	[78]
DLP	PANI	6	Polyurethane	9.3×10^{-7}	[79]
DLP	PANI	3	Acrylic resin	2.2×10^{-5}	[31]
DLP	PANI/MWCNT	2	Acrylic resin	7.4×10^{-4}	[32]
DLP	pTSA-PANI	5	Acrylic resin	1.7×10^{-7}	Present work
DLP	pTSA-PANI/LIGNIN	5/1	Acrylic resin	1.6×10^{-6}	Present work

In contrast, as far as the authors are concerned, there are no data on electrical conductivity of DLP 3D printed composites using pTSA-PANI filler, nor any study that employs unmodified lignin as a compatibilizer. Only a few works have been found in the literature about 3D-DLP printing of conductive light-curing acrylic resins based on polyanilines with other dopants and other conductive fillers. The electrical conductivity reached in the current study is higher than those reported for photocurable acrylic and polyurethane 3D-DLP printed composites loaded with HCl doped PANI between 1–6 wt.% (10^{-10}–10^{-7} S cm^{-1}) [78,79]. The data are slightly lower than those obtained for DLP 3D printed acrylic resins filled with 0.3–0.6 wt.% carbon nanotubes (10^{-4}–10^{-5} S cm^{-1}) [80–82] and similar contents of PANI-HCl (10^{-5} S cm^{-1}) [31] or a mixture of PANI/MWCNT (10^{-4} S cm^{-1}) [32]. Simultaneously, the conductivity values achieved are similar or higher than those reported for printable photocurable resins using 16 wt.% (10^{-6} S cm^{-1}) [83] and 1 wt.% (10^{-11} S cm^{-1}) [84] of Ag nanoparticles or 2 wt.% reduced graphene oxide (rGO) (10^{-7} S cm^{-1}) [85]. By contrast, some authors have reported conductivity values of the order of 10^{-2}–10^{-3} S cm^{-1} with 6 wt.% of rGO [86] or using 20 wt.% of PEDOT [60]. Nevertheless, the use of high amounts of filler is not desirable considering their cost and the negative effect on the printing process of the material [86].

4. Conclusions

This work reports on the systematic study of the structure–property relationship of 3D printed composites with increasing amounts of unmodified organosolv lignin in a light-curable acrylic matrix. Amounts of lignin below 3 wt.% can be successfully and reproducibly printed by LCD vat polymerization. This limit is due to the strong UV absorption at the critical wavelength of the UV lamp of the printer, which greatly decreases the rate of polymerization and the cure depth. The printed polymer composites have high degrees of conversion, rich acrylic surfaces, hardly any interactions between the filler and the matrix, and discrete lignin aggregates within the polymer matrix. For composites with lignin proportions below 3 wt.%, the bulk properties of the pristine resin are maintained, simultaneously improving the surface properties, resulting in smoother surfaces, increased Shore A hardness, and better wettability. These properties are beneficial from the point of view of 3D printing elements for applications involving interactions with fluids and facilitate subsequent finishing of the parts if necessary.

The effect of organosolv lignin as a dispersant of pTSA-PANI was assessed. A clear improvement in the dispersion of the conductive filler was achieved with as little as 1 wt.%

lignin in the formulation, resulting in more homogeneous samples with less surface roughness, better appearance, and electrical conductivity enhanced by one order of magnitude, up to 10^{-6} S cm^{-1}, without negatively affecting printability.

In brief, the present research opens the possibility of developing a range of novel solvent-free, eco-friendly photocurable nanocomposites for the fabrication of functional materials at low costs, valorizing a natural resource such as lignin.

Supplementary Materials: The following supporting information can be downloaded at: https://www.mdpi.com/article/10.3390/polym14194164/s1, Figure S1. Viscosity values of liquid formulations as a function of shear rate and Lignin amount at room temperature; Figure S2. TEM images of (A) and (B) PANI and (C) and (D) Lignin; Figure S3. Representative SEM images of (A) PANI and (B) Lignin with magnitude amplification of 10,000×; Figure S4. FTIR spectrum of p-TSA-PANI; Figure S5. Representative SEM images of (A) LG0PANI5 surface with magnitude amplification of 40×, (B) LG0PANI5 cryo-fractured cross-sections with magnitude amplification of 100× and (C) X5000. (D) LG1PANI5 surface with magnitude amplification of 40×, (E) LG1PANI5 cryo-fractured cross-sections with magnitude amplification of 100× and (F) X5000. Figure S6. SEM images of cryo-fractured cross-sections with magnitude amplification of 10,000× (A) LG0PANI5 and (B) LG1PANI5.

Author Contributions: G.A.-F.: conceptualization, methodology, validation, formal analysis, investigation, writing—original draft, visualization. A.L.-L.: methodology, investigation, validation, formal analysis, writing—original draft, visualization. A.A.-P.: writing—review and editing. M.S.D.-G.: conceptualization, methodology, validation, writing—review and editing, supervision. P.L.: investigation, resources. S.M.G.-G.: investigation, resources. M.-J.A.: conceptualization, writing—review and editing, supervision, project administration, Funding acquisition. All authors have read and agreed to the published version of the manuscript.

Funding: Goretti Arias-Ferreiro is thankful for the financial funding received from the Xunta de Galicia and the European Union (ED481A-2019/001). The authors are thankful for the financial support from the Ministerio de Ciencia e Innovacion/FEDER (reLiCom3D project ref PID2020-116976RB-I00) and Xunta de Galicia-FEDER (ED431C 2019/17 and ED431B 2019/44).

Institutional Review Board Statement: Not applicable.

Data Availability Statement: The data presented in this study are available on request from the corresponding author.

Conflicts of Interest: The authors declare no conflict of interest.

References

1. Fertier, L.; Koleilat, H.; Stemmelen, M.; Giani, O.; Joly-Duhamel, C.; Lapinte, V.; Robin, J.-J. The use of renewable feedstock in UV-curable materials—A new age for polymers and green chemistry. *Prog. Polym. Sci.* **2013**, *38*, 932–962. [CrossRef]
2. Wang, Q.; Sun, J.; Yao, Q.; Ji, C.; Liu, J.; Zhu, Q. 3D printing with cellulose materials. *Cellulose* **2018**, *25*, 4275–4301. [CrossRef]
3. Zhang, J.; Xiao, P. 3D printing of photopolymers. *Polym. Chem.* **2018**, *9*, 1530–1540. [CrossRef]
4. Mendes-Felipe, C.; Oliveira, J.; Etxebarria, I.; Vilas-Vilela, J.L.; Lanceros-Mendez, S. State-of-the-Art and Future Challenges of UV Curable Polymer-Based Smart Materials for Printing Technologies. *Adv. Mater. Technol.* **2019**, *4*, 1–16. [CrossRef]
5. El Magri, A.; Bencaid, S.E.; Vanaei, H.R.; Vaudreuil, S. Effects of Laser Power and Hatch Orientation on Final Properties of PA12 Parts Produced by Selective Laser Sintering. *Polymers* **2022**, *14*, 3674. [CrossRef]
6. Bártolo, P.J. (Ed.) *Stereolithography Materials, Processes and Applications*; Springer: New York, NY, USA, 2011; ISBN 9780387929033.
7. Ebers, L.S.; Arya, A.; Bowland, C.C.; Glasser, W.G.; Chmely, S.C.; Naskar, A.K.; Laborie, M.P. 3D printing of lignin: Challenges, opportunities and roads onward. *Biopolymers* **2021**, *112*, e23431. [CrossRef]
8. Wang, F.; Ouyang, D.; Zhou, Z.; Page, S.J.; Liu, D.; Zhao, X. Lignocellulosic biomass as sustainable feedstock and materials for power generation and energy storage. *J. Energy Chem.* **2021**, *57*, 247–280. [CrossRef]
9. Culebras, M.; Sanchis, M.J.; Beaucamp, A.; Carsí, M.; Kandola, B.K.; Horrocks, A.R.; Panzetti, G.; Birkinshaw, C.; Collins, M.N. Understanding the thermal and dielectric response of organosolv and modified kraft lignin as a carbon fibre precursor. *Green Chem.* **2018**, *20*, 4461–4472. [CrossRef]
10. Collins, M.N.; Nechifor, M.; Tanasă, F.; Zănoagă, M.; McLoughlin, A.; Stróżyk, M.A.; Culebras, M.; Teacă, C.A. Valorization of lignin in polymer and composite systems for advanced engineering applications—A review. *Int. J. Biol. Macromol.* **2019**, *131*, 828–849. [CrossRef]

11. Culebras, M.; Geaney, H.; Beaucamp, A.; Upadhyaya, P.; Dalton, E.; Ryan, K.M.; Collins, M.N. Bio-derived Carbon Nanofibres from Lignin as High-Performance Li-Ion Anode Materials. *ChemSusChem* **2019**, *12*, 4516–4521. [CrossRef]
12. Lai, C.; Zhou, Z.; Zhang, L.; Wang, X.; Zhou, Q.; Zhao, Y.; Wang, Y.; Wu, X.F.; Zhu, Z.; Fong, H. Free-standing and mechanically flexible mats consisting of electrospun carbon nanofibers made from a natural product of alkali lignin as binder-free electrodes for high-performance supercapacitors. *J. Power Sources* **2014**, *247*, 134–141. [CrossRef]
13. García-Negrón, V.; Chmely, S.C.; Ilavsky, J.; Keffer, D.J.; Harper, D.P. Development of Nanocrystalline Graphite from Lignin Sources. *ACS Sustain. Chem. Eng.* **2022**, *10*, 1786–1794. [CrossRef]
14. Grossman, A.; Wilfred, V. Lignin-based polymers and nanomaterials. *Curr. Opin. Biotechnol.* **2019**, *56*, 112–120. [CrossRef] [PubMed]
15. Thakur, V.K.; Thakur, M.K.; Raghavan, P.; Kessler, M.R. Progress in green polymer composites from lignin for multifunctional applications: A review. *ACS Sustain. Chem. Eng.* **2014**, *2*, 1072–1092. [CrossRef]
16. Wen, M.; Zhao, Y.; Li, Z.; Lai, S.; Zeng, Q.; Liu, C.; Liu, Y. Preparation of lignin-based carbon/polyaniline composites for advanced microwave absorber. *Diam. Relat. Mater.* **2021**, *111*, 108219. [CrossRef]
17. Bozó, É.; Ervasti, H.; Halonen, N.; Shokouh, S.H.H.; Tolvanen, J.; Pitkänen, O.; Järvinen, T.; Pálvölgyi, P.S.; Szamosvölgyi, Á.; Sápi, A.; et al. Bioplastics and Carbon-Based Sustainable Materials, Components, and Devices: Toward Green Electronics. *ACS Appl. Mater. Interfaces* **2021**, *13*, 49301–49312. [CrossRef]
18. Dianat, N.; Rahmanifar, M.S.; Noori, A.; El-Kady, M.F.; Chang, X.; Kaner, R.B.; Mousavi, M.F. Polyaniline-Lignin Interpenetrating Network for Supercapacitive Energy Storage. *Nano Lett.* **2021**, *21*, 9485–9493. [CrossRef]
19. Tanguy, N.R.; Wu, H.; Nair, S.S.; Lian, K.; Yan, N. Lignin Cellulose Nanofibrils as an Electrochemically Functional Component for High-Performance and Flexible Supercapacitor Electrodes. *ChemSusChem* **2021**, *14*, 1057–1067. [CrossRef]
20. Ye, W.; Li, X.; Luo, J.; Wang, X.; Sun, R. Lignin as a green reductant and morphology directing agent in the fabrication of 3D graphene-based composites for high-performance supercapacitors. *Ind. Crops Prod.* **2017**, *109*, 410–419. [CrossRef]
21. Wang, L.; Li, X.; Xu, H.; Wang, G. Construction of polyaniline/lignin composite with interpenetrating fibrous networks and its improved electrochemical capacitance performances. *Synth. Met.* **2019**, *249*, 40–46. [CrossRef]
22. Razaq, A.; Asif, M.H.; Kalsoom, R.; Khan, A.F.; Awan, M.S.; Ishrat, S.; Ramay, S.M. Conductive and electroactive composite paper reinforced by coating of polyaniline on lignocelluloses fibers. *J. Appl. Polym. Sci.* **2015**, *132*, 3–7. [CrossRef]
23. Arias-ferreiro, G.; Ares-pernas, A.; Lasagabáster-latorre, A.; Dopico-garcía, M.S.; Ligero, P.; Pereira, N.; Costa, P.; Lanceros-Mendez, S.; Abad, M. Photocurable Printed Piezocapacitive Pressure Sensor Based on an Acrylic Resin Modified with Polyaniline and Lignin. *Adv. Mater. Technol* **2022**, *7*, 2101503. [CrossRef]
24. Yang, D.; Huang, W.; Qiu, X.; Lou, H.; Qian, Y. Modifying sulfomethylated alkali lignin by horseradish peroxidase to improve the dispersibility and conductivity of polyaniline. *Appl. Surf. Sci.* **2017**, *426*, 287–293. [CrossRef]
25. Sutton, J.T.; Rajan, K.; Harper, D.P.; Chmely, S.C. Improving uv curing in organosolv lignin-containing photopolymers for stereolithography by reduction and acylation. *Polymers* **2021**, *13*, 3473. [CrossRef] [PubMed]
26. Zhang, S.; Li, M.; Hao, N.; Ragauskas, A.J. Stereolithography 3D Printing of Lignin-Reinforced Composites with Enhanced Mechanical Properties. *ACS Omega* **2019**, *4*, 20197–20204. [CrossRef]
27. Ibrahim, F.; Mohan, D.; Sajab, M.S.; Bakarudin, S.B.; Kaco, H. Evaluation of the Compatibility of Organosolv Lignin-Graphene Nanoplatelets with Photo-Curable Polyurethane in Stereolithography 3D Printing. *Polymers* **2019**, *11*, 1544. [CrossRef]
28. Sutton, J.T.; Rajan, K.; Harper, D.P.; Chmely, S.C. Lignin-Containing Photoactive Resins for 3D Printing by Stereolithography. *ACS Appl. Mater. Interfaces* **2018**, *10*, 36456–36463. [CrossRef]
29. Dopico-garcía, M.S.; Ares, A.; Lasagabáster-latorre, A.; García, X. Extruded polyaniline/EVA blends: Enhancing electrical conductivity using gallate compatibilizers. *Synth. Met.* **2014**, *189*, 193–202. [CrossRef]
30. Arias-Ferreiro, G.; Ares-Pernas, A.; Dopico-García, M.S.; Lasagabáster-Latorre, A.; Abad, M.-J. Photocured conductive PANI/acrylate composites for digital light processing. Influence of HDODA crosslinker in rheological and physicochemical properties. *Eur. Polym. J.* **2020**, *136*, 109887. [CrossRef]
31. Arias-ferreiro, G.; Ares-pernas, A.; Lasagabáster-latorre, A.; Aranburu, N.; Guerrica-echevarria, G.; Dopico-garcía, M.S.; Abad, M.J. Printability study of a conductive polyaniline/acrylic formulation for 3d printing. *Polymers* **2021**, *13*, 2068. [CrossRef]
32. Arias-Ferreiro, G.; Lasagabáster-Latorre, A.; Ares-Pernas, A.; Dopico-García, M.S.; Pereira, N.; Costa, P.; Lanceros-Mendez, S.; Abad, M.-J. Flexible 3D Printed Acrylic Composites based on Polyaniline/Multiwalled Carbon Nanotubes for Piezoresistive Pressure Sensors. *Adv. Electron. Mater.* **2022**, 2200590. [CrossRef]
33. *UNE-EN ISO 868*; Asociación Española de Normalización y Certificación. Determinación de la Dureza de Indentación por Medio de un Durómetro (Dureza Shore). AENOR: Madrid, Spain, 2003.
34. Voet, V.S.D.D.; Strating, T.; Schnelting, G.H.M.M.; Dijkstra, P.; Tietema, M.; Xu, J.; Woortman, A.J.J.J.; Loos, K.; Jager, J.; Folkersma, R. Biobased Acrylate Photocurable Resin Formulation for Stereolithography 3D Printing. *ACS Omega* **2018**, *3*, 1403–1408. [CrossRef] [PubMed]
35. Cortés, A.; Cosola, A.; Sangermano, M.; Campo, M.; González Prolongo, S.; Pirri, C.F.; Jiménez-Suárez, A.; Chiappone, A. DLP 4D-Printing of Remotely, Modularly, and Selectively Controllable Shape Memory Polymer Nanocomposites Embedding Carbon Nanotubes. *Adv. Funct. Mater.* **2021**, *31*, 2106774. [CrossRef]
36. Bennett, J. Measuring UV curing parameters of commercial photopolymers used in additive manufacturing. *Addit. Manuf.* **2017**, *18*, 203–212. [CrossRef] [PubMed]

37. Pezzana, L.; Malmström, E.; Johansson, M.; Sangermano, M. UV-curable bio-based polymers derived from industrial pulp and paper processes. *Polymers* **2021**, *13*, 1530. [CrossRef] [PubMed]
38. Li, X.; Zhong, H.; Zhang, J.; Duan, Y.; Li, J.; Jiang, D. Fabrication of zirconia all-ceramic crown via DLP-based stereolithography. *Int. J. Appl. Ceram. Technol.* **2020**, *17*, 844–853. [CrossRef]
39. Zhang, X.; Keck, S.; Qi, Y.; Baudis, S.; Zhao, Y. Study on Modified Dealkaline Lignin as Visible Light Macromolecular Photoinitiator for 3D Printing. *ACS Sustain. Chem. Eng.* **2020**, *8*, 10959–10970. [CrossRef]
40. Jafarzadeh, S.; Johansson, M.; Sundell, P.E.; Claudino, M.; Pan, J.; Claesson, P.M. UV-curable acrylate-based nanocomposites: Effect of polyaniline additives on the curing performance. *Polym. Adv. Technol.* **2013**, *24*, 668–678. [CrossRef]
41. Tomeckova, V.; Teyssandier, F.; Norton, S.J.; Love, B.J.; Halloran, J.W. Photopolymerization of acrylate suspensions. *J. Photochem. Photobiol. A Chem.* **2012**, *247*, 74–81. [CrossRef]
42. Ganewatta, M.S.; Lokupitiya, H.N.; Tang, C. Lignin biopolymers in the age of controlled polymerization. *Polymers* **2019**, *11*, 1176. [CrossRef]
43. Piccinino, D.; Capecchi, E.; Tomaino, E.; Gabellone, S.; Gigli, V.; Avitabile, D.; Saladino, R. Nano-structured lignin as green antioxidant and uv shielding ingredient for sunscreen applications. *Antioxidants* **2021**, *10*, 274. [CrossRef] [PubMed]
44. Goliszek, M.; Podkościelna, B.; Klepka, T.; Sevastyanova, O. Preparation, Thermal, and Mechanical Characterization of UV-Cured Polymer Biocomposites with Lignin. *Polymers* **2020**, *12*, 1159. [CrossRef] [PubMed]
45. Jacobs, P.F. *Rapid Prototyping & Manufacturing. Fundamentals of StereoLithography*, 1st ed.; Society of Manufacturing Engineers: Dearborn, MI, USA, 1992; ISBN 0-87263-425-6.
46. Gojzewski, H.; Guo, Z.; Grzelachowska, W.; Ridwan, M.G.; Hempenius, M.A.; Grijpma, D.W.; Vancso, G.J. Layer-by-Layer Printing of Photopolymers in 3D: How Weak is the Interface? *ACS Appl. Mater. Interfaces* **2020**, *12*, 8908–8914. [CrossRef]
47. Rojo, E.; Peresin, M.S.; Sampson, W.W.; Hoeger, I.C.; Vartiainen, J.; Laine, J.; Rojas, O.J. Comprehensive elucidation of the effect of residual lignin on the physical, barrier, mechanical and surface properties of nanocellulose films. *Green Chem.* **2015**, *17*, 1853–1866. [CrossRef]
48. Villaverde, J.J.; Li, J.; Ek, M.; Ligero, P.; De Vega, A. Native lignin structure of Miscanthus x giganteus and its changes during acetic and formic acid fractionation. *J. Agric. Food Chem.* **2009**, *57*, 6263–6270. [CrossRef] [PubMed]
49. Seo, J.H.; Choi, C.S.; Bae, J.H.; Jeong, H.; Lee, S.H.; Kim, Y.S. Preparation of a lignin/polyaniline composite and its application in Cr(VI) removal from aqueous solutions. *BioResources* **2019**, *14*, 9169–9182. [CrossRef]
50. Fodil Cherif, M.; Trache, D.; Brosse, N.; Benaliouche, F.; Tarchoun, A.F. Comparison of the Physicochemical Properties and Thermal Stability of Organosolv and Kraft Lignins from Hardwood and Softwood Biomass for Their Potential Valorization. *Waste Biomass Valoriz.* **2020**, *11*, 6541–6553. [CrossRef]
51. Tejado, A.; Peña, C.; Labidi, J.; Echeverria, J.M.; Mondragon, I. Physico-chemical characterization of lignins from different sources for use in phenol-formaldehyde resin synthesis. *Bioresour. Technol.* **2007**, *98*, 1655–1663. [CrossRef]
52. Wen, J.L.; Xue, B.L.; Sun, S.L.; Sun, R.C. Quantitative structural characterization and thermal properties of birch lignins after auto-catalyzed organosolv pretreatment and enzymatic hydrolysis. *J. Chem. Technol. Biotechnol.* **2013**, *88*, 1663–1671. [CrossRef]
53. Ke, J.; Singh, D.; Yang, X.; Chen, S. Thermal characterization of softwood lignin modification by termite Coptotermes formosanus (Shiraki). *Biomass Bioenergy* **2011**, *35*, 3617–3626. [CrossRef]
54. Sahoo, S.; Seydibeyoğlu, M.Ö.; Mohanty, A.K.; Misra, M. Characterization of industrial lignins for their utilization in future value added applications. *Biomass Bioenergy* **2011**, *35*, 4230–4237. [CrossRef]
55. Cocca, M.; D'Arienzo, L.; D'Orazio, L.; Gentile, G.; Martuscelli, E. Polyacrylates for conservation: Chemico-physical properties and durability of different commercial products. *Polym. Test.* **2004**, *23*, 333–342. [CrossRef]
56. Feng, X.; Yang, Z.; Chmely, S.; Wang, Q.; Wang, S.; Xie, Y. Lignin-coated cellulose nanocrystal filled methacrylate composites prepared via 3D stereolithography printing: Mechanical reinforcement and thermal stabilization. *Carbohydr. Polym.* **2017**, *169*, 272–281. [CrossRef] [PubMed]
57. Alsulami, Q.A.; Albukhari, S.M.; Hussein, M.A.; Tay, G.S.; Rozman, H.D. Biodegradable lignin as a reactive raw material in UV curable systems. *Polym. Technol. Mater.* **2020**, *59*, 1387–1406. [CrossRef]
58. Neukäufer, J.; Seyfang, B.; Grützner, T. Investigation of Contact Angles and Surface Morphology of 3D-Printed Materials. *Ind. Eng. Chem. Res.* **2020**, *59*, 6761–6766. [CrossRef]
59. Yang, N.; Zhang, D.D.; Li, X.D.; Lu, Y.Y.; Qiu, X.H.; Zhang, J.S.; Kong, J. Topography, Wettability, and Electrostatic Charge Consist Mayor Surface Properties of Intraocular Lenses. *Curr. Eye Res.* **2017**, *42*, 201–210. [CrossRef]
60. Scordo, G.; Bertana, V.; Scaltrito, L.; Ferrero, S.; Cocuzza, M.; Marasso, S.L.; Romano, S.; Sesana, R.; Catania, F.; Pirri, C.F. A novel highly electrically conductive composite resin for stereolithography. *Mater. Today Commun.* **2019**, *19*, 12–17. [CrossRef]
61. Serna, D.L.; Martínez, P.E.; González, M.Á.R.; Cadena, A.A.Z.; Contreras, E.A.Z.; Anguiano, M.G.S. Synthesis and characterization of a lignin-styrene-butyl acrylate based composite. *Polymers* **2019**, *11*, 1080. [CrossRef]
62. Vaidya, A.A.; Collet, C.; Gaugler, M.; Lloyd-Jones, G. Integrating softwood biorefinery lignin into polyhydroxybutyrate composites and application in 3D printing. *Mater. Today Commun.* **2019**, *19*, 286–296. [CrossRef]
63. Khalid, M.; Tumelero, M.A.; Brandt, I.S.; Zoldan, V.C.; Acuña, J.J.S.; Pasa, A.A. Electrical Conductivity Studies of Polyaniline Nanotubes Doped with Different Sulfonic Acids. *Indian J. Mater. Sci.* **2013**, *2013*, 718304. [CrossRef]

64. Li, S.; Tao, Y.; Maryum, P.; Wang, Q.; Zhu, J.; Min, F.; Cheng, H.; Zhao, S.; Wang, C. Bifunctional polyaniline electroconductive hydrogels with applications in supercapacitor and wearable strain sensors. *J. Biomater. Sci. Polym. Ed.* **2020**, *31*, 938–953. [CrossRef] [PubMed]
65. Usman, F.; Dennis, J.O.; Meriaudeau, F.; Ahmed, A.Y.; Seong, K.C.; Fen, Y.W.; Sadrolhosseini, A.R.; Abdulkadir, B.A.; Ayinla, R.T.; Daniyal, W.M.E.M.M.; et al. Dependence of the optical constant parameters of p-toluene sulfonic acid-doped polyaniline and its composites on dispersion solvents. *Molecules* **2020**, *25*, 4414. [CrossRef] [PubMed]
66. Beygisangchin, M.; Rashid, S.A.; Shafie, S.; Sadrolhosseini, A.R. Polyaniline synthesized by different dopants for fluorene detection via photoluminescence spectroscopy. *Materials* **2021**, *14*, 7382. [CrossRef]
67. Usman, F.; Dennis, J.O.; Ahmed, A.Y.; Seong, K.C.; Fen, Y.W.; Sadrolhosseini, A.R.; Meriaudeau, F.; Kumar, P.; Ayodele, O.B. Structural characterization and optical constants of p-toluene sulfonic acid doped polyaniline and its composites of chitosan and reduced graphene-oxide. *J. Mater. Res. Technol.* **2020**, *9*, 1468–1476. [CrossRef]
68. Abdiryim, T.; Xiao-Gang, Z.; Jamal, R. Comparative studies of solid-state synthesized polyaniline doped with inorganic acids. *Mater. Chem. Phys.* **2005**, *90*, 367–372. [CrossRef]
69. Horta-Romarís, L.; Abad, M.-J.; González-Rodríguez, M.V.; Lasagabáster, A.; Costa, P.; Lanceros-Méndez, S. Cyclic temperature dependence of electrical conductivity in polyanilines as a function of the dopant and synthesis method. *Mater. Des.* **2017**, *114*, 288–296. [CrossRef]
70. Horta Romarís, L.; González Rodríguez, M.V.; Huang, B.; Costa, P.; Lasagabáster Latorre, A.; Lanceros-Mendez, S.; Abad López, M.J. Multifunctional electromechanical and thermoelectric polyaniline–poly(vinyl acetate) latex composites for wearable devices. *J. Mater. Chem. C* **2018**, *6*, 8502–8512. [CrossRef]
71. Kulkarni, M.V.; Viswanath, A.K.; Aiyer, R.C.; Khanna, P.K. Synthesis, characterization, and morphology of p-toluene sulfonic acid-doped polyaniline: A material for humidity sensing application. *J. Polym. Sci. Part B Polym. Phys.* **2005**, *43*, 2161–2169. [CrossRef]
72. Makeiff, D.A.; Huber, T. Microwave absorption by polyaniline-carbon nanotube composites. *Synth. Met.* **2006**, *156*, 497–505. [CrossRef]
73. Surajit, S.; Sambhu, B.; Khastgir, D. Effect of Dopant Type on the Properties of Polyaniline Surajit. *J. Appl. Polym. Sci.* **2009**, *112*, 3135–3140. [CrossRef]
74. Zhang, Z.; Wei, Z.; Wan, M. Nanostructures of Polyaniline Doped with Inorganic Acids. *Macromolecules* **2002**, *35*, 5937–5942. [CrossRef]
75. Reig, M.; Farrokhzad, H.; Van der Bruggen, B.; Gibert, O.; Cortina, J.L. Synthesis of a monovalent selective cation exchange membrane to concentrate reverse osmosis brines by electrodialysis. *Desalination* **2015**, *375*, 1–9. [CrossRef]
76. Khandelwal, V.; Sahoo, S.K.; Kumar, A.; Manik, G. Electrically conductive green composites based on epoxidized linseed oil and polyaniline: An insight into electrical, thermal and mechanical properties. *Compos. Part B Eng.* **2018**, *136*, 149–157. [CrossRef]
77. Khandelwal, V.; Sahoo, S.K.; Kumar, A.; Manik, G. Study on the effect of carbon nanotube on the properties of electrically conductive epoxy/polyaniline adhesives. *J. Mater. Sci. Mater. Electron.* **2017**, *28*, 14240–14251. [CrossRef]
78. Han, H.; Cho, S. Fabrication of Conducting Polyacrylate Resin Solution with Polyaniline Nanofiber and Graphene for Conductive 3D Printing Application. *Polymers* **2018**, *10*, 1003. [CrossRef]
79. Joo, H.; Cho, S. Comparative studies on polyurethane composites filled with polyaniline and graphene for DLP-type 3D printing. *Polymers* **2020**, *12*, 67. [CrossRef]
80. Gonzalez, G.; Chiappone, A.; Roppolo, I.; Fantino, E.; Bertana, V.; Perrucci, F.; Scaltrito, L.; Pirri, F.; Sangermano, M. Development of 3D printable formulations containing CNT with enhanced electrical properties. *Polymer* **2017**, *109*, 246–253. [CrossRef]
81. Mu, Q.; Wang, L.; Dunn, C.K.; Kuang, X.; Duan, F.; Zhang, Z.; Qi, H.J.; Wang, T. Digital light processing 3D printing of conductive complex structures. *Addit. Manuf.* **2017**, *18*, 74–83. [CrossRef]
82. Mendes-Felipe, C.; Oliveira, J.; Costa, P.; Ruiz-Rubio, L.; Iregui, A.; González, A.; Vilas, J.L.; Lanceros-Mendez, S. Stimuli responsive UV cured polyurethane acrylated/carbon nanotube composites for piezoresistive sensing. *Eur. Polym. J.* **2019**, *120*, 109226. [CrossRef]
83. Fantino, E.; Chiappone, A.; Roppolo, I.; Manfredi, D.; Bongiovanni, R.; Pirri, C.F.; Calignano, F. 3D Printing of Conductive Complex Structures with in Situ Generation of Silver Nanoparticles. *Adv. Mater.* **2016**, *28*, 3712–3717. [CrossRef]
84. Sciancalepore, C.; Moroni, F.; Messori, M.; Bondioli, F. Acrylate-based silver nanocomposite by simultaneous polymerization–reduction approach via 3D stereolithography. *Compos. Commun.* **2017**, *6*, 11–16. [CrossRef]
85. Qian, C.; Xiao, T.; Chen, Y.; Wang, N.; Li, B.; Gao, Y. 3D Printed Reduced Graphene Oxide/Elastomer Resin Composite with Structural Modulated Sensitivity for Flexible Strain Sensor. *Adv. Eng. Mater.* **2022**, *24*, 2101068. [CrossRef]
86. Mendes-Felipe, C.; Costa, P.; Roppolo, I.; Sangermano, M.; Lanceros-Mendez, S. Bio-based Piezo- and Thermo-Resistive Photo-Curable Sensing Materials from Acrylated Epoxidized Soybean Oil. *Macromol. Mater. Eng.* **2022**, *307*, 2100934. [CrossRef]

polymers

MDPI

Article

Ion-Induced Polysaccharide Gelation: Peculiarities of Alginate Egg-Box Association with Different Divalent Cations

Anastasiya O. Makarova [1], Svetlana R. Derkach [2], Tahar Khair [3], Mariia A. Kazantseva [1,4], Yuriy F. Zuev [1,5,*] and Olga S. Zueva [3]

1 Kazan Institute of Biochemistry and Biophysics, FRC Kazan Scientific Center of RAS, Lobachevsky St., 2/31, 420111 Kazan, Russia
2 Institute of Natural Science and Technology, Murmansk State Technical University, Sportivnaya Str. 13, 183010 Murmansk, Russia
3 Institute of Electric Power Engineering and Electronics, Kazan State Power Engineering University, Krasnoselskaya St. 51, 420066 Kazan, Russia
4 HSE Tikhonov Moscow Institute of Electronics and Mathematics, Tallinskaya St., 34, 123458 Moscow, Russia
5 A. Butlerov Chemical Institute, Kazan Federal University, Kremlevskaya St. 18, 420008 Kazan, Russia
* Correspondence: yufzuev@mail.ru

Abstract: Structural aspects of polysaccharide hydrogels based on sodium alginate and divalent cations Ba^{2+}, Ca^{2+}, Sr^{2+}, Cu^{2+}, Zn^{2+}, Ni^{2+} and Mn^{2+} was studied using data on hydrogel elemental composition and combinatorial analysis of the primary structure of alginate chains. It was shown that the elemental composition of hydrogels in the form of freezing dried microspheres gives information on the structure of junction zones in the polysaccharide hydrogel network, the degree of filling of egg-box cells by cations, the type and magnitude of the interaction of cations with alginate chains, the most preferred types of alginate egg-box cells for cation binding and the nature of alginate dimers binding in junction zones. It was ascertained that metal–alginate complexes have more complicated organization than was previously desired. It was revealed that in metal–alginate hydrogels, the number of cations of various metals per C12 block may be less than the limiting theoretical value equal to 1 for completely filled cells. In the case of alkaline earth metals and zinc, this number is equal to 0.3 for calcium, 0.6 for barium and zinc and 0.65–0.7 for strontium. We have determined that in the presence of transition metals copper, nickel and manganese, a structure similar to an egg-box is formed with completely filled cells. It was determined that in nickel–alginate and copper–alginate microspheres, the cross-linking of alginate chains and formation of ordered egg-box structures with completely filled cells are carried out by hydrated metal complexes with complicated composition. It was found that an additional characteristic of complex formation with manganese cations is the partial destruction of alginate chains. It has been established that the existence of unequal binding sites of metal ions with alginate chains can lead to the appearance of ordered secondary structures due to the physical sorption of metal ions and their compounds from the environment. It was shown that hydrogels based on calcium alginate are most promising for absorbent engineering in environmental and other modern technologies.

Keywords: sodium alginate; divalent metal cations; hydrogel; metal sorption; association structure

Citation: Makarova, A.O.; Derkach, S.R.; Khair, T.; Kazantseva, M.A.; Zuev, Y.F.; Zueva, O.S. Ion-Induced Polysaccharide Gelation: Peculiarities of Alginate Egg-Box Association with Different Divalent Cations. *Polymers* **2023**, *15*, 1243. https://doi.org/10.3390/polym15051243

Academic Editors: Antonio M. Borrero-López, Concepción Valencia-Barragán, Esperanza Cortés Triviño, Adrián Tenorio-Alfonso and Clara Delgado-Sánchez

Received: 31 January 2023
Revised: 26 February 2023
Accepted: 27 February 2023
Published: 28 February 2023

1. Introduction

Many ionic polysaccharides have a strong tendency to bind metal ions [1,2]. Despite the variety of structural models, many studies marked out the dominant role of ions in their triggering of the polysaccharide structural transition and subsequent aggregation of polymer chains [3,4]. Ion binding may be involved in polysaccharide gelation as a part of their biological functions and the base for numerous technological applications. Low-toxic, biocompatible and biodegradable polysaccharides have found wide applications in food technologies, cosmetology, pharmaceutical and biomedical industries [5–8]. In addition to these advanced applications, polysaccharides have been shown to be useful in the adsorption and binding of harmful chemicals, heavy metals, antibiotics, pesticides and other contaminants from water and wastewater [9–14].

Alginate (alginic acid) is a copolymer of β–D–mannuronic acid (M) and α–L–guluronic (G) acid (Figure 1), generally extracted from brown algae or obtained using bacterial synthesis [15,16]. In recent years, progress has been made in the large-scale bacterial synthesis of alginates with desired G and M composition and sequences [17]. Alginate belongs to a group of polymers used in food and pharmaceutical technologies as a gelling agent. It also has a strong potential in the removal of heavy metals by biosorption [18].

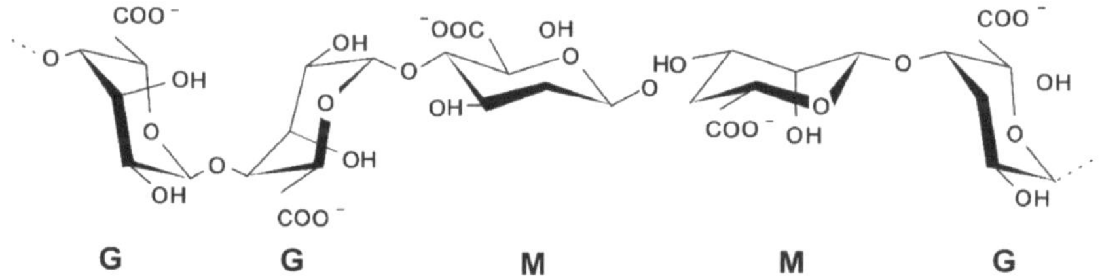

Figure 1. β–D–mannuronic (**M**) and α–L–guluronic (**G**) acid blocks in the alginate polymer chain.

One of the main alginate features is its ability to undergo the sol/gel transition in the presence of bivalent cations (Mg^{2+}, Ca^{2+}, Sr^{2+}, Ba^{2+}, etc.) [19]. According to the preparation procedure, such ionotropic gels occupy an intermediate position between chemical gels with the irreversible chemical crosslinking of polymer chains by covalent bonding and reversible physical gels, where polymer crosslinking is organized by electrostatics, hydrogen bonding, chain entanglement, hydrophobic interactions and crystallization [20]. In the process of ion-induced polysaccharide crosslinking, divalent metal ions take place in the polyelectrolyte complex formation due to an electrostatic interaction between the negatively charged carboxyl groups of polysaccharide molecules and the positively charged metal cations. Such kind of interaction can lead to strong chemical bonding of cations with certain groups of biopolymers, being fundamentally different in the case of alkaline earth and transition metal cations. The properties of alkaline earth ion-induced alginate gels are close to physical gels [21,22]. On the one hand, they are thermo-irreversible, similar to covalently cross-linked gels, but on the other hand, they can be formed under mild conditions at room temperatures and physiological pH. Furthermore, slight changes in physical-chemical parameters can lead to the rapid and reversible dissolution of some ionotropic gel networks, which also makes these systems close to physical gels.

The properties of the resulting gels strongly depend not only on the polysaccharide original structure but also on the type of metal cations as crosslinking agents. The mechanism of alginate gelation, induced by divalent ions, is generally described by the egg-box model proposed 50 years ago for calcium alginate [23]. However, different types of divalent ions have their own features in the complexation with polysaccharides, giving the difference in composition and microstructure of gels [24–30]. For example, with the help of density functional calculations, it was shown [29] that the binding of transition metals with carboxylates and chemical interactions in cation–alginate complexes differ from those of the alkaline earth metals. These authors have shown that the complexation between alkaline earth cations and alginate units occurs only due to ionic bonds, i.e., due to electrostatic interaction. In the case of transition metal cations, the long-range electrostatic interactions compete with stronger coordination-covalent bonding of cations with alginate units [29]. Thus, two different types of interactions lead to the equal macroscopic result, which is the formation of alginate gel in the presence of metal cations due to crosslinking of polysaccharide chains.

A similar point of view was expressed in [31], where the authors considered the coordination interactions between metal ions and functional polymer groups as crosslinking junctions. Here, metal coordination employs the intermediate state between covalent and ionic bonds, being weaker than covalent bonding and stronger than ionic bonding. Thus, for different types of ions, the ion–polysaccharide system might possess properties of either physical or chemical gels, leading to the appearance of various structures with different properties.

Alginate microspheres, produced using the conventional method of ion-induced alginate gelation, in the form of a hydrogel, aerogel or xerogel, are the most popular examples of alginate usage both as an adsorbent and immobilizing carrier for enzymes [32,33], dyes [34,35] or heavy metals [36,37]. The main goal of this work is to analyze how changing the crosslinking ion affects the internal structure and sorption ability of the resulting alginate system. It has been shown that different ions lead to different types of association of alginate dimers into junction zones, which change the morphology and properties of alginate microspheres. To research the structural features of gels crosslinked with divalent metals (Ba^{2+}, Sr^{2+}, Ca^{2+}, Zn^{2+}, Cu^{2+}, Ni^{2+} and Mn^{2+}), we investigated the elemental composition of these gels. We have shown how information on the filling of alginate blocks by ions makes it possible to obtain information on the structure of junction zones in the hydrogel network, on the degree of filling of egg-box cells by metal cations, on the type and magnitude of the interaction between cations and alginate chains and on the most preferred types of alginate egg-box cells for cations binding and to suggest the nature of alginate dimers binding in junction zones and to analyze the presence of sorption vacancies that can be occupied by heavy metal ions. The elemental composition of metal–alginate microspheres using specific cross-linking cations has not been previously discussed in the literature.

2. Materials and Methods

2.1. Materials

Sodium alginate (A2033) from Sigma-Aldrich, USA, was used to prepare polysaccharide solutions. The chemical formula of alginic acid $(C6H8O6)_n$ corresponds to both β–D–mannuronic (M) and α–L–guluronic (G) acids with M/G units ratio of 1.56 for sodium alginate [38,39].

Inorganic salts: barium chloride dihydrate ($BaCl_2{\cdot}2H_2O$), strontium chloride ($SrCl_2$), calcium chloride ($CaCl_2$), manganese chloride tetrahydrate ($MnCl_2{\cdot}4H_2O$), copper sulfate pentahydrate ($CuSO_4{\cdot}5H_2O$) and zinc sulfate heptahydrate ($ZnSO_4{\cdot}7H_2O$), from Tatchemproduct, Russia, and nickel sulfate hexahydrate ($NiSO_4{\cdot}6H_2O$), from Sigma, were used to prepare alginate microspheres.

MQ water purified with the "Arium mini" ultrapure water system (Sartorius, Gottingen, Germany) was used to prepare all solutions.

2.2. Preparation of Polysaccharide Microspheres

The concentrated aqueous solution of sodium alginate (2 wt.%) was prepared according to the standard procedure [23,40,41] by dissolving polysaccharide in water, preliminary swelling at room temperature, subsequent heating to 70 °C and cooling to room temperature. Then, samples were heated and exposed to ultrasound (35 kHz, 100 W) for 60 min at a temperature of 70 °C in a water bath of Bandelin SONOREX TK52 ultrasonic disperser (Germany). To prepare microspheres of alginate hydrogel, 0.5 mL of 2 wt.% alginate solution was added dropwise under constant stirring (500 rpm) to 1.5 mL of 1M solutions of Ba, Sr, Ca, Mn, Cu, Zn and Ni salts with a medical syringe (needle diameter of 0.63 mm). When droplets of sodium alginate solution enter the salt solution, the microspheres of composite hydrogel with a diameter of about 2 mm are instantly formed, in which monovalent sodium ions are replaced by divalent metal ions. Due to ionic and donor–acceptor interactions, divalent ions bind pairwise alginate chains, leading to the formation of a three-dimensional hydrogel structure. The subsequent elemental analysis has shown that there is an almost complete replacement of sodium ions by ions of other metals. The prepared microspheres were kept in solution for 20 min, then washed twice and frozen in liquid nitrogen for freeze drying with Martin Christ equipment. The microsphere preparation procedure and washing time were the same for all samples. Images of alginate microspheres in the presence of divalent metal cations are shown in Figure S1 in the Supporting Information.

2.3. Scanning Electron Microscopy

Scanning electron microscopy (SEM) was performed to control the structure of the freeze-dried hydrogel microspheres. Microspheres with Ba, Ca, Cu, Zn, Ni and Mn were examined using a field emission scanning electron microscope "Merlin" (Carl Zeiss, Germany) in the Interdisciplinary Center "Analytical Microscopy" (Kazan Federal University, Kazan). Samples using Sr were studied using an instrument Auriga Crossbeam Workstation (Carl Zeiss AG, Oberkochen, Germany) in the Shared Research Center of Kazan National Research Technical University "Applied Nanotechnology" (Kazan National Research Technical University).

2.4. Energy Dispersive X-ray Spectroscopy

Elemental analysis was performed using energy-dispersive X-ray spectrometry (EDX) with the X-Max setup (Oxford Instruments, Abingdon, UK) combined with SEM at an accelerating voltage of 20 kV. The analytical capabilities of the Merlin field emission scanning electron microscope were extended with additional attachments for X-ray microanalysis Oxford Instruments INCAx-act with the backscattered electron diffraction (EBSD) registration system Oxford Instruments CHANNEL5. The spectrometer, thanks to the INCASynergy package, is combined with the Oxford Instruments CHANNEL5 backscattered electron diffraction (EBSD) detection and analysis system, which makes it possible to study simultaneously the distribution of elemental composition and crystalline phases in the near-surface region of the sample. Samples using Sr were studied on the Auriga Crossbeam Workstation (Carl Zeiss AG, Oberkochen, Germany), equipped with INCA X-Max silicon drift detector for energy dispersive X-ray microanalysis (Oxford Instruments, Abingdon, UK).

The energy-dispersive X-ray spectroscopy is an analytical method for elemental analysis of solid matter based on the analysis of emission energy of its X-ray spectrum. With the help of an electron beam in an electron microscope, the atoms of a studied sample are excited to emit X-ray radiation, which is characteristic of each chemical element. Investigating the energy spectrum of such radiation, one can draw conclusions about the qualitative and quantitative composition of a sample, in our case, the freeze-dried alginate microspheres. When processing the obtained results, one should take into account the existing limitations of this method associated with the absence of characteristic X-ray radiation from hydrogen and lower accuracy in determining the quantitative composition of lightweight elements, such as carbon and oxygen.

3. Results

3.1. Theoretical Background

Anionic alginate solutions can form hydrogels by using metal cations with a valence of more than one as the crosslinking agents, which can provide the coordination of polysaccharide chains via alginate carboxylate groups. The replacement of monovalent sodium by divalent metal ions Me^{2+} results in the pairwise joining of adjacent alginate chains due to the formation of metal-dependent polyelectrolyte complexes. To describe the mechanism of the alginate cross-linking by divalent Ca^{2+} ions and their further association with the formation of junction zones in the form of flat sheets, Grant et al. proposed the egg-box model [23], which was repeatedly improved [2,17,42–47]. However, with some variations, this model is still appropriate to describe the crosslinking of alginate chains with alkaline earth and transition metal ions. Subsequently, it turned out that the crosslinking of alginate chains by alkaline earth metal ions proceeds in several stages [48]. The first stage is the formation of single crosslinks between biopolymers. The diaxial bond in the homopolymeric chain of guluronates determines cavities formed by the curved fiber structure, which facilitates the metal cation accommodation inside these cavities. Since MM and MG blocks do not form such cavities, Ca^{2+} ions prefer to bind to GG blocks, although the binding to other blocks are sometimes also observed. Thus, the most optimal binding site for Ca^{2+} in the egg-box structure is the cell made up of the GG block of one chain and the GG block of an adjacent chain. At present, such an egg-box model is more often recognized as a real cell, where the ion in the cavity holds both polysaccharide chains together. Such bond formation stimulates further chain "zipping", leading to the connection of two adjacent chains into a dimer.

It is convenient to represent such a junction as two closely spaced alginate chains consisting of series-connected GG blocks with Me^{2+} alkaline earth metal cations included in the formed cavities, as shown in Figure 2a. The junction of all blocks leads to the formation of a completely filled dimer (Figure 2b). The presence of blocks containing M units, that are not optimal for binding leads to the formation of dimers with unbound cells (Figure 2c). Following Grant's work [23], the alginate chains in Figure 2b,c are conventionally depicted as zigzag lines, and the absence of a bond is shown as the absence of a cation (blue ball) in the cell. Further lateral association of dimers can lead to the appearance of junction zones in the form of flat egg-box sheets. Fully cross-linked packing (Figure 2d), as will be shown later, corresponds to alginate gels induced by some transition metals, while the connected dimers bound by van der Waals interactions and hydrogen bonding are observed for calcium alginate (Figure 2e). Note that the egg-box model precisely corresponds to the scheme shown in Figure 2d. Subsequently, Sikorski [43], based on X-ray diffraction data for calcium alginate, showed that during the construction of junction zones, polymer chains are connected in the shape of already formed dimers and, therefore, half of the carboxyl groups do not participate in the bond formation. Therefore, in the case of a lateral interdimer association, the gaps between dimers contain sodium (or hydrogen) ions, which neutralize the excess charge of carboxyl groups. These ions are shown in Figure 2e as red balls. Nevertheless, the possibility of metal ions entering the zones of interdimer association exists even for calcium [28].

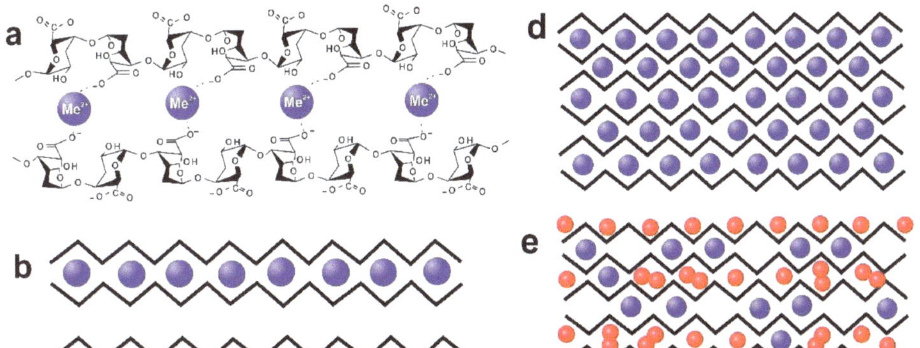

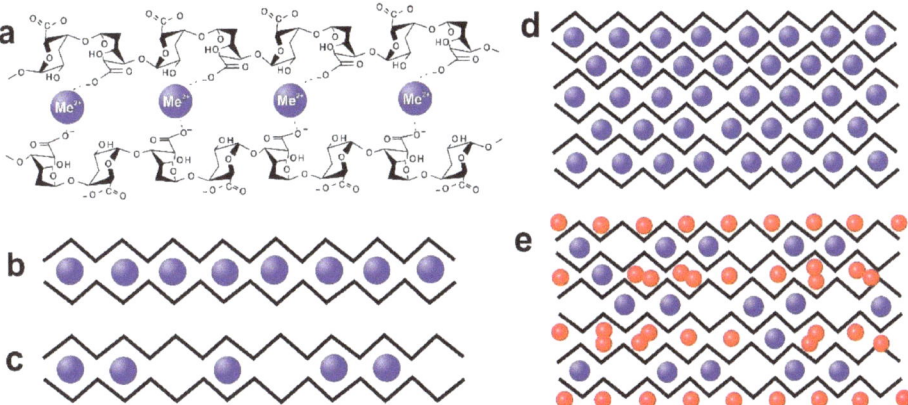

Figure 2. Integration of alginate chains into a dimer (**a**); schematic view of a dimer with completely (**b**) and partially (**c**) bound cells; types (**d**,**e**) of junction zones under the association of dimers. Blue balls show ions Me^{2+} and red balls correspond to sodium and hydrogen ions.

Since during the hydrogel formation, a divalent metal ion binds in pairs two monomer units of alginate chains, the chemical formula of sodium alginate is best considered for a block of two units as $(C_{12}H_{14}O_{12}Na_2)_n$, or for any divalent metal Me^{2+} in the form $(C_{12}H_{14}O_{12}Me_X)_n$, where the symbol X denotes the average number of divalent metal ions per block of two C_{12} monomeric units, i.e., the average block occupation number C_{12}. The limit value X = 1 corresponds to completely filled cells of an egg-box sheet from parallel-connected alginate chains (Figure 2d). In Figure 2d, each block of two monomeric units contains a Me^{2+} ion. For dimers formed by alkaline earth metals (primarily by calcium), as shown in Figure 2e, there is one metal cation (depicted by the blue ball) per egg-box of four monomeric units. Accordingly, in the absence of metal ions in interdimer space, the average number of Me^{2+} ions per C_{12} block should not exceed 0.5. Let us recall that this case corresponds to alginate dimers bound only by van der Waals interactions and hydrogen bonding. In Figure 2e, a case corresponding approximately to X = 0.35 is shown. If metal ions also form a bond in the interdimer space, this number can increase up to X = 1. Thus, the average number of metal ions per C_{12} block obtained using elemental analysis under the use of the egg-box model provides important information about the alginate gel structure and about the type of junction zones.

In addition, it should be noted that the block occupation number should be determined by the initial composition of the studied alginate, at least by the ratio of M and G blocks in the alginate chain. The importance of determining the chain composition and sequential structure has been noted by many authors [49]. Certainly, the knowledge of alginate primary monomeric composition does not suffice to determine its sequential structure. Certain attempts to describe the probabilistic distribution of monomeric units along the polymer chain were made in [50–52].

To assess the possible structures of junction zones, we applied combinatorial calculus. We took into account that in the alginate sample used, the M/G ratio is 1.56 [38,39], which is close to 1.5. For simplicity, we assumed that, on average, there are three M units for every two G units, i.e., M/G = 3:2. The approximate probabilities of block formation from two monomeric units can be calculated using the variant tabulation method (see Table S1 in SI). It turned out that the probabilities of the appearance of blocks GG, GM (together with MG) and MM, respectively, are equal to 16%, 48% and 36%. Here, the GM and MG blocks were considered equivalent, although the differences between the GM–GM and GM–MG cells must be taken into account when connecting in the egg-box.

When forming the Table of variants (Table S2 in SI) for the formation of egg-box cells of various types in the space between two alginate chains, the resulting proportion GG:GM:MG:MM = 4:6:6:9 was used for subsequent comparison with the elemental experiment. The results of calculations carried out using data in Tables S1 and S2 are presented in Table 1.

Table 1. Approximate probabilities of association of M and G units into different egg-box structures for alginate with M/G~1.5.

N	Type of Structure	Probability	N	Type of Structure	Probability
1	GG–GG	16/625 = 2.56%	4	GM–GM	72/625 = 11.52%
2	GG–GM	96/625 = 15.36%	5	GM–MG	72/625 = 11.52%
			6	GM–MM	216/625 = 34.56%
3	GG–MM	72/625 = 11.52%	7	MM–MM	81/625 = 12.96%

In this table, cells GM–GM and MG–MG were considered equivalent to each other, but not with the cells of GM–MG (MG–GM). The remaining structures containing GM blocks or MG blocks were considered equivalent to each other. It should be noted that taking into account nonequivalent positions almost does not complicate proposed method.

3.2. Experimental Results

Using the field emission scanning electron microscope, the SEM images of the surface, internal sections and internal cells of the freeze-dried microcapsules, the elemental composition of the near-surface and, in some cases, internal layers of microspheres were studied and the chemical formula corresponding to this composition was determined.

Since hydrogen does not show characteristic X-ray radiation, the method of energy-dispersive X-ray spectroscopy does not allow us to determine the quantitative composition of hydrogen in the structural formula. Therefore, all further formulas are given without hydrogen, the presence of which is simply implied in the above proportions. In addition, a characteristic feature of the energy-dispersive X-ray spectroscopy method is that lightweight elements are determined with a larger error, by which we explain the systematical mismatch of oxygen atoms by almost one per one C_{12} block.

The comparative results for seven samples of metal–alginate hydrogels are given below. The sequence of samples prepared on the basis of divalent metals Ba-Sr-Ca-Zn-Cu-Ni-Mn is considered in the descending order of their ionic radii 0.135-0.113-0.099-0.074-0.073-0.070-0.067 nm [27]. Traces of aluminum present in some samples are attributed to the result of sample preparation for scanning electron microscopy.

Barium–alginate microspheres. Cell images of the Ba–alginate microspheres and the elemental analysis data of the surface layers of freeze-dried barium–alginate microspheres are shown in Figure 3. For the above-mentioned reasons, hydrogen is absent.

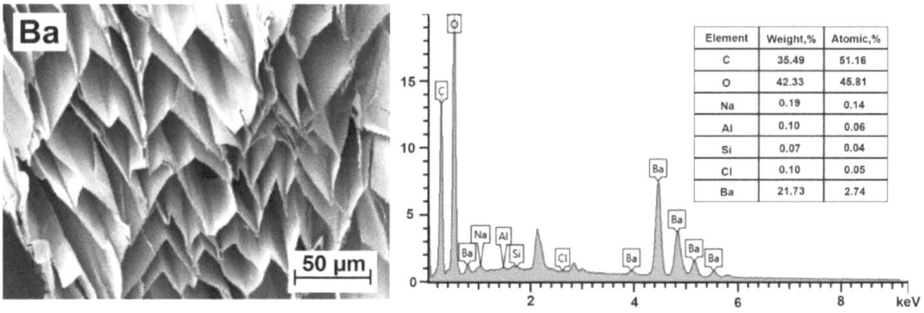

Figure 3. SEM image of transverse section of a Ba–alginate microsphere and its elemental content.

The obtained elemental composition of Ba–alginate microspheres corresponds to the formula $(C_{12}O_{11}Ba_{0.6})_n$. Here and below, the elements whose contribution is less than one tenth of an atom per cell were not included in the determined formula. For each C_{12} block composed by two monomeric units, there are on average X = 0.6 Ba atoms. This is less than the maximum possible theoretical value X = 1.0 but more than X = 0.5. There are traces of other metals (rest of Na and Al), but due to their small amount, they do not make a real contribution to a certain composition. Thus, the obtained data on the block occupation number indicate that barium cations bind alginate chains not only inside dimers but also in the interdimer space. This means that when joining in junction zones, the stage of formation of alginate dimers proceeds simultaneously with their joining into zones. Thus, in the junction zones of barium alginate, the dimers do not retain their individuality. This fact promotes the formation of strong Ba-based gels [53].

However, barium does not bind to all alginate blocks. In 40% of egg-box cells, the Ba^{2+} cross-linking cations are absent. The excess charge of the carboxyl groups remaining after sodium removal can be compensated in this case only by hydrogen. Due to the fact that, according to the available literature data, Ba interactions are more preferable with GG and MM blocks [27,28,52–59], some of the blocks, most likely MG, remain unbound by barium cations. Apparently, the shape and size of cells including MG blocks do not correspond to the barium cation. Thus, GG–GG, GG–MM, MM–MM and possibly GM–MM can be chosen as the most probable structures containing Ba^{2+}. The total probability of filling the structures, equal to $385/625 \approx 61.6\%$, allows us to determine the number of barium cations as X = 0.62, which is quite close to obtained value X = 0.6.

It should be also noted that after the preparation of barium alginate microspheres using barium chloride, all chloride ions remain in the washing solution after double flushing. This means that chloride ions do not interact with alginate chains. The cells of the obtained barium–alginate microspheres have a clear rhombic shape and a fairly homogeneous structure (Figure 3). The approximate cell size is 50×25 µm. According to [28], this gel is stable in acidic and neutral pH environments. If not for the toxicity of barium compounds, such a porous material would be ideal for drug delivery and other medical applications.

Strontium–alginate microspheres. In the table of elements, strontium occupies an intermediate position between barium and calcium. According to its properties, it is an analog of calcium. Natural strontium occurs as a mixture of four stable isotopes (mainly ^{88}Sr, 82.6%). It is a constituent of microorganisms, plants, animals and, unlike barium, is a low-toxic chemical.

The images of cells in Sr–alginate microspheres and the data from the elemental analysis of the near-surface layers of strontium–alginate microspheres obtained by treating sodium alginate with strontium chloride are shown in Figures S1 and 4. Despite the performed standard double washing, the obtained elemental composition, shown in Figure S1, corresponds to the chemical formula $(C_{12}O_{11.6}Cl_{2.3}Sr_{1.8}Na_{0.1})_n$. A similar composition was obtained for the inner regions of microcapsules. A distinctive feature of the obtained formula in comparison with the theoretical composition is the presence of chlorine and a significantly larger number of strontium atoms per block, as well as the existence of sodium residues. This composition indicates the presence of structures inside the sample not inherent to strontium alginate.

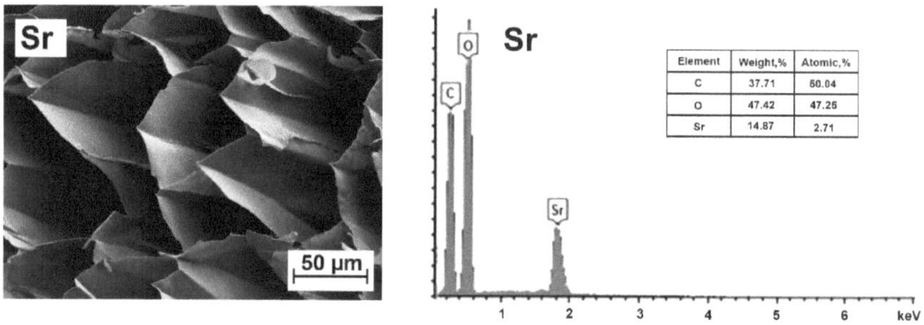

Figure 4. SEM image of a transverse section of a Sr–alginate microsphere and its elemental content after additional washing of the sample.

The obtained formula, in which there are more strontium atoms than possible binding sites, leads to a conclusion about the existence of strontium atoms in at least two fundamentally different nonequivalent positions: (a) connecting adjacent alginate chains, i.e., acting as cross-linking agents and (b) in the composition of $SrCl_2$ associates adsorbed by alginate chains. The excess amount of strontium atoms, the presence of chlorine and considerations of the electro-neutrality of the solution make one assume the formation of $SrCl_2$ associates. It should be noted that $SrCl_2$ structures exist precisely in the form of associates and only near alginate chains. Their possible dissociation in bulk solution would lead to the appearance of chloride ions, which are indifferent to the alginate structure and can be easily removed with washing. This is indicated by the results obtained for barium alginate gels. Such behavior of chloride ions was observed during ion-induced gelation with $BaCl_2$ when the absence of chloride ions was observed after a standard double washing. Separation of contributions from nonequivalent strontium atoms shows that approximately 0.65 of crosslinking Sr^{2+} ions per each C_{12} block carry out the spatial connection of chains, leading to the formation of egg-box cells.

In addition, for each C_{12} block of strontium alginate, there are on average 1.15 associates of strontium chloride $SrCl_2$. Taking into account that a distinctive feature of natural polysaccharides is the presence of sorption ability, we assumed that the binding of $SrCl_2$ by hydrogel structures observed in this case (physical adsorption) is most likely due to the presence of local energetically favorable positions for these associates near the alginate chains. Despite the fact that $SrCl_2$ associates do not form either covalent or ionic chemical bonds with alginate chains, they can be kept near them due to weaker (mainly van der Waals) interactions. To find the real elemental composition of strontium alginate, we subtracted 1.15 Sr atoms associated with 2.3 Cl atoms and obtained the resulting formula $(C_{12}O_{11.6}Sr_{0.65}Na_{0.1})_n$.

To test the hypothesis of the existence of various types of strontium binding with alginates, we assumed that the energy of the interaction of $SrCl_2$ associated with biopolymer chains corresponds to the physical adsorption. In this case, it will be not much more than the energy of its thermal motion in water. Therefore, an increase in the washing time should lead to the removal of weakly bound $SrCl_2$ associates. In the case of complexation based on ionic electrostatic interactions, the washing will not change this result.

Thus, we have conducted a study of newly prepared and thoroughly washed microcapsules of strontium alginate. The washing solution was changed five times with an interval of 2 h. The salt concentration in the wash solution was controlled using an inoLab Cond 7310 SET1 conductometer (Hungary). The washing took place at room temperature until the electrical conductivity of the washing solution reached the value of the electrical conductivity of distilled water (after 6 h of washing). The SEM images of Sr–alginate cells and elemental data for strontium–alginate microspheres after the additional washings are shown in Figure 4.

The elemental analysis carried out for two different cases of washed strontium–alginate microspheres gave new results: (a) $(C_{12}O_{11.3}Sr_{0.65})_n$ and (b) $(C_{12}O_{11.1}Sr_{0.70})_n$, which do not differ fundamentally from the result obtained earlier. This indicates that the used analysis of elemental composition was quite correct. In addition, this result can be regarded as an indirect confirmation of $SrCl_2$ physical adsorption by the hydrogel. It should be noted that with the standard procedure of microsphere preparation, described in the Materials and Methods, the existence of associates similar to $SrCl_2$ should also be expected in the case of cross-linking of alginates with other cations. It can even be assumed that the existence of such binding sites (physical adsorption) for emerging associates of metals (and other impurities) with hydrogel structure is explained by the adsorption capacity of alginate hydrogels and, therefore, can be numerically correlated with it.

According to the literature data, the structures corresponding to the egg-box model can appear as a result of the complex formation of strontium with alginate chains. Due to the known information that the interaction of strontium is more preferable with GG and MG blocks [27–29,58–60], some blocks, most likely MM, remain unbound by these cations. If we choose cells that do not contain MM blocks, namely, GG–GG, GG–GM, GM–GM and GM–MG, as the most probable structures containing Sr^{2+}, we shall obtain $X = 0.41$. The discrepancy between this number and the experimental data means that some cells containing MM blocks also can contain Sr^{2+} cations. In particular, the occupation of structures GG–GG, GG–GM, GG–MM and GM–MM gives the total probability of filling cells equal to $400/625 = 64\%$. This allows us to determine the average occupation number for strontium as $X = 0.64$, which is close to $X = 0.65 \div 0.7$ obtained from our experiments. The experimental value $X = 0.65 \div 0.7$ cations per C_{12} block exceeds the limiting value of 0.5, corresponding to the case of the association of dimers with cations according to the type of Figure 2e. This means that, as in the case of Ba^{2+}, the crosslinking occurs not only within dimers but also in the interdimer space and the dimers lose their individuality in resulting junction zones. In the case of strontium, the number of cations involved in the gel formation is still slightly larger than in the case of barium. However, a significant part of cells remains empty. Apparently, the shape and size of some cells do not correspond to the strontium cation. Thus, despite the fact that the experimental technique used does not allow us to draw conclusions about specific crosslinking sites, some considerations about the role of GG, MG and MM blocks in the binding of Me^{2+} ions by alginate chains can still be expressed.

The cells of strontium alginate microspheres obtained after intense washing turned out to be much more uniform in their composition and size (Figure 4). They are very similar in structure to the barium alginate microspheres. The Sr^{2+}-based microsphere cells, which are slightly larger, also have a rhombic shape with an approximate size of about 60×40 μm.

Strontium alginate forms a nontoxic gel with high chemical stability and strong mechanical performance. The Sr–alginate gels show great potential as biomaterials for bone regeneration based on enhanced cell proliferation and migration. It was noted in [60] that

Sr alginate is a suitable material for the immobilization of living cells in long-term perfusion studies compared with other ion-induced alginate gels. The Sr–alginate gel has higher chemical stability and effects on cells caused by Sr^{2+} are relatively mild compared to the other divalent cations.

Calcium–alginate microspheres. The most commonly used and studied divalent alkaline metal for the preparation of hydrogel microspheres is calcium. SEM images of the cells of the freeze-dried microspheres and data from the elemental analysis of near-surface layers of microspheres are shown in Figure 5.

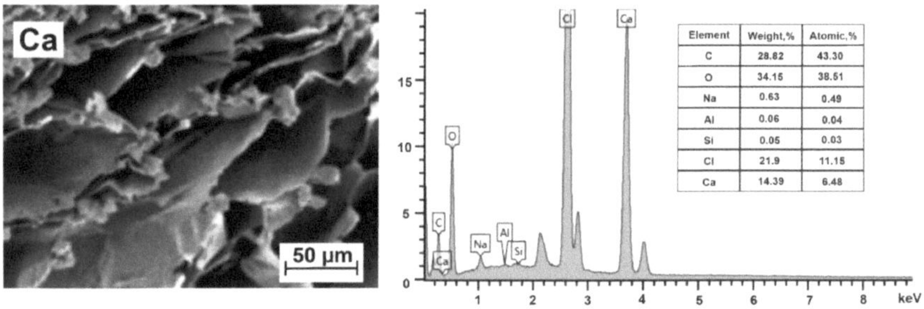

Element	Weight,%	Atomic,%
C	28.82	43.30
O	34.15	38.51
Na	0.63	0.49
Al	0.06	0.04
Si	0.05	0.03
Cl	21.9	11.15
Ca	14.39	6.48

Figure 5. SEM image of the transverse section of a Ca–alginate microsphere and its elemental content.

The obtained elemental composition, despite the standard washing of microspheres, corresponds to the formula $(C_{12}O_{11}Cl_3Ca_{1.8}Na_{0.14})_n$. A distinctive feature of this composition from the expected one is the presence of chlorine and a noticeably larger number of calcium atoms per block, as well as the presence of a small number of sodium atoms. Just as in the case of strontium alginate, the observed elemental composition leads to the conclusion that calcium atoms exist in two fundamentally different nonequivalent positions: (a) connecting adjacent alginate chains, i.e., acting as the cross-linking agents and (b) in the form of $CaCl_2$ associates adsorbed by alginate chains. The separation of these contributions can be carried out similarly to the previous case. Assuming that the adsorption binding of $CaCl_2$ associated with the hydrogel structure is due to the presence of energetically favorable local positions near the alginate chains, we subtracted 1.5 Ca atoms associated with 3 Cl atoms and obtained the following elemental composition $(C_{12}O_{11}Ca_{0.3}Na_{0,14})_n$. Thus, only 0.3 Ca^{2+} crosslinking ions in the structure of the hydrogels per C_{12} block provide spatial crosslinking of chains. This may be the reason why the strength of calcium-based gels is inferior to that of barium-based gels [53].

When carrying out theoretical calculations of the probability of the calcium cations binding in a hydrogel structure, we proceeded from the fact that bonds should occur in the cells containing GG blocks. There were no other preferences for Ca^{2+}. The probability to find calcium cations in cells GG–GG, GG–GM and GG–MM is 184/625 = 29.4%, which gives X = 0.29, being close to the value of 0.3 obtained in experiment. Note that the probability of GG–GG structures all over the sample is rather small (~2.5%), but their existence triggers the zipping mechanism of chains joining into dimers. There are also not so many cells that include one GG block. Therefore, the number of crosslinking ions turned out to be very small, and the type of junction zones correspond to the lateral association of dimers due to van der Waals interactions and hydrogen bonds (Figure 2e). The probability of occurrence of calcium ions in the interdimer space is low since it is determined by the probability of contact between the GG blocks of two dimers. This probability is about 1%. The uncompensated charges of carboxyl groups are partially compensated by the presence of small amounts of Na^+ and H^+ ions, which are mainly concentrated in the inter-dimer space. In general, such a structural organization of calcium alginate hydrogels makes its properties closer to those of physical hydrogels. Compared with Sr^{2+}, Ca^{2+} shows a smaller number of coordination sites and weaker binding with alginate molecules [29,60]. Here we are faced again with a situation where the shape and size of some cells do not correspond to the cation used. Therefore, Sr–alginate gels have a significantly higher chemical stability and stronger mechanical performance than Ca–alginate gels under the same concentration of alginate and ions. However, the total number of possible binding sites (both chemical and physical) in strontium and calcium alginates is approximately the same being about 1.8 per block of two monomeric alginate units. In comparison with strontium, the decrease in sites at which calcium ions can bind polysaccharide chains increases the number of sites suitable for physical adsorption. This fact allows us to conclude that calcium alginate can be used in the engineering of materials that are more effective in terms of sorption ability for their use in different technologies. It should also be noted that the adsorption of heavy metals by calcium alginate hydrogels leads to the strengthening of their structure [61].

The cells of calcium–alginate microspheres are heterogeneous not only in their composition but also in size (Figure 5), and their walls are rather thick compared to the walls of other metal–alginate microspheres. The approximate cell size is 70 × 30 μm. Calcium alginate gels are considered as safe and non-toxic.

Alginate systems based on transition metals Zn^{2+}, Cu^{2+}, Ni^{2+}. The alginate microspheres with transition metals were prepared with the help of their sulfate salts obtained in contrast to the case of alkaline earth metals, for which the chlorides were used. Since the comparison of elemental composition with the participation of Me^{2+} sulfates may not always be correct due to the difference in anions, we present below only the main features that characterize these hydrogels.

Zinc, which has the largest (comparable to alkaline earth metals) ionic radius (0.074 nm) of all transition metals we studied, produces ion-induced hydrogels similar to alkaline earth metal hydrogels. It is known that the Zn–alginate gel is always loose and weak [28,29,62–64]. This is due to the fact that Zn^{2+} can interact with the carboxylate of G blocks in a similar way to Ca^{2+}. However, the Zn^{2+}-mediated cross-linking of alginate shows a unidentate binding, which involves only one carboxylate oxygen atom [28].

The obtained elemental composition of Zn–alginate microspheres made it possible to determine the average occupation number of zinc ions per C_{12} block. It turned out that, on average, X = 0.6 crosslinking Zn^{2+} ions provide a spatial connection of polysaccharide chains. This number corresponds approximately to the number of Ba^{2+} cations that create bonds in the GG and MM blocks. It is possible that zinc cations bind to the same blocks, which may explain the similar result that is obtained. The incomplete use of all possible bonds (the number of zinc ions per block is less than one) casts doubt on the fact that covalent bonds predominate over the ionic bonds [29] in zinc alginate, although it may be possible when zinc ions interact with specific blocks. The total number of possible binding sites (both chemical and physical) for this alginate hydrogel turned out to be 1.5 per block of two monomeric alginate units, which is less than the value of 1.8 obtained for the strontium and calcium systems. Perhaps this fact also determines the weak structure of zinc alginate.

The cells of zinc–alginate microspheres have a disordered structure, in which elements of tetrahedral symmetry are still visible. In general, a fairly homogeneous porous structure is observed. The approximate cell size of 30 × 30 μm is noticeably smaller than that of alkaline earth metal alginate hydrogels but still larger than that of other transition metals. In the literature, good antibacterial properties of zinc-based gels are noted, however, exceeding certain concentrations for medical purposes can lead to toxic effects. Better preservation of enzyme activity was also noted when using zinc–alginate microcapsules [62].

The mechanism of Cu^{2+}- and Ni^{2+}-induced gelation of alginate remarkably differs from that of the alkaline earth metals [28,29,62,65–67]. It was concluded that the formation of coordination-covalent bonds in alginate gels with transition metal cations apparently prevails over electrostatic interactions in polyelectrolyte solutions [29]. The ions of these transition metals bind equally well to both M and G alginate units [66], forming much more ordered structures [68–70]. The observed elemental composition did not give grounds for assuming that the positions of various copper ions are not equivalent. This means that all copper cations are crosslinking ones (X = 1) leading to completely ordered junction zones described by the egg-box model (Figure 2d). Despite the fact that the size of the copper cation is one and a half times smaller than the calcium one, the egg-box cell formed by alginate chains and Cu^{2+} is always longer than that formed with Ca^{2+} [28]. The increased size in the egg-box dimer formed by alginate chains and Cu^{2+} may indicate a complex association that occurs when alginate chains are bound by copper cations, including the presence of hydration water molecules in egg-box cells. The calculations of optimized structures of Cu^{2+}−disaccharide complexes, which were studied with the density functional theory (DFT) [29], showed the presence of cavities that could accommodate more complicated copper complexes than single cations. Thus, it is possible that the cross-linking of alginate chains in copper–alginate microspheres and the formation of ordered egg-box structures with completely filled cells are provided by hydrated copper complexes of complicated composition.

The obtained elemental composition of Ni–alginate microspheres indicates a strong interaction of Ni^{2+} with polysaccharide chains, the average occupation number X = 1 for C_{12} blocks and the presence of hydration water which, along with the Ni^{2+} cation is responsible for the complex formation of alginate chains. In Ni–alginate microspheres, nickel ions bind equally well to all structural units of alginates, which makes it possible to speak about the formation of strong coordination-covalent bonding during complexation, leading to completely ordered junction zones in the egg-box model (Figure 2d). It should be noted that the calculations performed using the DFT method showed [29] that the presence of water molecules in the inner coordination shell of an ion provides a wide variety of stable hydrated structures. Moreover, these calculations show that the interaction energy of cations increases in hydrated complexes compared to corresponding anhydrous structures. Therefore, it is not surprising that some hydration water molecules can participate in the complex formation of alginates together with Ni^{2+}.

It should be noted that many nickel compounds are toxic and carcinogenic.

Manganese–alginate microspheres. SEM images of the cells and the elemental analysis data for Mn–alginate microspheres obtained by treating alginate with manganese chloride solution are shown in Figure 6.

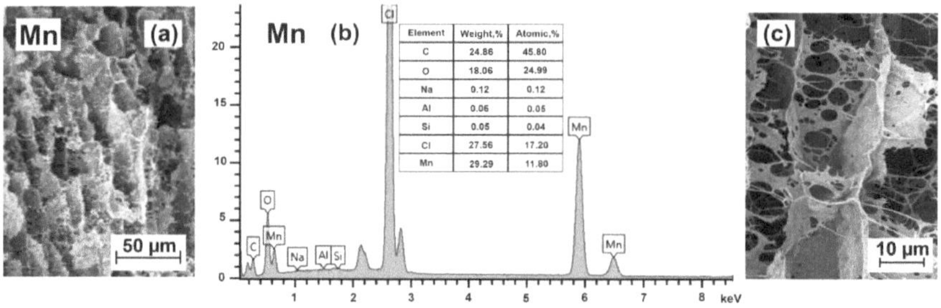

Figure 6. SEM image of cells in a Mn–alginate microsphere (**a**), elemental content of surface domains (**b**) and an enlarged image of cell walls of the interior (**c**).

The resulting elemental composition of the surface layers of Mn–alginate microspheres (Figure 6b) corresponds to the formula $(C_{12}O_6Cl_4Mn_3)_n$. A distinctive feature of this formula is an extremely small number of oxygen atoms per C_{12} block in the presence of three manganese atoms, which indicates very strong metal–alginate interactions. There was also a surprising openwork structure of the cell walls of manganese–alginate microspheres in the presence of strong Mn^{2+} interactions with all alginate blocks. Similar to calcium, we assumed that the addition of manganese salts leads not only to chain linking but also to the formation of physically adsorbed $MnCl_2$ complexes. The subtraction of atoms related to $MnCl_2$ associates gives the following elemental composition $(C_{12}O_6Mn)_n$. The cells of each block are completely filled (X = 1) with manganese cations indicating that manganese ions form complexes equally well with all structural units of alginate chains due to the appearance of covalent-coordination bonding.

The decrease in the number of oxygen atoms requires an additional explanation. It can be assumed that the presence of manganese leads to the appearance of processes during which H_2O and carbon dioxide CO_2 are released. Indeed, alginic acids exhibit a number of specific chemical reactions. Such reactions include, for example, dehydrogenation, decarboxylation and degradation up to oligomers in an acidic medium [26] or on heating in the presence of some metal salts [71]. The decomposition of alginate under physiological conditions can be caused by the partial oxidation of alginate chains [72]. Apparently, the manganese compounds contribute to the occurrence of such processes, being the catalysts for many organic reactions [73]. This can be facilitated by the fact that manganese chloride $MnCl_2$ is hydrated when dissolved in water, forming slightly acidic solutions with a pH of about 4. The instability of gels formed by manganese alginate has already been pointed out in [28,74].

The cells of freeze-dried Mn–alginate microspheres look very unusual. The cell partitions of the inner part of these microspheres at higher magnification are shown in Figure 6c. The openwork "petals" of cells resemble the structures left after the etching or some kind of chemical treatment, indirectly confirming by this picture the partial decomposition of alginate. According to the number of manganese atoms per block, a dense material should have been obtained, similar in properties to the nickel–alginate hydrogels. However, due to dehydration and decarboxylation, the hydrogel structure is loosened.

The approximate size of the cells can be estimated as 20 × 15 μm. The shape of the cells is closer to cubic. In our opinion, such a material could be of interest as a catalyst or catalyst carrier in biotechnological processes.

4. Discussion

We considered fundamental aspects of the influence of various ions on the morphology and elemental composition of alginate hydrogel microspheres obtained using various crosslinking metal ions (divalent cations Ba^{2+}, Sr^{2+}, Ca^{2+}, Cu^{2+}, Zn^{2+}, Ni^{2+} and Mn^{2+}). It should be noted that concentrated solutions of divalent metal salts were used to prepare hydrogel microcapsules. If the multi-staged binding of alginate chains is observed for weakly concentrated solutions [48], in highly concentrated solutions these stages occur almost simultaneously.

The result of the crosslinking of bivalent ions and polysaccharide chains is the formation of flat junction zones corresponding to egg-box structures with varying degrees of their cells being filled by divalent metal cations. The chemical formula of alginate with pairwise crosslinked chains underlying the egg-box structure, per the block of two monomers is $(C_{12}H_{14}O_{12}Me_X)_n$. The limiting value of the average number of Me^{2+} cations per C_{12} block (occupation number) X = 1 corresponds to the case of completely filled egg-box cells. We have shown that the analysis of elemental content of the near-surface zones of metal–alginate microspheres allows us to obtain information not only on the filling degree of junction cells by metal cations but also to draw conclusions about the type and magnitude of the interaction between cations and alginate chains, to clarify information about the composition of the most preferred egg-box cells and to suggest the nature of binding of alginate dimers in junction zones. In addition, the existence of opportunities for the sorption of metal ions and their compounds by metal–alginate hydrogels, primarily by the calcium–alginate system, has been established.

A cross-comparison of the obtained results, together with an analysis of the literature data, allowed us to conclude that the average degree of cell filling X correlates with the strength of cation binding to the alginate chains, i.e., with the relative contribution of the stronger than ionic coordination-covalent interaction, which is typical for alginate in the presence of transition metal cations [29]. Therefore, for cations such as Cu^{2+}, Ni^{2+} and Mn^{2+} where the contribution of the coordination-covalent interaction is sufficiently large, the average occupation number is X = 1 (Figure 6), and the junction zones have the form shown in Figure 2d. A slightly different situation is observed under the interaction of Zn^{2+}, which, although a transition metal, has a larger ionic radius comparable to that of alkaline earth metals. In the case of the zinc–alginate system, the experimentally observed average occupation number turned out to be less than 1 (X = 0.6), which indicates that the bonding of zinc with alginate is rather weak and cannot be of a coordination-covalent nature. The interaction of this cation with alginate, as well as the interaction with alkaline earth metal cations, is carried out using ionic bonds. Its additional features (unidentate binding [28]) make the Zn^{2+}-based hydrogel weak [62–64].

The result of crosslinking with alkaline earth cations, primarily with Ca^{2+}, is the pairwise association of alginate chains (chain-to-chain association) and the subsequent lateral association of cross-linked dimers, leading to the formation of flat junction zones, as shown in Figure 2e. In this case, cations fill the cells formed when alginate chains join into dimers. Due to van der Waals forces and hydrogen bonds, the dimers associate in the junction zone but there are no cations in the interdimer space. With the complete filling of all cells of each dimer with cations and their absence in the interdimer space, X = 0.5. A similar situation, but with partially unfilled cells of alginate dimers, corresponds to calcium alginate hydrogels, where the average number of crosslinking Me^{2+} ions per C_{12} block is even smaller and is approximately equal to X = 0.3. As we have shown, calcium cations can also enter the inter-dimer space, but the probability of such a process is low (about 1%). For other alkaline earth metals, the X number corresponds to 0.6 for barium and 0.65–0.7 for strontium (Figure 7). On the one hand, this fact is explained by the electrostatic binding of alkaline earth cations with alginate chains, which results in a different degree of complex formation of these cations with alginate blocks of various types (GG, MM, GM). This is manifested in a different probability of filling cells and the appearance of a certain number of unoccupied sites in the egg-box structures. On the other hand, the value X > 0.5 indicates the existence of Mn^{2+} cations in the inter-dimer association zone. Therefore, for barium and strontium, the stage of association into dimers occurs probably simultaneously with the stage of inter-dimer association, just as in the case of transition metals. The dependence, shown in Figure 7, reflects the average number of cations of various divalent metals per C_{12} block, which is equal to average occupation number in the cells of junction zones.

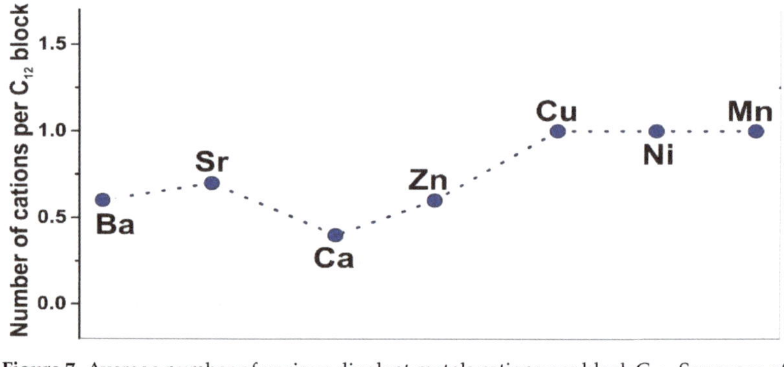

Figure 7. Average number of various divalent metals cations per block C_{12}. Sequence of chemical elements Ba-Sr-Ca-Zn-Cu-Ni-Mn is taken in descending order of their ionic radii 0.135-0.113-0.099-0.074-0.073-0.070-0.067 nm.

Molecular dynamics methods have established the existence of many optimal but still unequal binding sites for metal ions with alginate chains [17,29,75], some of which remain unoccupied. The presence of such places can lead to their occupation by various molecules which are bound by alginates as a result of physical sorption, the binding energy of which is less than the energy of the ionic bond of cations with alginate units and is comparable to the energy of thermal motion of molecules.

The interesting features of hydrogels can be revealed by studying the quantitative composition of the excess salts used to prepare microspheres and that remain in their structure after 20 min double washing. In principle, the components of the salt solution that have entered the sodium alginate solution are not equivalent. Some of the components can be incorporated into the structure of the hydrogel (Me^{2+} cations), the other part can be indifferent to gel structure and some parts can be in the zones of local energy minima, i.e., weakly interact with alginate structure. For example, when barium chloride is added, a certain amount of barium cations participates in crosslinking and all excess barium salt and chlorine anions are easily removed from the water in which hydrogel beads are located for 20 min during double washing. They are indifferent to the structure of barium alginate. On the contrary, when strontium chloride is added, almost the same amount of strontium cations is involved in crosslinking, however, not all $Me^{2+}Cl_2^-$ associates were eliminated by double washing. In particular, a 6h washing was required to remove excess salts (1.15 $SrCl_2$ per C_{12} block). It should be noted that washing leads to the removal of "excess" ions and associates, but not places for their possible binding, which determine the sorption capacity of alginates.

In the course of ion-induced gelation in the presence of transition metal cations, the egg-box structures of junction zones, i.e., the sheets of connected alginate chains, are formed almost instantly, so the excess salts can be located both above and below these sheets, which also form the ordered secondary structures. The elemental analysis data show that the maximum number of transition metal associates (except zinc) $Me^{2+}Cl_2^-$ per C_{12} block can be 2, and their location should be determined by the properties of the metal/alginate system. Thus, the number of physically adsorbed associates in this case is even greater than in the case of alkaline earth metal alginates. It is possible that the strength of their physical binding to alginate chains will also be greater. The time of the removal of "excess" associates by washing has not yet been established.

For zinc and strontium, the number of emerging secondary associates is 0.9 and 1.15 per C_{12} block, respectively. The large size of barium ions does not allow the participation in additional interactions with alginate chains.

According to the obtained structural models, the most optimal hydrogel in terms of sorption properties is the calcium–alginate hydrogel. A small number of filled cells of the egg-box structure in calcium alginate lead to the appearance of 1.5 additional possible sites for the interaction of ions and associates with alginates per each C_{12} block. Some of these sites can be located in the inter-dimer space of junction zones, which is associated with the electronegativity of weakly filled egg-box structures. Another part may be near the alginate chains, above and below these planes. Indirectly, the appearance of secondary structures is indicated by a decrease in the level of crystallinity upon soaking a sodium alginate film in a $CaCl_2$ solution, which was observed in [2].

5. Conclusions

In this work, we studied the elemental composition and structural features of the near-surface regions of freeze-dried microspheres obtained on the basis of the association of sodium alginate induced by divalent cations Ba^{2+}, Sr^{2+}, Ca^{2+}, Cu^{2+}, Zn^{2+}, Ni^{2+} and Mn^{2+}. It has been shown that in metal–alginate hydrogels, the average number of various Me^{2+} cations per C_{12} block with a limiting theoretical value equal to 1 is less than this number. In the case of alkaline earth metals and zinc, the average occupation number ranges from 0.3 to 0.35 for calcium to 0.65 to 0.7 for strontium. This fact points to the electrostatic binding of alkaline earth and zinc cations to alginate chains. A consequence of the relatively weak electrostatic interaction is a different degree in the complex formation of these cations with alginate blocks of various types (GG, MM, GM), which manifests itself in the presence of a certain number of unoccupied sites in the egg-box structure. Apparently, there is a discrepancy between the large size of the alkaline earth cation and the possibilities of its complex formation and the size and shape of certain egg-box cells. The use of combinatorial methods made it possible to calculate the probability that egg-box cells of various natures are formed during the association of alginate chains with the M/G unit ratio of 1.5.

The transition metal cations Cu^{2+}, Ni^{2+} and Mn^{2+} bind alginate chains through coordination-covalent bonding, which also results in the formation of structures similar to egg-boxes but with completely filled cells, i.e., the number of cations per C_{12} block is equal to 1. However, the completely filled cells are not the only structural feature of transition metal gels. The elemental analysis shows that in the copper–alginate microspheres, the crosslinking of alginate chains and the formation of ordered egg-box structures with completely filled cells are provided by hydrated copper complexes. A similar situation is observed for the nickel–alginate hydrogels. An additional distinctive feature of manganese–alginate systems is the partial destruction of alginate chains. Thus, the metal–alginate complexes may have a more complete organization than was assumed previously.

Using the example of strontium alginates, it has been established that the existence of unequal binding sites for metal ions with alginate chains can result in the appearance of an ordered secondary structure of the hydrogel due to the physical sorption of ions and other molecules from the environment. The study of the elemental composition of excess salts remaining in the gel structure after its double washing allows us to calculate the number of sites for their possible binding, which determines the sorption capacity of alginates.

From the point of view of sorption characteristics, calcium alginate shows the most optimal properties. The reduced number of sites where calcium ions link chains increases the number of sites available for physical adsorption. In addition, during adsorption from the environment, the vacancies in interdimer space and in the egg-box structures are occupied by heavy metal cations, which, as a rule, belong to transition metals and can be intruded into cells of any configuration. Their incorporation and bonding by prepotent coordination-covalent bonding led to hydrogel structural strengthening and to the appearance of additional sites for physical sorption. Therefore, we believe that on the basis of calcium alginate, the most effective materials in terms of sorption capacity can be obtained for their use in environmental and other modern technologies. The creation of innovative nanocomposite materials with improved mechanical properties and efficient adsorption capacity [76,77] based on calcium alginate holds a great promise for various ecological purposes.

Supplementary Materials: The following supporting information can be downloaded at: https://www.mdpi.com/article/10.3390/polym15051243/s1, Figure S1: Images of alginate microspheres in presence of divalent metal cations; Figure S2: SEM image of transverse section of Sr-alginate microsphere and its elemental content; Table S1: Calculation of probabilities of various blocks occurrence along alginate chain having a mannuronic acid to guluronic acid ratio M/G = 1.5; Table S2: Calculation of probabilities of various egg-box cells along alginate dimer having a mannuronic acid to guluronic acid ratio M/G = 1.5.

Author Contributions: Conceptualization, O.S.Z., Y.F.Z. and S.R.D.; methodology, O.S.Z. and A.O.M.; investigation, O.S.Z.; formal analysis, O.S.Z.; writing—original draft preparation, O.S.Z., Y.F.Z. and S.R.D.; writing—review and editing, Y.F.Z., O.S.Z. and S.R.D.; visualization, A.O.M., T.K. and M.A.K.; supervision and project administration, Y.F.Z. All authors have read and agreed to the published version of the manuscript.

Funding: This research received no external funding.

Institutional Review Board Statement: Not applicable.

Data Availability Statement: The data in this study are available on reasonable request from the corresponding author.

Acknowledgments: A.O.M., M.A.K. and Y.F.Z. give thanks for financial support from the government assignment for the FRC Kazan Scientific Center of RAS. The contribution of Y.F.Z. was partly supported by the Kazan Federal University Strategic Academic Leadership Program ("PRIORITY-2030"). O.S.Z. gives thanks for the support of the Kazan State Power Engineering University Strategic Academic Leadership Program ("PRIORITY-2030"). Electron microscopy investigation was performed using the equipment of the Interdisciplinary Center "Analytical Microscopy" (Kazan Federal University, Kazan) and the Shared Research Center of Kazan National Research Technical University "Applied Nanotechnology".

Conflicts of Interest: The authors declare no conflict of interest.

References

1. Diener, M.; Adamcik, J.; Bergfreund, J.; Catalini, S.; Fischer, P.; Mezzenga, R. Rigid, Fibrillar Quaternary Structures Induced by Divalent Ions in a Carboxylated Linear Polysaccharide. *ACS Macro Lett.* **2020**, *9*, 115–121. [CrossRef] [PubMed]
2. Braccini, I.; Pérez, S. Molecular Basis of Ca^{2+}—Induced Gelation in Alginates and Pectins: The Egg-Box Model Revisited. *Biomacromolecules* **2001**, *2*, 1089–1096. [CrossRef] [PubMed]
3. Smidsrød, O.; Grasdalen, H. Some physical properties of carrageenan in solution and gel state. *Carbohydr. Polym.* **1982**, *2*, 270–272. [CrossRef]
4. Makshakova, O.N.; Faizullin, D.A.; Zuev, Y.F. Interplay between secondary structure and ion binding upon thermoreversible gelation of κ-carrageenan. *Carbohydr. Polym.* **2020**, *227*, 115342. [CrossRef] [PubMed]
5. Choi, I.S.; Ko, S.H.; Lee, M.E.; Kim, H.M.; Yang, J.E.; Jeong, S.-G.; Lee, K.H.; Chang, J.Y.; Kim, J.-C.; Park, H.W. Production, Characterization, and Antioxidant Activities of an Exopolysaccharide Extracted from Spent Media Wastewater after *Leuconostoc mesenteroides* WiKim32 Fermentation. *ACS Omega* **2021**, *6*, 8171–8178. [CrossRef]
6. Arca, H.C.; Mosquera-Giraldo, L.I.; Bi, V.; Xu, D.; Taylor, L.S.; Edgar, K.J. Pharmaceutical Applications of Cellulose Ethers and Cellulose Ether Esters. *Biomacromolecules* **2018**, *19*, 2351–2376. [CrossRef]
7. Lawson, L.D.; Bauer, R. (Eds.) *Phytomedicines of Europe: Chemistry and Biological Activity*; ACS Symposium Series; American Chemical Society: Washington, DC, USA, 1998; Volume 691, pp. 74–82; ISBN 978-0-8412-3559-5.
8. Bogdanova, L.R.; Makarova, A.O.; Zueva, O.S.; Zakharova, L.Y.; Zuev, Y.F. Encapsulation of diagnostic dyes in the polysaccharide matrix modified by carbon nanotubes. *Russ. Chem. Bull.* **2020**, *69*, 590–595. [CrossRef]
9. Ali, I. New Generation Adsorbents for Water Treatment. *Chem. Rev.* **2012**, *112*, 5073–5091. [CrossRef]
10. Yagub, M.T.; Sen, T.K.; Afroze, S.; Ang, H.M. Dye and its removal from aqueous solution by adsorption: A review. *Adv. Colloid Interface Sci.* **2014**, *209*, 172–184. [CrossRef]
11. Basheer, A.A. New generation nano-adsorbents for the removal of emerging contaminants in water. *J. Mol. Liq.* **2018**, *261*, 583–593. [CrossRef]
12. Ali, M.E.; Hoque, M.E.; Safdar Hossain, S.K.; Biswas, M.C. Nanoadsorbents for wastewater treatment: Next generation biotechnological solution. *Int. J. Environ. Sci. Technol.* **2020**, *17*, 4095–4132. [CrossRef]
13. Wan Ngah, W.S.; Teong, L.C.; Hanafiah, M.A.K.M. Adsorption of dyes and heavy metal ions by chitosan composites: A review. *Carbohydr. Polym.* **2011**, *83*, 1446–1456. [CrossRef]
14. Ching, S.H.; Bansal, N.; Bhandari, B. Alginate gel particles–A review of production techniques and physical properties. *Crit. Rev. Food Sci. Nutr.* **2017**, *57*, 1133–1152. [CrossRef] [PubMed]
15. Aarstad, O.A.; Tøndervik, A.; Sletta, H.; Skjåk-Bræk, G. Alginate Sequencing: An Analysis of Block Distribution in Alginates Using Specific Alginate Degrading Enzymes. *Biomacromolecules* **2012**, *13*, 106–116. [CrossRef] [PubMed]
16. Mørch, Ý.A.; Donati, I.; Strand, B.L.; Skjåk-Bræk, G. Molecular Engineering as an Approach to Design New Functional Properties of Alginate. *Biomacromolecules* **2007**, *8*, 2809–2814. [CrossRef]
17. Hecht, H.; Srebnik, S. Structural Characterization of Sodium Alginate and Calcium Alginate. *Biomacromolecules* **2016**, *17*, 2160–2167. [CrossRef]
18. Davis, T.A.; Volesky, B.; Mucci, A. A review of the biochemistry of heavy metal biosorption by brown algae. *Water Res.* **2003**, *37*, 4311–4330. [CrossRef]
19. Dodero, A.; Pianella, L.; Vicini, S.; Alloisio, M.; Ottonelli, M.; Castellano, M. Alginate-based hydrogels prepared via ionic gelation: An experimental design approach to predict the crosslinking degree. *Eur. Polym. J.* **2019**, *118*, 586–594. [CrossRef]
20. Maitra, J.; Shukla, V.K. Cross-linking in Hydrogels—A Review. *Cross-Link. Hydrogels Rev.* **2014**, *4*, 25–31. [CrossRef]
21. Hu, H.; Xu, F.-J. Rational design and latest advances of polysaccharide-based hydrogels for wound healing. *Biomater. Sci.* **2020**, *8*, 2084–2101. [CrossRef]
22. Duceac, I.A.; Stanciu, M.-C.; Nechifor, M.; Tanasă, F.; Teacă, C.-A. Insights on Some Polysaccharide Gel Type Materials and Their Structural Peculiarities. *Gels* **2022**, *8*, 771. [CrossRef]
23. Grant, G.T.; Morris, E.R.; Rees, D.A.; Smith, P.J.C.; Thom, D. Biological interactions between polysaccharides and divalent cations: The egg-box model. *FEBS Lett.* **1973**, *32*, 195–198. [CrossRef]
24. Smidsrød, O. Molecular basis for some physical properties of alginates in the gel state. *Faraday Discuss. Chem. Soc.* **1974**, *57*, 263–274. [CrossRef]
25. Russo, R.; Malinconico, M.; Santagata, G. Effect of Cross-Linking with Calcium Ions on the Physical Properties of Alginate Films. *Biomacromolecules* **2007**, *8*, 3193–3197. [CrossRef]
26. Wang, B.; Wan, Y.; Zheng, Y.; Lee, X.; Liu, T.; Yu, Z.; Huang, J.; Ok, Y.S.; Chen, J.; Gao, B. Alginate-based composites for environmental applications: A critical review. *Crit. Rev. Environ. Sci. Technol.* **2019**, *49*, 318–356. [CrossRef]
27. Brus, J.; Urbanova, M.; Czernek, J.; Pavelkova, M.; Kubova, K.; Vyslouzil, J.; Abbrent, S.; Konefal, R.; Horský, J.; Vetchy, D.; et al. Structure and Dynamics of Alginate Gels Cross-Linked by Polyvalent Ions Probed via Solid State NMR Spectroscopy. *Biomacromolecules* **2017**, *18*, 2478–2488. [CrossRef]
28. Hu, C.; Lu, W.; Mata, A.; Nishinari, K.; Fang, Y. Ions-induced gelation of alginate: Mechanisms and applications. *Int. J. Biol. Macromol.* **2021**, *177*, 578–588. [CrossRef] [PubMed]
29. Agulhon, P.; Markova, V.; Robitzer, M.; Quignard, F.; Mineva, T. Structure of Alginate Gels: Interaction of Diuronate Units with Divalent Cations from Density Functional Calculations. *Biomacromolecules* **2012**, *13*, 1899–1907. [CrossRef] [PubMed]

30. Bhowmik, S.; Ghosh, B.N.; Rissanen, K. Transition metal ion induced hydrogelation by amino-terpyridine ligands. *Org. Biomol. Chem.* **2014**, *12*, 8836–8839. [CrossRef]
31. Li, H.; Yang, P.; Pageni, P.; Tang, C. Recent Advances in Metal-Containing Polymer Hydrogels. *Macromol. Rapid Commun.* **2017**, *38*, 1700109. [CrossRef] [PubMed]
32. Bogdanova, L.R.; Zelenikhin, P.V.; Makarova, A.O.; Zueva, O.S.; Salnikov, V.V.; Zuev, Y.F.; Ilinskaya, O.N. Alginate-Based Hydrogel as Delivery System for Therapeutic Bacterial RNase. *Polymers* **2022**, *14*, 2461. [CrossRef] [PubMed]
33. Bogdanova, L.R.; Rogov, A.M.; Zueva, O.S.; Zuev, Y.F. Lipase enzymatic microreactor in polysaccharide hydrogel: Structure and properties. *Russ. Chem. Bull.* **2019**, *68*, 400–404. [CrossRef]
34. Enayatzamir, K.; Alikhani, H.A.; Yakhchali, B.; Tabandeh, F.; Rodríguez-Couto, S. Decolouration of azo dyes by Phanerochaete chrysosporium immobilised into alginate beads. *Environ. Sci. Pollut. Res.* **2010**, *17*, 145–153. [CrossRef] [PubMed]
35. Daâssi, D.; Rodríguez-Couto, S.; Nasri, M.; Mechichi, T. Biodegradation of textile dyes by immobilized laccase from Coriolopsis gallica into Ca-alginate beads. *Int. Biodeterior. Biodegrad.* **2014**, *90*, 71–78. [CrossRef]
36. Arıca, M.Y.; Kaçar, Y.; Genç, Ö. Entrapment of white-rot fungus Trametes versicolor in Ca-alginate beads: Preparation and biosorption kinetic analysis for cadmium removal from an aqueous solution. *Bioresour. Technol.* **2001**, *80*, 121–129. [CrossRef]
37. Bayramoğlu, G.; Tuzun, I.; Celik, G.; Yilmaz, M.; Arica, M.Y. Biosorption of mercury(II), cadmium(II) and lead(II) ions from aqueous system by microalgae Chlamydomonas reinhardtii immobilized in alginate beads. *Int. J. Miner. Process.* **2006**, *81*, 35–43. [CrossRef]
38. Gómez-Ordóñez, E.; Rupérez, P. FTIR-ATR spectroscopy as a tool for polysaccharide identification in edible brown and red seaweeds. *Food Hydrocoll.* **2011**, *25*, 1514–1520. [CrossRef]
39. Park, J.; Lee, S.J.; Lee, H.; Park, S.A.; Lee, J.Y. Three dimensional cell printing with sulfated alginate for improved bone morphogenetic protein-2 delivery and osteogenesis in bone tissue engineering. *Carbohydr. Polym.* **2018**, *196*, 217–224. [CrossRef]
40. Zueva, O.S.; Makarova, A.O.; Zuev, Y.F. Carbon Nanotubes in Composite Hydrogels Based on Plant Carbohydrates. *MSF* **2019**, *945*, 522–527. [CrossRef]
41. Paques, J.P.; van der Linden, E.; van Rijn, C.J.M.; Sagis, L.M.C. Preparation methods of alginate nanoparticles. *Adv. Colloid Interface Sci.* **2014**, *209*, 163–171. [CrossRef]
42. Morris, E.R.; Rees, D.A.; Thom, D.; Boyd, J. Chiroptical and stoichiometric evidence of a specific, primary dimerisation process in alginate gelation. *Carbohydr. Res.* **1978**, *66*, 145–154. [CrossRef]
43. Sikorski, P.; Mo, F.; Skjåk-Bræk, G.; Stokke, B.T. Evidence for Egg-Box-Compatible Interactions in Calcium−Alginate Gels from Fiber X-ray Diffraction. *Biomacromolecules* **2007**, *8*, 2098–2103. [CrossRef]
44. Li, L.; Fang, Y.; Vreeker, R.; Appelqvist, I.; Mendes, E. Reexamining the Egg-Box Model in Calcium−Alginate Gels with X-ray Diffraction. *Biomacromolecules* **2007**, *8*, 464–468. [CrossRef]
45. Borgogna, M.; Skjåk-Bræk, G.; Paoletti, S.; Donati, I. On the Initial Binding of Alginate by Calcium Ions. The Tilted Egg-Box Hypothesis. *J. Phys. Chem. B* **2013**, *117*, 7277–7282. [CrossRef]
46. Wang, H.; Wan, Y.; Wang, W.; Li, W.; Zhu, J. Effect of calcium ions on the III steps of self-assembly of SA investigated with atomic force microscopy. *Int. J. Food Prop.* **2018**, *21*, 1995–2006. [CrossRef]
47. Cao, L.; Lu, W.; Mata, A.; Nishinari, K.; Fang, Y. Egg-box model-based gelation of alginate and pectin: A review. *Carbohydr. Polym.* **2020**, *242*, 116389. [CrossRef]
48. Fang, Y.; Al-Assaf, S.; Phillips, G.O.; Nishinari, K.; Funami, T.; Williams, P.A.; Li, L. Multiple Steps and Critical Behaviors of the Binding of Calcium to Alginate. *J. Phys. Chem. B* **2007**, *111*, 2456–2462. [CrossRef]
49. Donati, I.; Paoletti, S. Material Properties of Alginates. In *Alginates: Biology and Applications*; Microbiology Monographs; Rehm, B.H.A., Ed.; Springer: Berlin/Heidelberg, Germany, 2009; Volume 13, pp. 1–53; ISBN 978-3-540-92678-8.
50. Painter, T.; Smidsrød, O.; Larsen, B.; Haug, A.; Paasivirta, J. A Computer Study of the Changes in Composition-Distribution Occurring during Random Depolymerization of a Binary Linear Heteropolysaccharide. *Acta Chem. Scand.* **1968**, *22*, 1637–1648. [CrossRef]
51. Smidsrød, O.; Whittington, S.G. Monte Carlo Investigation of Chemical Inhomogeneity in Polymers. *Macromolecules* **1969**, *2*, 42–44. [CrossRef]
52. Larsen, B.; Painter, T.J. The periodate-oxidation limit of alginate. *Carbohydr. Res.* **1969**, *10*, 186–187. [CrossRef]
53. Montanucci, P.; Terenzi, S.; Santi, C.; Pennoni, I.; Bini, V.; Pescara, T.; Basta, G.; Calafiore, R. Insights in Behavior of Variably Formulated Alginate-Based Microcapsules for Cell Transplantation. *BioMed Res. Int.* **2015**, *2015*, 965804. [CrossRef] [PubMed]
54. Mørch, Ý.A.; Donati, I.; Strand, B.L.; Skjåk-Bræk, G. Effect of Ca^{2+}, Ba^{2+} and Sr^{2+} on Alginate Microbeads. *Biomacromolecules* **2006**, *7*, 1471–1480. [CrossRef] [PubMed]
55. Bajpai, S.K.; Sharma, S. Investigation of swelling/degradation behaviour of alginate beads crosslinked with Ca^{2+} and Ba^{2+} ions. *React. Funct. Polym.* **2004**, *59*, 129–140. [CrossRef]
56. Zimmermann, U.; Mimietz, S.; Zimmermann, H.; Hillgärtner, M.; Schneider, H.; Ludwig, J.; Hasse, C.; Haase, A.; Rothmund, M.; Fuhr, G. Hydrogel-Based Non-Autologous Cell and Tissue Therapy. *BioTechniques* **2000**, *29*, 564–581. [CrossRef] [PubMed]
57. Huang, S.-L.; Lin, Y.-S. The Size Stability of Alginate Beads by Different Ionic Crosslinkers. *Adv. Mater. Sci. Eng.* **2017**, *2017*, 9304592. [CrossRef]

58. Hassan, R.M. Prospective and comparative Novel technique for evaluation the affinity of alginate for binding the alkaline-earth metal ions during formation the coordination biopolymer hydrogel complexes. *Int. J. Biol. Macromol.* **2020**, *165*, 1022–1028. [CrossRef]
59. Santhanes, D.; Teng, L.Y.; Sheng, F.S.; Coombes, A.G.A. Exploiting the versatility of oral capsule formulations based on high M-alginate for targeted delivery of poorly water soluble drugs to the upper and lower GI tract. *J. Drug Deliv. Sci. Technol.* **2018**, *46*, 384–391. [CrossRef]
60. Zhang, X.; Wang, L.; Weng, L.; Deng, B. Strontium ion substituted alginate-based hydrogel fibers and its coordination binding model. *J. Appl. Polym. Sci.* **2020**, *137*, 48571. [CrossRef]
61. Kong, C.; Zhao, X.; Li, Y.; Yang, S.; Chen, Y.M.; Yang, Z. Ion-Induced Synthesis of Alginate Fibroid Hydrogel for Heavy Metal Ions Removal. *Front. Chem.* **2020**, *7*, 905. [CrossRef]
62. Rezaii, N.; Khodagholi, F. Evaluation of Chaperone-like Activity of Alginate: Microcapsule and Water-soluble Forms. *Protein. J.* **2009**, *28*, 124–130. [CrossRef]
63. Maire du Poset, A.; Lerbret, A.; Boué, F.; Zitolo, A.; Assifaoui, A.; Cousin, F. Tuning the Structure of Galacturonate Hydrogels: External Gelation by Ca, Zn, or Fe Cationic Cross-Linkers. *Biomacromolecules* **2019**, *20*, 2864–2872. [CrossRef] [PubMed]
64. Plazinski, W.; Drach, M. Binding of bivalent metal cations by α- L -guluronate: Insights from the DFT-MD simulations. *New J. Chem.* **2015**, *39*, 3987–3994. [CrossRef]
65. Cheng, X.; Guan, H.; Su, Y. Polymerization of vinyl acetate initiated by a copper alginate coordination polymer film/Na_2SO_3/H_2O system. *J. Inorg. Organomet. Polym.* **2000**, *10*, 115–126. [CrossRef]
66. Lu, L.; Liu, X.; Tong, Z. Critical exponents for sol–gel transition in aqueous alginate solutions induced by cupric cations. *Carbohydr. Polym.* **2006**, *65*, 544–551. [CrossRef]
67. Boguń, M.; Mikołajczyk, T. Sorption and tensile strength properties of selected fibres of cupric alginate. *Fibres Text. East. Eur.* **2008**, *16*, 39–42.
68. Caccavo, D.; Ström, A.; Larsson, A.; Lamberti, G. Modeling capillary formation in calcium and copper alginate gels. *Mater. Sci. Eng. C* **2016**, *58*, 442–449. [CrossRef] [PubMed]
69. Haug, A.; Smidsrød, O.; Högdahl, B.; Øye, H.A.; Rasmussen, S.E.; Sunde, E.; Sørensen, N.A. Selectivity of Some Anionic Polymers for Divalent Metal Ions. *Acta Chem. Scand.* **1970**, *24*, 843–854. [CrossRef]
70. Rui Rodrigues, J.; Lagoa, R. Copper Ions Binding in Cu-Alginate Gelation. *J. Carbohydr. Chem.* **2006**, *25*, 219–232. [CrossRef]
71. Moore, A. (Ed.) *Alginic Acid: Chemical Structure, Uses and Health Benefits (Chemistry Research and Applications)*; Nova Publishers: Hauppauge, NY, USA, 2015; ISBN 978-1-63463-224-9.
72. Lee, K.Y.; Mooney, D.J. Alginate: Properties and biomedical applications. *Prog. Polym. Sci.* **2012**, *37*, 106–126. [CrossRef]
73. Khusnutdinov, R.I.; Bayguzina, A.R.; Dzhemilev, U.M. Manganese compounds in the catalysis of organic reactions. *Russ. J. Org. Chem.* **2012**, *48*, 309–348. [CrossRef]
74. Emmerichs, N.; Wingender, J.; Flemming, H.-C.; Mayer, C. Interaction between alginates and manganese cations: Identification of preferred cation binding sites. *Int. J. Biol. Macromol.* **2004**, *34*, 73–79. [CrossRef]
75. Perry, T.D.; Cygan, R.T.; Mitchell, R. Molecular models of alginic acid: Interactions with calcium ions and calcite surfaces. *Geochim. Cosmochim. Acta* **2006**, *70*, 3508–3532. [CrossRef]
76. Liu, H.; Pan, B.; Wang, Q.; Niu, Y.; Tai, Y.; Du, X.; Zhang, K. Crucial roles of graphene oxide in preparing alginate/nanofibrillated cellulose double network composites hydrogels. *Chemosphere* **2021**, *263*, 128240. [CrossRef]
77. Liu, C.; Liu, H.; Xiong, T.; Xu, A.; Pan, B.; Tang, K. Graphene oxide reinforced alginate/PVA double network hydrogels for efficient dye removal. *Polymers* **2018**, *10*, 835. [CrossRef]

MDPI AG
Grosspeteranlage 5
4052 Basel
Switzerland
Tel.: +41 61 683 77 34

Polymers Editorial Office
E-mail: polymers@mdpi.com
www.mdpi.com/journal/polymers

www.ingramcontent.com/pod-product-compliance
Lightning Source LLC
LaVergne TN
LVHW072314090526
838202LV00019B/2285
* 9 7 8 3 7 2 5 8 3 6 8 0 2 *